Problem Solving in Computational Molecular Science

NATO ASI Series

Advanced Science Institutes Series

A Series presenting the results of activities sponsored by the NATO Science Committee, which aims at the dissemination of advanced scientific and technological knowledge, with a view to strengthening links between scientific communities.

The Series is published by an international board of publishers in conjunction with the NATO Scientific Affairs Division

A Life Sciences **B Physics**	Plenum Publishing Corporation London and New York
C Mathematical and Physical Sciences **D Behavioural and Social Sciences** **E Applied Sciences**	Kluwer Academic Publishers Dordrecht, Boston and London
F Computer and Systems Sciences **G Ecological Sciences** **H Cell Biology** **I Global Environmental Change**	Springer-Verlag Berlin, Heidelberg, New York, London, Paris and Tokyo

PARTNERSHIP SUB-SERIES

1. Disarmament Technologies	Kluwer Academic Publishers
2. Environment	Springer-Verlag / Kluwer Academic Publishers
3. High Technology	Kluwer Academic Publishers
4. Science and Technology Policy	Kluwer Academic Publishers
5. Computer Networking	Kluwer Academic Publishers

The Partnership Sub-Series incorporates activities undertaken in collaboration with NATO's Cooperation Partners, the countries of the CIS and Central and Eastern Europe, in Priority Areas of concern to those countries.

NATO-PCO-DATA BASE

The electronic index to the NATO ASI Series provides full bibliographical references (with keywords and/or abstracts) to more than 50000 contributions from international scientists published in all sections of the NATO ASI Series.
Access to the NATO-PCO-DATA BASE is possible in two ways:

– via online FILE 128 (NATO-PCO-DATA BASE) hosted by ESRIN, Via Galileo Galilei, I-00044 Frascati, Italy.

– via CD-ROM "NATO-PCO-DATA BASE" with user-friendly retrieval software in English, French and German (© WTV GmbH and DATAWARE Technologies Inc. 1989).

The CD-ROM can be ordered through any member of the Board of Publishers or through NATO-PCO, Overijse, Belgium.

Series C: Mathematical and Physical Sciences – Vol. 500

Problem Solving in Computational Molecular Science

Molecules in Different Environments

edited by

S. Wilson

Rutherford Appleton Laboratory,
Oxfordshire, United Kingdom

and

G. H. F. Diercksen

Max-Planck Institut für Astrophysik,
Garching, Germany

Kluwer Academic Publishers

Dordrecht / Boston / London

Published in cooperation with NATO Scientific Affairs Division

Proceedings of the NATO Advanced Study Institute on
Problem Solving in Computational Molecular Science:
Molecules in Different Environments
Bad Windsheim, Germany
12–22 August 1996

A C.I.P. Catalogue record for this book is available from the Library of Congress

ISBN 978-94-010-6506-1 e-ISBN-13:978-94-009-0039-4

DOI: 10.1007/978-94-009-0039-4

Published by Kluwer Academic Publishers,
P.O. Box 17, 3300 AA Dordrecht, The Netherlands.

Sold and distributed in the U.S.A. and Canada
by Kluwer Academic Publishers,
101 Philip Drive, Norwell, MA 02061, U.S.A.

In all other countries, sold and distributed
by Kluwer Academic Publishers Group,
P.O. Box 322, 3300 AH Dordrecht, The Netherlands.

Printed on acid-free paper

TABLE OF CONTENTS

Preface vii

Molecular properties in different environments
B.T. Sutcliffe 1

Quantum-chemical models
J. Karwowski 37

Practical *ab initio* methods for molecular electronic structure studies
I. An overview 85
II. Finite basis sets and the algebraic approximation 109
III. Molecular integrals over gaussian-type functions 159
IV. Relativistic many-body perturbation theory 185
S. Wilson

Models for simulating molecular properties in condensed systems
M. Karelson and G.H.F. Diercksen 215

On calculating the electronic spectroscopy of very large molecules
M.C. Zerner 249

Fine and hyperfine structure: Spin properties of molecules
C.M. Marian 291

Molecular simulation - A primer
F. Müller-Plathe 389

Chemical reactions in bulk and on surfaces 413
R. Nieminen

The following contributions were not available at the date of publication:

Semi-empirical quantum chemical methods
M.C. Zerner

Theoretical nuclear magnetic resonance studies
J. Gauss

Heterogeneous catalysis
C.R.A. Catlow

PREFACE

Preface

Recent methodological and software advances have made a great variety of problems in the molecular sciences amenable to computational study. These problems are often of immediate industrial relevance; for example, in the design of novel, more efficient and selective homogenous petrochemical catalysts, in the development of new synthetic photodynamic clinical agents for cancer therapy and treatment of diseases caused by human immunodeficiency viruses, and in the construction of molecular switches and optically non-linear substances.

For all practical purposes the basic physical equations governing the behaviour of a system at the molecular level can only be solved approximately. The key issue, therefore, in any reliable and accurate computational study in molecular physics and quantum chemistry is the adoption of a suitable model which contains the essential physics and chemistry, is computationally tractable, and preferably amenable to systematic refinement. The provision of advice on the choice of an appropriate model which encapsulates the essential physics relevant to a particular problem has so far received scant attention. This issue is becoming acute as *standard* software packages are becoming widely available. These packages are being increasingly heavily used in both the academic and industrial sectors by researchers who have received no special training in the theoretical physics and chemistry underpinning these *standard applications*.

This volume records the proceedings of a NATO Advanced Study Institute, held at the Kurhotel Residenz, Bad Windsheim, 12th-22nd August, 1996, which provided researchers, whose background is not in the computational molecular sciences, with the necessary background to make intelligent use of the methods available by performing reliable calculations of an appropriate accuracy and making a considered interpretation of the data so obtained. In pursuit of this objective the traditional *methodology-orientated* approach to the subject has been abandoned in favour of *application-*

orientated lecture courses. These lectures have been of two types: *(i)* descriptions of recipes for practical investigations, *(ii)* reviews of the physical content of tractable computational methods.

The lectures describing recipes in the computational molecular sciences consisted of explicit and detailed expositions of the method (or methods) most appropriate to a particular problem and often included discussion of the associated operators and parameters, such as the Hamiltonian, basis set and wave function. In general, the recipes were very specific to the target application; the type of property required or class of molecule considered. Each lecture course, therefore, addressed a single property, or a series of closely related properties, for similar molecular species that can be examined by use of the same recipe; for example, fine and hyperfine structure for small molecules in the gas phase, or the electronic structure of large molecules in the liquid phase. Each lecture course discussed the important physics and chemistry which determine the property or related properties of interest. An outline of the different models, which have been developed to facilitate the calculation of a particular property, was given and the approximations which they involve discussed. Attention was focused on the consequences of the approximations employed in a particular model in terms of the accuracy and reliability achieved. Explicit recipes were given for the calculation of a property for a class of molecular species to a specified accuracy. This included, for example, the choice of the computational method and associated operators and parameters. In evolving fields, the recipes presented were critically assessed. Finally, the lectures demonstrated explicitly the reliability of the methods and accuracy they achieve by presenting the results of applications to carefully selected examples. The emphasis was based on the detailed analysis of the results by comparison with other work. A critical comparison of the computational efficiency of the competing methods has been attempted.

The review lectures presented computational methods from a unified point of view. The lectures discussed the physical model being considered in each case. They outlined the approximations introduced in the mathematical derivation with respect to the physical model solved. The lectures discussed the reliability and accuracy of different properties calculated by the chosen method in view of the physical model and the approximations implemented in the method. The strengths as well as the weaknesses of the method were demonstrated by judiciously selected examples. The results were discussed, again, with respect to the physical model adopted and compared with the results obtained by other computational methods and also by laboratory experiment. An informed assessment of the computational efficiency of the various methods for selected examples was provided.

Most of important industrial chemical processes and all biochemical reactions occur in condensed phases. It is well known that the structure and properties of a molecular system can be dramatically affected by the surrounding medium. Therefore, much of the Institute focused on recipes for computational studies of the structure, properties and dynamics of molecules in different environments and on the review of the physical ingredients of the appropriate methods. Not only have established recipes been described but evolving procedures in developing fields have been discussed and assessed.

We are most grateful to the lecturers at the Institute not only for the high standard that they achieved in their lectures and accompanying notes but also for the stimulating atmosphere which they helped create. It is our duty and privilege to thank the management and staff of the Kurhotel Residenz, Bad Windsheim, for their warm hospitality and attention to detail. We thank Erich Schreiner for taking care of much of the local organization of the Institute. Finally, we are most grateful to NATO Division of Scientific Affairs for the generous support afforded to the Institute.

S. Wilson & G.H.F. Diercksen

MOLECULAR PROPERTIES IN DIFFERENT ENVIRONMENTS

B. T. SUTCLIFFE
Department of Chemistry,
University of York,
York YO1 5DD, England

1. Introduction

It is the aim of this set of lectures to identify clearly and coherently the context for computational problem solving. They will not be concerned with computational strategies in detail but will nevertheless be presented in the context of what is computationally possible.

When we speak of molecules in different environments we sometimes mean that we are thinking about an isolated molecule in an environment provided by an electromagnetic field and sometimes we mean that we are thinking about the molecule in a material environment, for example in a dilute gas of other molecules or as a part of a liquid or as a solute molecule in a bulk solvent or as part of a molecular crystal.

The notion of field and matter as separate entities is rather difficult to sustain within the theory of relativity, so the idea of the field as an environment for a molecule is somewhat problematic in modern relativity theory. The idea on an isolated molecule in contradistinction from its material environment, is problematic within classical quantum mechanics, since a single equation describes both the molecule and the environment. Because there is no relativistic quantum theory it is not possible to deal with both these groups of problems simultaneously in an adequate way. It is the custom in dealing with chemical phenomena however, to assume that classical quantum mechanics is the correct theory to use and to treat field and hence relativistic effects as perturbations. This is the line that we shall take in these lectures. In particular we shall not consider the radiation field as such but simply the perturbations that are effected by its interaction with the molecule. But the first problem that we need to solve is that of identifying the isolated molecule and we shall begin by explaining why it is that even that idea is a tricky one.

S. Wilson and G. H. F. Diercksen (eds.), Problem Solving in Computational Molecular Science, 1–36.

1.1. CLASSICAL QUANTUM MECHANICS

From a quantum mechanical perspective, the idea of a molecule in an environment provided by other molecules, is one that is not quite obvious. Both the notion of a *molecule* and that of an *environment* are classical ideas and have to be realised within quantum mechanics according to particular conventions. If we assume that the ordinary Schrödinger equation is sufficient, at least to first order, to describe the systems of interest to us, then we can specify the time-independent Schrödinger Hamiltonian describing a system of N charged particles in a coordinate frame fixed in the laboratory as

$$\hat{H}(\mathbf{x}) = -\frac{\hbar^2}{2}\sum_{i=1}^{N} m_i^{-1}\nabla^2(\mathbf{x}_i) + \frac{e^2}{8\pi\epsilon_o}\sum_{i,j=1}^{N}{}' \frac{Z_i Z_j}{x_{ij}} \tag{1}$$

where the separation between particles is defined by

$$x_{ij}^2 = \sum_{\alpha}(x_{\alpha j} - x_{\alpha i})^2 \tag{2}$$

It is convenient to regard $\mathbf{x}_i$ as a column matrix of three cartesian components $x_{\alpha i}$, $\alpha = x, y, z$ and to regard $\mathbf{x}_i$ collectively as the 3 by N matrix $\mathbf{x}$. Each of the particles has mass m_i and charge $Z_i e$. The charge-numbers Z_i are positive for a nucleus and minus one for an electron. In a neutral system the charge-numbers sum to zero. The Schrödinger equation for the (constant energy) system is then

$$\hat{H}(\mathbf{x})\psi(\mathbf{x}) = E\psi(\mathbf{x}) \tag{3}$$

This form is appropriate for any number of particles and there is nothing in it that would give us any clue about which bits of the operator correspond to molecules and which to environment. The form might also represent a chemical reaction at constant energy and it is equally uninformative about which bits are the products or the reactants. Or the form might represent an isolated neutral molecule but if it did, it could equally well represent all the products into which the molecule could dissociate while the system remains electrically neutral. We could go on: the form could represent a solid or a liquid, the number of particles could be Avogadro's number; some thinkers suppose that the appropriate value for N is always Eddington's number, the total number of particles in the universe.

It is clear that if we are to understand the outcomes of experiments in terms of quantum mechanics, we must make choices and judgments about how to divide the equation up to produce the most tractable and informative way possible of dealing with the problem and also to yield results that make sense of our experience. But it should always be kept in mind that we are making choices and judgments, admittedly ones which are heavily circumscribed by the press of experiment, but about which, nevertheless, there can be honest and sensibly argued differences of opinion.

We begin these lectures by considering the symmetry properties of the equation: then we consider the extent to which information implicit in the equations, such as the masses and charges of the particles, can be used to suggest possible solution forms and in the light of these two sets of considerations, which we will call *intrinsic*, we shall try to identify the molecule and, by contrast, its environment. The considerations here will be *extrinsic* to the quantum mechanical problem as such and will involve clothing a particular model of experience in quantum mechanics. Of course not any old model will do: it has to be one that leads to results that are congruent with our experience, so the procedure is not a mere matter of taste. However more than one model may fill the bill and it is very improbable that one model alone will suffice to account for all of our experience of a particular type.

2. The free system

2.1. THE INTRINSIC SYMMETRY ASPECTS OF THE PROBLEM

The Hamiltonian (1) is invariant under uniform translations in the frame fixed in the laboratory. This means that the centre of molecular mass moves through space like a free particle and the states of a free particle are not quantised and eigenfunctions are not square integrable. The centre of mass motion must therefore be separated out to disentangle any bound states from the translational continuum.

The Hamiltonian is also invariant under all orthogonal transformations (rotation-reflections) of the particle variables in the frame fixed in the laboratory. It is therefore possible to separate, at least to some extent, the orientational motions of the system from its purely internal motions. The internal motions comprise dilations, contractions and deformations of a specified configuration of particle variables. Put colloquially, the internal motions simply change the system geometry. But in any case solutions of the equation must form basis functions for irreducible representations of the orthogonal group in three dimensions, $O(3)$.

The Hamiltonian is also invariant under the permutation of the variable sets of all identical particles. So the solutions of the equation must form a basis for the permutation group of the system. The permutation group of the system is formed as the direct product of the permutation groups for each of the sets of identical particles. The fact that the particles involved are either fermions or bosons restricts the kind of representations that can arise and we shall discuss these restriction later and in context.

If we stick to the time-independent Schrödinger equation in the absence of external fields or other perturbations, then this is all the symmetry that there is. Subsequently we shall see that in certain contexts these symmetries are either lowered or even broken. An example is the point group symmetry that arises naturally in accounting for the vibration-rotation spectra of isolated molecules. That this happens must always be a consequence of the particular model that is being treated quantum mechanically and the account so provided is appropriate only in so far as the model yields results in line with experiment.

2.1.1. *Translational symmetry*

The easiest way to consider the translational symmetry of the problem is to consider the coordinate transformation that removes the centre of mass motion from the full Hamiltonian. This is

$$(\mathbf{t}\ \mathbf{X}_T) = \mathbf{x}\ \mathbf{V} \tag{4}$$

In (4) $\mathbf{t}$ is a 3 by $N-1$ matrix and $\mathbf{X}_T$ is a 3 by 1 matrix, so that the combined (bracketed) matrix on the left of (4) is 3 by N. $\mathbf{V}$ is an N by N matrix which, from the structure of the left side of (4), has a special last column whose elements are

$$V_{iN} = {M_T}^{-1} m_i, \qquad\qquad M_T = \sum_{i=1}^{N} m_i. \tag{5}$$

Hence $\mathbf{X}_T$ is the standard centre of mass coordinate.

$$\mathbf{X}_T = {M_T}^{-1} \sum_{i=1}^{N} m_i \mathbf{x}_i \tag{6}$$

As the coordinates $\mathbf{t}_j, j = 1, 2, ...N-1$ are to be translationally invariant, we require on each remaining column of $\mathbf{V}$

$$\sum_{i=1}^{N} V_{ij} = 0, \qquad\qquad j = 1, 2, ...N-1 \tag{7}$$

and it is easy to see that (7) forces $\mathbf{t}_j \to \mathbf{t}_j$ as $\mathbf{x}_i \to \mathbf{x}_i + \mathbf{a}$, all i.

The $\mathbf{t}_i$ are independent if the inverse transformation

$$\mathbf{x} = (\mathbf{t}\,\mathbf{X}_T)\mathbf{V}^{-1} \tag{8}$$

exists. The structure of the right side of (8) shows that the bottom row of $\mathbf{V}^{-1}$ is special and, without loss of generality, we may require its elements to be

$$(\mathbf{V}^{-1})_{Ni} = 1 \qquad\qquad i = 1, 2.....N \tag{9}$$

The inverse requirement on the remainder of $\mathbf{V}^{-1}$ implies that

$$\sum_{i=1}^{N}(\mathbf{V}^{-1})_{ji}m_i = 0 \qquad j = 1, 2, ...N-1 \tag{10}$$

When we write the column matrix of the cartesian components of the partial derivative operator as $\partial/\partial\mathbf{x}_i$, the coordinate change (4) gives

$$\frac{\partial}{\partial\mathbf{x}_i} = \sum_{j=1}^{N-1} V_{ij}\frac{\partial}{\partial\mathbf{t}_j} + m_i {M_T}^{-1}\frac{\partial}{\partial\mathbf{X}_T} \tag{11}$$

Hence the Hamiltonian (1) in the new coordinates becomes

$$\hat{H}(\mathbf{t}, \mathbf{X}_T) = -\frac{\hbar^2}{2}\sum_{i,j=1}^{N-1}\mu_{ij}^{-1}\vec{\nabla}(\mathbf{t}_i).\vec{\nabla}(\mathbf{t}_j) + \frac{e^2}{8\pi\epsilon_o}\sum_{i,j=1}^{N}{}' \frac{Z_i Z_j}{r_{ij}(\mathbf{t})} - \frac{\hbar^2}{2M_T}\nabla^2(\mathbf{X}_T). \tag{12}$$

Here

$$\mu_{ij}^{-1} = \sum_{k=1}^{N} {m_k}^{-1} V_{ki} V_{kj} \qquad i, j = 1, 2, ...N-1 \tag{13}$$

The $N-1$ dimensional square matrix composed of all the μ_{ij}^{-1} is denoted as $\boldsymbol{\mu}^{-1}$. It is inverse to a matrix of reduced masses $\boldsymbol{\mu}$ but in our approach it is straightforward to construct the elements of the inverse directly. The operator r_{ij} is just x_{ij} as given by (2) but expressed as a function of $\mathbf{t}_i$. Thus

$$r_{ij}(\mathbf{t}) = (\sum_{\alpha}(\sum_{k=1}^{N-1}((\mathbf{V}^{-1})_{kj} - (\mathbf{V}^{-1})_{ki})t_{\alpha k})^2)^{1/2} \tag{14}$$

In (12) the $\vec{\nabla}(\mathbf{t}_i)$ are the usual grad operators expressed in the cartesian components of $\mathbf{t}_i$ and the last term represents the centre of mass kinetic energy. Since the centre of mass variable does not enter the potential term, the centre of mass problem may be separated off completely so that the full solution is of the form

$$T(\mathbf{X}_T)\Psi(\mathbf{t}) \tag{15}$$

where $\Psi(\mathbf{t})$ is a solution to the problem specified by the first two terms in (12) which will be denoted collectively by $\hat{H}(\mathbf{t})$ and referred to as the *translationally invariant* Hamiltonian.

We shall need the angular momentum operator in terms of X_T and the $\mathbf{t}_i$; the total angular momentum operator may be written as

$$\hat{\mathbf{L}}(\mathbf{x}) = \frac{\hbar}{i}\sum_{i=1}^{N}\hat{\mathbf{x}}_{\mathrm{i}}\frac{\partial}{\partial \mathbf{x}_i} \tag{16}$$

where $\hat{\mathbf{L}}(\mathbf{x})$ and $\frac{\partial}{\partial \mathbf{x}_i}$ are column matrices of cartesian components and the skew-symmetric matrix $\hat{\mathbf{x}}_i$ is

$$\hat{\mathbf{x}}_{\mathrm{i}} = \begin{pmatrix} 0 & -x_{zi} & x_{yi} \\ x_{zi} & 0 & -x_{xi} \\ -x_{yi} & x_{xi} & 0 \end{pmatrix} \tag{17}$$

The matrix $\hat{\mathbf{x}}_{\mathrm{i}}$ can also be written in terms of the infinitesimal rotation generators

$$\mathbf{M}^x = \begin{pmatrix} 0 & 0 & 0 \\ 0 & 0 & 1 \\ 0 & -1 & 0 \end{pmatrix} \quad \mathbf{M}^y = \begin{pmatrix} 0 & 0 & -1 \\ 0 & 0 & 0 \\ 1 & 0 & 0 \end{pmatrix} \quad \mathbf{M}^z = \begin{pmatrix} 0 & 1 & 0 \\ -1 & 0 & 0 \\ 0 & 0 & 0 \end{pmatrix} \tag{18}$$

so that

$$\hat{\mathbf{x}}_i = \sum_{\alpha} x_{\alpha i}\mathbf{M}^{\alpha T} \tag{19}$$

A variable symbol with a caret over it will, from now on, be used to denote a skew-symmetric matrix as defined by (19). The matrices are simply representations of the antisymmetric tensor, usually written as $e_{\alpha\beta\gamma}$.

Transforming to coordinates $\mathbf{X}_T, \mathbf{t}_i$ gives

$$\hat{\mathbf{L}}(\mathbf{x}) \rightarrow \frac{\hbar}{i}\sum_{i=1}^{N-1}\hat{\mathbf{t}}_i\frac{\partial}{\partial \mathbf{t}_i} + \frac{\hbar}{i}\hat{\mathbf{X}}_T\frac{\partial}{\partial \mathbf{X}_T} \tag{20}$$

and in future the first term will be denoted as $\hat{\mathbf{L}}(\mathbf{t})$ and called the translationally invariant angular momentum.

Whether or not the Schrödinger equation has any bound state solutions even when the translational continuum has been removed, is actually a rather hard question to answer in general. An outline of what is at present known together with some references can be found in [1]. Of course this emphasis on bound states should not obscure the fact that even using the translationally invariant Hamiltonian, there will be among its solutions, continuum states resulting from the relative motion of two or more fragments. The energies of the continuum states will be above the energies of the bound states (if any) and thus will not give rise to the formal problems that the translation continuum does. Such states can be treated, formally at least in an expansion, by stipulating that the sum includes integration over the continuum. In future expansions this interpretation will be assumed where appropriate but in practice, of course, such integration cannot be achieved. These continuum states are of the greatest interest in any discussions of scattering and hence of reactions.

The separation involves an essentially arbitrary choice of translationally invariant coordinates and there is always one less such coordinate than the original number because of the centre of mass coordinate. After this separation is made, it is clearly a matter of opinion and/or convention how the translationally invariant coordinates should be identified. So the role of the coordinates in specifying either electronic or nuclear motions becomes problematic. Of course the exact solutions are indifferent to the translationally invariant coordinate choice made and we shall always make a choice that is convenient for the purposes in hand.

2.1.2. *Rotation-reflection symmetry*

For a system with more than two particles one can transform the coordinates $\mathbf{t}$ such that rotational motion can be expressed in terms of three orientation variables, with the remaining motions expressed in terms of variables (commonly called internal coordinates) which are invariant under all orthogonal transformations of the $\mathbf{t}$. For $N = 2$ only two orientation variables are required and this case is rather special and is excluded from all subsequent discussion. The coordinate transformation is often referred to as constructing a frame fixed in the body and it is supposed that the three orientation variables are specified by means of an orthogonal matrix $\mathbf{C}$, the elements of which are expressed as functions of three eulerian angles $\phi_m, m = 1, 2, 3$ which are orientation variables and so there will be just $3N - 6$ internal variables for the particles.

The cartesian coordinates $\mathbf{t}$ are are considered related to a set $\mathbf{z}$ by

$$\mathbf{t} = \mathbf{C}\mathbf{z} \tag{21}$$

Since $\mathbf{z}$ are fixed in the body, not all their $3N - 3$ components are independent, for there must be three relations between them. Hence components of $\mathbf{z}_i$ must be writable in terms of $3N - 6$ independent internal coordinates $q_i, i = 1, 2, \ldots, 3N-6$. Some of the q_i maybe components of $\mathbf{z}_i$ but generally q_i are expressible in terms of scalar products of the $\mathbf{t}_i$ (and equally of the $\mathbf{z}_i$) since scalar products are the most general constructions that are invariant under orthogonal transformations of their constituent vectors. If only proper orthogonal transformations are considered the scalar triple products are also invariants but they change sign under improper operations.

The transformation of the Schrödinger Hamiltonian for these new coordinates cannot be achieved as easily as was possible for translationally invariant coordinates and we shall not consider it further here but see [1]. We shall quote, with an appropriate reference, whatever results we need in future. However it will be useful to have explicit expressions for the angular momentum operators and eigen functions.

The translationally invariant angular momentum operator becomes

$$\hat{\mathbf{L}}(\mathbf{t}) = -\frac{\hbar}{i}|\mathbf{C}|\mathbf{C}\mathbf{D}\frac{\partial}{\partial\boldsymbol{\phi}} \tag{22}$$

where $|\mathbf{C}|$ is either plus or minus one according to whether $\mathbf{C}$ corresponds to a proper rotation or to an improper rotation. This term arises because on the formal variable change (21)

$$\hat{\mathbf{t}}_i = |\mathbf{C}|\mathbf{C}\hat{\mathbf{z}}_i\mathbf{C}^T \tag{23}$$

The matrix $\mathbf{D}$ is a function of the eulerian angles alone.

There is at this stage an element of choice for the definition of the angular momentum in the frame fixed in the body: we choose

$$\hat{\mathbf{L}}(\boldsymbol{\phi}) = \frac{\hbar}{i}\mathbf{D}\frac{\partial}{\partial\boldsymbol{\phi}} \tag{24}$$

The components of this operator obey the *standard* commutation conditions, so that the angular momentum eigenfunctions, $|JMk>$, have the standard properties of those defined in Brink and Satchler [2] or in Biedenharn and Louck [3]. Explicitly, if $\mathbf{C}$ is parameterised by the standard eulerian angle choice made in [2] or [3] then

$$|JMk> = \left(\frac{2J+1}{8\pi^2}\right)^{\frac{1}{2}} (-1)^k \mathcal{D}^{J^*}_{M-k}(\boldsymbol{\phi}) \tag{25}$$

where $\mathcal{D}^J$ is the standard Wigner matrix as defined in [2] or [3]. The functions $|JMk>$ are often called *symmetric-top* eigenfunctions.

It is somewhat more usual to chose the negative of (24) as the definition, in which case its components obey the celebrated anomalous commutation conditions and the relevant symmetric top functions are proportional to $\mathcal{D}^{J^*}_{Mk}$. Whatever the choice, however the components of the body-fixed angular momentum operator are functions of the ϕ_m alone. With our choice it may be imagined that there are angular momentum eigenfunctions $|JMk>$, just as in Brink and Satchler [2] or in Biedenharn and Louck [3] such that

$$\hat{L}^2(\mathbf{t})|JMk> = \hat{L}^2(\boldsymbol{\phi})|JMk> = \hbar^2 J(J+1)|JMk>$$

$$\hat{L}_z(\mathbf{t})|JMk> = \hbar M|JMk>$$

$$\hat{L}_z(\boldsymbol{\phi})|JMk> = \hbar k|JMk> \tag{26}$$

Furthermore defining the step-up and -down operators in the usual way as $L_\pm = L_x \pm iL_y$ then

$$\hat{L}_\pm(\mathbf{t})|JMk> = \hbar C^\pm_{JM}|JM \pm 1k>$$

$$\hat{L}_\pm(\phi)|JMk> = \hbar C^\pm_{Jk}|JMk \pm 1> \tag{27}$$

where the phase conventions are chosen as the standard Condon and Shortley ones (see *eg* [4]) so that

$$C^\pm_{Jj} = [J(J+1) - j(j \pm 1)]^{\frac{1}{2}} \tag{28}$$

A more extended discussion of these matters can be found in Section (3.8) of [3] and also in [5] and [6].

The full rotation-reflection symmetry of the problem is that of the group $O(3)$ but the symmetric top functions (25) are a basis only for representations of the rotation group $SO(3)$, that is parity is ignored. To introduce parity would however complicate the discussion somewhat without adding much of interest.

In fact it can be shown [3], [7] that, whatever the parameterisation of $\mathbf{C}$,

$$|1Mk> = \left(\frac{3}{8\pi^2}\right)^{\frac{1}{2}} (\mathbf{X}^T\mathbf{C}\mathbf{X})_{Mk} \tag{29}$$

with

$$\mathbf{X} = \begin{pmatrix} -1/\sqrt{2} & 0 & 1/\sqrt{2} \\ -i/\sqrt{2} & 0 & -i/\sqrt{2} \\ 0 & 1 & 0 \end{pmatrix} \tag{30}$$

provided that $C_{\alpha\beta}$ is ordered $\alpha, \beta = x, y, z$ and the indices M, k run $+1, 0, -1$ across each row and down each column. The general angular momentum eigenfunction $|JMk>$ can now be constructed from (29) by vector-coupling (see Appendix 2 of [7]).

It is clear from what has gone before and indeed from general group theoretical arguments, that the eigenfunctions $\Psi(\mathbf{t})$ from (15) can be written in the form

$$\Psi(\mathbf{t}) \rightarrow \Psi^{J,M}(\phi, \mathbf{q}) = \sum_{k=-J}^{+J} \Phi^J_k(\mathbf{q})|JMk> \tag{31}$$

where the internal coordinate function on the right side cannot depend on M because, in the absence of a field, the energy of the system does not depend on M.

Since the angular momentum eigenfunctions are of known form and with known properties, it is always possible to eliminate the angular motion from the problem by integrating out the angular part of any expected value and hence producing an effective Hamiltonian within a given (J, M, K) rotational manifold that depends only on the internal coordinates. Such a Hamiltonian will contain the angular quantum numbers as parameters and only a portion of it will actually be independent of them. This is to be expected, since the group $SO(3)$ (unlike the translation group in three dimensions) is not abelian and so rotation-internal motion interactions cannot be eliminated. We shall always make a specific choice of the effective Hamiltonian for internal motion motivated by the particular problem considered.

In the discussion of internal coordinates made below (21) it was noted that internal coordinates composed from scalar triple products of translationally invariant coordinates would not be used in the present work since it was desired that the internal coordinates should be invariant not only under rotations but under rotation-reflections too. This seems the proper thing to do because it makes the most complete separation possible between the internal and orientation parts of the wave function, but it is not a choice that is universally made.

Before going further, it is necessary to say something about the jacobian of the transformation and its domain of validity.

As the transformation (4) is linear its jacobian is simply a constant that can be ignored. The transformation from the $\mathbf{t}_i$ to the eulerian angles and the internal coordinates is non-linear and has a jacobian $|\mathbf{J}|^{-1}$ where $\mathbf{J}$ is the matrix constructed from the derivatives of ϕ_m and q_k with respect to the $t_{\alpha i}$. The non-linearity is a topological consequence of any transformation that allows rotational motion to be separated [8] and there is always some conformation of the particles that causes the jacobian to vanish. Clearly where the jacobian vanishes, the transformation is undefined. This failure manifests itself in the Hamiltonian by the presence of terms which diverge unless, acting on the wavefunction, they vanish. This can occur either by cancelation or by the wavefunction itself being vanishingly small in the divergent region.

The origin of these divergences is not physical. They arise simply as a consequence of the choice of coordinates. A particular choice can obviously preclude the description of a possible physical state of a system. Thus suppose that a triatomic is described according to Eckart's approach [9], [10] with the equilibrium geometry specified as bent. In this case the jacobian vanishes when the internal coordinates correspond to a linear geometry. The problem then becomes ill conditioned for states with large amplitude angular deformations. Such states are physically reasonable, merely they cannot be described according to this formulation.

The important point is that the non-linear transformation cannot be globally valid. As it has only local validity, one can at most derive a local

Hamiltonian that is valid within a particular domain. According to general topological considerations [8] one can construct a sequence of transformations which have common ranges of validity sufficient for passage from one to another to cover the whole space.

The volume element for integration is

$$dt = |\mathbf{J}|^{-1} d\phi d\mathbf{q} \tag{32}$$

2.1.3. *Permutation symmetry*

A permutation of the variables describing a set of identical particles will naturally leave the centre of mass coordinate unchanged but it will induce a linear transformation among the translationally invariant coordinates and that transformation will certainly not have the form of a standard permutation matrix, neither will it be orthogonal. In general too, a permutation will induce changes in the orientation variables and in the internal coordinates. It can be shown that a permutation of identical particles can produce from a given set of orientation variables, a new set of orientation variables which may be functions of the original orientation variables and the internal coordinates, However such a permutation can change an internal coordinate only into a function of internal coordinates. It is possible in principle to choose a set of orientational variables that are invariant under permutations of identical particles, but the commonly made choices are not invariant. Similarly the commonly made choices of internal coordinates are not invariant. Unless choice of invariant orientation variables is made, however, the form of the internal motion Hamiltonian within any rotational manifold, cannot be invariant either. In this case the idea of vibration-rotation separation becomes a problematic one.

It is difficult and, for present purposes unnecessary, to give a general account of what is going on. Instead, what happens will be considered in each particular case.

2.2. MASSES AND CHARGES

Historically in quantum mechanics, the nuclei have always been treated on a different footing from the electrons. This difference is a hangover from the old quantum theory but it is usually justified now by appeal to the pioneering work of Born and Oppenheimer [11]. The argument of that paper, put informally, is that because the nuclei are so much heavier than the electrons, they must move much more slowly. So, to a first approximation it is sensible to hold (or clamp) them at a sequence of fixed positions and to solve the problem for the fast-moving electrons as a function of the nuclear positions. The nuclear motions can then be accommodated as perturbations. In fact the perturbation argument of the original paper is now seldom quoted but rather the later work of Born which is most readily accessible as Appendix VIII in [12]. (See also [13].)

In this approach the natural thing to do is to regard the nuclei as specifying a coordinate frame and to solve the electronic problem in this frame. To distinguish between electrons and nuclei, the variables are split up into two sets, one set consisting of L variables, $\mathbf{x}_i^e$, describing the electrons and the other set of H variables, $\mathbf{x}_i^n$, describing the nuclei and $N = L + H$. When it is necessary to emphasise this split, (1) will be denoted $\hat{H}(\mathbf{x}^n, \mathbf{x}^e)$. The electronic wave function is then provided by a solution of the clamped nuclei electronic Hamiltonian

$$\hat{H}^{cn}(\mathbf{a}, \mathbf{x}^e) = -\frac{\hbar^2}{2m}\sum_{i=1}^{L}\nabla^2(\mathbf{x}_i^e) - \frac{e^2}{4\pi\epsilon_0}\sum_{i=1}^{H}\sum_{j=1}^{L}\frac{Z_i}{|\mathbf{x}_j^e - \mathbf{a}_i|} + \frac{e^2}{8\pi\epsilon_0}{\sum_{i,j=1}^{N}}'\frac{1}{|\mathbf{x}_i^e - \mathbf{x}_j^e|} \tag{33}$$

This Hamiltonian is obtained from the original one (1) by assigning the values $\mathbf{a}_i$ to the nuclear variables $\mathbf{x}_i^n$, hence the designation *clamped nuclei* for this form. Within the electronic problem each nuclear position $\mathbf{a}_i$ is treated as a parameter. It has eigenfunctions of the form $\phi(\mathbf{a}, \mathbf{x}^e)$ and energies of the form $E(\mathbf{a})$. Thus both the eigenfunctions and the eigenvalues contain the nuclear variables as parameters. The energy here is commonly called the electronic energy and the eigenfunction, the electronic wave function. For solution of the entire problem, the electronic wave function must be available for all values of the parameters.

If it were the case that the solution to the full problem could be written as a product of the solution to the clamped nuclei problem and a nuclear motion function

$$\psi(\mathbf{x}^n, \mathbf{x}^e) = \Phi(\mathbf{x}^n)\phi(\mathbf{x}^n, \mathbf{x}^e)$$

the function, $\Phi(\mathbf{x}^n)$ could be determined as a solution of an effective nuclear motion equation, obtained from the full equation (3) by multiplying from the left by $\phi^*(\mathbf{x}^n, \mathbf{x}^e)$ and integrating over all $\mathbf{x}_i^e$. Ignoring the effects of the nuclear derivative operators on the electronic wave function, the effective nuclear motion problem so constructed, has a potential that is just the sum of the electronic energy and the nuclear repulsion energy. This potential is usually referred to as the potential energy surface.

It is clear that there is something a little fishy about this approach, for the electronic energy is invariant under all uniform translations and all rigid rotation-reflections of the geometrical figure specified by the particular choice of $\mathbf{a}$. So It cannot be a function of all the nuclear coordinates. In fact it can be shown that it is a function of only $3H - 6$ internal coordinates formed from the nuclear variables (excluding the case where $H = 2$). There are unfortunately, formidable technical problems in the way of making a precise argument here and connecting the solutions of the clamped nuclei electronic Hamiltonian in an unambiguous fashion with solutions to the full problem.

Even at a qualitative level it is clear enough that any argument made here must depend on the magnitude of the electronic energy and the details of its behaviour as a function of geometry and the details of the separation between electronic energy levels. To some extent however it is possible to justify the simple product approach for an appropriately small collection of nuclei at suitably low energies near sufficiently deep minima on the potential surface, if this surface is sufficiently well separated from any other potential surface. Assuming it justified, then a molecule is identified as occurring at where there is a local minimum on the potential energy surface. This minimum is then thought of as specifying an equilibrium geometry for a molecule.

This apparent simplification of the problem is not, however, without its price. In the clamped nuclei Hamiltonian (33), the nuclei are identified in order to define the coordinate frame for the electronic calculation. A typical "point" on a potential energy surface occurs at an identifiable nuclear geometry. But if some of the nuclei are identical then geometries resulting from permutations of such nuclei require equal consideration. In practice, as first noted in print by Berry [14], they do not receive it. An attempt to justify this was made by Longuet-Higgins [15]. He postulated that any permutation that could be realised only by dis-assembling the geometrical figure of a molecule, could be achieved only by surmounting exceptionally high barriers on the potential energy surface and would, therefore be unimportant. Such a permutation was therefore called *un-feasible.* Only permutations that could realised by orthogonal transformations of the rigid geometrical figure were important or *feasible.* This approach has been extremely influential and detailed expositions of it can be found in [7] and [16]. Although this way of looking at things does allow us to understand why point group symmetry is such an effective means of classification in molecular spectroscopy, it still leaves some puzzles. Suppose that the potential surface argument is appropriate. In the absence of a calculation, whether there is a barrier or not in realising a permutation, is largely a matter of opinion. It may well be one on which there can be widespread agreement but that may simply be because those agreeing, recognise the need to achieve a particular group symmetry to explain observations. It is not clear either that the potential argument is appropriate. The idea of un-feasibility rests on the notion that the permutation is a real motion of particles in the potential computed from a clamped nuclei calculation. But the idea that a permutation is a real motion of particles is an incongruous one, from a mathematical point of view, as is the idea of un-feasibility, outside the single product approximation for the wave function and hence a single uncoupled potential function.

It remains unclear how best to deal with these difficulties. They add to the problems arising from permutations which are adumbrated above. Neither are they the only problems. It is not clear whether this separation is the best way to consider systems consisting of large numbers of nuclei and electrons, neither is it clear that it is the best way of dealing with systems at high energies or at small separations between or even crossings of, potential energy surfaces.

In other words, although number of particles, together with their masses and charges are intrinsic properties of the system, it is not really possible to use them to identify the molecule *a priori*. It is necessary to use extrinsic criteria. We shall begin our discussion of such criteria by considering the isolated molecule.

3. The isolated molecule

If it is assumed that the clamped nuclei electronic problem has been solved to yield a potential for nuclear motion but that otherwise the electrons are not involved in the nuclear motion, then rather detailed arguments may be used to show that for a general choice of orientation coordinates and internal coordinates the kinetic energy operator for nuclear motion in such a potential is of the form

$$\hat{K}(\boldsymbol{\phi}, \mathbf{q}) + \hat{K}(\mathbf{q}) \tag{34}$$

where

$$\hat{K}(\boldsymbol{\phi}, \mathbf{q}) = \frac{1}{2}\Big(\sum_{\alpha\beta} \kappa_{\alpha\beta}\hat{L}_\alpha \hat{L}_\beta + \hbar \sum_\alpha \lambda_\alpha \hat{L}_\alpha\Big) \tag{35}$$

and

$$\hat{K}(\mathbf{q}) = -\frac{\hbar^2}{2}\Big(\sum_{k,l=1}^{3H-6} g_{kl}\frac{\partial^2}{\partial q_k \partial q_l} + \sum_{k=1}^{3H-6} h_k \frac{\partial}{\partial q_k}\Big) \tag{36}$$

The matrix $\boldsymbol{\kappa}$ is an 3 by 3 inverse generalised inertia tensor and $\mathbf{g}$ is a $3H-6$ by $3H-6$ metric matrix The term linear in the angular momentum is

$$\lambda_\alpha = \frac{1}{i}\Big(\nu_\alpha + 2\sum_{k=1}^{3H-6} \tau_{k\alpha}\frac{\partial}{\partial q_k}\Big) \tag{37}$$

with the $3H-6$ by 3 matrix $\boldsymbol{\tau}$ is a coupling tensor between the internal and angular motions All of these matrices are functions of the internal coordinates alone. The quantities ν_α and h_k arise because the derivative operators do not generally commute with the constituent parts of these matrices. Only the angular momentum operators depend on the eulerian angles and these depend upon the eulerian angles alone.

Using the form (31) one can eliminate angular motion from the problem and write an effective Hamiltonian within any (J, M, k) rotational manifold, that depends only on the internal coordinates. The matrix elements with respect to the angular variables of the operators that depend only on the q_k are trivial. Thus

$$< J'M'k' \mid \hat{K}(\mathbf{q}) + V \mid JMk >= \delta_{J'J}\delta_{M'M}\delta_{k'k}(\hat{K}_I + V) \tag{38}$$

in which V is the potential which arises as the sum of the electronic and nuclear repulsion energy. In what follows explicit allowance for the diagonal requirement on J and M will be assumed and the indices suppressed to save writing. Similarly the fact that the integration implied is over ϕ only will be left implicit.

To treat the operator containing the angular terms is much more complicated and best done by re-expressing the components of $\hat{\mathbf{L}}$ in terms of $\hat{L}_\pm(\phi)$ and $\hat{L}_z(\phi)$ and using (26) and (27). When this is done

$$< JMk' \mid \hat{K}(\phi, \mathbf{q}) \mid JMk >=$$

$$\frac{\hbar^2}{4}(b_{+2}C^+_{Jk+1}C^+_{Jk}\delta_{k'k+2} + b_{-2}C^-_{Jk-1}C^-_{Jk}\delta_{k'k-2})$$

$$+\frac{\hbar^2}{4}(C^+_{Jk}(b_{+1}(2k+1) + \lambda_+)\delta_{k'k+1} + C^-_{Jk}(b_{-1}(2k-1) + \lambda_-)\delta_{k'k-1})$$

$$+\frac{\hbar^2}{2}((J(J+1) - k^2)b + b_0k^2 + \lambda_0 k)\delta_{k'k} \tag{39}$$

In this expression

$$b_{\pm 2} = (\kappa_{xx} - \kappa_{yy})/2 \pm \kappa_{xy}/i$$

$$b_{\pm 1} = \kappa_{xz} \pm \kappa_{yz}/i$$

$$b = (\kappa_{xx} + \kappa_{yy})/2 \qquad\qquad b_0 = \kappa_{zz} \tag{40}$$

and in terms of the λ_α in (37) λ_0 is λ_z and the $\lambda_\pm$ are

$$\lambda_\pm = (\lambda_x \pm \lambda_y/i) \tag{41}$$

3.1. THE SPECTROSCOPIC MOLECULE

When thinking of a system containing at most a dozen or so nuclei it is usual to realise this nuclear motion Hamiltonian in terms of the approach begun by Eckart [9] and brought to full fruition in the work of Watson [10]. The resulting quantum mechanical form is particularly appropriate for interpreting molecular spectra and so will be thought of as defining the spectroscopic molecule. The physical picture is one in which there is an isolated deep minimum on the potential surface at a particular nuclear geometry, the *equilibrium* geometry or structure. To describe this geometry Eckart begins from the H redundant nuclear coordinates

$$\mathbf{t}_i = \mathbf{x}_i^n - \mathbf{X}$$

where $\mathbf{X}$ is the centre of nuclear mass coordinate. Thus

$$\sum_{i=1}^{H} m_i \mathbf{t}_i = 0$$

and a matrix $\mathbf{C}$ is chosen as in (21) to define a set of cartesians in the frame fixed in the body such that the equilibrium structure is specified by

$$\mathbf{z}_i = \mathbf{a}_i$$

where the $\mathbf{a}_i$ are constant matrices. Naturally, the $\mathbf{z}_i$ (and, in consequence, the $\mathbf{a}_i$) are a redundant set, summing to zero in the manner of the $\mathbf{t}_i$ above. The internal coordinates are expressed in terms of displacements from the equilibrium geometry by

$$q_k = \sum_{i=1}^{H} \sum_{\alpha} b_{k\alpha i} (\mathbf{z}_i - \mathbf{a}_i)_\alpha, \quad k = 1, 2, \cdots 3H - 6$$

where the elements $b_{k\alpha i}$ are simply constants which may be regarded as components of a column matrix $\mathbf{b}_{ki}$. The range of these coordinates is $(-\infty, \infty)$. The precise conditions specifying $\mathbf{C}$ are

$$\sum_{i=1}^{H} m_i (\hat{\mathbf{a}}_i (\mathbf{z}_i - \mathbf{a}_i)) \equiv \sum_{i=1}^{H} m_i \vec{a}_i \times (\vec{z}_i - \vec{a}_i) = 0$$

provided that the $\mathbf{a}_i$ do not define a line. These constraint conditions are on the components of the mass weighted sum over all the vectors, of the vector products of the equilibrium geometry vectors with the displacement vectors. In classical mechanics the vanishing of these components would be interpreted as the system having no internal angular momentum at the equilibrium geometry.

If the displacements are sufficiently small then the potential for nuclear motion can be expanded about the equilibrium geometry as a power series in the q_k

$$V(\mathbf{q}) = V(\mathbf{a}) + \frac{1}{2} \sum_{kl=1}^{3H-6} F_{kl} q_k q_l + \cdots$$

The first term in this expression is the sum of the electronic and the nuclear repulsion energy at the equilibrium geometry. The linear term is absent because the first derivatives of the potential vanish at the equilibrium geometry, by definition. The matrix $\mathbf{F}$ is composed of the second derivatives of the potential evaluated at the equilibrium geometry. The matrix of second derivatives is, in the context of power series often called the Hessian matrix and so $\mathbf{F}$ is often called the Hessian at the minimum. It is a symmetric positive definite matrix and its elements are usually called the *force constants* for the problem. This usage arises because they have units of [force][distance]$^{-1}$ and in the classical texts on spectroscopy are always quoted in millidynes per Ångström.

Internal coordinates must be linearly independent and so the $\mathbf{b}_{ki}$ must be linearly independent and in order for an inverse transformation to exist between the internal coordinates and the coordinates in the frame fixed in the laboratory, it is required that

$$\sum_{i=1}^{H} \mathbf{b}_{ki} = 0 \qquad \sum_{i=1}^{H} \hat{\mathbf{a}}_i \mathbf{b}_{ki} \equiv \sum_{i=1}^{H} \vec{a}_i \times \vec{b}_{ki} = 0$$

The Hamiltonian derived as outlined above using these general internal coordinates is really very cumbersome and extremely difficult to use, but Watson [10] showed (see also Louck [17] for an approach rather more like ours) showed that it could be simplified by a particular choice of orthogonal internal coordinates, and by incorporating the internal coordinate part of the jacobian into the operator. This last trick is analogous to the fairly familiar process when working in spherical polars, for example, where the radial volume element $r^2 dr$ can be reduced to dr by writing the trial wavefunction $\psi(r)$ as $r^{-1}P(r)$ and modifying the Hamiltonian to refer to $P(r)$. This modification changes the derivative terms in the operator by $\partial/\partial r \rightarrow (\partial/\partial r - 1/r)$ and so on but alters none of the multiplicative or $\partial/\partial\theta$ terms. The resulting Hamiltonian is often said to be in *manifestly hermitian* form. Particular examples of this kind of construction can be found in [10] and in [17] while a general account is given in Section 35 of Kemble [18].

With Watson's choice the $\mathbf{b}_{ki}$ are written $\sqrt{m_i}\mathbf{l}_{ki}$ and it is required that

$$\sum_{i=1}^{H} \mathbf{l}_{ki}^T \mathbf{l}_{mi} = \delta_{km}, \qquad \sum_{i=1}^{H} \sqrt{m_i}\mathbf{l}_{ki} = 0, \qquad \sum_{i=1}^{H} \sqrt{m_i}\hat{\mathbf{a}}_i \mathbf{l}_{ki} = 0$$

The internal coordinates so defined are usually written as Q_k where, explicitly

$$Q_k = \sum_{i=1}^{H} \sqrt{m_i}\mathbf{l}_{ki}^T(\mathbf{z}_i - \mathbf{a}_i)$$

Coordinates satisfying the above conditions can always be chosen to diagonalise a quadratic approximation to the potential as well. If they are so chosen, then they are called *normal* coordinates and are the basis of the simplest first-order approach to describing the vibration-rotation motion of the molecule.

The kinetic energy operators for the spectroscopic molecule can be written as

$$\hat{K}(\phi, \mathbf{Q}) = \frac{1}{2}\sum_{\alpha\beta} \mu_{\alpha\beta}(\hat{L}_\alpha + \hat{\pi}_\alpha)(\hat{L}_\beta + \hat{\pi}_\beta)$$

$$\hat{K}(\mathbf{Q}) = -\frac{\hbar^2}{2}\sum_{k=1}^{3H-6} \frac{\partial^2}{\partial Q_k^2} - \frac{\hbar^2}{8}\sum_{\alpha} \mu_{\alpha\alpha} \qquad (42)$$

in which $\boldsymbol{\mu}$ is a special form of the inverse generalised inertia tensor $\boldsymbol{\kappa}$ in (35) defined as

$$\boldsymbol{\mu} = \mathbf{I}''^{-1}\mathbf{I}^0\mathbf{I}''^{-1}$$

Here $\mathbf{I}^0$ is the inertia tensor for the molecule at the equilibrium geometry.

$$\mathbf{I}^0 = \sum_{i=1}^{H} m_i \hat{\mathbf{a}}_i^T \hat{\mathbf{a}}_i$$

and so is a constant matrix, while

$$\mathbf{I}'' = \sum_{i=1}^{H} m_i \hat{\mathbf{z}}_i^T \hat{\mathbf{a}}_i = -\sum_{i=1}^{H} m_i \hat{\mathbf{z}}_i \hat{\mathbf{a}}_i$$

and is a matrix which allows for the instantaneous geometry changes. The operator $\hat{\pi}_\alpha$ is defined as

$$\hat{\pi}_\alpha = \frac{\hbar}{i}\sum_{k,l=1}^{3H-6} \zeta_{kl}^\alpha Q_k \frac{\partial}{\partial Q_l}$$

in which

$$\zeta_{kl}^\alpha = \sum_{i=1}^{H} (\hat{\mathbf{l}}_{ki}\mathbf{l}_{li})_\alpha$$

This operator is often called a component of internal angular momentum but this is rather a misnomer, for it has none of the properties of angular momentum. It is actually a component of the Coriolis coupling operator and thus the constants ζ_{kl}^α are quite properly given their usual designation as Coriolis coupling constants.

In the original presentation by Watson, the components of the total angular momentum operators are denoted $\hat{\Pi}_\alpha$ and obey the anomalous commutation conditions. Here we have followed Louck and used angular momentum operators that obey the standard commutation conditions. The Watson form is achieved from the present one by making the substitution

$$(\hat{L}_\alpha + \hat{\pi}_\alpha) \rightarrow -(\hat{\Pi}_\alpha - \hat{\pi}_\alpha)$$

If the displacements are small, then the Q_k are vanishingly small and so the Coriolis coupling operator vanishes and the inverse generalised inertia tensor becomes $\boldsymbol{\mu}^0$, the inverse of the equilibrium inertia tensor. In these circumstances one can choose the frame fixed in the body so that the equilibrium inertia tensor is diagonal (the principal axis choice). Also an expansion of the potential up to quadratic terms in the coordinates, will be adequate. If this is the case then the equations for the kinetic energy above simplify to yield a Hamiltonian

$$\hat{H}(\boldsymbol{\phi}, \mathbf{Q}) = \hat{K}(\boldsymbol{\phi}) + \hat{H}(\mathbf{Q})$$

in which

$$\hat{K}(\boldsymbol{\phi}) = \frac{1}{2}\sum_\alpha \mu^0_{\alpha\alpha}\hat{L}^2_\alpha \tag{43}$$

$$\hat{H}(\mathbf{Q}) = -\frac{\hbar^2}{2}\sum_{k=1}^{3H-6}\frac{\partial^2}{\partial Q_k^2} + \frac{1}{2}\sum_{k=1}^{3H-6}\lambda_k Q_k^2 \tag{44}$$

where the zero of the potential energy in (44) has been chosen to incorporate the constant terms. It has also been assumed that the coordinates are normal ones, chosen to diagonalise the quadratic approximation to the potential. Thus the λ_i are the eigenvalues of the quadratic form $\mathbf{F}$. All the eigenvalues must be positive for a stable molecule because the matrix of second derivatives of the internal coordinates evaluated at a minimum (the Hessian at the minimum) must be positive definite. If for any reason the expansion is made not about the equilibrium geometry, but rather at some other geometry, then it is possible to get negative values of λ_i and hence imaginary vibration frequencies.

Because the elements of $\boldsymbol{\mu}^0$ are constants then (43) is just the Hamiltonian for an asymmetric top. The rotational matrix element in (39) simplifies to

$$< JMk' \mid \hat{K}(\boldsymbol{\phi}) \mid JMk >=$$

$$\frac{\hbar^2}{4}\left(\frac{(\mu^0_{xx} - \mu^0_{yy})}{2}\left(C^+_{Jk+1}C^+_{Jk}\delta_{k'k+2} + C^-_{Jk-1}C^-_{Jk}\delta_{k'k-2}\right)\right)$$

$$+\frac{\hbar^2}{2}\left(\frac{(\mu^0_{xx} + \mu^0_{yy})}{2}\left(J(J+1) - k^2\right) + \mu^0_{zz}k^2\right)\delta_{k'k} \tag{45}$$

The $2(J+1)$ dimensional secular problem composed of these matrix elements cannot generally be solved to give an energy expression in closed form but the rotational wave functions solutions are of the form

$$^M\chi_\tau^J(\phi) = \sum_{k=-J}^{k=J} c_{\tau k}^J |JMk>, \quad \tau = -J, -J+1, \cdots, J$$

The $c_{\tau k}^J$ are constant coefficients and each rotational wave function is associated with an energy $E_{J\tau}$.

If two of the equilibrium moments of inertia are the same (the symmetric top case) then these may be designated as x and y and the first term in (45) vanishes. The energy is then given by the last term in (45) and the $|JMk>$ are individually angular eigenfunctions. Thus for the symmetric top k is a good quantum number. These matters are treated in standard texts on molecular spectroscopy.

The Hamiltonian (44) simply represents a sum of non-interacting Harmonic oscillators, each with a wavefunction of the form

$$\psi_{n_i}(Q_i) = N_i e^{\frac{\alpha_i Q_i^2}{2}} H_{n_i}(\sqrt{\alpha_i} Q_i)$$

where

$$\alpha_i = \frac{\sqrt{\lambda_i}}{\hbar} \equiv \frac{\omega}{\hbar}$$

and the energy of the oscillator is

$$\epsilon_{n_i} = (n_i + \frac{1}{2})\hbar\omega \equiv (n_i + \frac{1}{2})h\nu$$

The full vibrational wave function is then

$$\Psi(\mathbf{Q}) = \prod_{i=1}^{3H-6} \psi_{n_i}(Q_i)$$

and the total vibrational energy of the system is just

$$E_v = \sum_{i=1}^{3H-6} \epsilon_{n_i}$$

The wavefunction for the nuclear motion part of the problem arises from the simplification of (31) and is a product

$$^M\chi_\tau^J(\phi)\Psi(\mathbf{Q})$$

and the wavefunction for the full problem in the single product approximation is

$$\psi(\mathbf{x}^n, \mathbf{x}^e) \Rightarrow T(\mathbf{X}_T)\, {}^M\chi_\tau^J(\boldsymbol{\phi})\Psi(\mathbf{Q})\phi(\mathbf{a}, \mathbf{z}^e)$$

in which $\phi(\mathbf{a}, \mathbf{z}^e)$ is the electronic wavefunction taken at the equilibrium geometry and with the electronic coordinates expressed in the frame fixed in the body denoted $\mathbf{z}^e$. The total energy of the molecule in this approximation is

$$E = E_T + E_{J\tau} + E_v + V(\mathbf{a})$$

The translational energy E_T is usually ignored as is the translational wavefunction and the fact that in this approximation the energy is the sum of an electronic and a rotational and a vibrational part is often (quite wrongly) said to specify the Born Oppenheimer approximation.

So to compute the properties of isolated spectroscopic molecule in quantum mechanics, the following separate problems need to be solved. We must first perform an electronic structure calculation and identify the equilibrium geometry. At that geometry we must calculate, in a convenient coordinate system, the Hessian for the nuclear variables. The electronic structure calculation must therefore be good enough to distinguish the lowest energy minimum from any others that might occur locally and at the lowest energy minimum, be good enough to yield a decent equilibrium geometry and a potential surface that is parallel to the exact one for small displacements about this geometry. For a closed-shell system an LCAO-MO-SCF calculation is generally believed to be adequate if a suitably large basis is chosen. If it is necessary to go further, as it sometimes is to make sure that the true minimum has been found, perturbation techniques such as MP2 are used.

The Hessian determined from the electronic structure calculation can be diagonalised to yield the eigenvalues λ_i and hence the harmonic the vibrational frequencies of the molecule. The eigenvectors express the normal coordinates in terms of the coordinates used to evaluate the Hessian. The equilibrium inertia tensor can be calculated from the nuclear coordinates that specify the equilibrium geometry. This tensor can then be diagonalised to give the moments of inertia and the secular problem with matrix elements given by (45) can be solved to yield the rotational energy levels. The Hessian is usually constructed using the cartesian coordinates in which the clamped nuclei calculations are performed. In this case then, the Hessian is of dimension $3H$ but six of its eigenvalues are zero corresponding to the three possible uniform translations and the three possible rigid rotations of the equilibrium geometry.

Provided that this simple picture allows a reasonable interpretation of the molecular spectrum then it is worthwhile refining it by passing beyond the harmonic approximation and allowing for a more realistic representation of the potential. Also by allowing for some non-rigidity in the rotor. All the corrections at this level can be treated by first and second order perturbation theory. At this level then, the isolated molecule problem can clearly be solved. However, if vibration-rotation interaction is important then perturbation theoretic treatments are often inadequate. Also if the electronic motion interacts with the nuclear motion, a state of affairs usually called the breakdown of the Born Oppenheimer approximation, it is less clear how best to deal with it. Indeed in these circumstances the isolated molecule problem must be considered still unsolved.

However it should be remembered that the picture given above predated any possibility of decent calculations. Following Eckart's paper [9] the scheme outlined above was developed and used to interpret spectra, that is, it was assumed to be valid and experiment was used to parametrise it. In this process it was discovered that spectra could be interpreted in terms of molecular properties that were, in good approximation, transferable. Thus the rotational spectrum of a pair of molecules that would be thought of as having chemically similar bonds can be interpreted by means of an equilibrium geometry in which the common bonds are the same length. Similarly the vibrational spectrum of both can be interpreted by constructing a potential quadratic in terms of bond lengths and bond angles as internal coordinates with characteristic force constants for each quadratic term. For example, it is found that C–H stretching force constant (one quadratic in the displacement along the bond) can be given the same value in whatever molecule it occurs. This picture is the one that experimentalists use and it is in terms of this picture that you will be expected to present any results that you have calculated. This does not necessarily mean that you must calculate within this scheme, but it does mean that you must present your results in terms of it.

The approach outlined in this section is probably not going to be helpful when considering really large molecules. Once the usual state of the system is a liquid or a solid without a significant gas phase component, then looking at its wave function in the manner outlined above, is rather a waste of time. However it is still quite useful to assume that the internal motions can be separated off and to consider an aspect of the behaviour of such molecules as arising from their tendency to achieve a minimum energy geometry on a potential energy surface with many minima in it.

4. The electromagnetic field as an environment

The problem to be solved here is how to treat fields in quantum mechanics in order to describe the results of the many interesting experiments that result from molecules responding to applied fields. Among the effects that result from the presence of fields are the Stark effect, the Zeeman effect, the Kerr effect and the Cotton-Mouton and the Faraday effects. Intuitively we tend to believe that fields usually have only a small effect on molecules. So we tend to believe that in the esr or nmr experiment we are essentially looking at signals resulting from the molecule, perturbed only to a small extent. To examine these assumptions it is not necessary to pick out the isolated molecule initially and we can mount a slightly more general discussion.

4.1. THE SYSTEM IN AN ELECTROMAGNETIC FIELD

In quantum mechanics the electromagnetic operators (see [19], [20]) are multiplicative and are the scalar potential ϕ and the vector potential $\mathbf{A}$ associated with the electric field strength $\mathbf{E}$ and the magnetic flux density $\mathbf{B}$ according to

$$\mathbf{E} = -\vec{\nabla}\phi - \frac{\partial \mathbf{A}}{\partial t}$$

$$\mathbf{B} = \vec{\nabla} \times \mathbf{A}$$

The electromagnetic operators in the Hamiltonian represent interactions between the field and the particles and so are written as $\phi(\mathbf{x}_i)$ and $\mathbf{A}(\mathbf{x}_i)$ and associated with the $i-$th particle. In the presence of a field, it can be shown that the Hamiltonian (1) becomes

$$\hat{H}(\mathbf{x},\phi,\mathbf{A}) = \sum_{i=1}^{N} \frac{1}{2m_i}(\frac{\hbar}{i}\vec{\nabla}(\mathbf{x}_i) - eZ_i\mathbf{A}(\mathbf{x}_i)) \cdot (\frac{\hbar}{i}\vec{\nabla}(\mathbf{x}_i) - eZ_i\mathbf{A}(\mathbf{x}_i))$$

$$+\sum_{i=1}^{N} eZ_i\phi(\mathbf{x}_i) + \frac{e^2}{8\pi\epsilon_o}\sum_{i,j=1}^{N}{}' \frac{Z_iZ_j}{x_{ij}} \tag{46}$$

Expanding the first term gives

$$\hat{H}(\mathbf{x},\phi,\mathbf{A}) = \hat{H}(\mathbf{x}) + \hat{H}'(\mathbf{x},\phi,\mathbf{A})$$

where $\hat{H}(\mathbf{x})$ is given by (1) and

$$\hat{H}'(\mathbf{x},\phi,\mathbf{A}) =$$

$$\sum_{i=1}^{N}\frac{1}{2m_i}(-eZ_i\frac{\hbar}{i}\vec{\nabla}(\mathbf{x}_i)\cdot\mathbf{A}(\mathbf{x}_i) - eZ_i\mathbf{A}(\mathbf{x}_i)\cdot\frac{\hbar}{i}\vec{\nabla}(\mathbf{x}_i) + e^2Z_i^2|\mathbf{A}(\mathbf{x}_i)|^2) + \sum_{i=1}^{N} eZ_i\phi(\mathbf{x}_i) \tag{47}$$

It is usual to write

$$\boldsymbol{\pi}(\mathbf{x}_i) = (\frac{\hbar}{i}\vec{\nabla}(\mathbf{x}_i) - eZ_i\mathbf{A}(\mathbf{x}_i))$$

and to call it either the mechanical momentum or the gauge invariant momentum of the particle. The reason for the latter name arises from the observation that replacement of the potentials by the new potentials according to

$$\mathbf{A} \to \mathbf{A}' = \mathbf{A} + \vec{\nabla} f$$

$$\phi \to \phi' = \phi - \frac{\partial f}{\partial t}$$

where f is an arbitrary function of position and time, leads to exactly the same fields. These equations define a change of gauge.

In what follows, the potentials will be assumed not to depend explicitly on time, so the time partial derivatives will not be considered further. In this case it is easy to see that $\boldsymbol{\pi}$ has an invariant expectation value under gauge change provided that the original function undergoes a phase change.

$$\psi \to \psi' = e^{i\lambda}\psi$$

with

$$\lambda = \frac{Ze}{\hbar} f$$

It follows therefore that the physical quantities calculated from the Hamiltonian will be gauge invariant as is required.

The freedom that gauge change allows can be used to recast the Hamiltonian in a way that is particularly useful for some purpose or other. Thus because

$$\vec{\nabla} \cdot \mathbf{A}\psi = \mathbf{A} \cdot (\vec{\nabla}\psi) + \psi(\vec{\nabla} \cdot \mathbf{A})$$

the second term can be made to vanish if f is chosen to make $\vec{\nabla} \cdot \mathbf{A}$ zero. This choice of gauge is often called coulomb gauge but a number of other choices are commonly made too.

In the case of a uniform electric field and of a uniform magnetic flux density it is easily seen that

$$\mathbf{A} = \frac{1}{2}\mathbf{B} \times \vec{\mathbf{x}}$$

and

$$\phi = -\mathbf{E} \cdot \vec{\mathbf{x}}$$

where $\mathbf{E}$ and $\mathbf{B}$ are constant vectors. Notice that both these quantities depend upon a particular choice of origin for the position vector $\vec{\mathbf{x}}$ and that any constant translation (shift of origin) leaves the derived fields invariant.

Notice too that **A** satisfies the coulomb gauge condition. In this case (47) may be re-written as

$$\sum_{i=1}^{N} \frac{1}{2m_i}(-eZ_i\mathbf{B}\cdot\frac{\hbar}{i}\vec{\mathbf{x}}_i \times \vec{\nabla}(\mathbf{x}_i) + e^2 Z_i^2 |\frac{\mathbf{B}}{2} \times \vec{\mathbf{x}}_i|^2) - e\mathbf{E}\cdot\sum_{i=1}^{N} Z_i\vec{\mathbf{x}}_i \qquad (48)$$

It might be prudent here to say a little about units and dimensions. The SI unit of electric field strength is the volt metre^{-1} (V m^{-1}) and the unit of magnetic flux density is the tesla (T). The unit of scalar potential is therefore the volt which in the present context is most helpfully regarded as having units of joule coulomb^{-1} (J C^{-1}). The vector potential has units of tesla metre (T m).

In (48) the term linear in the magnetic field is usually interpreted as a field-dipole interaction with the magnetic dipole arising from the vector product term with a coefficient of the form $e\hbar m_i^{-1}$. If m_i is taken as the electron mass then this coefficient has the approximate value of 1.8546 10^{-23} J T^{-1} and if m is taken as the proton mass then it is about 2000 times smaller. Commonly occurring **B** fields are all less than 1 T. The value for the earth's field is about 5 10^{-5} T, that at the centre of a circular coil of 1000 turns each of radius 10^{-1} m and carrying 1.5 A is about 10^{-2} T. However that between the pole pieces of a large electromagnet is about 2 T while the latest in superconducting magnets for nmr machines have a maximum field of about 20 T. In our simple molecular picture we should expect low-lying electronic states to be separated by about 10^{-18} J, low-lying vibrational states by about 10^{-21} J, and low-lying rotational states by about 10^{-23} J.

The electric field term is usually interpreted as an electric dipole interaction and the basic unit is ea_o where $a_o = \frac{4\pi\epsilon_o\hbar^2}{me^2}$. This unit is approximately 8.4778 10^{-30} C m (about 2.54 Debye) and thus the electric field term yields only small effects for most commonly occurring cases. However it can be shown [21] that an electric field makes a qualitative difference to the problem. In the presence of an electric field, no matter how small, the system becomes metastable and has no discrete spectrum at all. It can also be shown that to treat the electric field as a perturbation to the field-free problem gives perturbation series which are divergent. Nevertheless such series seem to give the correct answer in low order and we shall assume that we can use them in what follows.

Under the change to translationally invariant coordinates the term linear in $\mathbf{B}$ in (48) becomes

$$-\frac{e\hbar\mathbf{B}}{2i}\cdot\sum_{i=1}^{N}\left(\frac{Z_i}{m_i}\vec{\mathbf{x}}_i\times\vec{\nabla}(\mathbf{x}_i)\right)=$$

$$-\frac{e\hbar\mathbf{B}}{2i}\cdot\Big(Z_T {M_T}^{-1}\vec{\mathbf{X}}_T\times\vec{\nabla}(\mathbf{X}_T)+$$

$$\vec{\mathbf{X}}_T\times\sum_{j=1}^{N-1}\tilde{p}_j\vec{\nabla}(\mathbf{t}_j)+{M_T}^{-1}\sum_{j=1}^{N-1}\tilde{Z}_j\vec{\mathbf{t}}_j\times\vec{\nabla}(\mathbf{X}_T)+\sum_{j,k=1}^{N-1}\tilde{q}_{jk}\vec{\mathbf{t}}_j\times\vec{\nabla}(\mathbf{t}_k)\Big) \quad (49)$$

where Z_T is the total charge of the system and

$$\tilde{p}_j=\sum_{i=1}^{N}\frac{Z_i}{m_i}V_{ij},\qquad \tilde{Z}_j=\sum_{i=1}^{N}Z_i(\mathbf{V}^{-1})_{ji},\qquad \tilde{q}_{jk}=\sum_{i=1}^{N}\frac{Z_i}{m_i}(\mathbf{V}^{-1})_{ji}V_{ik}$$

The term quadratic in $\mathbf{B}$ is

$$\frac{e^2}{8}\sum_{i=1}^{N}\frac{Z_i^2}{m_i}\left(|\mathbf{B}|^2|\vec{\mathbf{x}}_i|^2-|\mathbf{B}\cdot\vec{\mathbf{x}}_i|^2\right)=$$

$$\frac{e^2}{8}\sum_{i=1}^{N}\frac{Z_i^2}{m_i}\Big(((|\mathbf{B}|^2|\vec{\mathbf{X}}_T|^2-|\mathbf{B}\cdot\vec{\mathbf{X}}_T|^2)$$

$$+2\sum_{j=1}^{N-1}(\mathbf{V}^{-1})_{ji}\vec{\mathbf{t}}_j\cdot(\vec{\mathbf{X}}_T|\mathbf{B}|^2-\mathbf{B}(\mathbf{B}\cdot\vec{\mathbf{X}}_T))$$

$$+\sum_{j,k=1}^{N-1}(\mathbf{V}^{-1})_{ji}(\mathbf{V}^{-1})_{ki}(|\mathbf{B}|^2\vec{\mathbf{t}}_j\cdot\vec{\mathbf{t}}_k-(\mathbf{B}\cdot\vec{\mathbf{t}}_j)(\mathbf{B}\cdot\vec{\mathbf{t}}_k))\Big) \quad (50)$$

The term in the electric field becomes

$$-e\mathbf{E}\cdot(\vec{\mathbf{X}}_T Z_T+\sum_{i=1}^{N}Z_i\sum_{j=1}^{N-1}(\mathbf{V}^{-1})_{ji}\vec{\mathbf{t}}_j) \quad (51)$$

or

$$-e\mathbf{E}\cdot(\vec{\mathbf{X}}_T Z_T+\sum_{j=1}^{N-1}\tilde{Z}_j\vec{\mathbf{t}}_j)$$

$\tilde{Z}_j$ is now seen to be an effective charge for the pseudo-particle with coordinate $\mathbf{t}_j$. Some idea of its order of magnitude may be gained by means of an example. If we consider a neutral atom with all electronic variables measured from the nucleus then the $\mathbf{t}_j$ become the electronic pseudo-particle variables and with effective charges,in this case, of precisely -1. It seems to be the case then that a physically motivated choice of the translationally invariant coordinates can lead to physically reasonable values of the effective charge. In this context it is plausible to suppose that the third term in (49) will be small and that it may be neglected, as it usually is. Using the same example the p_j turn out to be $-(m_1 m)^{-1} M_T$ where m_1 is the nuclear and m the electronic mass. So they are approximately of order m^{-1} and certainly not small. In the same approximation the q_{jk} are negative and of order m^{-1} when $j = k$ but are of order m_1^{-1} when $j \neq k$. This last observation makes it reasonable to regard the fourth term in (49) as approximately the electronic angular momentum of the atom and the approximation would be exact if m_1 were really infinite. This is certainly comforting as a justification of what we usually do but the fact that p_j is large means that we are not, at least *a priori*, justified in neglecting the second term in (49), which we also usually do.

Looking at the terms in (50) in the same way, all of them, including the first one on the right, have coefficients which, in the limit of infinite nuclear mass, are of order m^{-1}. So again it is not clear that we are free to neglect them.

Now it would obviously be much neater if it were possible to perform a gauge transformation so as to remove from (49) and (50) the terms corresponding to interactions between the translational motion and the internal motion but this can be done only with a rather particular choice of phase factor f (see for example [22] and for a general discussion [23]) and is not helpful in the present discussion. In any case the interaction between the moving system and the fields is a real physical phenomenon whose effects cannot be transformed away.

If the system is not electrically neutral then the first term on the right in (49) gives rise to cyclotron motion for the centre of mass while the first term in (51) gives rise to a centre of mass Stark effect. If the system is electrically neutral ($Z_T = 0$) then the first term on the right in both (49) and (51) vanishes. The centre of mass motion then occurs in the second and third term on the right in (49) and in the first two terms on the right in (50). To estimate the size of the effects due to these terms we would expect to calculate their expectation values between translation functions of the type $T(\mathbf{X}_T)$ from (15). However such functions, if chosen to be eigenfunctions of the free centre of mass problem, are

$$T(\mathbf{X}_T) = A e^{i\vec{K}\cdot\vec{\mathbf{X}}_T}$$

and are not square integrable. There is no agreed method for dealing with this problem. One possibility is to treat the centre of mass motion as occurring in a box subject to periodic boundary conditions. This is a widely used trick. It is used, for example, when the quantization of the radiation field is considered and in solid state theory. If the periodic choice is used then the translational wavefunction becomes

$$T(\mathbf{X}_T) = D^{-3/2} e^{i\vec{K}\cdot\vec{\mathbf{X}}_T}, \quad \vec{K} = \frac{2\pi}{D}(n_x\hat{\mathsf{i}} + n_y\hat{\mathsf{j}} + n_z\hat{\mathsf{k}}) = \frac{2\pi}{D}\vec{n}$$

in which D is the length of the side of a cubic box treated as the repeating unit and the n_α are $0, \pm 1, \pm 2, \ldots$. The integration of $\mathbf{X}_T$ is in the range $(0, D)$.

This wave function is a momentum eigenfunction so

$$\vec{\nabla}(\mathbf{X}_T)\, T(\mathbf{X}_T) = i\vec{K}\, T(\mathbf{X}_T)$$

and the expected value of the centre of mass vector is

$$< T|\vec{\mathbf{X}}_T|T > = \frac{\vec{D}}{2}, \quad \vec{D} = D\hat{\mathsf{d}}, \quad \hat{\mathsf{d}} = (\hat{\mathsf{i}} + \hat{\mathsf{j}} + \hat{\mathsf{k}})$$

and the expected value of its square is

$$< T|\, |\vec{\mathbf{X}}_T|^2\, |T > = |\vec{D}|^2 \equiv D^2$$

Assuming that this is a satisfactory wave function it follows that, on integrating over $\mathbf{X}_T$, the expected values of the first three terms on the right in (49) are

$$-\frac{eZ_T h}{4M_T}\mathbf{B}\cdot\hat{\mathsf{d}}\times\vec{n}$$

$$-\frac{e\hbar}{4i}\mathbf{B}\cdot\vec{D}\times\sum_{j=1}^{N-1}\tilde{p}_j\vec{\nabla}(\mathbf{t}_j) - \frac{eh}{2M_T D}\mathbf{B}\cdot\sum_{j=1}^{N-1}\tilde{Z}_j\vec{\mathbf{t}}_j\times\vec{\mathbf{n}}$$

The interaction terms depend on the length D of the box side and this is a very unsatisfactory outcome since this parameter is entirely at our disposal and may be assigned at will. The larger we choose D to be, the larger the first interaction term and the smaller the second, other things being equal. But as we have seen in the atom example, the first term is unlikely to be intrinsically small and the second term is, so our difficulties are actually compounded by our field choice. At present there seems to be no satisfactory way round this problem.

Turning to (50), the expected value of first two terms on the right is

$$\frac{e^2}{8}\sum_{i=1}^{N}\frac{Z_i^2}{m_i}\Big((|\mathbf{B}|^2|\vec{D}|^2 - |\mathbf{B}\cdot\vec{D}|^2)$$

$$+\sum_{j=1}^{N-1}(\mathbf{V}^{-1})_{ji}\vec{\mathbf{t}}_j\cdot(\vec{D}|\mathbf{B}|^2 - \mathbf{B}(\mathbf{B}\cdot\vec{D}))\Big)$$

and again the position here is unsatisfactory as it obviously is too for the expression arising from the first term in (51).

In exceptionally accurate work on atoms (see for example [24] but also see [25]) the energy arising from the collective motion alone is subtracted from the total energy and this appears yield quantities that are independent of the choices made with respect to the translational function. However the cited work does not approach the problem as has been done here. The traditional approach in molecular physics is simply to ignore these terms for neutral systems and to use rather heroic approximations in the case of charged systems. For the present we must simply flag up this area as one in which there are many unsolved problems which we shall ignore and turn our attention to the internal motion part of the problem.

The last term in (49) is an internal motion term which arises from the angular momentum of the pseudo-particles and is generally regarded as a paramagnetic term. It is the term responsible for the non-spin part of the Zeeman splitting. The last term in (50) is also an internal motion term and is reminiscent in form of a classical dipole interaction term. It is always positive and so is a diamagnetic term. To consider these matters further we need now to re-incorporate the molecule.

4.2. THE MOLECULE IN AN ELECTROMAGNETIC FIELD

We need here to expand a little on the discussion given earlier of the isolated molecule and the position of the clamped nuclei approximation. Although as was said earlier, there are formidable technical problems in the way of making a precise argument about the status of the clamped nuclei electronic Hamiltonian, it is generally agreed that to get an electronic Hamiltonian anything at all like it, one must choose the centre of nuclear mass coordinate as origin for the translationally invariant electronic coordinates and choose L such variables. Thus we can have only $H-1$ translationally invariant nuclear coordinates and these must be expressed entirely in terms of of the original nuclear coordinates so that

$$\mathbf{x}_i^e = \mathbf{t}_i^e + \mathbf{X} \tag{52}$$

and

$$\mathbf{x}_i^n = \sum_{j=1}^{H-1} \mathbf{t}_j (\mathbf{V}^n)_{ji}^{-1} + \mathbf{X} \tag{53}$$

in which

$$\mathbf{X}_T = \mathbf{X} + M_T^{-1} m \sum_{i=1}^{L} \mathbf{t}_i^e$$

The derivative operators are then

$$\vec{\nabla}(\mathbf{x}_i^e) = \vec{\nabla}(\mathbf{t}_i^e) + mM_T{}^{-1}\vec{\nabla}(\mathbf{X}_T) \tag{54}$$

$$\vec{\nabla}(\mathbf{x}_i^n) = \sum_{j=1}^{H-1} V_{ij}^n \vec{\nabla}(\mathbf{t}_j) - m_i M^{-1} \sum_{j=1}^{L} \vec{\nabla}(\mathbf{t}_j^e) + m_i M_T{}^{-1}\vec{\nabla}(\mathbf{X}_T) \tag{55}$$

The terms from (49) that involve only the internal coordinates then are

$$-m^{-1} \sum_{i=1}^{L} \mathbf{t}_i^e \times \vec{\nabla}(\mathbf{t}_i^e) + M_T^{-1}(1 + Z_H m M^{-1}) \sum_{i,k=1}^{L} \mathbf{t}_i^e \times \vec{\nabla}(\mathbf{t}_k^e)$$

$$+ \sum_{k=1}^{L} \sum_{j=1}^{H-1} (M^{-1} \overline{Z}_j \mathbf{t}_j \times \vec{\nabla}(\mathbf{t}_k^e) - M_T^{-1} m p_j \mathbf{t}_k^e \times \vec{\nabla}(\mathbf{t}_j))$$

$$+ \sum_{j,k=1}^{H-1} q_{jk} \vec{\mathbf{t}}_j \times \vec{\nabla}(\mathbf{t}_k) \tag{56}$$

in which

$$p_j = \sum_{i=1}^{H} Z_i m_i^{-1} V_{ij}^n, \qquad q_{jk} = \sum_{i=1}^{H} Z_i m_i^{-1} (\mathbf{V}^n)_{ji}^{-1} V_{ik}^n$$

and so are restrictions to the nuclear variables alone of the terms defined earlier belowIn fact force constants quoted in these units have exactly the same value when they are quoted in the proper SI unit of Newton metre^{-1}. (49). Similarly $\overline{Z}_j$ is defined in terms of $(\mathbf{V}^n)^{-1}$ and so is an effective charge for the j−th nuclear pseudo-particle and Z_H is the total nuclear charge. In a neutral molecule this will just be L, making the coefficient of the second term in (56) M^{-1}.

The terms from (50) that involve only the internal coordinates are

$$m^{-1}\sum_{i=1}^{L}(|\mathbf{B}|^2|\vec{\mathbf{t}}_i^e|^2-(\mathbf{B}\cdot\vec{\mathbf{t}}_i^e)^2))+M'^{-1}(|\mathbf{B}|^2|\sum_{i=1}^{L}\vec{\mathbf{t}}_i^e|^2-(\mathbf{B}\cdot\sum_{i=1}^{L}\vec{\mathbf{t}}_i^e)^2)$$

$$-2mM_T^{-1}\sum_{j=1}^{H-1}\sum_{k=1}^{L}\left(\sum_{i=1}^{H}Z_i^2m_i^{-1}(\mathbf{V}^{n-1})_{ji}\right)\vec{\mathbf{t}}_j\cdot(|\mathbf{B}|^2\vec{\mathbf{t}}_k^e-(\mathbf{B}\cdot\vec{\mathbf{t}}_k^e))$$

$$+\sum_{j,k=1}^{H-1}\left(\sum_{i=1}^{H}Z_i^2m_i^{-1}(\mathbf{V}^{n-1})_{ji}(\mathbf{V}^{n-1})_{ki}\right)(|\mathbf{B}|^2\vec{\mathbf{t}}_j\cdot\vec{\mathbf{t}}_k-(\mathbf{B}\cdot\vec{\mathbf{t}}_j)(\mathbf{B}\cdot\vec{\mathbf{t}}_k))\quad(57)$$

in which

$$M'^{-1}=M_T^{-2}(Lm+m^2\sum_{i=1}^{H}Z_i^2m_i^{-1})-2M_T^{-1}$$

If $Z_im_i^{-1}$ were a constant then p_j would vanish and q_{jk} would become diagonal and the last term in (56) would be proportional to the total nuclear angular momentum. However $Z_im_i^{-1}$ cannot really be a constant, for different isotopes have the same charge. However if m_i is regarded as the relative atomic mass, a thing commonly done in interpreting spectra, it being justified as allowing for the electron following motion in the Born Oppenheimer approximation, then the ratio is, in suitable units, roughly a half at least as far up in the periodic table as sulphur. Thus we might expect to be able to ignore the second and third terms in (56) and to concentrate on dealing with the two angular momentum-like terms remaining. It is difficult to make even such a crude analysis as this for the terms in (57) but in general here too, it seems as if the first and last terms are likely to be the most important.

The internal motion part of the electric field term is

$$-e\mathbf{E}\cdot((mM_T^{-1}Z_T-1)\sum_{i=1}^{L}\vec{\mathbf{t}}_i^e+\sum_{j=1}^{H-1}\overline{Z}_j\vec{\mathbf{t}}_j)\qquad(58)$$

It must be stressed again that these results depend intrinsically on the internal coordinate choice. They would not be the same in any other choice.

To transform these operators further to give their forms in terms of orientation coordinates and rotationally invariant coordinates we need, as above, to expand a little on our earlier discussion. The transformation to orientation and internal coordinates is achieved via

$$\mathbf{t}_i^e=\mathbf{C}\mathbf{z}_i^e,\qquad \mathbf{t}_i=\mathbf{C}\mathbf{z}_i$$

in which the matrix $\mathbf{C}$ is defined entirely in terms of the translationally invariant nuclear coordinates $\mathbf{t}_i$. Thus the eulerian angles, which are defined by $\mathbf{C}$ are functions of the translationally invariant nuclear coordinates only. The $3H-6$ rotationally invariant coordinates q_k are now defined entirely in terms of the $\mathbf{t}_i$.

The derivative operators are

$$\frac{\partial}{\partial \mathbf{t}_i} = \mathbf{C}(\boldsymbol{\Omega}^i\mathbf{D}\frac{\partial}{\partial \boldsymbol{\phi}} + \mathbf{Q}^i\frac{\partial}{\partial \mathbf{q}} + \boldsymbol{\Omega}^i\sum_{j=1}^{L}\hat{\mathbf{z}}_j^e\frac{\partial}{\partial \mathbf{z}_j^e}) \tag{59}$$

and

$$\frac{\partial}{\partial \mathbf{t}_i^e} = \mathbf{C}\frac{\partial}{\partial \mathbf{z}_i^e} \tag{60}$$

where $\partial/\partial\boldsymbol{\phi}$ and $\partial/\partial\mathbf{q}$ are column matrices of 3 and $3N-6$ partial derivatives respectively and $\partial/\partial\mathbf{t}_i$ and $\partial/\partial\mathbf{z}_i^e$ are column matrices of 3 partial derivatives. Substituting these into $\hat{H}(\mathbf{t})$ yields an electronic part that can be written as

$$\hat{H}^{cn}(\mathbf{a},\mathbf{z}^e) = -\frac{\hbar^2}{2m}\sum_{i=1}^{L}\nabla^2(\mathbf{z}_i^e) - \frac{e^2}{4\pi\epsilon_0}\sum_{i=1}^{H}\sum_{j=1}^{L}\frac{Z_i}{|\mathbf{z}_j^e-\mathbf{a}_i|} + \frac{e^2}{8\pi\epsilon_0}\sum_{i,j=1}^{N}{}'\frac{1}{|\mathbf{z}_i^e-\mathbf{z}_j^e|} \tag{61}$$

and so is essentially the same as the standard clamped nucleus Hamiltonian (33) but expressed in coordinates defined in a frame which, roughly speaking, rotates with the molecule. They also yield the nuclear motion parts given in (35) and (36). It should be stressed here again, as it was earlier, that to reach these results it is necessary to neglect some electron-nucleus coupling terms. The neglected terms it is hoped will be small since they all involve inverse powers of either the total mass M_T or the total nuclear mass M. It is thus perfectly consistent of us to ignore similar terms arising from (56) and (57) and we shall ignore them without further comment.

It might be hoped that in view of the standard approach to the molecular Zeeman effect (see for example [26]) in which the molecule is treated as a magnetic dipole arising from the total angular momentum interacting with the magnetic field, it would be possible to eliminate the first and third terms from (56) in favour of a term involving just the total angular momentum operator. But this does not appear to be possible.

Transforming the first term and reincorporating the field gives

$$\frac{e\hbar\mathbf{B}'}{2mi}\cdot\sum_{i=1}^{L}\vec{\mathbf{z}}_i^e\times\vec{\nabla}(\mathbf{z}_i^e)$$

in which

$$B'_\alpha = |\mathbf{C}|\sum_\beta B_\beta C_{\beta.\alpha}$$

The components of $\mathbf{B}'$ are those of the magnetic induction expressed in the axis system defined from the laboratory fixed one by means of $\mathbf{C}$. They can be regarded as components expressed in terms of the axis frame rotating with the molecule, as outlined above. This term is essentially proportional to the electronic angular momentum expressed in the rotating frame. The last term, however, is somewhat more problematic. It is written as

$$\sum_{j,k=1}^{H-1} q_{jk}\vec{\mathbf{t}}_j \times \vec{\nabla}(\mathbf{t}_k) = |\mathbf{C}|\mathbf{C}\left(\boldsymbol{\Omega}\,(\mathbf{D}\frac{\partial}{\partial \boldsymbol{\phi}} + \sum_{j=1}^{L} \hat{\mathbf{z}}_j^e \frac{\partial}{\partial \mathbf{z}_j^e}) + \mathbf{Q}\frac{\partial}{\partial \mathbf{q}}\right)$$

where

$$\boldsymbol{\Omega} = \sum_{j,k=1}^{H-1} q_{jk}\hat{\mathbf{z}}_j\boldsymbol{\Omega}^k, \quad \mathbf{Q} = \sum_{j,k=1}^{H-1} q_{jk}\hat{\mathbf{z}}_j\mathbf{Q}^k$$

To get explicit forms for $\boldsymbol{\Omega}$ and $\mathbf{Q}$, the $\mathbf{z}_j$ must be expressed in terms of the q_k and the precise form of these expressions will depend on that relationship. In what follows it is assumed that we know appropriate expression for $\mathbf{z}_j$.

Combining the first and the third terms and incorporating the field gives

$$-\frac{e\hbar\mathbf{B}'}{2i}\left(\boldsymbol{\Omega}\,\mathbf{D}\frac{\partial}{\partial \boldsymbol{\phi}} + \mathbf{Q}\frac{\partial}{\partial \mathbf{q}} + (\boldsymbol{\Omega} - m^{-1})\sum_{j=1}^{L} \hat{\mathbf{z}}_j^e \frac{\partial}{\partial \mathbf{z}_j^e}\right) \tag{62}$$

where now $\mathbf{B}'$ is to be treated as a row matrix of three components and where there is an implicit unit matrix multiplying m^{-1}. If the matrix with elements q_{jk} was a constant diagonal matrix then $\boldsymbol{\Omega}$ would be a negative unit matrix multiplied by that constant and $\mathbf{Q}$ would vanish. Were the constant in this case to be m^{-1}, then so would the electronic term too. This would leave just an angular momentum term (compare (24)) of the kind that is usually used to interpret the molecular Zeeman effect.

The first and third terms in (57) undergo an almost trivial change to give

$$m^{-1}\sum_{i=1}^{L}(\,|\mathbf{B}'|^2|\vec{\mathbf{z}}_i^e|^2 - (\mathbf{B}'\cdot\vec{\mathbf{z}}_i^e)^2)\,)$$

$$+\sum_{j,k=1}^{H-1}\left(\sum_{i=1}^{H} Z_i^2 m_i^{-1}(\mathbf{V}^{n-1})_{ji}(\mathbf{V}^{n-1})_{ik}\right)(\,|\mathbf{B}'|^2\vec{\mathbf{z}}_j\cdot\vec{\mathbf{z}}_k - (\mathbf{B}'\cdot\vec{\mathbf{z}}_j)(\mathbf{B}'\cdot\vec{\mathbf{z}}_k)\,) \tag{63}$$

The internal motion part of the electric field term becomes

$$-e\mathbf{E}'((mM_T^{-1}Z_T - 1)\sum_{i=1}^{L}\vec{\mathbf{z}}_i^e + \sum_{j=1}^{H-1}\tilde{Z}_j\vec{\mathbf{z}}_j) \tag{64}$$

where now $\mathbf{E}'$, which is defined just like $\mathbf{B}'$, is treated as a row matrix of three components. To calculate matrix elements of $\mathbf{C}$ over the rotational part of the wave function over, we note that by an extension of the arguments that led to (29) we can write the matrix in terms of the appropriate standard Wigner matrix $\mathcal{D}^1$ as

$$\mathbf{C} = \mathbf{X}\mathcal{D}^1\mathbf{X}^\dagger$$

and the resulting integrals are expressible in terms of products of $3-j$ symbols.

4.3. THE SPECTROSCOPIC MOLECULE IN AN ELECTROMAGNETIC FIELD

Making Watson's choice for the internal coordinates the derivative operator like (59) for a (redundant) coordinate is

$$\frac{\partial}{\partial \mathbf{t}_i} = \mathbf{C}(\boldsymbol{\Omega}^i(\mathbf{D}\frac{\partial}{\partial \phi} + \sum_{j=1}^{L} \hat{\mathbf{z}}_j^e \frac{\partial}{\partial \mathbf{z}_j^e} + \frac{i}{\hbar}\boldsymbol{\pi}) + m_i^{\frac{1}{2}} \sum_{k=1}^{3H-6} \mathbf{l}_{ki} \frac{\partial}{\partial \mathbf{Q}_k} +) \tag{65}$$

and the coordinate $\mathbf{z}_i$ is given in terms of the Q_k by

$$\mathbf{z}_i = \mathbf{a}_i + m_i^{-\frac{1}{2}} \sum_{k=1}^{3H-6} \mathbf{l}_{ki} Q_k \tag{66}$$

To be strictly compatible with our earlier discussion of the isolated molecule we should add to the derivative term a part to allow for the incorporation of the jacobian. This part is just a multiplicative term and it will be ignored as it is likely to be small except in regions where the jacobian is close to zero. Details of its calculation (which is rather tricky) can be found in [17].

These expressions only make proper sense if the terms on the left are regarded as being defined by means of the right hand side, on which all the terms are properly defined and none are redundant. With this proviso, it turns out that it is perfectly proper in (63) and (64), to replace according to

$$\sum_{j=1}^{H-1} (\mathbf{V}^{n-1})_{ji} \mathbf{z}_j \rightarrow \mathbf{z}_i$$

where $\mathbf{z}_i$ on the right is given by (66). In these coordinates (62) becomes

$$-\frac{e\hbar\mathbf{B}'}{2i}\left(\boldsymbol{\Omega}^m(\mathbf{D}\frac{\partial}{\partial\phi} + \frac{i}{\hbar}\boldsymbol{\pi}) + \mathbf{Q}^m\frac{\partial}{\partial\mathbf{Q}} + (\boldsymbol{\Omega}^m - m^{-1})\sum_{j=1}^{L}\hat{\mathbf{z}}_j^e\frac{\partial}{\partial\mathbf{z}_j^e}\right) \tag{67}$$

in which

$$\boldsymbol{\Omega}^m = \sum_{i=1}^{H} Z_i \hat{\mathbf{z}}_i \hat{\mathbf{a}}_i \mathbf{I}''^{-1}, \quad \mathbf{Q}^m_{\alpha k} = \sum_{i=1}^{H} Z_i m_i^{-\frac{1}{2}} (\hat{\mathbf{z}}_i \mathbf{l}_{ki})_\alpha$$

We thus have expressions for the molecule-field interactions in the standard coordinate system.

5. Conclusions

What I hope that we have seen is that the Schrödinger equation for an isolated molecule has to be quite carefully constructed within ordinary quantum mechanics. However we have seen that such an equation can be constructed and that the molecule can be understood in terms on an equilibrium geometry defined by a minimum in a potential energy surface, calculated in principle from the ordinary clamped nuclei electronic Hamiltonian. The full molecular Hamiltonian based on this and as realised in the Eckart-Watson approach, certainly seems to to describe the isolated spectroscopic molecule very well.

Although the description of a collection of molecules has not been dealt with, a context has been provided for any such description. It is hoped that it is now clear why any such description is liable to be tricky and open to disagreement (even more so than in the isolated molecule case) because there is nothing in the Schrödinger Hamiltonian as such, that is responsive to such a division. So extrinsic considerations will be of the essence of such descriptions. Furthermore the exigencies of calculation make it certain that in any account semi-empirical, even empirical and frankly model methods will have to be used.

When we come to consider placing the molecule in an electromagnetic field then even greater care is necessary than it was in the case of the isolated molecule. But we have seen how it can be done, at least for constant fields and we have indicated what the deep intrinsic problems are. The importance of the gauge choice and hence the coordinate choice is a matter of the first importance here. If we wanted to calculate the electronic contribution to the field-internal motion coupling terms arising from (62) or (67) using wave functions obtained from a clamped nuclei calculation, then we *must* choose the origin of any electronic coordinate as the centre of nuclear mass. It is only with such a choice that the clamped nuclei Hamiltonian can be placed properly in the present context and in which the coupling expressions are valid. But this is not as easy as it seems because the standard orbital forms take as their origin the nucleus on which they happen to be centred. This is complication has been studied (see for example [27] and also the article by J. Gauss in this volume) and can be dealt with but it requires care.

We have not attempted to incorporate the radiation field into our discussion and it would not have been too easy to do it. A recent paper [28] and a book [29] cover much of the relevant ground, albeit without considering the problem of translation and rotation separation. Neither have we attempted to deal with collections of molecules in a field. This is an area in which there are few investigations but one which seems fertile ground for future work.

References

1. Sutcliffe, B. T. (1993), *J. Chem. Soc., Faraday Transactions*, **89**, 2321.
2. Brink, D.M.and Satchler, G.R. (1968) *Angular Momentum* , 2nd ed. Clarendon Press, Oxford.
3. Biedenharn, L.C. and Louck, J.C. (1982) *Angular Momentum in Quantum Physics* , Addison- Wesley, Reading, Mass.
4. Condon E.U.and Shortley, G.H. (1935) *The theory of atomic spectra.*, Cambridge University Press, Cambridge.
5. Brown, J. M. and Howard, B. J. (1976) *Mol. Phys.*, **31**, 1517.
6. Zare, R. N. (1988) *Angular Momentum*, Wiley, New York.
7. Ezra, G. (1982) *Symmetry properties of molecules* , Lecture Notes in Chemistry **28**, Springer-Verlag, Berlin.
8. Schutz, B. (1980) *Geometrical methods of mathematical physics*, Cambridge University Press, Cambridge.
9. Eckart, C. (1935) *Phys. Rev.*, **47**, 552.
10. Watson, J.K.G. (1968) *Mol. Phys.*, **15**, 479.
11. Born, M. and Oppenheimer, J.R. (1927) *Ann.der Phys.*, **84**, 457.
12. Born, M. and Huang, K. (1955) *Dynamical Theory of Crystal Lattices* , Oxford University Press, Oxford..
13. Slater, J.C. (1927) *Proc. Nat. Acad. Sci.*, **13**, 423.
14. Berry, R. S. (1960) *Rev. Mod. Phys.*, **32**, 447.
15. Longuet-Higgins, H. C. (1963) *Molec. Phys.*, **6**, 445.
16. Bunker, P. R. (1979) *Molecular Symmetry and Spectroscopy*, Academic Press, London.
17. Louck, J.C. (1976) *J. Mol. Spec.*, **61**, 107.
18. Kemble, E.C. (1937) *The fundamental principles of Quantum Mechanics* , McGraw-Hill, New York.
19. McWeeny, R. (1989) *Methods of Molecular Quantum Mechanics*, 2nd edn., Academic Press, London.
20. Moss, R. E. (1973) *Advanced Molecular Quantum Mechanics*, Chapman and Hall, London.
21. Thirring, W. (1981) *A Course in Mathematical Physics, 3, Quantum Mechanics of Atoms and Molecules*, tr. E. M. Harrell, Springer-Verlag, New York.
22. Moss, R. E. and Perry, A. J. (1992) *Mol. Phys*, **23**, 957.
23. Johnson, B. R., Hirschfelder J. O. and Yang, K-H (1983) *Rev. Mod. Phys*, **55**, 109.
24. Baye, D. and Vincke, M. (1990) *J. Phys B: At. Mol Opt. Phys*, **23**, 246.
25. Kravchenko, Yu. P., Lieberman, M. A. and Johansson, B. (1996) *Phys. Rev A* **54**, 287.
26. Townes, C. H. and Schawlow, A. L. (1975) *Microwave Spectroscopy*, Dover, New York.
27. Hall, G. G. (1973) *Int. journ. Quant. Chem.*, **7**, 15.
28. Woolley, R. G. (1996) *Mol. Phys*, **88**, 291.
29. Craig, D. P. and Thirunamachandran, T. (1984) *Molecular quantum electrodynamics: an introduction to radiation-molecule interactions*, Academic Press., London.

QUANTUM-CHEMICAL MODELS

JACEK KARWOWSKI
Instytut Fizyki Uniwersytetu Mikołaja Kopernika
ul. Grudziądzka 5, 87-100 Toruń, Poland

1. Introduction

Over centuries of development of science, it gradually became clear that our theories are unable to reflect the "physical reality" in all its complexity but can only supply models which describe certain aspects of some phenomena. The Newtonian mechanics, the Maxwellian electrodynamics, thermodynamics, statistical physics, the special and the general theory of relativity - all supply us with a consistent set of models describing various aspect of the macroscopic world. The apparent completeness of this description has been broken by a failure to account for processes in which elementary (and, hence, non-distinguishable) particles take part. Systems ranging from single elementary particles to their aggregates such as atoms and their nuclei, molecules, polymers and crystals cannot be described by any of the above mentioned *classical theories.* Properties of these systems, in a surprisingly successful way, are described by *quantum mechanics*, probably the most strange and unusual theory ever invented. Its basic notions such as wavefunctions, observables, the Hilbert space, do not correspond to anything measurable and its conclusions are frequently contradictory to the "common sense". However, in spite of this, all experimental facts from the area of its applicability are in a perfect agreement with its predictions. In particular, all of quantum chemistry may be considered as a branch of applied quantum mechanics.

The aim of this presentation is a brief description of the basic ideas which are derived from quantum mechanics and form a base for most important quantum chemical theories. The models are presented without considering one of their most important features: converting them into computational algorithms. This subject is addressed in detail in other presentations. Special care has been taken to stress the similarities of various approaches rather than their differences. For this reason the presentation

S. Wilson and G. H. F. Diercksen (eds.), Problem Solving in Computational Molecular Science, 37–84.

has been constructed from a rather general perspective, starting from the fundamental postulates of quantum mechanics. However it seems that only in this way one can show a simple path leading from the basic assumptions of our model to its final conclusions. The paper is divided into five sections. In the first two sections, a review of concepts leading to formulation of the Schrödinger equation and of the Dirac equation, their properties and relations between them are discussed. In the remaining three sections different quantum chemical models, ranging from the ones describing a single electron in the field of an external "core" to the approaches derived from the notion of the N-electron antisymmetric model space, are briefly reviewed.

There are many textbooks and review articles in which subjects addressed in this article are presented in all details. A reader willing to extend his understanding of relations between different branches of physics is referred to a particularly clear and broad presentation given by Feynman *et al.* (1977) in their famous "The Feynman Lectures on Physics". The fundamental concepts of quantum mechanics are presented, among others, in an excellent monograph by Sakurai (1984) and in the classical textbooks by Messiah (1961), Davydov (1965), Schiff (1968), Sakurai and Tuan (1985). Quantum chemical methods aimed at solving a Hamiltonian eigenvalue problem in the finite-dimensional model space (which constitute most of the "classical" quantum chemistry), are reviewed in the monographs by McWeeny (1973, 1989), in books edited by Wilson (1987) and in the proceedings of the previous Bad Windsheim NATO ASI (Wilson and Diercksen 1992). A detailed discussion of the density functional theory (only briefly mentioned in this paper) may be found in the monographs by Gross and Dreizler (1995) and by Kryachko and Ludeña (1990).

2. The Fundamentals

2.1. THE POSTULATES

The mathematical framework in which quantum mechanics may be conveniently formulated is the Hilbert space theory. Details of this formulation are given in standard textbooks, as e.g. in the ones by Dirac (1948) and by Messiah (1961), Sakurai and Tuan (1985). For our purposes it is sufficient to recall the following postulates:

- *A quantum system is associated with a linear vector space $\mathcal{H}$.*
  The elements of $\mathcal{H}$ are called *kets* and denoted by the symbol $|\ \rangle$. Different kets are labeled by letters or by letters and numbers, as for example $|A_1\rangle$, $|A_2\rangle$, etc. The space is spanned by an orthonormal, in general infinite, basis. The basis may be discrete or continuous or both.

Then, an arbitrary vector $|A\rangle$ belonging to $\mathcal{H}$ may be represented as

$$|A\rangle = \sum_n A_n|n\rangle + \int_C A(\xi)|\xi\rangle d\xi, \tag{1}$$

where the sum is extended over the discrete part of the basis and the integral – over the continuous one. The vectors dual to the kets are called *bras* and denoted $\langle \quad |$ with the appropriate identifying symbol inserted. Then, the vector dual to $|A\rangle$ reads

$$\langle A| = \sum_n A_n\langle n| + \int_C A(\xi)\langle \xi|d\xi. \tag{2}$$

The scalar product of A by B is the complex conjugate of the scalar product of B by A: $\langle A|B\rangle = \langle B|A\rangle^*$. The orthonormality of the basis means that

$$\langle n|n'\rangle = \delta_{mn'}, \qquad \int_C \langle \xi|\xi'\rangle = \delta(\xi - \xi'). \tag{3}$$

The basis is assumed to be complete, i.e. it fulfils the closure relation:

$$\hat{I} = \sum_n |n\rangle\langle n| + \int_C |\xi\rangle d\xi\langle \xi|, \tag{4}$$

where $\hat{I}$ is the identity operator and $|n\rangle\langle n| \equiv \hat{P}_n$ denotes the projector on the one dimensional space defined by the basis vector $|n\rangle$.

– *Each state of the quantum system is described by a normalized ket in $\mathcal{H}$.*
In other words, a quantum state may be identified with a point on a unit sphere in $\mathcal{H}$ or, alternatively, with a direction in $\mathcal{H}$. The linearity of $\mathcal{H}$ implies that a superposition of two different states is also a state of the quantum system, i.e., if $|\psi_1\rangle$ and $|\psi_2\rangle$ describe quantum states, then also

$$|\psi\rangle = c_1|\psi_1\rangle + c_2|\psi_2\rangle$$

is a state of this system. This means that equations which determine the evolution of a quantum system are linear.

– *Each dynamical variable Ω is attached to a hermitian operator $\hat{\Omega}$ in $\mathcal{H}$.*
The relationship between the dynamical variables and the operators is established through the *correspondence principle*. This means that if $\hat{x}$ and $\hat{p}_x$ are the operators corresponding, respectively, to the coordinate x and to the momentum p_x then, for example, the operator of the non-relativistic kinetic energy is given by $\hat{T} = \frac{\hat{p}^2}{2m}$ and the operator of the potential energy of a harmonic oscillator is $k\hat{x}^2$. Operators

corresponding to dynamical variables are, after Dirac (1948), referred to as *observables.* Each observable has a complete set of eigenvectors and real eigenvalues. Its eigenvalue problem

$$\hat{\Omega}|\omega_n\rangle = \omega_n|\omega_n\rangle, \tag{5}$$

where, for simplicity, the operator is assumed to have a discrete spectrum only, defines a complete basis $|\omega_n\rangle$, which without any loss of generality, may be taken as orthonormal. The set of eigenvectors of $\hat{\Omega}$ may be taken as the basis in $\mathcal{H}$. We say that this set forms a basis of the representation $\{\Omega\}$.
Since the closure relation gives the resolution of the identity

$$\hat{I} = \sum_n |\omega_n\rangle\langle\omega_n|, \tag{6}$$

an arbitrary ket (i.e. an arbitrary quantum state) may be expressed in terms of the eigenvectors of $\hat{\Omega}$:

$$|A\rangle = \sum_n |\omega_n\rangle\, \langle\omega_n|A\rangle. \tag{7}$$

Similarly, an arbitrary observable may be written as

$$\hat{\Xi} = \sum_{mn} |\omega_m\rangle\Xi_{mn}\langle\omega_n|, \tag{8}$$

where

$$\Xi_{mn} = \langle\omega_n|\hat{\Xi}|\omega_m\rangle \tag{9}$$

is referred to as a matrix element of the operator $\hat{\Xi}$. In particular, the operator $\hat{\Omega}$ may be expressed in terms of its eigenvalues:

$$\hat{\Omega} = \sum_n |\omega_n\rangle\omega_n\langle\omega_n|. \tag{10}$$

The last equation is known as the spectral decomposition of $\hat{\Omega}$ in its own representation. As we see, in a given representation $\{\Omega\}$, a vector $|A\rangle$ is represented by a set of numbers $A_n = \langle\omega_n|A\rangle$ labelled by one index (a one-column matrix, an analog of components of a vector in a Cartesian finite-dimensional space) and an operator $\hat{\Xi}$ is represented by a set of numbers $\Xi_{mn} = \langle\omega_m|\hat{\Xi}|\omega_n\rangle$ labelled by two indices, i.e. by a square matrix. These matrices are referred to as the *representatives* of the vectors or of the operators. When the vector space $\mathcal{H}$ is spanned by the basis of the representation $\{\Omega\}$, it is frequently referred to as *the Ω space.* Then, if the basis vectors are taken as eigenvectors of the coordinate operator we have the *coordinate space*, if they are the eigenvectors of the momentum operator, we have the *momentum space*, etc. A transition from one representation to another is equivalent to a transition from one orthonormal basis to another orthonormal basis in the same space. Hence, such a transition is described by a unitary transformation performed on all the representatives of vectors and operators. Such a transformation changes the numerical values of the matrix elements, however it does not change the physical content of the theory.

- *Each result of a measurement of* Ω *is equal to one of the eigenvalues of* $\hat{\Omega}$.

- *The probability* $P(\omega_n)$ *that* ω_n *is obtained as the result of a measurement of* Ω *performed on a quantum system in a state* $|A\rangle$ *is equal to*

$$P(\omega_n) = \langle A|\hat{P}_n|A\rangle = \langle A|\omega_n\rangle\langle\omega_n|A\rangle. \tag{11}$$

The last two postulates are concerned with measurements and define links between the formalism and the physical reality modelled by this formalism. A very important conclusion is that the observable $\{\Omega\}$ has a precisely defined value in its eigenstate only. Indeed, $P(\omega_n) = \delta_{kn}$ only if $|A\rangle = |\omega_k\rangle$. Consequently, two different dynamical variables may have precisely defined values in a given state, only if this state is simultaneously an eigenstate of the corresponding observables. As it is known, this is possible if these two operators commute. If the quantum system is in an arbitrary state, then the average (or expectation) value of $\hat{\Omega}$ is equal to

$$\langle\hat{\Omega}\rangle_{av} = \sum_n P(\omega_n)\omega_n = \sum_n \langle A|\omega_n\rangle\omega_n\langle\omega_n|A\rangle$$

or, according to Eq. (10),

$$\langle\hat{\Omega}\rangle_{av} = \langle A|\hat{\Omega}|A\rangle. \tag{12}$$

Then, the average value of Ω in a state $|A\rangle$ is equal to the appropriate diagonal matrix element of $\hat{\Omega}$. In particular, if $|A\rangle = |\omega_k\rangle$ then, as one should expect, $\langle\hat{\Omega}\rangle_{av} = \omega_k$.

2.2. THE EQUATIONS OF MOTION

Let us assume that the state vectors $|\psi, t\rangle$ depend upon time t. In order to derive the equation of motion for a quantum system we introduce the *evolution operator* $\hat{T}(t, t_0)$, such that

$$|\psi, t\rangle = \hat{T}(t, t_0)|\psi, t_0\rangle. \tag{13}$$

Then, the evolution operator moves the state vector from t_0 to t. If the displacement in time is infinitesimal, we can write

$$\hat{T} \equiv \hat{T}(t_0 + \delta t, t_0) = \hat{I} - \frac{i}{\hbar}\hat{H}\delta t, \tag{14}$$

where $\hat{H}$ is referred to as the generator of the infinitesimal displacement in time. This operator is hermitian if $\hat{T}$ is unitary. By analogy to the classical mechanics we may identify this operator with the Hamiltonian of the quantum system. Then, Eqs (13) and (14) yield

$$\frac{|\psi, t_0 + \delta t\rangle - |\psi, t_0\rangle}{\delta t} = -\frac{i}{\hbar}\hat{H}|\psi, t_0\rangle. \tag{15}$$

In the limit of $\delta t \to 0$ we get

$$i\hbar\frac{d|\psi\rangle}{dt} = \hat{H}|\psi\rangle. \tag{16}$$

This equation is known as the *Schrödinger equation* in a form which is representation-independent. It describes the time evolution of a time-dependent state vector $|\psi\rangle$. Let us note that at this stage $|\psi\rangle$ depends on time only. In particular it depends upon neither coordinates nor momenta. A description based on Eq. (16) is known as the *Schrödinger picture* of quantum mechanics.

Using Eq. (13) and remembering that the initial state $|\psi, t_0\rangle$ is an arbitrary normalized vector in $\mathcal{H}$, we may rewrite Eq. (16) in an operator form as

$$i\hbar\frac{d\hat{T}}{dt} = \hat{H}\hat{T}. \tag{17}$$

In particular, if $\hat{H}$ is time-independent then the last equation may be easily solved and

$$\hat{T}(t, t_0) = e^{-\frac{i}{\hbar}\hat{H}(t-t_0)}. \tag{18}$$

If we select the representation in which $\hat{H}$ is diagonal, i.e

$$\hat{H}|E_n\rangle = E_n|E_n\rangle, \tag{19}$$

then

$$\hat{T}(t, t_0)|E_n\rangle = e^{-\frac{i}{\hbar}E_n(t-t_0)}|E_n\rangle, \tag{20}$$

This means, that if $\hat{H}$ is time-independent, the quantum system would remain permanently in this eigenstate. Transitions between different eigenstates of the Hamiltonian are possible as a consequence of time-dependent perturbations (e.g. due to an external electromagnetic field).

For an arbitrary state we have

$$|\psi, t_0\rangle = \sum_n c_n(0)|E_n\rangle. \tag{21}$$

Its time evolution (for a time-independent Hamiltonian) is described by

$$|\psi, t\rangle = \sum_n c_n(t)|E_n\rangle = \hat{T}(t, t_0)|\psi, t_0\rangle. \tag{22}$$

Then,

$$c_n(t) = c_n(0)e^{-\frac{i}{\hbar}E_n(t-t_0)}. \tag{23}$$

Eq. (23) gives us a general solution of the Schrödinger equation in the case of a time-independent Hamiltonian. As we see, the state vector changes in time. Also time-dependent are distributions of the probabilities $P(\omega_n)$ and, consequently, the expectation values of obseravables, except for the energy. The probability distribution $P(E_n)$ of the energy values remains constant since $|c_n(t)|^2 = |c_n(0)|^2$.

One may obtain another, equivalent, mode of description of the evolution of a quantum system by performing a unitary transformation on the vectors and operators of the Schrödinger picture. In particular, by performing the time-dependent transformation $\hat{T}(t, t_0)^\dagger$ one defines the *Heisenberg picture.* Then, we have

$$|\psi_H\rangle = \hat{T}(t, t_0)^\dagger|\psi_S\rangle = \hat{T}(t, t_0)^{-1}|\psi_S\rangle \tag{24}$$

and

$$\hat{\Xi}_H = \hat{T}(t, t_0)^\dagger \hat{\Xi}_S \hat{T}(t, t_0), \tag{25}$$

where the subscripts S and H refer to the Schrödinger and the Heisenberg pictures, respectively. By comparing Eqs, (13) and (24) one can see that $|\psi_H\rangle = |\psi, t_0\rangle$, i.e. it is time-independent. On the other hand the operators, which for an isolated system are in the Schrödinger picture time-independent, in the Heisenberg picture are time-dependent. By differentiating Eq. (25), after some simple manipulations, one obtains the Heisenberg equation of the motion:

$$i\hbar\frac{d\hat{\Xi}_H}{dt} = [\hat{\Xi}_H, \hat{H}] + i\hbar\frac{\partial\hat{\Xi}_H}{\partial t}. \tag{26}$$

From the last equation one can draw a very important conclusion: if the quantum system is isolated (i.e. if $\frac{\partial \hat{\Xi}_H}{\partial t} = 0$), then an observable is time-independent (i.e. it describes a constant of the motion) if it commutes with the Hamiltonian.

2.3. COORDINATE AND MOMENTUM REPRESENTATIONS

The basis vectors of the coordinate representation are defined as

$$\hat{x}|x\rangle = x|x\rangle, \tag{27}$$

where $\hat{x}$ is the coordinate operator and x is its eigenvalue, i.e. the value of the coordinate. The spectrum is continuous and extends from $-\infty$ to $+\infty$. The orthonormality condition and the closure relation read:

$$\langle x|x'\rangle = \delta(x - x'), \quad \hat{I} = \int_{-\infty}^{+\infty} |x\rangle dx \langle x|. \tag{28}$$

The representative of the state vector $|\psi, t\rangle$ depends upon the basis vector index x, i.e. it is a function of x. It is called the *wavefunction*:

$$\psi(x, t) \equiv \langle x|\psi, t\rangle. \tag{29}$$

Similarly,

$$\psi(x, t)^* = \langle \psi, t|x\rangle. \tag{30}$$

The normalization condition may be written in two equivalent ways:

$$1 = \langle \psi, t|\psi, t\rangle = \int_{-\infty}^{+\infty} \langle \psi, t|x\rangle dx \langle x|\psi, t\rangle = \int_{-\infty}^{+\infty} |\psi(x, t)|^2 dx. \tag{31}$$

It is known from the electron diffraction experiments that the coordinate representative of an eigenvector of the momentum operator is a plane wave:

$$\langle x|p\rangle = \frac{1}{\sqrt{2\pi\hbar}} e^{\frac{i}{\hbar}xp}, \tag{32}$$

where

$$\hat{p}|p\rangle = p|p\rangle \tag{33}$$

and $\hat{p}$ is the momentum operator. Eq. (32) may also be considered as a direct consequence of the de Broglie hypothesis on the particle-wave duality of matter. By differentiating Eq. (32) with respect to x one gets a useful relation

$$-\frac{\hbar}{i}\frac{\partial}{\partial x}\langle p|x\rangle = p\langle p|x\rangle. \tag{34}$$

An arbitrary observable which depends upon the coordinate operator only, is diagonal in the coordinate representation:

$$\langle x|f(\hat{x})|x'\rangle = f(x')\delta(x - x'). \tag{35}$$

Using Eqs. (32), (34) and the resolution of the identity in the basis of the momentum operator eigenvectors, one gets

$$\langle x|\hat{p}|x'\rangle = -i\hbar\frac{\partial}{\partial x}\delta(x - x') \tag{36}$$

and

$$\langle x|\hat{p}|\psi, t\rangle = -i\hbar\frac{\partial}{\partial x}\psi(x, t). \tag{37}$$

The last equation gives the coordinate representative of $\hat{p}|\psi, t\rangle$. In this way we have derived the well known formula for the momentum operator in the coordinate representation:

$$\hat{\vec{p}} = -i\hbar\,\nabla, \tag{38}$$

(Eq. (38) has been generalized in a trivial way in order to account for three components of $\vec{p}$ and $\vec{x}$).

By multiplying Eq. (16) by $\langle x|$ from the left-hand side and inserting between $\hat{H}$ and $|\psi, t\rangle$ the resolution of the identity, we get

$$i\hbar\frac{\partial\psi(x, t)}{\partial t} = \int_{-\infty}^{+\infty} H(x, x')\psi(x', t)dx', \tag{39}$$

where

$$H(x, x') = \langle x|\hat{H}|x'\rangle. \tag{40}$$

In particular, if

$$\hat{H} = \frac{\hat{p}^2}{2m} + V(\hat{x}), \tag{41}$$

then, according to Eqs. (35) and (36),

$$\langle x|\hat{H}|x'\rangle = -\left[\frac{\hbar^2}{2m}\Delta_{x'} + V(x')\right]\delta(x - x') \tag{42}$$

and Eq. (39) assumes the familiar form:

$$i\hbar\frac{\partial\psi(x, t)}{\partial t} = \left(-\frac{\hbar^2}{2m}\Delta + V(x)\right)\psi(x, t). \tag{43}$$

The momentum representation is defined in a way analogous to that of the coordinate representation. Then, the basis in $\mathcal{H}$ is formed by the eigenvectors of the momentum operator (Eq. (33)). The representative of the state vector in the momentum representation is given by

$$\chi(p,t) \equiv \langle p|\psi,t\rangle. \tag{44}$$

The coordinate and the momentum representatives of the state vector are related by a Fourier transformation:

$$\psi(x,t) = \langle x|\psi,t\rangle = \int_{-\infty}^{+\infty} \langle x|p\rangle\chi(p,t)dp = \frac{1}{\sqrt{2\pi\hbar}}\int_{-\infty}^{+\infty} e^{\frac{i}{\hbar}xp}\chi(p,t)dp, \tag{45}$$

where we have used Eqs. (44), (32) and the resolution of the identity in the $|p\rangle$ basis. The analogs of Eqs. (35) and (36) in the momentum space are, respectively,

$$\langle p|f(\hat{p})|p'\rangle = f(p)\delta(p-p') \tag{46}$$

and

$$\langle p|\hat{x}|p'\rangle = -i\hbar\nabla_p\delta(p-p'). \tag{47}$$

The Schrödinger equation in the momentum representation reads

$$i\hbar\frac{\partial\chi(p,t)}{\partial t} = \int_{-\infty}^{+\infty} H(p,p')\chi(p',t)dp', \tag{48}$$

where

$$H(p,p') = \langle p|\hat{H}|p'\rangle. \tag{49}$$

If the Hamiltonian is given by Eq. (41), the Schrödinger equation becomes

$$i\hbar\frac{\partial\chi(p,t)}{\partial t} = \frac{p^2}{2m}\chi(p,t) + \int_{-\infty}^{+\infty} V(p,p')\chi(p',t)dp', \tag{50}$$

where

$$\begin{aligned} V(p,p') &= \langle p|V(\hat{x})|p'\rangle \\ &= \int_{-\infty}^{+\infty}\int_{-\infty}^{+\infty} \langle p|x\rangle\langle x|V(\hat{x})|x'\rangle\langle x'|p'\rangle dx\, dx' \\ &= \frac{1}{2\pi\hbar}\int_{-\infty}^{+\infty} V(x)e^{\frac{i}{\hbar}(p-p')x}dx. \end{aligned} \tag{51}$$

In the case of the harmonic oscillator with

$$\hat{H} = \frac{\hat{p}^2}{2m} + \frac{k^2}{2}\hat{x}^2 \tag{52}$$

the Schrödinger equation in the coordinate and in the momentum representations has the same mathematical structure. In most of other cases, the

Schrödinger equation in the momentum representation is an integral equation in which the potential function is the kernel of the integral operator.

3. The Schrödinger and the Dirac Equations

3.1. INVARIANCE PROPERTIES OF THE HAMILTONIAN

Let us assume that **S** describes a transformation of the reference frame. The results given by a quantum theory are independent of this transformation if the Hamiltonian is invariant with respect to **S**, i.e. if $\mathbf{S}\hat{H}\mathbf{S}^{-1} = \hat{H}$. In such a case we say that **S** is a symmetry transformation. Since a quantum system is fully defined by its Hamiltonian, we may say that the transformation **S** is a symmetry transformation if one cannot design an experiment which would allow to detect whether this transformation has been performed or not. For example, a translation in three-dimensional space of a free particle or a rotation of a free atom belong the symmetry operations of these systems.

The most universal symmetry is related to transformations between two inertial (i.e. moving with a constant velocity relative to each other) reference frames. It had been recognized already by Newton that *"The motions of bodies included in a given space are the same among themselves, whether that space is at rest or moves uniformly forward in a straight line"*. A more refined formulation of this *principle of relativity* is due to Poincaré: *"The laws of physical phenomena must be the same for a fixed observer as for an observer who has a uniform motion of translation relative to him, so that we have not, nor can we possibly have, any means of discerning whether or not we are carried along in such a motion"*. Then, all physical theories, including quantum mechanics, should be formulated in such a way that the results of all measurements are the same independent of whether they are taken by an observer in rest or by an observer moving with a constant velocity. This implies that the Hamiltonian must be invariant with respect to a transformation from one inertial frame to another. The Newtonian mechanics is invariant with respect to the Galilean transformations:

$$\begin{aligned} x_1' &= x_1 - ut, \\ x_2' &= x_2, \quad x_3' = x_3, \\ t' &= t, \end{aligned} \tag{53}$$

where the axes of both frames are parallell, the unprimed quantities refer to the reference frame "in rest" and the primed ones to the frame moving with a constant velocity u along the x_1 axis.

In 1887, Michelson and Morley performed an experiment which demonstrated that the velocity of light measured from different inertial reference frames is always the same. This result is clearly inconsistent with the

Galilean transformation. In 1905 Einstein noticed that the transformation between two inertial frames which leads to the conclusion that the velocity of light is a universal constant is the Lorentz transformation (it was discovered earlier by Lorentz that the Maxwell equations are invariant with respect to this transformation):

$$\begin{aligned} x_1' &= \frac{x_1 - ut}{\sqrt{1-u^2/c^2}}, \\ x_2' &= x_2, \quad x_3' = x_3, \\ t' &= \frac{t - ux_1/c^2}{\sqrt{1-u^2/c^2}}, \end{aligned} \tag{54}$$

where c is the velocity of light. As one can see, the Lorentz transformation approaches the Galileo one when $u/c \to 0$. The transformation becomes singular if $u = c$. Its analysis leads to the *causality principle* which says that in a Lorentz-invariant theory no signal can travel with a velocity larger than that of light.

A requirement of invariance with respect to either Galilean or Lorentzian transformation imposes very strong restrictions on the form of the Hamiltonian. A theory or a model which is Galileo-invariant is usually referred to as the *non-relativistic* one. In a non-relativistic model three coordinates of a point, momentum, velocity, are three-component vectors, while time or energy are invariants (scalars). Therefore the equation of motion (Eq. (16)) is invariant with respect to the Galilean transformation if the Hamiltonian is Galileo-invariant. The non-relativistic (Galileo-invariant) Hamilton function of a free particle is:

$$H^0 = \frac{p^2}{2m}. \tag{55}$$

If the particle moves in an external electromagnetic field characterized by the scalar potential V and by the vector potential $\vec{A}$, the non-relativistic Hamilton function becomes

$$H = \frac{1}{2m}\left(\vec{p} - \frac{e}{c}\vec{A}\right)^2 + V, \tag{56}$$

where e is the electric charge of the particle. The correspondence principle applied to the nonrelativistic Hamilton function leads to the nonrelativistic Hamilton operator and opens a way to formulation of the nonrelativistic (Schrödinger) quantum mechanics (cf. Eqs. (41), (43), (50)).

Models which are Lorentz-invariant, are called *relativistic*. In a relativistic model three coordinates: x_1, x_2, x_3 and time, $x_0 = ct$, form a four-component vector. Also momentum and energy are components of the same four-vector. Therefore, in each Lorentz-invariant model, the space coordinates of a particle and the time corresponding to this particle should appear on an equal footing, contrary to non-relativistic models, where time is an absolute quantity, the same for all particles. For example, if an equation is of the first-order in time, it must be of the first-order in all coordinates, etc. In the equation describing the evolution of a state vector (Eq. (16)), time plays a special role. We say that the equation is written in a non-covariant form, i.e. its possible Lorentz-invariance is hidden and may be discovered only after analyzing simultaneously the form of the Hamiltonian and the structure of the equation. As we shall see, a Lorentz-invariant quantum mechanical equation (the Dirac equation) results from Eq. (16) if the Hamilton operator is derived, through the correspondence principle, from the relativistic Hamilton function. In the case of a free particle this function is given by the following expression:

$$H^0 \equiv E_p = mc^2\sqrt{1+\left(\frac{p}{mc}\right)^2}. \tag{57}$$

In its resolution into a power series of $\delta = \left(\frac{p}{mc}\right)^2$:

$$H^0 = mc^2 + \frac{p^2}{2m} - \frac{p^4}{8m^3c^2} + O\left(\delta^6\right) + \cdots, \tag{58}$$

one can recognize the rest energy of the particle, the non-relativistic kinetic energy term and the correction due to the relativistic change of the mass. If the particle moves in an external field, then

$$H = mc^2\sqrt{1+\frac{1}{m^2c^2}\left(\vec{p}-\frac{e}{c}\vec{A}\right)^2} + V. \tag{59}$$

However, no Lorentz-invariant equation describing more than one particle can be written in the form of Eq. (16) - in the case of N particles such an equation would have to depend on N different time coordinates.

3.2. THE DIRAC EQUATION

Historically, the first Lorentz-invariant equation describing the evolution of a quantum state was obtained by Schrödinger, Klein and Gordon. By taking a square of the operators on both sides of Eq. (16) and substituting the free-particle relativistic Hamiltonian as it is given by Eq. (57), we have:

$$-\hbar^2\frac{\partial^2\psi}{\partial t^2} = (c^2\hat{p}^2 + m^2c^4)\psi. \tag{60}$$

Now, putting $\hat{p}^2\psi = -\hbar^2\Delta\psi$ we get

$$\left[\Box - (mc/\hbar)^2\right]\psi = 0, \tag{61}$$

where

$$\Box = \Delta - \frac{1}{c^2}\frac{\partial^2}{\partial t^2}. \tag{62}$$

This equation is known as the Klein-Gordon equation. One can easily show its relativistic invariance. It may also be generalized to account for an external field. However, equally easily, one can demonstrate that it does not describe electrons. After some careful studies one can prove that the reason of the failure of the Klein-Gordon equation to describe electrons is that it cannot accomodate the spin-$\frac{1}{2}$ nature of electrons (it describes spinless particles). In order to derive a relativistic equation describing the behaviour of an electron, let us consider first the non-relativistic free-particle Hamiltonian for a spin-$\frac{1}{2}$ particle.

Usually the free particle Hamiltonian corresponding to the Hamilton function given by Eq. (55) is written as

$$\hat{H}^0 = \frac{\hat{p}^2}{2m}. \tag{63}$$

However, equally well we may take the expression

$$\hat{H}^0 = \frac{(\vec{\sigma}\cdot\hat{\vec{p}})(\vec{\sigma}\cdot\hat{\vec{p}})}{2m}, \tag{64}$$

where $\vec{\sigma}$ are the Pauli spin matrices. Due to the identity

$$(\vec{\sigma}\cdot\hat{\vec{a}})(\vec{\sigma}\cdot\hat{\vec{b}}) = \hat{\vec{a}}\cdot\hat{\vec{b}} + i\vec{\sigma}\cdot[\hat{\vec{a}}\times\hat{\vec{b}}], \tag{65}$$

where $\hat{\vec{a}}$ and $\hat{\vec{b}}$ are operators, Eqs. (63) and (64) are equivalent. However there is a difference if the particle moves in an external field. After the substitution $\vec{p} \to \hat{\vec{p}} - \frac{e}{c}\hat{\vec{A}}$ to Eq. (64) we get

$$\hat{H} = \frac{1}{2m}\left(\hat{\vec{p}} - \frac{e}{c}\hat{\vec{A}}\right)^2 - \frac{e\hbar}{2mc}\left(\vec{\sigma}\cdot\hat{\vec{B}}\right), \tag{66}$$

where $\hat{\vec{B}} = \nabla\times\hat{\vec{A}}$ is the magnetic field operator. The $\vec{\sigma}$-dependent term in the last equation is equal to the term added by Pauli to the Schrödinger equation, on a purely phenomenological basis, in order to account for the interaction of the electron spin magnetic moment with the external magnetic field. This term would not appear if we took Eq. (63) instead.

Now let us consider a relativistic spin-$\frac{1}{2}$ free particle. The correspondence principle applied to Eq. (57) taken to the second power gives

$$(\hat{E} - c\vec{\sigma} \cdot \hat{\vec{p}})(\hat{E} + c\vec{\sigma} \cdot \hat{\vec{p}})\psi = m^2c^4\psi, \tag{67}$$

where instead of H, we introduced E (the component of the energy-momentum four-vector) and defined ψ as a two-component function (because $\vec{\sigma}$ is a 2×2 matrix). Taking $\hat{E} = i\hbar c\frac{\partial}{\partial x_0} \equiv i\hbar\frac{\partial}{\partial t}$ and $\hat{\vec{p}} = -i\hbar\nabla$ we may rewrite Eq. (67) in the form

$$\hat{\Omega}_+\hat{\Omega}_-\psi = m^2c^4\psi, \tag{68}$$

where

$$\hat{\Omega}_\pm = i\hbar\left(\frac{\partial}{\partial t} \pm c\vec{\sigma} \cdot \nabla\right). \tag{69}$$

Defining $\psi^R = \frac{1}{mc^2}\hat{\Omega}_-\psi$ and $\psi^L = \psi$ we get

$$\hat{\Omega}_-\psi^L = mc^2\psi^R, \tag{70}$$

$$\hat{\Omega}_+\psi^R = mc^2\psi^L. \tag{71}$$

This pair of equations may be rewritten as

$$i\hbar\frac{\partial\psi}{\partial t} = \hat{H}_D^0\psi \tag{72}$$

where

$$\psi = \begin{pmatrix}\psi^G \\ \psi^F\end{pmatrix} = \begin{pmatrix}\psi^R + \psi^L \\ \psi^R - \psi^L\end{pmatrix} \tag{73}$$

is a four-component wavefunction and

$$\hat{H}_D^0 = c\vec{\alpha} \cdot \vec{p} + \beta mc^2 \tag{74}$$

with

$$\vec{\alpha} = \begin{pmatrix}0 & \vec{\sigma} \\ \vec{\sigma} & 0\end{pmatrix}, \quad \beta = \begin{pmatrix}I_2 & 0 \\ 0 & -I_2\end{pmatrix} \tag{75}$$

and I_2 is the 2×2 unit matrix. Eq. (72) is known as the Dirac equation. It is written in the Hamiltonian form (cf. Eq. (16)), however it is Lorentz invariant. In the presence of an external electromagnetic field the free-particle Dirac Hamiltonian is replaced by

$$\hat{H}_D = c\vec{\alpha} \cdot \left(\hat{\vec{p}} - \frac{e}{c}\hat{\vec{A}}\right) + \beta mc^2 + \hat{V}. \tag{76}$$

It is important to note that Eqs. (43) and (72) are defined in different Hilbert spaces. The Hilbert space of Eq. (43) is $\mathcal{H}$. In this case it is spanned by the set of eigenvectors of the coordinate operator. Eq. (76) is represented in a space $\mathcal{H}_D$ which is a Cartesian product of $\mathcal{H}$ and a 4-dimensional spin-space $\mathcal{H}_\Sigma$ in which operators $\vec{\alpha}$ and β are represented as 4×4 matrices:

$$\mathcal{H}_D = \mathcal{H} \otimes \mathcal{H}_\Sigma. \tag{77}$$

It has already been noticed by Levy-Leblond (1970) that the non-relativistic Schrödinger equation may be written as a first-order, four-component equation:

$$(c\vec{\alpha} \cdot \hat{\vec{p}} + \beta^+ V - 2\beta^- mc^2)\psi = \beta_+ E, \tag{78}$$

where

$$\beta^+ = \begin{pmatrix} I_2 & 0 \\ 0 & 0 \end{pmatrix} \text{ and } \beta^- = \begin{pmatrix} 0 & 0 \\ 0 & I_2 \end{pmatrix} \tag{79}$$

are projection operators on two-dimensional subspaces of $\mathcal{H}_\Sigma$. Eq. (78) may be rewritten as

$$\begin{pmatrix} V & c(\vec{\sigma} \cdot \hat{\vec{p}}) \\ c(\vec{\sigma} \cdot \hat{\vec{p}}) & -2mc^2 \end{pmatrix} \begin{pmatrix} \psi^G \\ \psi^F \end{pmatrix} = E \begin{pmatrix} \psi^G \\ 0 \end{pmatrix}.$$

Elimination of ψ^F gives

$$\left[\frac{(\vec{\sigma} \cdot \hat{\vec{p}})(\vec{\sigma} \cdot \hat{\vec{p}})}{2m} + V \right] \psi^G = E\psi^G,$$

i.e. the Schrödinger equation. Then, Eq. (78) may be interpreted as a representation of the Schrödinger equation (Eq. (43)) in the augmented space. As we shall see, this equation constitutes a starting point to formulation by Rutkowski (1986) of the *direct perturbation theory* (see also Kutzelnigg 1989).

3.3. SIMPLE SOLUTIONS

From now on, we shall be concerned with solutions of the Schrödinger and of the Dirac equations corresponding to some specific energies. In this case the time-dependent evolution equations reduce to the eigenvalue problems (19) of the appropriate Hamiltonians. First let us consider the case of a free particle. Both Schrödinger and Dirac Hamiltonians commute in this case with the momentum operator. Therefore in both cases the wavefunctions may be represented by the plane waves. Then the eigenfunction of the free-particle Dirac equation, corresponding to the momentum $\vec{p}$, may be expressed as

$$\psi = \begin{pmatrix} \psi_1 \\ \psi_2 \\ \psi_3 \\ \psi_4 \end{pmatrix} \frac{1}{\sqrt{2\pi\hbar}^3} e^{\frac{i}{\hbar}(\vec{x}\cdot\vec{p})}, \tag{80}$$

where ψ_k, $k = 1,2,3,4$ are the components of a four-component *spinor*. According to Eqs. (72), (74) and (75) the components of the spinor may be obtained from the Dirac equation, which in this case may be written as

$$\begin{array}{rcrcl} (mc^2 - E)\psi_G & + & c(\vec{\sigma}\cdot\vec{p})\psi_F & = & 0, \\ c(\vec{\sigma}\cdot\vec{p})\psi_G & - & (mc^2 + E)\psi_F & = & 0, \end{array} \tag{81}$$

with $\psi_G = \binom{\psi_1}{\psi_2}$ and $\psi_F = \binom{\psi_3}{\psi_4}$. If $E = mc^2 + \varepsilon$ and $\varepsilon << mc^2$ (i.e. in a non-relativistic approximation) the second of Eqs. (81) gives

$$|\psi_F| \approx \left|\frac{\vec{\sigma}\cdot\vec{p}}{2mc}\psi_G\right| << |\psi_G|. \tag{82}$$

Therefore ψ_G and ψ_F are, respectively, known as the large and the small components of the Dirac wavefunction ψ.

The free-particle Schrödinger Hamiltonian commutes with the orbital angular momentum operator $\hat{\vec{L}} = \hat{\vec{r}} \times \hat{\vec{p}}$, but the Dirac one does not. This means that, for a free particle, $\hat{\vec{L}}$ is a constant of the motion for the Schrödinger electron but not for the Dirac one. As one can easily check, the free-particle Dirac Hamiltonian commutes with an operator

$$\hat{\vec{J}} = \hat{\vec{L}} + \hat{\vec{\Sigma}}, \tag{83}$$

where

$$\vec{\Sigma} = \frac{\hbar}{2}\begin{pmatrix} \vec{\sigma} & 0 \\ 0 & \vec{\sigma} \end{pmatrix}. \tag{84}$$

Eq. (83) may be interpreted as a definition of the total angular momentum operator of the electron, composed of the orbital part $\hat{\vec{L}}$ and the spin part $\hat{\vec{\Sigma}}$.

Solving Eqs. (81) one gets a most surprising result of the Dirac theory:

$$E = \pm mc^2\sqrt{1 + \left(\frac{p}{mc}\right)^2}, \tag{85}$$

i.e. the spectrum of the Dirac Hamiltonian is not limited from below (a Dirac electron does not have any lowest energy state). One should expect that an electron in a positive energy state should (due to any kind of interaction with an external field) fall into a negative energy state and emit a photon with an energy larger than $2mc^2$. This energy may be infinite, since the spectrum is unlimited. In order to solve this difficulty Dirac proposed that all the negative-energy states are filled under normal conditions and the Pauli exclusion principle prevents transitions to these states. The state with no positive-energy electron and all negative-energy states fully occupied is referred to as the *Dirac vacuum*. An excitation of one negative-energy electron results in creation of a hole in the Dirac vacuum and of one positive-energy electron. This corresponds to creation of an electron-positron pair. This interpretation contains a deep self-contradiction: the Dirac model describes, in principle, a single electron. At the same time, it is unable even to explain the behaviour of a free particle without assuming that it is surrounded by an infinite number of particles occupying the negative-energy states. This contradiction may be solved on the ground of quantum electrodynamics (c.f. Dirac 1948, Sakurai 1984).

An interesting, paradoxial, result is obtained when studying the velocity of a free particle. In the Schrödinger case, as one should expect,

$$\left(\frac{d\hat{\vec{x}}}{dt}\right)_S = \frac{i}{\hbar}\left[\hat{H}_S^0, \vec{x}\right] = \frac{\hat{\vec{p}}}{m}. \tag{86}$$

However in the case of the Dirac Hamiltonian we have

$$\left(\frac{d\hat{\vec{x}}}{dt}\right)_D = \frac{i}{\hbar}\left[\hat{H}_D^0, \hat{\vec{x}}\right] = c\hat{\vec{\alpha}}. \tag{87}$$

This result is very strange in many aspects. First, the eigenvalues of α_k are ± 1. Therefore the eigenvalues of $\left(\frac{d\hat{\vec{x}}}{dt}\right)_D$ are equal to $\pm c$. Besides, different components of $\left(\frac{d\hat{\vec{x}}}{dt}\right)_D$ do not commute. Then, a measurement of one component of the Dirac velocity is incompatible with a measurement of another one while different components of the momentum operator do commute. Moreover, $\left(\frac{d\hat{\vec{x}}}{dt}\right)_D$ does not commute with $\hat{H}_D^0$ and therefore the velocity is not a constant of the motion despite the fact that the particle is free.

An explanation of this facts involves an analysis of the time-dependence of $\vec{\alpha}$ and of the coordinate operator $\hat{x}$. It results that both of them execute very rapid oscillations with an angular frequency $\frac{2mc^2}{\hbar} = 1.5 \cdot 10^{21} sec^{-1}$. This motion, named by Schrödinger *Zitterbewegung*, is due to an interference between the positive- and negative-energy components of the wave-packet describing the electron. Intuitively it may be interpreted as a consequence of a permanent creation and annihilation of the so called *virtual* electron-positron pairs.

In order to explore this problem in more detail and also for later use it is convenient to introduce the *even* and *odd* operators. An even operator acting on a function which belongs to the positive (negative) energy subspace of a free electron transforms it into a function which belongs to the same subspace. An odd operator transforms a positive (negative)-energy-subspace function into a function which belongs to the complementary subspace. Operators

$$\Pi_{\pm} = \frac{1}{2}(\hat{I} \pm \hat{\Lambda}), \tag{88}$$

where

$$\hat{\Lambda} = \frac{\hat{H}_D}{E_p} = \frac{c\vec{\alpha}\cdot\vec{p} + \beta mc^2}{\sqrt{m^2c^4 + c^2p^2}} \tag{89}$$

is the operator of the sign of energy (its eigenvalues are $+1$ for the positive- and -1 for the negative-energy states), are projection operators to the positive (Π_+) and to the negative (Π_-) energy subspace. For example, an operator $\Pi_+\hat{\Omega}\Pi_+$ acts in the free-electron positive-energy space only. In the free particle space each operator, say $\hat{\Omega}$, may be decomposed into its even part $[\hat{\Omega}]$ and its odd part $\{\hat{\Omega}\}$:

$$\hat{\Omega} = [\hat{\Omega}] + \{\hat{\Omega}\}, \tag{90}$$

where

$$[\hat{\Omega}] = \frac{1}{2}(\hat{\Omega} + \hat{\Lambda}\hat{\Omega}\hat{\Lambda}), \tag{91}$$

$$\{\hat{\Omega}\} = \frac{1}{2}(\hat{\Omega} - \hat{\Lambda}\hat{\Omega}\Lambda). \tag{92}$$

Theories in which only even operators are used may be formulated in the positive energy part of the Hilbert space. In such theories creation of the virtual electron-positron pairs is not possible. Therefore they are frequently referred to as *no-pair* theories. In the case of the Dirac velocity operator (87), its even (i.e. the "no-pair") part is equal to

$$\left[\left(\frac{d\hat{x}}{dt}\right)_D\right] = c\,[\vec{\alpha}] = \frac{c^2p}{E_p}\hat{\Lambda}. \tag{93}$$

Then, the "no-pair velocity" corresponds to the standard one-particle velocity - the *Zitterbewegung* disappears. However another difficulty remains: if the particle is described by a wavefunction belonging to the negative energy space then $\Lambda = -1$ and the direction of velocity is opposite to that of momentum.

A more detailed discussion of this and of related paradoxes of the Dirac theory may be found, for example, in textbooks of Messiah (1961), Baym

(1969), Sakurai (1984).

3.4. ONE-ELECTRON ATOM

An exact quantum-mechanical treatment of a hydrogen-like atom is equivalent to solving a two-body problem with a Coulomb interaction. In the nonrelativistic case it may be solved exactly. By the separation of the center of mass the six-dimensional Schrödinger equation is split into two three-dimensional equations. One of them describes a free motion of the center of mass and the other one – the relative motion of the electron and the nucleus. The Hamiltonian which governs the relative motion reads

$$\hat{H}_S = \frac{\hat{p}^2}{2\mu} - \frac{Ze^2}{r}, \tag{94}$$

where Ze is the nuclear charge, $\mu = \frac{mM}{m+M}$ is the reduced mass; m is the mass of electron and M is the mass of the nucleus. The Hamiltonian commutes with the orbital angular momentum operators $\hat{L}^2$ and $\hat{L}_z$. Its eigenvalue spectrum consists of the discrete part (describing the bound states) and the continuous part (describing the ionized states). The discrete part of the eigenvalue problem may be written as

$$\hat{H}_S \psi_{nlm}(r, \theta, \phi) = E_n^S \psi_{nlm}(r, \theta, \phi) \tag{95}$$

where r, θ and ϕ are the spherical coordinate system variables and

$$\begin{aligned} n &= 1, 2, \ldots, \\ l &= 0, 1, \ldots, n-1, \\ m &= -l, -l+1, \ldots, l \end{aligned}$$

are the quantum numbers describing, respectively, the energy (the principal quantum number), the orbital angular momentum and its projection. The eigenvalues (i.e. the energies of the bound states of the electron) in the Hartree atomic units (a.u.: $m = 1$, $e = 1$, $\hbar = 1$) are given by

$$E_n^S = -\frac{Z^2}{2n^2}\left(1 + \frac{m}{M}\right)^{-1} \text{a.u.} = -\left(\frac{Z}{n}\right)^2 R_M, \tag{96}$$

where

$$R_M = R_\infty \left(1 + \frac{m}{M}\right)^{-1} \tag{97}$$

and

$$R_\infty = \frac{1}{2} mc^2 \alpha^2 \tag{98}$$

are the Rydberg constants for the finite and infinite nuclear mass, respectively, and

$$\alpha = \frac{e^2}{\hbar c} \approx \frac{1}{137} \tag{99}$$

is the fine structure constant.

The eigenvalues depend upon the principal quantum number only while the eigenfunctions are labelled by three quantum numbers (n, l and m). The degree of the degeneracy is, in this case, equal to n^2. There are two origins for this degeneracy. One, with respect to the quantum number m, is a consequence of the spherical symmetry of the problem (in all spherically symmetric one-electron systems this kind of degeneracy would appear). The second one, with respect to the quantum number l, is a consequence of a *dynamical symmetry* connected with some specific properties of the interaction potential. In the case of the non-relativistic one-electron atom there is an additional constant of the motion arising from the commutation of the so called *Runge-Lenz* vector with the Hamiltonian (for more details see e.g. Schiff 1968). The l-degeneracy is related to the existence of this constant of the motion and is removed if the potential is deformed so that it loses its Coulombic character.

The eigenfunctions of the hydrogen-like Hamiltonian are prototypes for most of approximate one-electron functions (orbitals) used in quantum chemical calculations. They may be expressed as

$$\psi_{nlm}(r, \theta, \phi) = \frac{1}{r} R_{nl}(r) Y_{lm}(\theta, \phi), \tag{100}$$

where Y_{lm} (the spherical harmonics) are the eigenfunctions of $\hat{L}^2$ and $\hat{L}_z$ and R_{nl} are the radial functions. The radial functions are of the form

$$R_{nl}(r) = W_n^l(r) e^{-Zr/n}, \tag{101}$$

where W_n^l is a polynomial of the degree n with l-dependent coefficients. The number of nodes of this polynomial is equal to $n - l$. At the origin the radial part of the wavefunction $r^{-1}R_{nl}$ behaves as r^l, i.e. it vanishes for $r = 0$ if $l \neq 0$. The spherical harmonics are, for $m \neq 0$, complex and:

$$Y_{l,m}^* = (-1)^m Y_{l-m}. \tag{102}$$

Their real combinations, i.e $Y_{lm} + (-1)^m Y_{l,-m}$ and $i(Y_{lm} - (-1)^m Y_{l,-m})$ may be expressed as $\sum_{abc} C_{abc} x^a y^b z^c / r^l$, where $a + b + c = l$. For $l = 0, 1, 2, 3, \ldots$ they are referred to as $s, p, d, f, \ldots$-type functions, respectively. The discrete states of the hydrogen-like atom are, customarily, labelled by the quantum numbers n and l as $1s$, $2s$, $2p$, $3s$, $3p$, $3d$, ...

The Schrödinger model of a hydrogen-like atom does not contain spin. Spin may be introduced by using the phenomenological Pauli model. Then the wavefunction ψ_{nlm} has to be multiplied by a two-component spin function α or β corresponding, respectively, to the eigenvalues $+\frac{\hbar}{2}$ and $-\frac{\hbar}{2}$ of a projection $\hat{S}_z$ of the spin operator. These functions are referred to as *spinorbitals*. The spinorbitals may be combined according to the rules of coupling the angular momenta to form eigenfunctions of the total angular momentum operator which, of course, also commutes with the Hamiltonian. However introducing spin at this level does not influence the energy of the atom, unless the atom is placed into an external magnetic field when the Pauli term (c.f. Eq. (66) influences the energy (do not confuse with the Pauli relativistic corrections which result from a reduction of the relativistic Dirac equation).

In the relativistic case a two-body Coulomb problem cannot be solved exactly. Therefore the hydrogen-like atom in the Dirac theory is modelled by an electron moving in an external Coulomb field. We solve the corresponding Dirac equation in the reference frame in which the singularity of the potential is placed at the origin, i.e. with $V(r) = -\frac{Ze^2}{r}$. A projection of the orbital angular momentum does not commute with this Hamiltonian, but its square does. Also the total angular momentum operator commutes with it. It is customary to introduce, apart of the total angular momentum quantum number $j = l \pm 1/2$ and its projection $m_j = -j, -j+1, \ldots, j$, the relativistic angular momentum quantum number

$$\kappa = \pm(j + \frac{1}{2}) = \begin{cases} l & \text{if } j = l - 1/2 \\ -(l+1) & \text{if } j = l + 1/2. \end{cases} \tag{103}$$

The spectrum of the Dirac Hamiltonian is unbounded from below. It consists of two continua (one extending from the energy $+mc^2$ up and another one extending from the energy $-mc^2$ down) and the discrete part located below the ionization threshold $+mc^2$. The Dirac energy depends on two quantum numbers: n and j (or, equivalently, on n and κ). Then, the eigenvalue equation reads:

$$\hat{H}_D \psi_{nljm_j} = E^D_{nj} \psi_{nljm_j}, \tag{104}$$

where

$$E^D_{nj} - mc^2 = -\frac{2Z^2}{N(N+\tilde{n})} \cdot R_\infty \tag{105}$$

with

$$N = \sqrt{\alpha^2 Z^2 + \tilde{n}^2}, \tag{106}$$

$$\tilde{n} = n + s - |\kappa| \tag{107}$$

and

$$s = \sqrt{\kappa^2 - \alpha^2 Z^2}. \tag{108}$$

The Dirac states are labeled by three quantum numbers: n, l and j. Then, the consecutive discrete states of the Dirac atom are: $1s_{1/2}$, $2s_{1/2}$, $2p_{1/2}$, $2p_{3/2}$, $3s_{1/2}$, $3p_{1/2}$, $3p_{3/2}$, $3d_{3/2}$, $3d_{5/2}$, ... As one can easily see, for $\alpha \to \infty$, i.e. in the nonrelativistic limit, $s \to \kappa$, $\tilde{n} \to n$, $N \to n$ and $(E^D_{nj} - mc^2) \to E^S_n$. By expanding the rhs of Eq. (105) into a power series of $\alpha^2 Z^2$ one gets

$$E^D_{nj} = mc^2 \left[1 - \frac{(\alpha Z)^2}{2n^2} - \frac{(\alpha Z)^4}{2n^3} \left(\frac{1}{|\kappa|} - \frac{3}{4n} \right) + O((\alpha Z)^6) \right], \tag{109}$$

where the first term is the rest energy of the electron, the second one is the non-relativistic (Schrödinger) energy and the third one is the relativistic (Pauli) correction. In the Dirac spectrum, apart of shifting the energy levels relative to the Schrödinger ones, some of the degeneracies have been removed. Now the energy levels depend upon the total angular momentum quantum number j but, also in the Dirac theory, they do not depend upon l. Then, for example, the energy of the states $2p_{1/2}$ and $2p_{3/2}$ are different while the energies of $2p_{1/2}$ is the same as that of $2s_{1/2}$. The splitting of the energy levels due to j is called the *fine structure splitting.* It is relatively small for small Z but it grows very fast with increasing nuclear charge (as one can see from Eq. (109) it is proportional to Z^4). Experimental measurements show that the energies of a real hydrogen-like atom depend upon l. In particular the energies of $2s_{1/2}$ and $2p_{1/2}$ are different. For the first time this splitting was measured for the hydrogen atom by Lamb and Retherford (Lamb 1951) and the effect is called the *Lamb shift.* It may be explained on the ground of quantum electrodynamics. Its value grows rapidly with increasing Z. For the hydrogen atom it is equal 1060 Mc $= 4{\cdot}10^{-6}$ eV; for the uranium like atoms it is about 70 eV, i.e. by 7 orders of magnitude larger than in hydrogen!

The Dirac wavefunctions are four-component spinors. They may be represented as

$$\psi_{nl\kappa m_j} = \frac{1}{r} \begin{pmatrix} G_{nl\kappa}(r)\Phi(\theta.\phi)_{nl\kappa m_j} \\ iF_{nl\kappa}(r)\Phi(\theta,\phi)_{nl-\kappa m_j} \end{pmatrix}, \tag{110}$$

where κ is related to j according to Eq. (103), G and F are, respectively, the large and and the small radial components of the wavefunction, and $\Phi(\theta,\phi)$ are combinations of products of the spherical harmonics and the two-component spin functions constructed according to the rules of coupling the angular momenta. The radial functions fulfil the following eigenvalue equation (c.f. Eq. (81) for the free particle):

$$(mc^2 - E + V(r))G(r) - c\left(\frac{d}{dr} - \frac{\kappa}{r}\right)F(r) = 0, \tag{111}$$

$$c\left(\frac{d}{dr} + \frac{\kappa}{r}\right)G(r) - (mc^2 + E - V(r))F(r) = 0, \tag{112}$$

where $V(r) = -\frac{Ze^2}{r}$. Then, the small component is related to the large one:

$$F(r) = c(mc^2 + E - V(r))^{-1}\left(\frac{d}{dr} + \frac{\kappa}{r}\right)G(r). \tag{113}$$

Both large and small components of the radial wavefunctions may be expressed as products of $e^{-\zeta r}$ and a polynomial, where $\zeta = \sqrt{m^2c^2 - E^2/c^2}$. At the origin $R(r)$ behaves as r^s. Then, if $|\kappa| > 1$ it vanishes for $r = 0$. However if $\kappa = \pm 1$ then $s < 1$ and $r^{-1}R(r)$ is singular. However this singularity is weak and for $Z\alpha < 1$ does not obstruct the normalizability of the wavefunctions. The relativistic radial electron density is contracted relative to the non-relativistic one (the Dirac hydrogen atom is "smaller" than the Schrödinger one). Besides, it is nodeless since the nodes of the large and small radial components (except the $r = 0$ node) never coincide.

Singularity of the Dirac wavefunctions at the origin is a consequence of the Coulomb singularity in the potential. A more general potential may be taken in the form

$$V(r) = -Z(r)\frac{e^2}{r}.$$

If Z(r) is constant, then we have the point nucleus and singular potential. One of the simplest and commonly used is a model in which a uniform nuclear charge distribution is assumed. In this case

$$V(r) = \begin{cases} -\frac{3Z}{2A}\left(1 - \frac{r^2}{3A^2}\right)e^2, & \text{if } 0 \le r \le A \\ -\frac{Ze^2}{r} & \text{if } r > A. \end{cases} \tag{114}$$

Taking a finite nuclear model we get the wavefuncion without any singularity: $r^{-1}G(r)$ behaves as r^l and $r^{-1}F(r)$ – as r^{l+1}. A detailed discussion of this problem has been given by Grant (1996)

4. Approximations

Exact solutions of both Schrödinger and Dirac equations are possible for very few systems. In most cases one has to construct approximate methods of solving the corresponding eigenvalue equation. It is customary to divide the approximate methods of solving eigenvalue problems into two classes: variational methods and perturbational methods.

4.1. VARIATIONAL METHODS

Most of quantum chemical models originate from variational approaches. The variational methods are based on a very simple property of an eigenvalue problem known as the variational principle. Let us assume that we have a discrete eigenvalue problem

$$\hat{H}|\psi_k\rangle = E_k|\psi_k\rangle, \quad k = 1, 2, \ldots \tag{115}$$

with $E_1 \leq E_2 \leq E_3 \leq \cdots$. The basic theorem of the variational principle says that for an arbitrary $|\Phi\rangle$ which belongs to the space spanned by the eigenvectors of $\hat{H}$,

$$W[\Phi] \equiv \frac{\langle\Phi|\hat{H}|\Phi\rangle}{\langle\Phi|\Phi\rangle} \geq E_1. \tag{116}$$

The equality can be achieved only if $|\Phi\rangle = |\psi_1\rangle$. The functional $W[\Phi]$ is often referred to as the *Rayleigh quotient*. From this theorem one can conclude that

$$\frac{\langle\Phi^{(q)}|\hat{H}|\Phi^{(q)}\rangle}{\langle\Phi^{(q)}|\Phi^{(q)}\rangle} \geq E_q \tag{117}$$

if $|\Phi^{(q)}\rangle$ is orthogonal to $|\psi_i\rangle$ for $i = 1, 2, \ldots, q-1$.

Applications of the variational principle result in modifying a trial function Φ so that the value of $W[\Phi]$ becomes as small as it is possible within the constraints imposed upon Φ. Taking different kinds of the trial functions one gets different variational models. A precondition for a straightforward applicability of the variational method is that the eigenvalue problem we are dealing with is bounded from below, i.e. that the lowest energy state exists. This condition is fulfiled in the case of the Schrödinger equation but not in the case of the Dirac equation. Therefore, in order to apply the variational principle to the Dirac equation we have to take some preliminary steps. In general there are two classes of approaches to this problem. In the first approach one restricts the space of the trial functions by imposing the boundary conditions which would force the trial functions to be orthogonal to the negative-energy solutions. In such a space the Dirac eigenvalue problem is bounded from below and, consequently, the variational method may be applied in a way similar to that in the non-relativistic case (for a detailed discussion see Grant 1996). In the second approach one modifies the Dirac Hamiltonian in order to transform it into an operator which is bounded from below for all square-integrable trial functions retaining the interesting part of its spectrum unchanged (for a general discussion of this topic see Chang, Pélissier and Durand, 1986). There are many ways to accomplish this goal. For example, Sucher (1980) proposed to project the Hamiltonian onto the space of the positive-energy states (c.f. Eqs. (88) - (92)). Then the resulting eigenvalue problem is bounded from below but the formalism becomes prohibitively complicated. One can transform the Hamiltonian in such a way that the negative-energy continuum would overlap with the positive-energy continuum (Kutzelnigg 1984, Karwowski and Kobus 1986). Finally, using one of variants of the Foldy and Wouthuysen (1950) od Douglas and Kroll (1974) transformation one can tranform the Hamiltonian so that the large and small components are approximately decoupled. The most successful approaches in this cathegory are the *no-pair* approximation of Hess (1985, 1986) and the *regular approximation* of van Lenthe *et al.* (1993, 1994, 1996).

4.2. CLASSICAL PERTURBATIONAL METHODS

The perturbational methods are applicable if the Hamiltonian may be split into the unperturbed part $\hat{H}_0$ and a perturbation $\hat{H}'$:

$$\hat{H} = \hat{H}_0 + \hat{H}'. \tag{118}$$

If solutions of the eigenvalue problem of $\hat{H}_0$ are known then solutions of the eigenvalue problem of $\hat{H}$ may be expanded into a power series of the perturbation using the solutions of the unperturbed problem. The non-relativistic perturbation techniques are usually related to either Brillouin-Wigner or to Rayleigh-Schrödinger scheme (see e.g. Thouless 1961). In the relativistic case, the most successful is the direct perturbation method (Rutkowski 1986, Kutzelnigg 1989)).

Let us assume that

$$(\hat{H}_0 + \hat{H}')|\Psi\rangle = E|\Psi\rangle \tag{119}$$

and

$$\hat{H}_0|\Psi^0\rangle = E^0|\Psi^0\rangle. \tag{120}$$

Then, assuming that $\langle\Psi^0|\Psi^0\rangle = 1$ and $\langle\Psi|\Psi^0\rangle = 1$ (the so called *intermediate normalization*), after some simple algebra, we get

$$E = E^0 + \langle\Psi^0|\hat{H}'|\Psi\rangle \tag{121}$$

and

$$|\Psi\rangle = |\Psi^0\rangle + \frac{1}{E - \hat{H}_0}\hat{P}\hat{H}'|\Psi\rangle, \tag{122}$$

where

$$\hat{P} = \hat{I} - |\Psi^0\rangle\langle\Psi^0| \tag{123}$$

is the projection operator onto the space orthogonal to $|\Psi^0\rangle$. The last two equations may be iterated giving in effect expansions of the energy and of the wavefunction. In particular,

$$E = E^0 + \langle\Psi^0|\hat{H}'|\Psi^0\rangle + \langle\Psi^0|\hat{H}'\frac{\hat{P}}{E - \hat{H}_0}\hat{H}'|\Psi^0\rangle + \cdots. \tag{124}$$

Methods of constructing the perturbation series based on this expansion are known as the *Brillouin-Wigner perturbation theory.* Let us note that in this expansion the *exact* energy E appears on both left- and right-hand-side of the equation. Therefore it cannot be considered as a power series expansion of neither E nor $|\Psi\rangle$ in terms of powers of the perturbation operator. Besides one can show (see e.g. Thouless 1961, Wilson 1987) that the method, when restricted to a given order and applied to an N-electron problem, is not *size consistent*, i.e. its N-behaviour is incorrect.

Another approach, known as the *Rayleigh-Schrödinger perturbation theory* is obtained as an expansion into a power series of the perturbation of both energy and the wavefunction. The Rayleigh-Schrödinger expansion may be derived from the following expressions (equivalent to Eqs. (119) and (120)):

$$E = E^0 + \langle\Psi^0|\hat{H}'|\Psi\rangle \tag{125}$$

and

$$|\Psi\rangle = |\Psi^0\rangle + \frac{1}{E^0 - \hat{H}_0}\hat{P}(\hat{H}' - E + E^0)|\Psi\rangle. \tag{126}$$

The energy expansion in this case is

$$E = E^0 + \langle\Psi^0|\hat{H}'|\Psi^0\rangle + \langle\Psi^0|\hat{H}'\frac{\hat{P}}{E_0 - \hat{H}_0}\hat{H}'|\Psi^0\rangle + \cdots. \tag{127}$$

In the third term of the Rayleigh-Schrödinger expansion there appears the unperturbed energy E^0, while in the corresponding term of the Brillouin-Wigner expansion - the exact one, E. The higher terms of the Rayleigh-Schrödinger expansion are entirely different from the ones of the Brillouin-Wigner ones. Their classification may be conveniently performed by means of diagrammatic techniques (the Goldstone and the Hugenholtz diagrams are the most commonly used). The operator

$$\hat{R}_0 = \frac{\hat{P}}{E^0 - \hat{H}_0} \tag{128}$$

is referred to as the *reduced resolvent.* In the basis of eigenvectors of $\hat{H}_0$ it may be represented as

$$\hat{R}_0 = \sum_{i\neq 1} \frac{|\Psi_i^0\rangle\langle\Psi_i^0|}{E_0^{(i)} - E_1^0}, \tag{129}$$

where $\Psi_1^0 \equiv \Psi^0$ and $E_1^0 \equiv E^0$. The N-dependence of the Rayleigh-Schrödinger expansion is correct in every order of approximation (we say that this formulation is size-consistent).

4.3. DIRECT PERTURBATION THEORY

The direct perturbation method of Rutkowski (1986) proved to be the most appropriate way for performing relativistic perturbational calculations. It is based on the partition of the Dirac equation which defines the unperturbed problem in the same Hilbert space as the exact one. Then, we write the Dirac equation in the form (c.f. Eq. (78))

$$\left[\hat{H}_0 - \beta^+ E + \beta^-(V - E)\right]\psi = 0, \tag{130}$$

where E has been substituted for $E - mc^2$ (i.e. the origin of the energy scale has been chosen at the rest energy of the electron),

$$\psi = \begin{pmatrix}\psi^G\\ \psi^F\end{pmatrix} = \begin{pmatrix}\psi^G\\ 0\end{pmatrix} + \begin{pmatrix}0\\ \psi^F\end{pmatrix} = \psi^+ + \psi^- \tag{131}$$

and

$$\hat{H}_0 = c\vec{\alpha}\cdot\hat{\vec{p}} + \beta^+ V - 2\beta^- mc^2. \tag{132}$$

The four-component spinors ψ^+ and ψ^- may be obtained as appropriate projections of ψ:

$$\psi^+ = \beta^+\psi, \quad \psi^- = \beta^-\psi \tag{133}$$

and are related, respectively, to the large and to the small components of ψ. As one can easily see (c.f. section 3. 2.), equation

$$(\hat{H}_0 - \beta^+ E)\psi = 0 \tag{134}$$

is equivalent to the Schrödinger equation. In the direct perturbation method Eq. (134) is regarded as the unperturbed problem and

$$\hat{H}' = \beta^-(V - E) \tag{135}$$

as the perturbation. Both the energy and the wavefunction are expanded into power series with respect to the perturbation (the perturbation parameter is not written explicitly). Then, we have

$$\psi = \psi^{(0)} + \psi^{(1)} + \psi^{(2)} + \cdots \tag{136}$$

and

$$E = E^{(0)} + E^{(1)} + E^{(2)} + \cdots . \tag{137}$$

Substituting these expansions into Eq. (130) and separating different orders of perturbation we get an infinite set of equations:

$$(\hat{H}_0 - \beta^+ E^{(0)})\psi^{(0)} = 0, \tag{138}$$

$$(\hat{H}_0 - \beta^+ E^{(0)})\psi^{(1)} + (\hat{H}' - \beta^+ E^{(1)})\psi^{(0)} = 0, \tag{139}$$

$$\cdots \quad \cdots \quad \cdots .$$

From here the consecutive corrections may be derived in the usual way:

$$E^{(1)} = \langle\psi^{(0)-}|V - E^{(0)}|\psi^{(0)-}\rangle, \tag{140}$$

$$E^{(2)} = \langle\psi^{(0)-}|V - E^{(0)}|\psi^{(1)-}\rangle - E^{(1)}\left(\langle\psi^{(0)-}|\psi^{(0)-}\rangle + \langle\psi^{(0)+}|\psi^{(1)+}\rangle\right), \tag{141}$$

$$\cdots \quad \cdots \quad \cdots \quad \cdots .$$

In this theory the perturbation parameter is equal to the square of the fine-structure constant α. Therefore the k-th order of the perturbation corresponds to a term proportional to α^{2k}.

A review of different modifications of this method and of other perturbational schemes suitable in the case of a relativistic theory have recently been presented by Kutzelnigg (1989, 1990).

4.4. RELATIVISTIC TWO-COMPONENT METHODS

4.4.1. *Elimination of the small components*

The simplest and the most commonly used approach to an approximation describing relativistic effects is the method in which the small components of the wavefunctions are expressed by the large ones using the Dirac equation (as e.g. in Eq. (112)) and then eliminated. As a result a relativistic description based on a two-component wavefunction is obtained. By expanding the resulting equation into a power series of $\frac{E-V}{mc^2}$ one obtains the well known Pauli approximation (see e.g. Davydov 1965, Schiff 1968). The resulting Hamiltonian is strongly singular and the relativistic terms can only be used as the first order perturbations.

Works directed towards development of a two-component relativistic theory free of the deficiencies of the Pauli approach resulted in formulation of numerous quasirelativistic theories. Particularly, the method of Chang, Pélissier and Durand (1986), derived from the theory of effective Hamiltonians, is based on a regular, i.e. suitable to variational calculations, Hamiltonian. A similar regular two-component Hamiltonian has been recently obtained by van Lenthe *et al.* (1993, 1994, 1996) by expanding the relativistic Hamiltonian in terms of $(2mc^2 - V)^{-1}$.

A large family of two-component relativistic equations may be obtained by using different modifications of unitary transformations which decouple the large and small components in the Dirac equation. The best known is the Foldy-Wouthuysen (1950) transformation. However the most successful in applications appeared to be the Douglas-Kroll (1974) transformation in its modification due to Hess (1985, 1986). The resulting from this transformation *no-pair* approach belongs to the most frequently used relativistic quantum chemical models.

The Dirac equation for an electron moving in the field of an external potential V may be written as

$$\begin{aligned} V\psi_G + c(\vec{\sigma}\cdot\hat{\vec{p}})\psi_F &= E\psi_G, \\ c(\vec{\sigma}\cdot\hat{\vec{p}})\psi_G + (V - 2mc^2)\psi_F &= E\psi_F, \end{aligned} \tag{142}$$

where, similarly as in the previous section, the origin of the energy scale has been chosen at mc^2. Eliminating the small component gives

$$\psi_F = \frac{1}{2mc}\left(1 + \frac{E - V}{2mc^2}\right)^{-1} (\vec{\sigma}\cdot\hat{\vec{p}})\,\psi_G \equiv \hat{X}\,\psi_G, \tag{143}$$

and

$$\hat{H}^{lc}\psi_G = E\psi_G. \tag{144}$$

The Hamiltonian

$$\hat{H}^{lc} = \frac{1}{2m}(\vec{\sigma}\cdot\hat{\vec{p}})\left(1 + \frac{E - V}{2mc^2}\right)^{-1} (\vec{\sigma}\cdot\hat{\vec{p}}) + V \tag{145}$$

is energy-dependent and acts on the large component ψ_G of the Dirac spinor. At this stage Eq. (144) is equivalent to the original Dirac equation. In order to convert Eq. (144) into a quasirelativistic equation of a theory in which only two-component wavefunctions appear one has to renormalize ψ_G (see e.g. Davydov 1965 for more details). Then, we have

$$1 = \langle\psi|\psi\rangle = \langle\psi_G|\psi_G\rangle + \langle\psi_F|\psi_F\rangle = \langle\psi_G|\hat{I} + \hat{X}^{\dagger}\hat{X}|\psi_G\rangle = \langle\psi_{qr}|\psi_{qr}\rangle, \tag{146}$$

where the quasirelativistic wavefunction is defined as

$$|\psi_{qr}\rangle = \hat{O}|\psi_G\rangle \tag{147}$$

with

$$\hat{O} = \left(\hat{I} + \hat{X}^\dagger \hat{X}\right)^{\frac{1}{2}}. \tag{148}$$

Consequently, the corresponding Hamiltonian is

$$\hat{H}_{qr} = \hat{O}\hat{H}^{lc}\hat{O}^{-1}. \tag{149}$$

Expanding $\hat{H}^{lc}$ and $\hat{O}$ into a power series of $\frac{E-V}{mc^2}$ and retaining the lowest-order terms, one obtains the well known Pauli Hamiltonian

$$\hat{H}_P = \frac{\hat{p}^{\,2}}{2m} + V - \frac{\hat{p}^{\,4}}{8m^3c^2} - \frac{\vec{s}\cdot(\nabla V \times \hat{\vec{p}})}{2m^2c^2} - \frac{\hbar^2}{8m^2c^2}\Delta V, \tag{150}$$

where $\vec{s} = \frac{\hbar}{2}\vec{\sigma}$ is the electron spin operator of the Pauli theory. The non-relativistic Schrödinger Hamiltonian is given by the first two terms in the rhs of the last equation and the subsequent terms describe, respectively, the effect of change of the electron mass with velocity, the spin-orbit interaction and a correction due to *Zitterbewegung* (known as the Darwin correction). All these terms are highly singular. In effect the eigenvalue problem of $\hat{H}_P$ does not have any square-integrable solutions. However this operator may be (and has been since more than half a century) used to estimate the relativistic corrections as the first-order perturbations. By properly restricting its domain one can also use this operator in some variational approaches. However one should remember that for a Coulomb potential there always exists an area around the nucleus in which the expansion is invalid because $\left|\frac{E-V}{mc^2}\right| > 1$. On the other hand, this is the area in which the kinetic energy of the electron is the largest, i.e. the relativistic effects are the most important.

There exist many other ways of reducing the Dirac equation to a quasirelativistic (two-component) form. The subject is well covered in the literature (e.g by Kutzelnigg 1984, by Karwowski and Kobus 1985, 1986, by Chang, Pélissier and Durand 1986). For pedagogical reasons a recent approach by van Lenthe *et al.* (1993, 1994, 1996) deserves special attention. We can rewrite Eq. (145) as

$$\hat{H}^{lc} = (\vec{\sigma}\cdot\hat{\vec{p}})\,\frac{c^2}{2mc^2 - V}\left(1 + \frac{E}{2mc^2 - V}\right)^{-1}(\vec{\sigma}\cdot\hat{\vec{p}}) + V. \tag{151}$$

From this form of $\hat{H}^{lc}$ one can see that instead of expanding it in terms of the ill-behaving $\frac{E-V}{mc^2}$, one can use as the expansion parameter $E\tau$, where

$$\tau = \frac{1}{2mc^2 - V}.$$

For an attractive Coulomb-type potential and $|E| < 2mc^2$, the value of $|E\tau|$ never exceeds 1. Therefore the corresponding expansion is convergent in the whole space, also in the area of the singularity. Then, one gets

$$\hat{H}^{lc} = V + c^2(\vec{\sigma}\cdot\hat{\vec{p}})\,\tau\,(\vec{\sigma}\cdot\hat{\vec{p}}) - Ec^2(\vec{\sigma}\cdot\hat{\vec{p}})\,\tau^2\,(\vec{\sigma}\cdot\hat{\vec{p}}) + O\left[(E\tau)^3\right]. \quad (152)$$

In a similar way one can expand the normalization operator $\hat{O}$. In effect, in the lowest order of the expansion, one obtains the so called *zeroth-order regular approximated* (ZORA) Hamiltonian:

$$\hat{H}^{zora} = (\vec{\sigma}\cdot\hat{\vec{p}})\,\frac{c^2}{2mc^2 - V}\,(\vec{\sigma}\cdot\hat{\vec{p}}) + V. \quad (153)$$

identical to the relativistic regular effective Hamiltonian of Chang, Pélissier and Durand (1986). ZORA Hamiltonian is regular, bounded from below and can be used in variational calculations without any special restrictions. In the higher order one gets the Hamiltonian in the *first-order regular approximation* (FORA):

$$\hat{H}^{fora} = \hat{H}^{zora} - \frac{1}{2}\left[\hat{B}\hat{H}^{zora} + \hat{H}^{zora}\hat{B}\right], \quad (154)$$

where

$$\hat{B} = (\vec{\sigma}\cdot\hat{\vec{p}})\frac{c^2}{(2mc^2 - V)^2}(\vec{\sigma}\cdot\hat{\vec{p}}).$$

This Hamiltonian, similarly as the Pauli one, does not have square-integrable eigenfunctions and is not limited from below. Therefore it cannot be used in variational calculations, unless the space of the trial function is properly constrained.

4.4.2. *Foldy-Wouthuysen and related transformations*

Foldy and Wouthuysen (1950) introduced a systematic procedure for decoupling the large and the small component parts of the Dirac equation to any fixed order in the fine structure constant. Numerous modifications and generalizations of their approach have been discussed by Kutzelnigg (1990) and by van Lenthe (1996). Let us consider a unitary transformation performed on the Dirac Hamiltonian

$$\hat{H}^{fw} = U\hat{H}_D U^\dagger \quad (155)$$

with

$$U = (\hat{I} + \hat{W}^\dagger\hat{W})^{-\frac{1}{2}}\begin{pmatrix} \hat{I} & \hat{W}^\dagger \\ -\hat{W} & \hat{I} \end{pmatrix}. \quad (156)$$

This transformation brings the Dirac Hamiltonian into a block-diagonal form (i.e. decouples the large and the small components) if $\hat{W}$ fulfils the following condition (Kutzelnigg 1990):

$$c\,(\vec{\sigma}\cdot\hat{\vec{p}}) - c\,\hat{W}\,(\vec{\sigma}\cdot\hat{\vec{p}})\,\hat{W} - \hat{W}\,V + (V - 2mc^2)\,\hat{W} = 0. \tag{157}$$

The "large component" part of the transformed equation reads

$$\hat{H}^{FW}\psi^{FW} = E\psi^{FW}. \tag{158}$$

The Foldy-Wouthuysen Hamiltonian $\hat{H}^{FW}$ is bounded from below and its spectrum is the same as the positive-energy spectrum of $\hat{H}_D$. Expanding $\hat{W}$ into a power series of $\frac{E-V}{mc^2}$ and retaining the lowest-order terms we obtain the Pauli Hamiltonian; if the expansion is performed in terms of $(2mc^2 - V)^{-1}$ instead then, depending upon the degree of approximation, we end up with ZORA or with FORA Hamiltonians.

A very important modification of the Foldy-Wouthuysen transformation is known as the Douglas-Kroll (1974), or Douglas-Kroll-Hess (1985, 1986) transformation. If we assume that $\hat{W}$ commutes with V, Eq. (157) may be solved giving

$$\hat{W} = \frac{c\,(\vec{\sigma}\cdot\hat{\vec{p}})}{mc^2 + \sqrt{m^2c^4 + \hat{p}^2c^2}}. \tag{159}$$

In the case of a free particle this is the exact Foldy-Wouthuysen transformation. Otherwise the transformation may be expanded in terms of powers of the external potential V. The resulting large-component part $\hat{H}^{DC}$ of the Douglas-Kroll-transformed Dirac Hamiltonian, in the first order in the external potential V, reads

$$\hat{H}^{DC} = \hat{E}_p + \hat{A}(\hat{V} + \hat{R}\hat{V}\hat{R})\hat{A}, \tag{160}$$

where

$$\hat{E}_p = \sqrt{c^2\hat{p}^2 + m^2c^4}, \tag{161}$$

$$\hat{A} = \sqrt{\frac{\hat{E}_p + mc^2}{2\hat{E}_p}}, \tag{162}$$

and

$$\hat{R} = \frac{c\,(\vec{\sigma}\cdot\hat{\vec{p}})}{\hat{E}_p + mc^2}. \tag{163}$$

Using an expansion method one can derive formulae which are exact to higher orders in V. This Hamiltonian, frequently referred to as the *no-pair* Hamiltonian, is regular also for Coulomb-like potentials and may easily be applied in all kinds of variational methods.

5. N-electron Systems: The Hamiltonian and the Model Space

5.1. THE INDEPENDENT PARTICLE MODEL

Construction of a quantum theory of a system of N-electrons meets many new conceptual problems, when compared to the the theory of a single particle. This can be seen already in the simplest case of non-interacting particles, which may be described using the *independent-particle model.* In this case, the non-relativistic Hamiltonian may be written as

$$\hat{H}_S^0(1,2,\ldots,N) = \sum_{j=1}^{N} \hat{h}_1^S(j), \tag{164}$$

where

$$\hat{h}_1^S(1) = \hat{T}(1) + \hat{V}(1) \tag{165}$$

is the one-electron non-relativistic Hamiltonian composed of the operators describing the kinetic energy of the electron ($\hat{T}$) and its interaction with the external potential ($\hat{V}$). If we are concerned with isolated atoms or molecules, the external potential is generated by the clamped nuclei. The eigenvalue equation

$$\hat{H}_S^0 \Psi_n(1,2,\ldots,N) = E_n^0 \Psi_n(1,2,\ldots,N) \tag{166}$$

may be separated into N one-electron equations. Since electrons are non-distinguishable, the N-electron Hamiltonian has to be invariant with respect to their permutations. Therefore all one-electron equations resulting from the separation are the same. If

$$\hat{h}_1^S(1)\psi_q(1) = \varepsilon_q \psi_q(1) \tag{167}$$

with $\varepsilon_1 \leq \varepsilon_2 \leq \cdots$, then a product

$$\Psi_\lambda^0(1,2,\ldots,N) = \prod_{j=1}^{N} \psi_{\lambda_j}(j) \tag{168}$$

is a solution of Eq. (166) to the eigenvalue

$$E_\lambda^0 = \sum_{j=1}^{N} \varepsilon_{\lambda_j}. \tag{169}$$

The one-electron functions ψ_j are usually assumed to be orthonormal (for simplicity we make this assumption as well) and are referred to as *orbitals.*

The set of indices $\lambda = \{\lambda_1, \lambda_2, \dots, \lambda_N\}$ is called the *electronic configuration* and the product Ψ^0_λ is the *Hartree product* corresponding to the electronic configuration λ. If we do not refer to a specific representation, we use the symbol $|\lambda\rangle$ to denote the Hartree product corresponding to the orbital configuration λ.

The eigenvalue problem (166) is highly degenerate since all products of orbitals which form the same configuration correspond to the same eigenvalue. An important restriction on the form of the wavefunction results from the fact that the electrons are indistinguishable: the electron density must be symmetric with respect to any permutation of electrons. Hence the wavefunction should either be symmetric or antisymmetric with respect to a transposition of two electrons. A theorem, first proved by Pauli on the ground of quantum electrodynamics (more details about the proof are given by Sakurai 1984), states that systems of particles that have half-integer spins (the fermions) are described by antisymmetric wavefunctions, while the systems of integer-spin particles (the bosons) – by the symmetric ones. Then, an N-electron wavefunction must be antisymmetric with respect to a transposition of electrons. The antisymmetry is related to the half-integer electron spin. Therefore introducing spin, optional in the case of a single Schrödinger electron in the absence of magnetic fields, is necessary when we have more than one particle, even if the particles do not interact. Thus, in the non-relativistic theory on N-electron systems each orbital is associated with two mutually orthogonal spin functions, customarily denoted α and β. They are eigenfunctions of the one electron operator $\hat{S}^2$ of the total spin to the eigenvalue $\frac{1}{2}(1+\frac{1}{2})\hbar^2$ and of its projection $\hat{S}_z$ to the eigenvalues $+\frac{1}{2}\hbar$ and $-\frac{1}{2}\hbar$, respectively. They form a basis in a two-dimensional one-particle spin space and may be represented as

$$\alpha = \begin{pmatrix} 1 \\ 0 \end{pmatrix}, \quad \beta = \begin{pmatrix} 0 \\ 1 \end{pmatrix}. \tag{170}$$

Then, with each orbital ψ one may associate two *spinorbitals*: ψ_α and ψ_β. The N-electron product of spinorbitals corresponding to a Hartree product Ψ^0_λ is denoted $\tilde{\Psi}^0_\lambda$. The antisymmetric combination of the spinorbital products corresponding to a given λ may be obtained by acting on $\tilde{\Psi}^0_\lambda$ by the antisymmetrization operator

$$\hat{\mathcal{A}} = \frac{1}{N!}\sum_P (-1)^p \hat{P}, \tag{171}$$

where $\hat{P}$ is a permutation operator and p is its parity. The antisymmetrization operator may easily be recognized as a projection operator onto the antisymmetric subspace of the space spanned by all products of spinorbitals.

Then, the antisymmetric wavefunction corresponding to the configuration λ reads

$$\tilde{\Psi}_{\lambda} = \sqrt{N!}\hat{\mathcal{A}}\tilde{\Psi}^{0}_{\lambda}, \tag{172}$$

where $\sqrt{N!}$ is the normalization factor. The antisymmetric function $\tilde{\Psi}_{\lambda}$ given by Eq. (172) may be represented by a determinant built of N spinorbitals (the *Slater determinant*). From the basic properties of determinants one may easily deduce that if any two spinorbitals in λ are the same, the corresponding wavefunction $\tilde{\Psi}_{\lambda}$ vanishes. In other words, no two electrons can occupy the same one-electron state. This is the famous *Pauli exclusion principle* being a base for understanding a multitude of chemical and physical phenomena.

The set of all Slater determinants which may be constructed out of a given set of spinorbitals forms a discrete basis in an antisymmetric, N-electron Hilbert space. One may easily prove that if the orbital basis is complete in the one-electron Hilbert space, then the basis of all N-electron Slater determinants is complete in the antisymmetric N-electron space (Löwdin 1955). In all models directed towards practical applications the number of orbitals is finite. If the dimension of the orbital space is K then we have $2K$ spinorbitals and $\binom{2K}{N}$ Slater determinants. If $K >> N$, the dimension of the N-electron space $\binom{2K}{N} \sim \frac{(2K)^N}{N!}$, grows exponentially with the number of electrons. The finite-dimensional antisymmetric N-electron space is referred to as the *model space.* It is also called the *full configuration interaction* (FCI) space.

Let us note that the lowest eigenvalue of $\hat{H}^{0}_{S}(1, 2, \ldots, N)$ is equal to $N\varepsilon_1$ (see Eq. (169)) while the lowest eigenvalue of this Hamiltonian in the FCI space, is equal to

$$E^{0}_{1} = \sum_{j=1}^{N/2} 2\varepsilon_j. \tag{173}$$

It is larger than the lowest eigenvalue $N\varepsilon_1$ of the unconstrained eigenvalue problem. Therefore, when solving the eigenvalue problem using variational methods, one should include the condition of antisymmetry of the space as a constraint in the variational procedure. This is an analog of the boundary conditions which have to be imposed on an unbounded from below variational eigenvalue problem in order to prevent the variational collapse. In the case of a variational problem in a FCI space, imposing this constraint is easy. However in some approaches, as for example in the ones in which the variational principle is applied to a density matrix, a proper taking into account the antisymmetry of the model space results in a very difficult and still not completely solved *N-representability problem* (Coleman 1965, Erdahls and Smith 1987). Also in the density functional theory, this problem is far from being trivial (Kryachko and Ludeña 1990, Gross and Dreizler 1995).

In the case of N non-interacting Dirac particles moving in a stationary external field, one can write a Lorentz-invariant equation as a generalization of Eq. (72). Then, we have

$$\sum_{j=1}^{N}\left[i\hbar\frac{\partial}{\partial t_j}-\hat{h}_1^D(j)\right]\psi=0, \tag{174}$$

where

$$\hat{h}_1^D=c\vec{\alpha}_1\cdot\hat{\vec{p}}(1)+\beta_1 mc^2+\hat{V}(1) \tag{175}$$

is the one-electron Dirac Hamiltonian. Eq. (174) may be separated into N independent Dirac equations and each of them may be converted into an one-electron eigenvalue problem. The further reasoning goes in a way similar to that in the non-relativistic case except that now each one-electron eigenfunction of $\hat{h}_1^D$ is a four-component spinor which is not separable into an orbital and a spin part. The set of Slater determinants constructed from these spinors span the model space in which the N-electron Hamiltonian eigenvalue problem may be represented, similarly as in the nonrelativistic case. There is, however, an important difference between the relativistic and non-relativistic description. In the relativistic case the energy of every N-electron discrete state is degenerate with an infinite number of continuum states. This may easily be seen in the case of two particles. Let us say that $E_{12}=\varepsilon_1+\varepsilon_2$ corresponds to a Slater determinant built from two Dirac spinors corresponding to one-electron energies ε_1 and ε_2. However the same energy is obtained if we combine $\overline{\varepsilon}_1=\varepsilon_1+mc^2+\delta$ and $\overline{\varepsilon}_2=\varepsilon_1-mc^2-\delta$, where $\delta>0$. Obviously, $\overline{\varepsilon}_1$ correspnds to a positive continuum energy state and $\overline{\varepsilon}_2$ - to a negative continuum energy state. This property of the Dirac spectrum is known as the Brown-Ravenhall disease (Bethe and Salpeter 1957, Wilson 1987, Grant 1996). If the one-electron equations are exactly solvable, one can easily separate many-electron discrete states from the continuum ones: the discrete N-electron states are products of the Dirac spinors belonging to the discrete parts of the one-electron spectra. However, when approximate methods are used, one has to carefully avoid any contamination of the spinors by the continnum states. This may either be achieved by properly imposing the boundary conditions (Grant 1996) or by other techniques, already discussed in section 4.

5.2. INTERACTIONS

Though very few systems may be described reasonably well by the model of non-interacting particles, the model space defined within this approach is a base for formulation of a vast majority of quantum chemical methods. An arbitrary Hamiltonian may be represented in this space as a matrix and since, at least in principle, the space can be complete, the model may supply a formal solution of the eigenvalue problem of this Hamiltonian. A Hamiltonian of an N electron system may be written as

$$\hat{H}(1,2,\ldots,N) = \sum_{j=1}^{N} \hat{h}_1(j) + \sum_{i<j}^{N} \hat{h}_2(i,j) \tag{176}$$

where $\hat{h}_1$ are the one-electron Hamiltonians and $\hat{h}_2$ describe the interactions between electrons. In the nonrelativistic case, $\hat{h}_1 = \hat{h}_1^S$ and $\hat{h}_2(1,2) = \frac{e^2}{r_{12}}$ is the inter-electron Coulomb interaction. In the relativistic theory defining an N-electron Hamiltonian is a major fundamental problem. A Lorentz-invariant quantum mechanics of many particles cannot be formulated. A rigorous formulation of such a theory is only possible at the level of quantum electrodynamics. Also an approximate form of the many-electron Hamiltonian may be derived from there. However, using the correspondence principle, one may obtain the same result from classical electrodynamics. Classically, the scalar and the vector potentials generated at the point $\vec{r}_j$ by a charge e located at the point $\vec{r}_i$ and moving with velocity $\vec{v}_i$ (Lienard-Wiechert potentials) are given by

$$V_i^{LW}(\vec{r}_j) = \frac{e}{r_{ij}\left(1 + \frac{(\vec{v}_i \cdot \vec{r}_{ij})}{c r_{ij}}\right)} \tag{177}$$

and

$$\vec{A}_i^{LW}(\vec{r}_j) = \frac{\vec{v}_i}{c} V_i^{LW}(\vec{r}_j) \tag{178}$$

respectively. By substituting these potentials into the relativistic Hamilton function (Eq. 59)) and expanding into a power series with respect to $\frac{\vec{v}}{c}$ we get

$$H = \sum_{j=1}^{N} H_1(j) + \sum_{i<j}^{N} \frac{e^2}{r_{ij}} \left[1 - \frac{\vec{v}_i \cdot \vec{v}_j}{2c^2} - \frac{(\vec{v}_i \cdot \vec{r}_{ij})(\vec{v}_j \cdot \vec{r}_{ij})}{2c^2 {r_{ij}}^2}\right] + O\left(\left(\frac{v}{c}\right)^2\right), \tag{179}$$

where

$$H_1(j) = mc^2 \sqrt{1 + \frac{1}{m^2c^2}\left(\vec{p}_j - \frac{e}{c}\vec{A}(\vec{r}_j)\right)^2} + V(\vec{r}_j). \tag{180}$$

Now, according to the correspondence principle, we substitute:

$$H_1(j) \Rightarrow \hat{h}_1^D(j) = c\vec{\alpha}_j \left(\hat{\vec{p}}_j - \frac{e}{c}\hat{\vec{A}}(\vec{r}_j)\right) + \hat{V}(\vec{r}_j) \tag{181}$$

and

$$\vec{v} \Rightarrow c\vec{\alpha}. \tag{182}$$

The resulting Hamiltonian reads:

$$\hat{H}(1,2,\ldots,N)_D = \sum_{j=1}^{N} \hat{h}_1^D(j) + \sum_{i<j}^{N} \left[\hat{h}_2(i,j) + \hat{h}_2^{mag}(i,j) + \hat{h}_2^{ret}(i,j)\right], \tag{183}$$

where

$$\hat{h}_2^{mag}(i,j) = -\frac{e^2}{2r_{ij}}(\vec{\alpha}_i \cdot \vec{\alpha}_j) \tag{184}$$

and

$$\hat{h}_2^{ret}(i,j) = -\frac{e^2}{2{r_{ij}}^3}(\vec{\alpha}_i \cdot \vec{r}_{ij})(\vec{\alpha}_j \cdot \vec{r}_{ij}) \tag{185}$$

are, respectively, corrections due to two-electron magnetic interactions and due to retardation resulting from the finite velocity of the interaction. A relativistic Hamiltonian in which the two-electron terms are approximated by $\hat{h}_2(i,j) = \frac{e^2}{r_{ij}}$ is known as the *Dirac-Coulomb Hamiltonian.* If the magnetic and retardation corrections are included (referred to as the *Breit interactions*) then the Hamiltonian is callled the *Dirac-Breit Hamiltonian.* Since the Hamilton function (179) is correct up to $(\frac{v}{c})^2$ terms, also the quantum-mechanical Hamiltonian is correct up to this accuracy and the formulation based on this Hamiltonian is only approximately Lorentz invariant.

Each Hamiltonian $\hat{H}$ is represented in the model space by a matrix. Its matrix elements are

$$H_{LM} = \langle L|\hat{H}|M\rangle, \tag{186}$$

where $|L\rangle$ and $|M\rangle$ are the kets corresponding to the pertinent Slater determinants. The operator eigenvalue problem

$$(\hat{H} - E)|\Psi\rangle = 0 \tag{187}$$

is represented by the corresponding matrix secular equation

$$\sum_M \left(\langle L|\hat{H} - E|M\rangle\right) \langle M|\Psi\rangle = 0 \tag{188}$$

or alternatively by

$$\sum_M (H_{LM} - E\delta_{LM})C_M = 0, \tag{189}$$

where $C_M = \langle M|\Psi\rangle$ is the M-th component of the representative of the wavefunction. The matrix elements H_{LM} are linear combinations of one- and two-electron integrals defined, respectively, as

$$(p|q) = \langle\psi_p(1)|h_1(1)|\psi_q(1)\rangle \tag{190}$$

and

$$(pq|rs) = \langle \psi_p(1)\psi_r(2)|h_2(1,2)|\psi_q(1)\psi_s(2)\rangle. \tag{191}$$

The corresponding formulae may be found in textbooks of molecular quantum mechanics, e.g. in the ones by Weissbluth (1978) or by McWeeny (1989). Attempts to solve this eigenvalue problem in very large model spaces stimulated development of many ingenious models and computational approaches. Some of them will be discussed in the next section.

5.3. THE FOCK SPACE

The direct sum of model spaces corresponding to all possible numbers of electrons is referred to as the *Fock space.* In a representation-independent formulation the Fock space is spanned by kets corresponding to all Slater determinants built from a given set of spinorbitals or, in the case of a relativistic theory, from a given set of Dirac spinors. Each basis ket in this space is labelled by a string of 0's and 1's showing the distribution of electrons occupying particular spinorbitals (or spinors). For example, in the simplest case of $K = 1$ the Fock space has 2 dimensions and the basis vectors are $|0\rangle$ (0 electrons) and $|1\rangle$ (1 electron), if $K = 2$ then the Fock space is 4-dimensional with the following basis: $|00\rangle$, $|01\rangle$, $|10\rangle$ and $|11\rangle$ corresponding, respectively, to 0, 1, 1 and 2 electrons. The state with 0 electrons is called the *vacuum state* and is usually denoted $|0\rangle$. The number of electrons is a well defined measurable quantity. Therefore we may define an operator $\hat{N}$ with eigenvalues equal to the number of electrons N. All basis vectors in the Fock space are eigenvectors of this operator. Then, $\hat{N}$ is diagonal in the Fock space. The representation defined by its eigenvectors is referred to as the *occupation number representation.* A subspace of the Fock space corresponding a given N is identical to the FCI space.

All basis vectors in the Fock space may be obtained from the vacuum state by consecutively acting on it by the one-electron *creation operators* $a_k^\dagger$ corresponding to the k-th spinorbital (or to the k-th spinor). A creation operator acting on an unoccupied spinorbital changes its occupation number to 1 and acting on a spinorbital which is occupied produces zero. Similarly one defines the *annihilation operators* a_k changing the occupation number of the k-th spinorbital from 1 to 0 and producing zero when acting on an empty spinorbital. In the case of the Fock space with $K = 1$ the representatives of the basis vectors are:

$$|0\rangle \Rightarrow \begin{pmatrix} 1 \\ 0 \end{pmatrix}, \quad |1\rangle \Rightarrow \begin{pmatrix} 0 \\ 1 \end{pmatrix}. \tag{192}$$

As one can check, the creation and annihilation operators are represented, in this case, by the following 2×2 matrices

$$a^\dagger \Rightarrow \begin{pmatrix} 0 & 0 \\ 1 & 0 \end{pmatrix}, \quad a \Rightarrow \begin{pmatrix} 0 & 1 \\ 0 & 0 \end{pmatrix}. \tag{193}$$

These matrices are mutually adjoint. One can show that also in a general case $a_k^\dagger$ and a_k are mutually adjoint. The operator $\hat{N}$ is diagonal and has to be represented by a diagonal 2×2 matrix with eigenvalues 0 and 1. Then,

$$\hat{N} \Rightarrow \begin{pmatrix} 0 & 0 \\ 0 & 1 \end{pmatrix}. \tag{194}$$

Hence,

$$\hat{N} = a^\dagger a. \tag{195}$$

Another equation which can be deduced from (193) is

$$aa^\dagger + a^\dagger a = 1. \tag{196}$$

It is not difficult to generalize Eqs. (195) and (196). Then,

$$\hat{N}_k = a_k^\dagger a_k, \tag{197}$$

where $\hat{N}_k$ is the occupation number operator of spinorbital k, and

$$a_k a_k^\dagger + a_k^\dagger a_k = 1. \tag{198}$$

The operator of the total number of electrons is

$$\hat{N} = \sum_k \hat{N}_k = \sum_k a_k^\dagger a_k. \tag{199}$$

Eq. (198) and the antisymmetry of the kets imply that

$$a_k a_l^\dagger + a_l^\dagger a_k = \delta_{kl}. \tag{200}$$

and

$$a_k a_l + a_l a_k = a_k^\dagger a_l^\dagger + a_l^\dagger a_k^\dagger = 0. \tag{201}$$

If the spinorbital basis is the eigen-basis of $\hat{h}_1$, the Hamiltonian of the non-interacting particles may be written as

$$\hat{H}_0 = \sum_k \hat{N}_k \varepsilon_k = \sum_k a_k^\dagger a_k \varepsilon_k. \tag{202}$$

Correctness of this equation may be immediately checked by acting with the rhs of Eq. (202) on an arbitrary basis ket of the Fock space – for obvious reasons one obtains the same ket multiplied by the eigenvalue of $\hat{H}_0$ as it is given by Eq. (169). On can show that in general

$$\hat{H} = \sum_{pq} (p|q) a_p^\dagger a_q + \sum_{pqrs} (pq|rs) a_p^\dagger a_r^\dagger a_q a_s. \tag{203}$$

Correctness of Eq. (203) may be checked by showing that the matrix representing operator given by Eq. (176) in the basis of Slater determinants is the same as the matrix representing operator (203) in the basis of the corresponding kets in the occupation number representation.

The formalism based on using creation and annihilation operators is frequently called the *second quantization* formalism. Let us note that in the second quantization formalism the Hamiltonian, i.e. also the wavefunction and, consequently, all properties of the quantum system, are derived from a knowledge of the set of one- and two-electron integrals $(p|q)$ and $(pq|rs)$. Though this observation could be made also by analyzing the structure of the matrix representing $\hat{H}$ in the FCI space (Eq. (186)), here the dependence on the one- and two-electron integrals is introduced already at the level of the Hamilton operator. It is important to remember that this formalism is restricted to the Fock space only. Then, it is not applicable if the wavefunction is not expressible as a combination of Slater determinants (as e.g. in the case of explicitly correlated wavefunctions).

If the Hamiltonian is spin-independent, one can execute the integration over spin in the one- and two-electron intergrals transforming them into the corresponding integrals in the *orbital* rather than *spinorbital* basis. Then, after some algebra, one may show (Paldus 1976) that

$$\hat{H} = \sum_{pq}^{K} (p|q) E_{pq} + \sum_{pqrs}^{K} (pq|rs)(E_{pq} E_{rs} - \delta_{qr} E_{ps}), \tag{204}$$

where

$$E_{pq} = a_{p\alpha}^\dagger a_{q\alpha} + a_{p\beta}^\dagger a_{q\beta}, \tag{205}$$

α and β correspond to the one-electron spin functions and the spinorbital indices of the creation and annihilation operators have been split into their orbital and spin parts. Then, in Eqs. (204) and (205) labels p, q, r and s refer to *orbitals.* One should remember that the same indices in Eq. (203) refer to either spinorbitals or to spinors. We also retain the same symbols to denote the integrals in the spinor, spinorbital and orbital bases.

The operators E_{pq} are spin-independent (in Eq. 205 the "integration" over spin variables has been performed). When they act on a product of orbitals they replace ψ_q by ψ_p. Therefore they are called *replacement operators* (Duch and Karwowski 1985). They are also referred to as the *shift operators, excitation operators* (see a review paper by Karwowski, 1992, for references and for more discussion of this subject) and appear to be the *generators of the unitary group* $U(K)$ (Paldus 1976). The replacement operators act in the orbital space only and

$$E_{pq} = \mathcal{I}_\Sigma \otimes \sum_{l=1}^{N} |\psi_p(l)\rangle\langle\psi_q(l)|, \tag{206}$$

where $\mathcal{I}_\Sigma$ is the identity operator in the spin space. The equivalence of Eqs. (205) and (206) may be demonstrated by showing that their right-hand sides give the same results when acting on an orbital product multiplied by an arbitrary spin function (one should remember, however, that Eq. (205) is defined in the Fock space while Eq. (206) in its N-electron subspace). If $p = q$ then

$$E_{pp}|\lambda\rangle = n_p^\lambda|\lambda\rangle, \tag{207}$$

where $n_p^\lambda = 0, 1, 2$ is the occupation number of ψ_p in λ. Then, E_{pp} is the *orbital* occupation number operator.

5.4. GRAPHICAL REPRESENTATIONS OF THE MODEL SPACES

Model spaces in which the basis vectors are labelled by sequences of one-electron occupation numbers may be conveniently represented in a graphical way. The model space is described by a graph composed of a network of paths. Each of the paths describes one N-electron basis function. The path is composed of either K or $2K$ arcs, depending upon the mode of the description. The slope of the arc describes the one-electron occupation number. Then, in the spinorbital description we have the *two-slope* graphs (since the occupation numbers may be equal to either 0 or 1). In the orbital description the occupation numbers may be equal to 0, 1 and 2. Therefore in this case we have *three-slope* orbital graphs. In the case of spin-independent Hamiltonians the total spin operators are constants of the motion. Therefore one can split the FCI space into spin-adapted subspaces. The basis in each space is composed of spin-adapted combinations of Slater determinants (i.e. of eigenfunctions ot the total spin operators $\hat{S}^2$ and $\hat{S}_z$). Such spaces may be represented by four-slope graphs (Shavitt 1977, 1978), in which both orbital and spin structure of the basis functions are described by each path. This representation of the model spaces is specific for the so called *unitary group approach* (UGA) to theory of many-electron systems (Paldus 1976). Alternatively, one may represent the model space as an antisymmetrized direct product of the orbital space and of the spin space, each of them being represented by an independent graph (Duch and Karwowski 1985). The orbital space is then described by a three-slope orbital graph and the spin space – by a two-slope spin graph also known as the *van Vleck's branching diagram* (Pauncz 1979). A detailed analysis of properties of the model spaces and a description of different ways of their graphical representations may be found in a monograph by Duch (1986).

6. The Eigenvalue Problem in the Model Space

Constructing the secular equation (188) and solving the matrix eigenvalue problem constitutes one of the most important tasks of computational quantum chemistry. The problem, in principle very simple conceptually, triggered development of most ingenious and sophisticated approaches as a consequence of a drive towards optimising the space and making it as large as possible. If the secular equation is formulated in a complete one-electron basis then, independent of the specific choice of the basis, the solution is the same. However, for practical reasons, we try to select the one-electron basis in such a way that a good approximation to the exact wavefunction may be obtained by using not only finite, but small model space.

6.1. HARTREE-FOCK METHODS

Particularly important is an optimization of the one-electron basis under the assumption that the N-electron model space is one-dimensional. This leads to the variational problem in which, depending upon the Hamiltonian and upon the specific approach, we optimize a set of orbitals, spinorbitals or Dirac spinors which are occupied in the electronic configuration which approximates the N-electron state we are interested in. The resulting equations are known as the *restricted Hartree-Fock* equations, *unrestricted Hartree-Fock equations* and *Dirac-Fock equations*, respectively. In each case we insist that the one-electron functions are selected in such a way that the Rayleigh quotient (Eq. 116)) in which the trial function Φ has been selected as a Slater determinant (in some cases a spin- and/or symmetry-adapted combination of Slater determinants) assumes its minimum. The expression which is subject to the minimization procedure is, in the simplest case, equal to the diagonal matrix element of the Hamiltonian matrix in the determinantal basis. Then

$$\langle\Phi|\hat{H}|\Phi\rangle = \sum_{p=1}^{N}(p|p) + \sum_{p<q}^{N}[(pp|qq) - (pq|qp)], \tag{208}$$

whre the indices p and q refer to the one-electron functions the Slater determinant Φ is composed of, i.e. to either spinorbitals or to the Dirac spinors (but *not* to orbitals). The two-electron integrals $(pp|qq)$ and $(pq|qp)$ are referred to, respectively as the Coulomb and the exchange integrals.

Eq. (208) corresponds to the Rayleigh quotient for a determinantal wavefunction under condition that all spinorbitals are orthonormal. i.e.

$$\langle p|q\rangle = \delta_{pq}. \tag{209}$$

Therefore the minimization procedure has to be performed by constrained variations of the spinorbitals with the constraints given by Eq. (209). Besides, in the case of the Dirac-Coulomb or Dirac-Breit Hamiltonian, the Dirac spinors have to fulfil very strict boundary conditions in order to prevent the variational collapse or the so callled *continuum dissolution* resulting from the "Brown-Ravenhall disease" (Grant 1987). Among these conditions particularly important are the correct behaviour of the spinors at the nuclei and the so called *kinetic balance*, i.e. assuring correct relations between large and small components of the Dirac spinors. A detailed discussion of this subject has been performed by Grant (1987, 1996).

A constrained variational problem leads to a set of equations for the functions being subject of the variations. The equations are known as the Euler-Lagrange equations of the variational problem. Their fulfilment implies that the minimized functional reaches its minimum. In the case of minimization of the functional given by Eq. (208) with the constraints given by Eq. (209), the Euler-Lagrange equations are known as the *Hartree-Fock equations* (in the non-relativistic case) or the *Dirac-Fock equations* (in the relativistic case). Symbolically they may be written as a one-electron pseudo-eigenvalue problem:

$$\hat{F}(1)\psi_k(1) = \varepsilon_k\psi_k(1), \tag{210}$$

where

$$\hat{F} = \hat{h}_1 + \sum_{j=1}^{N} [\hat{J}_j(1) - \hat{K}_j(1)] \tag{211}$$

with

$$\hat{J}_j(1)\psi_k(1) = \int_{v_2} \psi_j(2)^\dagger \hat{h}_2(1,2)\psi_j(2)\, dv_2\, \psi_k(1) \tag{212}$$

and

$$\hat{K}_j(1)\psi_k(1) = \int_{v_2} \psi_j(2)^\dagger \hat{h}_2(1,2)\psi_k(2)\, dv_2\, \psi_j(1). \tag{213}$$

Operators $\hat{J}$ and $\hat{K}$ are referred to, respectively, as *Coulomb* and *exchange* operators. In their definition a general two-electron operator $\hat{h}_2(1,2)$ has been used. However usually this operator is simply equal to $\frac{e^2}{r_{12}}$.

There exist numerous versions of these equations. The most commonly known are the ones which correspond to the non-relativistic *closed-shell* case, i.e. to the case when each orbital appears twice in the configuration defining the trial function Φ. A detailed description of this and other versions of the Hartree-Fock method may be found in the works of McWeeny (1989), of McWeeny and Sutcliffe (1985) and in references given therein. The Dirac-Fock method has been discussed in detail by Grant (1987, 1996).

6.2. THE ELECTRON CORRELATION

All Hartree-Fock and Dirac-Fock methods reduce the N-electron problem to a one-electron one, i.e. they may be viewed as the optimum one-electron approximations. In this way one can obtain an exact solution of the corresponding Schrödinger or Dirac-Coulomb equation if the N-electron Hamiltonian eigenvalue equation is separable. In the case of electrons interacting by the Coulomb forces it is not. Therefore the Hartree-Fock (Dirac-Fock) models may give us only an approximate (though in some cases very good) description of the physical system. The difference between the exact eigenvalue of the many-electron Hamiltonian and the corresponding Hartree-Fock (Dirac-Fock) energy is referred to as the *correlation energy* (Löwdin 1955). Effects associated with differences between the "exact", i.e. either Schrödinger or Dirac-Coulomb, and, respectively, Hartree-Fock or Dirac-Fock description on an atom or a molecule are called the *electron correlation effects.* In order to estimate these effects one has to go beyond the one-electron model, i.e. one has to consider the Hamiltonian eigenvalue problem in a multi-dimensional model space. By solving Eq. (210) one obtains a set of one-electron functions $\psi_1, \psi_2, \ldots, \psi_N$ which give us the optimum one-determinantal wavefunction Φ and define the Fock operator $\hat{F}$. Once the one-electron functions are known, $\hat{F}$ becomes a well defined Hermitian operator. Its eigenfunctions form a complete set and may be used as the one-electron basis for construction of the N-electron model space. Then, spinorbitals which fulfil Eq. (210) may be divided into two classes: the ones which are occupied in Φ and the ones which are empty in Φ. The empty orbitals ones are referred to as the *virtual orbitals.*

The method in which the energies and the wavefunctions are obtained by diagonalization of the Hamiltonian matrix in the FCI space is called *full configuration interaction method.* If the diagonalization is performed in a subspace of the FCI space, the approach is simply called *configuration interaction* (CI) method and the subspace of the FCI space is called the *CI space.* Numerous ways of restricting the space in which the CI procedure is performed are described by McWeeny (1989), Karwowski (1992) and in references given in these works.

Relations between results of a full and a restricted CI calculations are given by the McDonald theorem (Löwdin 1955):
Let the CI space be spanned by J basis functions and let the CI energies be

$$E_1^{(J)} \leq E_2^{(J)} \leq \cdots \leq E_J^{(J)}; \tag{214}$$

as the dimension of the CI space is increased by adding an additional basis function, eigenvalues $E_p^{(J+1)}$ *of the* $(J+1)$*-dimensional problem satisfy the inequalities:*

$$E_{p-1}^{(J)} \leq E_p^{(J+1)} \leq E_p^{(J)} \tag{215}$$

and as the set of one-electron functions approaches completeness, the FCI solutions approach the exact solutions of the pertinent eigenvalue equation. From this theorem results that each eigenvalue of the CI matrix is an upper bound to the corresponding eigenvalue of the Hamiltonian.

Approximate methods of solving the FCI problem comprise a large variety of models ranging from different kinds or restricted CI calculations to methods based on the perturbational expansions, such as the many-body perturbation theory (MBPT) and the coupled clusters (CC) method. Broad reviews of all these approaches may be found in books edited by Wilson (1987) and by Wilson and Diercksen (1992).

7. Acknowledgements

The support by the Polish KBN, Project No. 2 P03B 011 08 is gratefully acknowledged. Sincere thanks are due to Brian G. Wybourne for his most useful comments. Helpful remarks of Marcus Schweizer are also appreciated.

References

1. Baym, G., 1969 *Lectures on Quantum Mechanics*, Benjamin, London.
2. Bethe, H. A. and Salpeter, E. E., 1957, *Quantum Mechanics of One- and Two-Electron Atoms*, Springer, Berlin.
3. Chang, Ch., Pélissier, M. and Durand, Ph., 1986, Phys. Scripta **34**, 394.
4. Coleman, A. J., 1965, Rev. Mod. Phys. **35**, 668.
5. Davydov, A. S., 1965, *Quantum Mechanics*, Pergamon.
6. Dirac, P. A. M., 1948, *The Principles of Quantum Mechanics*, Claredon, Oxford.
7. Douglas, M. and Kroll, N. M., 1974, Ann. Phys. **82**, 89
8. Duch, W., 1986, *GRMS or Graphical Representation of Model Spaces*, Lecture Notes in Chemistry, Vol. 42, Springer, Berlin.
9. Duch, W. and Karwowski, J., 1985, Comp. Phys. Reports, **2**, 94.
10. Erdahl, R. and Smith, V., 1987, Editors, *Density Matrices and Density Functionals*, Reidel, Dordrecht.

11. Feynman, R. P., R. B. Leighton, R. B. and Sands, M., 1977, *The Feynman Lectures on Physics*, Addison-Wesley, Reading MA.
12. Foldy, L. L. and Wouthuysen, S. A., 1950, Phys. Rev. **78**, 29.
13. Gould, H. and Munger, C. T., 1987, Phys. Scripta **36**, 476.
14. Grant, I. P., 1996, *Relativistic Atomic Structure*, in *Atomic, Molecular and Optical Physics Reference Book*, ed. G. W. F. Drake, p. 258, American Institute of Physics, New York.
15. Grant, I. P., 1987 in *Methods in Computational Chemistry*, ed. S. Wilson, vol. 2, p. 1, Plenum, New York.
16. Gross, E. K. U. and Dreizler, R. M., Editors, 1995, *Density Functional Theory*, Plenum, New York.
17. Hess, B. A., 1985, Phys. Rev. A **32**, 756.
18. Hess, B. A., 1986, Phys. Rev. A **33**, 3742.
19. Karwowski, J., 1992, in *Methods in Computational Molecular Physics*, ed. S. Wilson and G. H. F. Diercksen, NATO ASI Series, Series B: Physics Vol. 293, p. 65, Plenum, New York.
20. Karwowski, J. and Kobus, J., 1985, Int. J. Quantum Chem. **28**, 741.
21. Karwowski, J. and Kobus, J., 1986, Int. J. Quantum Chem. **30**, 809.
22. Kryachko, E. S. and Ludeña, E. V., 1990, *Energy Density Functional Theory of Many-Electron Systems*, Kluwer, Dodrecht.
23. Kutzelnigg, W., 1984, Int. J. Quantum Chem. **25**, 107.
24. Kutzelnigg, W., 1989, Z. Physik D, **11**, 15.
25. Kutzelnigg, W., 1990, Z. Phys. D **15**, 27.
26. Lamb, W. E., 1951, Rept. on Progr. Phys. **14**, 19.
27. Levy-Leblond, J. M., 1970, Ann. Phys. **57**, 481.
28. Löwdin, P. -O., 1955, Phys. Rev. **97**, 1509.
29. McWeeny, R., 1973, *Quantum Mechanics: Methods and Basic Applications*, Pergamon, Oxford.
30. McWeeny, R., 1989, *Methods of Molecular Quantum Mechanics*, Academic, New York.
31. McWeeny, R and Sutcliffe, B. T., 1985, Comp. Phys. Reports, **5**, 218.
32. Messiah, A., 1961, *Quantum Mechanics*, North Holland, Amsterdam.
33. Paldus, J., 1976, in *Theoretical Chemistry. Advances and Perspectives*, ed. H. Eyring and D. Henderson, vol. 2, p. 131, Academic, New York.
34. Pauncz, R., 1979, *Spin Eigenfunctions: Construction and Use*, Plenum, New York.
35. Rutkowski, A., 1986, J. Phys. B, **19**, 149; 3431; 3443.
36. Sakurai, J. J., 1984 (Tenth printing), *Advanced Quantum Mechanics*, Addison-Wesley, Reading, MA.
37. Sakurai, J. J. and Tuan, S. F., 1985, *Modern Quantum Mechanics*, Benjamin/Cummings, Menlo Park, CA
38. Schiff, L. I., 1968, *Quantum Mechanics*, McGraw-Hill, New York
39. Shavitt, I., 1977, Int. J. Quantum Chem. **S11**, 131.
40. Shavitt, I., 1978, Int. J. Quantum Chem. **S12**, 5.
41. Sucher, J., 1980, Phys. Rev. A **22**, 348.
42. Thouless, D. J., 1961, *The Quantum Mechanics of Many-Body Systems*, Academic, New York.
43. van Lenthe, E., 1996, *The ZORA Equation*, Vrije Universiteit te Amsterdam (Thesis).
44. van Lenthe, E., Baerends, E. J. and Snijders, J. G., 1993, J. Chem. Phys. **99**, 4597.
45. van Lenthe, E., Baerends, E. J. and Snijders, J. G., 1994, J. Chem. Phys. **101**, 9783.
46. Weissbluth, M., 1978, *Atoms and Molecules*, Academic, New York.
47. Wilson, S., Editor, 1987, *Methods in Computational Chemistry*, Plenum, New York.
48. Wilson, S. and Diercksen, G. H. F., Editors, 1992, *Methods in Computational Molecular Physics*, NATO ASI Series, Series B: Physics Vol. 293, Plenum, New York.

PRACTICAL *AB INITIO* METHODS FOR MOLECULAR ELECTRONIC STRUCTURE STUDIES. I. AN OVERVIEW

S. WILSON
Rutherford Appleton Laboratory
Chilton, Oxfordshire OX11 0QX, England

Abstract. An overview of practical *ab initio* methods for molecular electronic structure studies is given. A graphical interface, the UNICHEM computational chemistry package, is used to emphasize the various choices made in performing an electronic structure calculation and, in particular, to describe the way in which these choices affect both the utility of the results obtained and the tractability of the computation.

1. Introduction

The development of *ab initio* quantum chemistry and its dependence on the computer can be well illustrated by quoting two Nobel Laureates. In 1929, P.A.M. Dirac[1] wrote

> *The underlying physical laws necessary for the mathematical theory of a large part of physics and the whole of chemistry are ... completely known, and the difficulty is only that the exact application of these laws leads to equations much too complicated to be solved.*

Recently, in 1990, K.G. Wilson[2] observed that

> *There are roughly ten million classified chemical compounds at the present time. Each individual molecule has many properties to compute and/or measure: binding energy, electron density, atomic structure, spectra (vibrational, rotational and electronic), reaction rates, electron and molecular scattering cross sections. However, the spectacular opportunity for the future lies in compounds not yet synthesized or classified. ... Quantum chemistry research can in principle start with ANY collection of atoms so the huge number of unexplored forms of matter defines the number of possible research projects for quantum chemists.*

S. Wilson and G. H. F. Diercksen (eds.), Problem Solving in Computational Molecular Science, 85–108.

Ab initio quantum chemistry is concerned with the approximate solution of the basic equations of quantum mechanics to predict the properties of matter at a molecular level. The term "*ab initio*" implies that our methods are rigorous. They can be refined. It does not imply that our approximations are accurate.

Neglecting relativity, the equation to be solved for molecules in a stationary state is the time-independent Schrödinger equation

$$\hat{H}\Psi = E\Psi \tag{1}$$

with hamiltonian operator

$$\begin{aligned}\hat{H} &= \sum_{A=1}^{N} -\frac{1}{2M_A}\nabla_A^2 + \sum_{i=1}^{n} -\frac{1}{2}\nabla_i^2 + \sum_{A>B=1}^{N} \frac{Z_A Z_B}{R_{AB}} \\ &\quad - \sum_{A=1}^{N}\sum_{i=1}^{n} \frac{Z_A}{r_{Ai}} + \sum_{i\geq j=1}^{n} \frac{1}{r_{ij}}\end{aligned} \tag{2}$$

where the first two terms are associated with the kinetic energies of the nuclei and the electrons, respectively, the third term describes the repulsive effects between the nuclei, the penultimate term arises from the attractions between the nuclei and the electrons and the final term describes interelectronic repulsions. So that for the nitrogen molecule (N_2), for example, we have a second order partial differential equation in 48 variables [14 electrons plus two nuclei, each with 3 Cartesian coordinates], whilst for C_{60}, the number of variables is 1260 [60 × 6 electrons together with 60 nuclei each with 3 Cartesian coordinates].

The large disparity of the masses of the electrons and the nuclei allows the problem to be separated, to a very good approximation in the vast majority of cases, into an electronic problem, which is the subject of these lectures, and a nuclear problem. The electrons accommodate almost instantaneously nuclear motion. The electronic Schrödinger equation, which may be written

$$\hat{H}_e\Phi = \mathcal{E}\Phi \tag{3}$$

with the electronic hamiltonian operator

$$\hat{H}_e = -\frac{1}{2}\sum_{i=1}^{n}\nabla_i^2 - \sum_{A=1}^{N}\sum_{i=1}^{n}\frac{Z_A}{r_{Ai}} + \sum_{i\geq j=1}^{n}\frac{1}{r_{ij}}, \tag{4}$$

describes the motion of the electrons in the field of clamped nuclei and is still a formidable problem involving, for example, 42 variables for N_2 and 1080 for C_{60}. This is a many-fermion problem. Even for this simplified problem it is unlikely that a standard library program for solving differential equations would even predict the existence of molecules!

The next step is to assume that each electron moves in an effective field generated by the other electrons in the molecule; this is termed the independent electron model. It leads to a second order partial differential equation in just 3 variables but, except in the case of the very simplest model, the bare-nucleus model in which all electron-electron interactions are neglected, contains an unknown effective potential describing the averaged interactions between the electrons. In the Hartree-Fock method this effective potential is determined self-consistently; that is, we guess the solutions, construct an effective potential from these guesses, solve the resulting eigenvalue equation, use the solutions thus obtained to construct an improved effective potential, and then iterate this process until the solutions are self-consistent, i.e. are unchanged to within some chosen tolerance. Independent electron models, of which the Hartree-Fock methods is by far the most widely used, often provide an excellent approximation.

Typically, total electronic energies are given by the Hartree-Fock model to an accuracy of about 1%, but unfortunately this error is of the same order of magnitude as most energies of chemical interest; i.e. bond energies, the differences in energy between different conformers, etc. It is, therefore, usually necessary to consider the corrections to independent electron models. These corrections are associated with the instantaneous interactions between the electrons. For the Hartree-Fock model these corrections are usually termed electron correlation effects. In the majority of contemporary computations electron correlation is described by perturbation theoretic methods because of their well established theoretical and computational advantages.

2. Toward practical *ab initio* methods for molecular quantum mechanics

The dependence of progress in applied *ab initio* quantum chemistry on developments in computing is widely recognized. (The influence of atomic and molecular electronic structure calculations on progress in computing is, perhaps, less widely recognized. Throughout the history of computing, theoretical chemistry has been a major source of challenging computational tasks.) Indeed, although much progress on the molecular electronic structure problem was made before the advent of the electronic computer, the prospects changed radically in the mid-twentieth century with the dawn of the computer age. The late twenties and thirties saw pioneering calculations by Hylleraas[3] on the He ground state and by James and Coolidge[4] on the hydrogen molecule ground state. Coulson[5] reported a molecular orbital study of the hydrogen molecule whilst Hartree, Hartree and Swirles[6] developed the multiconfiguration self-consistent field approach. Swirles[7] also introduced the Dirac-Hartree-Fock self-consistent field method for systems containing heavy atoms where relativistic effects are important. Although many qualitative developments were made in quantum chemistry during the late twenties and thirties, progress in quantitative applications awaited the development of the electronic digital computer. Schaefer[8] describes how

> *The 1940s were, in a certain sense, the 'dark ages' of ab initio electronic structure theory. It was not apparent how to extend the highly reliable results of Hylleraas for He and James and Coolidge for H_2 to larger atoms and molecules. Although the Hartree-Fock and multiconfiguration Hartree-Fock formalisms were available, the application of these ab initio methods even to molecules as small as water seemed impossible at the time.*

The importance of computers in electronic structure theory was recognized at an early stage. D.R. Hartree visited V. Bush at M.I.T. who had built, in 1931, the first class of machines which have become know as mechanical differential analyzers. The precision of the solutions obtained from the mechanical differential analyzer is dependent on a variety of factors, *e.g.* slip at the friction wheel of an integrator, the quality of the gears and bearings. The mechanical nature of the device would impose serious limitations on the speed of operation. At Manchester University, D.R. Hartree and A. Porter constructed a small prototype machine in the early 1930's largely from standard Meccano parts which achieved an accuracy of the order of 2%. Hartree[9] was later to remark

> *This (the construction of the Meccano prototype) was originally done more for amusement than with any serious purpose, but the first results were successful beyond my expectations, and suggested that it would be practical to build such models to do some serious work on problems for which high accuracy was not required in the results.*

The Metropolitan-Vickers Electrical Company Ltd. were contracted to build a full scale mechanical differential analyzer at the University of Manchester and this was completed in 1935. A similar machine was installed in the Mathematical Laboratory of Cambridge University in 1939 by M.V. Wilkes and his associates. As early as 1936, theoretical chemists (backed by J. Lennard-Jones) provided the main impetus for the setting up of a Central Mathematics Laboratory in Cambridge equipped with a few Brunsvigas, a collection of mathematical tables, a differential analyzer and other special purpose machines.

Hartree was one of the first scientists to combine an interest in electronic structure theory with an interest in automatic computation. During the summer of 1945, Hartree visited the Moore School of Electrical Engineering at the University of Pennsylvania, to become acquainted with electronic digital systems and, in particular, the ENIAC (Electronic Numerical Integrator and Computer). He became convinced that digital computation was superior to analog for scientific applications. In 1947, Hartree left Manchester for Cambridge where the EDSAC1 (Electronic Delay Storage Automatic Calculator) became operational in 1949.

Even in these early days, Hartree[10] cautioned that

> *Use of the machine is no substitute for the thought of organizing the computations, only for the labour of carrying them out. This point seems to me of great importance, and to be missed entirely by those who speak of a machine of this kind as an "electronic brain". Even if the organization of the calculations is done by the machine, as is possible in future developments, the operator will still have to think about the sequence of operating instructions which will enable the machine to do this organization.*

The prospects for *ab initio* molecular structure calculations changed radically with the electronic digital computer. McWeeny[11] describes the situation

> *When, forty years ago, as a student of Charles Coulson in Oxford, I began work in Theoretical Chemistry, I was provided with a Brunsviga calculator — a small mechanical device with a handle for propulsion, metal levers for setting the numbers, and a bell that rang to indicate overflow.*
>
> *After the Brunsviga came the electromechanical machines of the late 1940s and early 1950s, and a great reduction in the burden of calculating molecular wave functions. We were now happy. At least for systems containing a few electrons it was possible to make fully ab initio calculations, even though semiempirical remained indispensable for most molecules of everyday interest. The 1950 papers by Hall and Roothaan represented important milestone along the road to larger-scale non-empirical calculations, extending the prewar work of Hartree and Fock from many-electron atoms to many-electron molecules — and thus onto "real chemistry". But in a practical sense perhaps the most dramatic event was the appearance of the automatic digital computer — which ended an era and opened a new one.*
>
> *The first electronic computer I used was the "Whirlwind" at MIT in 1954; it solved my large* (20 × 20!) *system of secular equations in only a few minutes, but high-level languages were not then available*

and "integral packages" were nonexistent. Integrals over Slater orbitals were evaluated using the Barnett-Coulson expansion technique (and desk machines) and a single three-centre two-electron integral might take a few hours of tedious work.

However, it was to take some time for the potential of the computer to be realized. Hall[12] and, independently, Roothaan[13] presented the matrix Hartree-Fock equations in a form suitable for implementation on a digital computer. The use of Gaussian basis functions in contemporary molecular electronic structure calculations is ubiquitous. Their use in this context was first suggested by McWeeny [14] [15] and, independently, by Boys[16]. However, as related by Shavitt [17] in a recently published historical review, for some years after the introduction of Gaussian basis functions, the supposedly "more physical" Slater (exponential)-type orbitals remained the basis function of choice. A few pioneering calculations, most notably by Boys[18], Matsen[19], Nesbet[20], Pople[21] and Shavitt[22], were made in the fifties and those that were were for small molecules using quite drastic approximations. By 1959, Mulliken and Roothaan[23] were able to write a paper entitled *"Broken bottlenecks and the future of molecular quantum mechanics"* describing a program for completely automatic self-consistent field studies for diatomic molecules. In this paper, Mulliken and Roothaan[23] write

Looking toward the future, it seems certain that colossal rewards lie ahead from large-scale quantum-mechanical calculations of the structure of matter.

The development of packages began in earnest in the sixties. Programs such as POLYATOM[24] and IBMOL[25] put down the foundations of modern practical *ab initio* quantum chemistry and by the end of the decade the self-consistent field problem could be solved routinely for small molecules and a start had been made on the correlation problem[26] [27] [28] [29] [30]. Work had also begun on the development of techniques for the calculation of energy derivatives[31] [32].

The seventies saw the introduction of methods derived from quantum field theory[33] [30] into the study of the molecular electronic structure problem and the many-body perturbation theory[34] [35] [36], formulated within the algebraic approximation and employing the Møller-Plesset partition of the hamiltonian[37] [38], was set to become the method of choice in dealing with the correlation problem.

The late seventies and early eighties saw the introduction of more powerful computers - supercomputers, vector processors and then parallel vector processors, which changed quite radically the size of systems that was amenable to *ab initio* study and the accuracy with which such studies could be conducted[39]. The relative increase in computing speed compared with the speed of disk access spurred the introduction of direct methods[40], first as direct self-consistent field algorithms and then as low order perturbation theoretic correlation treatments. Here the numerous two-electron integrals are computed "on the fly" rather than being stored on disk. The use of direct approaches has spurred a renewed interest in the development of more efficient integral evaluation and integral approximation techniques[41].

During the 1980s there was a growing interest in the practical realization of relativistic molecular electronic structure theory. This interest was kindled by the need to extended the methods of quantum chemistry to systems containing heavy elements. Relativistic quantum chemistry is discussed in more detail in Part IV[42].

3. Practical *ab initio* methods for molecular quantum mechanics

3.1. HARDWARE AND SOFTWARE

The choice of theoretical model and computational algorithms is inextricably linked with the hardware facilities and software capabilities which can be devoted to the problem. Contemporary quantum chemistry calculations are executed on a wide spectrum of computers ranging from supercomputers to personal computers. At the supercomputing end of the spectrum, parallel vector processors, such as the CRAY C-90, with 16 processors each delivering a peak processing speed of 1 Gflop/s giving an aggregate peak performance of 16 Gflop/s, offer the highest performance, whilst machines such as the CRAY J-90, with 32 processors each delivering a peak processing speed of 200 Mflop/s and, therefore, an aggregate peak performance of some 6.4 Gflop/s, offer a cost-effective solution. Meanwhile, developments at the personal computer end of the spectrum facilitate calculations on desk-top machines which, less than a decade ago, would have necessitated access to a mainframe machine. Attempts to harness the potential power of massively parallel processors continue, but with different stages of a molecular electronic structure calculations exhibiting differing levels of parallelism, a heterogeneous system would appear to provide the most general purpose solution.

Much quantum chemical software is now commercially available. Software costs now represent a significant fraction of the total cost of making a calculation. One of the most popular of the software products for performing *ab initio* molecular electronic structure studies is the GAUSSIAN system which is a multifeatured suite of programs written by a team of 35 authors in the latest (GAUSSIAN94) version[43].

3.2. GRAPHICAL USER INTERFACES AND VISUALIZATION

The use of quantum chemical methods has become more widespread with the development of graphical user interfaces which make such methods more readily available, particularly, to the non-specialist.

The UNICHEM system[44], originally developed by Cray Research Inc. but acquired in 1996 by the Oxford Molecular Group, runs on a heterogeneous system consisting of a Silicon Graphics workstation, on which the interactive graphics is carried out, and a Cray supercomputer, on which the numerically intensive molecular electronic structure calculation is performed. From the workstation the user can define the molecular structure to be considered, choose the theoretical model to be employed, decided on the various approximations to be made in the computational implementation, initiate and monitor the execution of the quantum chemical program, and, finally, examine the results of the calculation, display and manipulate the orbitals, the electron density, electrostatic potential, *etc.*

In Figure 1, the main window of the UNICHEM computational chemistry environment is shown. It consists of a periodic table and a choice of coordination types. By clicking on *File* the user opens a new window on which to construct a model of the target molecule or opens a window containing a previously constructed molecular system. This window is referred to as the "canvas". To construct the target molecule, the user first clicks on an atom in the Periodic Table in the main window and the required coordination. Then, to introduce the new atom at the appropriate point on the canvas the user moves the cursor to the bond to which the new atom is to be attached and clicks the left hand mouse button. The partial construction of a histidine molecule obtained in this fashion is shown in Figure 2. Note the "open" bond in this Figure to which the rest of the molecule will be attached. The completed histidine molecule is displayed in Figure 3. The user also has access to libraries of molecular structures by means of the *Show Library* button.

Whilst constructing the molecule, the user can manipulate the molecular structure in a number of ways. For example, the molecule can be rotated in any direction, translated, and moved towards or away from the user. By clicking on *Build* the user can change bond lengths, delete atoms, etc. Clicking on *View* allows the user to change various characteristics of the display:- 'ball and stick' model, 'line' model, 'stick' model or 'atoms only'. The bond lengths are taken, by default, to be the sum of the covalent radii of the component atoms but this can be modified (to use, for example, van der Waals radii) from the main UNICHEM window.

At this point the molecule is completely defined. It is defined by the Cartesian coordinates of the nuclei together with the associated nuclear

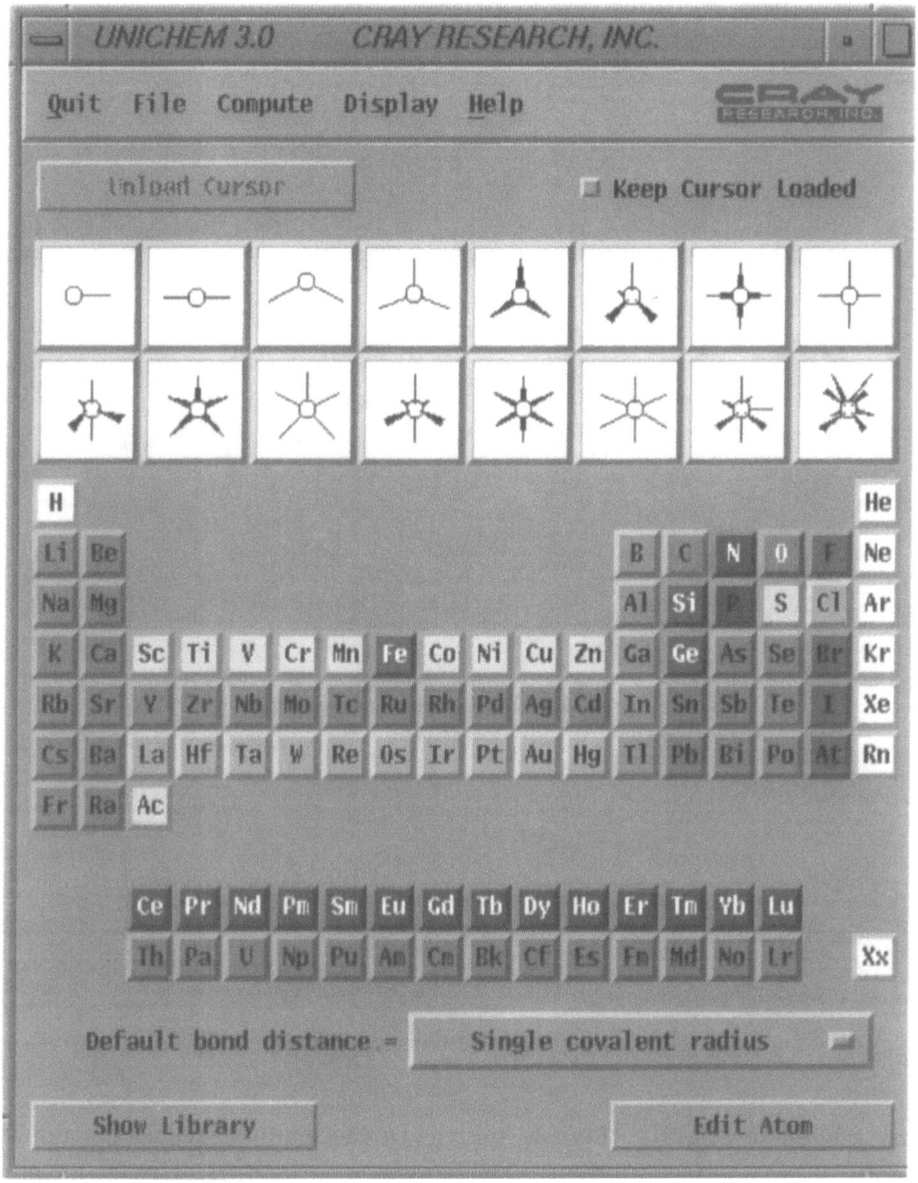

Figure 1. The main window of the UNICHEM computational chemistry environment.

charges. The molecule is assumed, at this stage, to be electrically neutral so that the total number of electrons is just the sum of the numbers of electrons associated with each of the neutral component atoms.

Having constructed the target molecular structure and saved it by click-

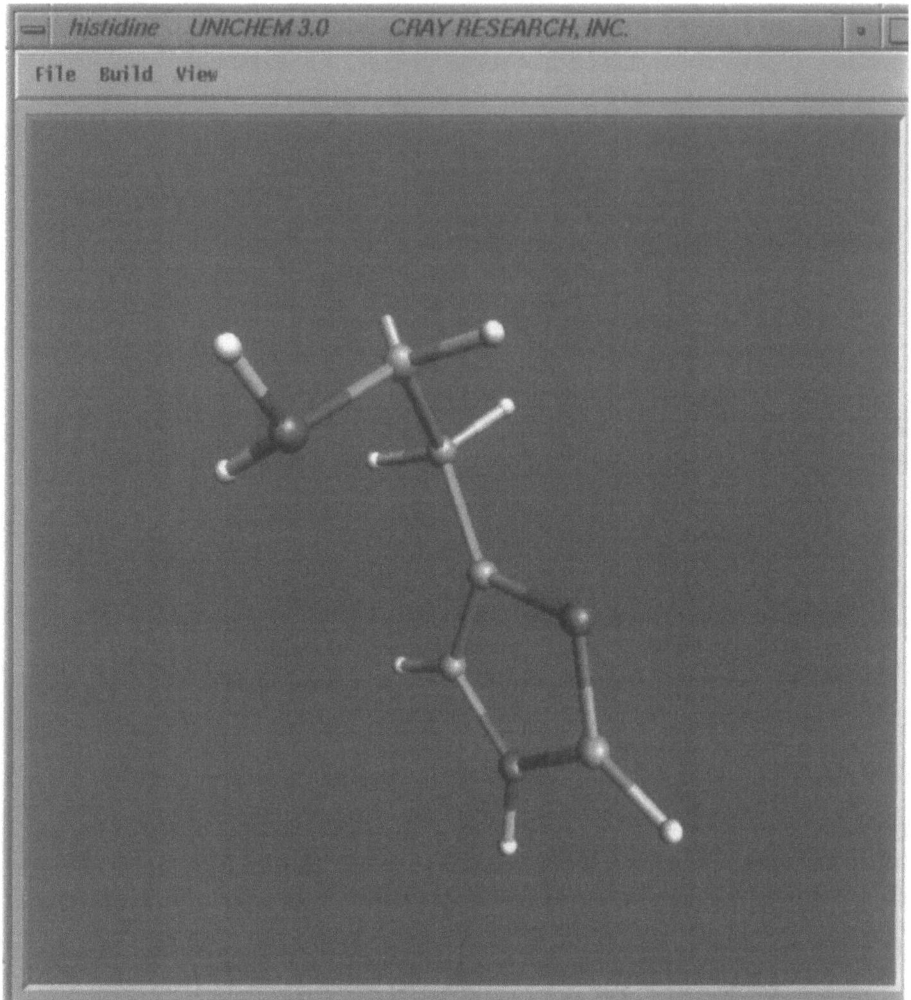

Figure 2. A partially constructed histidine molecule on the UNICHEM canvas.

ing *File* on the canvas, the user returns to the main UNICHEM window, clicks on *Compute* and then selects *Setup and Launch.* The *Setup and Launch* window is shown in Figure 4. Here the user selects the particular chemistry code to be employed (the GAUSSIAN program in our example), confirms the structure for which the calculation is to be carried out (histidine in our example) and defines the computer to be used (a CRAY J-90 in the example) and the computer resources required:- the job time, the memory required, *etc.*.

At this point, the user could click the *Launch* button and a job would be

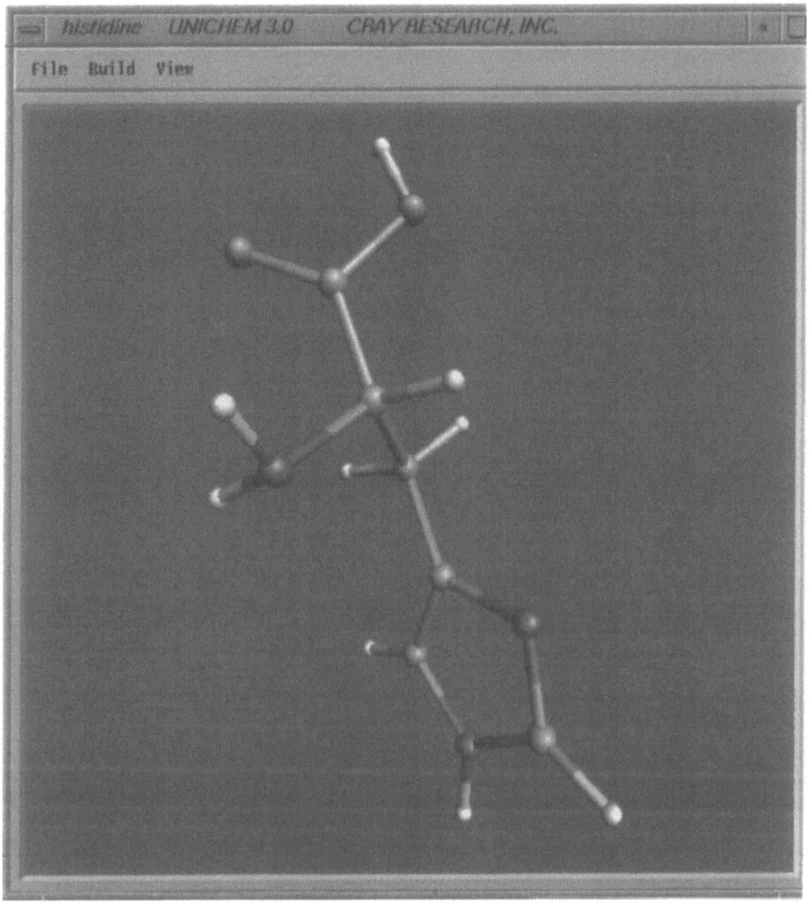

Figure 3. A completed histidine molecule on the UNICHEM canvas.

initialized and executed using default choices; a default theoretical model and default approximation schemes. In the case of the histidine example, this would be a restricted Hartree-Fock (RHF) calculation with a small (STO-3G) basis set of contracted Gaussian functions. However, the efficacy of a particular quantum chemic3al study depends critically both on the assumed theoretical model and in the details of the approximation schemes invoked in its practical realization.

Clicking on the *Setup* button in the *Setup and Launch* window enables a more detailed specification of the method and approximations to be made. The *Gaussian Setup* window is illustrated in Figure 5. Here, the assumed electronic state is specified, the Hartree-Fock method defined and the approach used in the treatment of electron correlation effects selected. Choices made here will affect not only the utility of the molecular electronic struc-

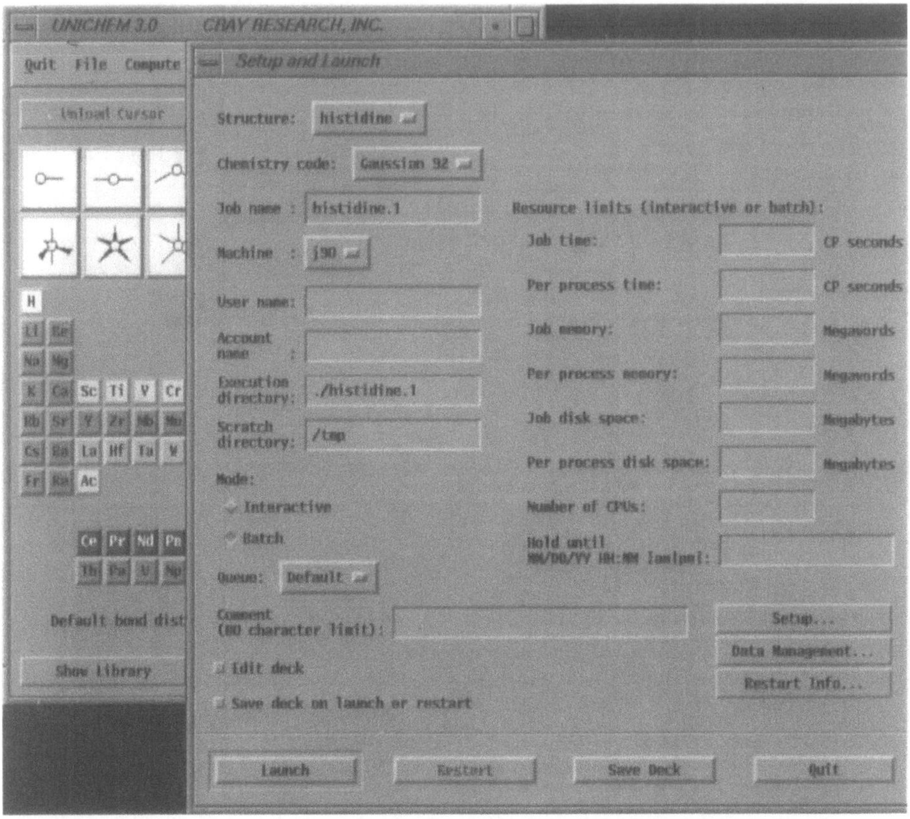

Figure 4. The *Setup and Launch* window in the UNICHEM computational chemistry environment.

ture calculation but also the computational resources required for its successful expedition. In the *Gaussian Setup* window, the user can choose between an energy calculation at a single geometry, a complete geometry optimization or a search for a transition state.

Greater control over a geometry optimization procedure can be obtained by clicking the top button of the group of five buttons in the lower right hand corner of the Gaussian Setup window and thereby obtaining the *Optimization Options* window. The *Optimization Options* window is displayed in Figure 6.

Various options available during the self-consistent field procedure are changed via the *SCF Options* window, which is opened by clicking on the second button of the group of five buttons in the lower right hand corner of the *Gaussian Setup* window. The *SCF Options* window is illustrated in Figure 7.

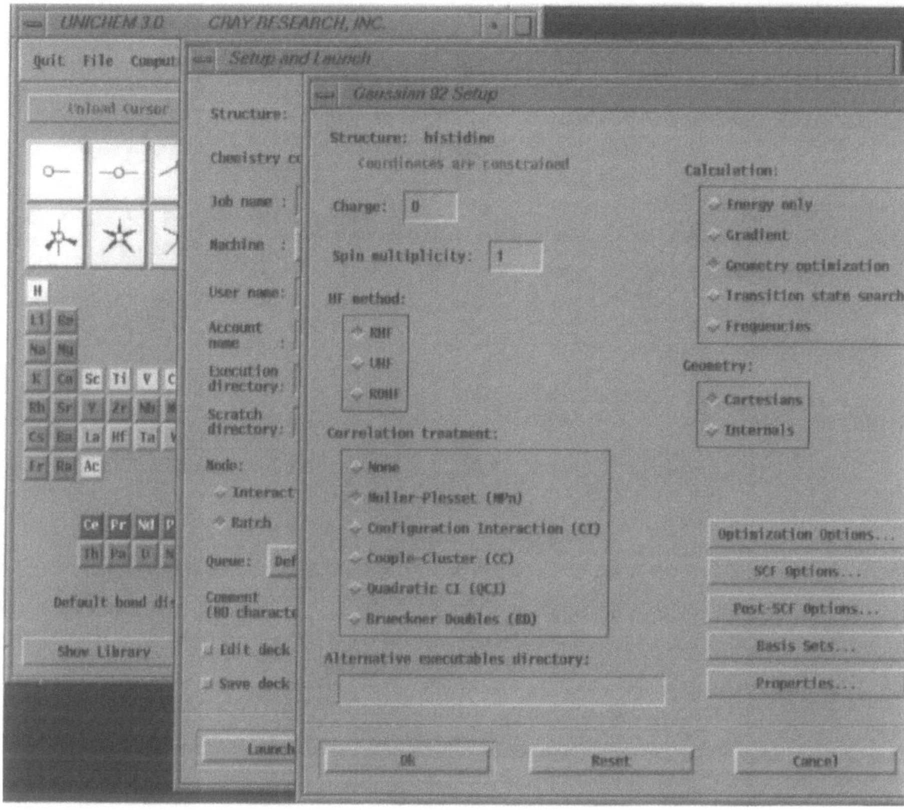

Figure 5. The *Gaussian Setup* in the UNICHEM computational chemistry environment.

The variety of options available during the electron correlation calculation are changed via the *Post-SCF Options* window, which is opened by clicking on the third button of the group of five buttons in the lower right hand corner of the *Gaussian Setup* window. The *Post-SCF Options* window is illustrated in Figure 8.

The selection of a basis set is one of the most critical elements of a quantum chemical calculation. The various pre-defined basis sets available to the UNICHEM/GAUSSIAN92 system are accessed via the *Basis Sets* window, which is opened by clicking on the fourth button of the group of five buttons in the lower right hand corner of the *Gaussian Setup* window. The *Basis Sets* window is illustrated in Figure 9.

The *Properties* window, which is opened by clicking on the final button of the group of five buttons in the lower right hand corner of the *Gaussian Setup* window, is illustrated in Figure 10. The properties selected in this window are calculated at the final geometry achieved in any calculation

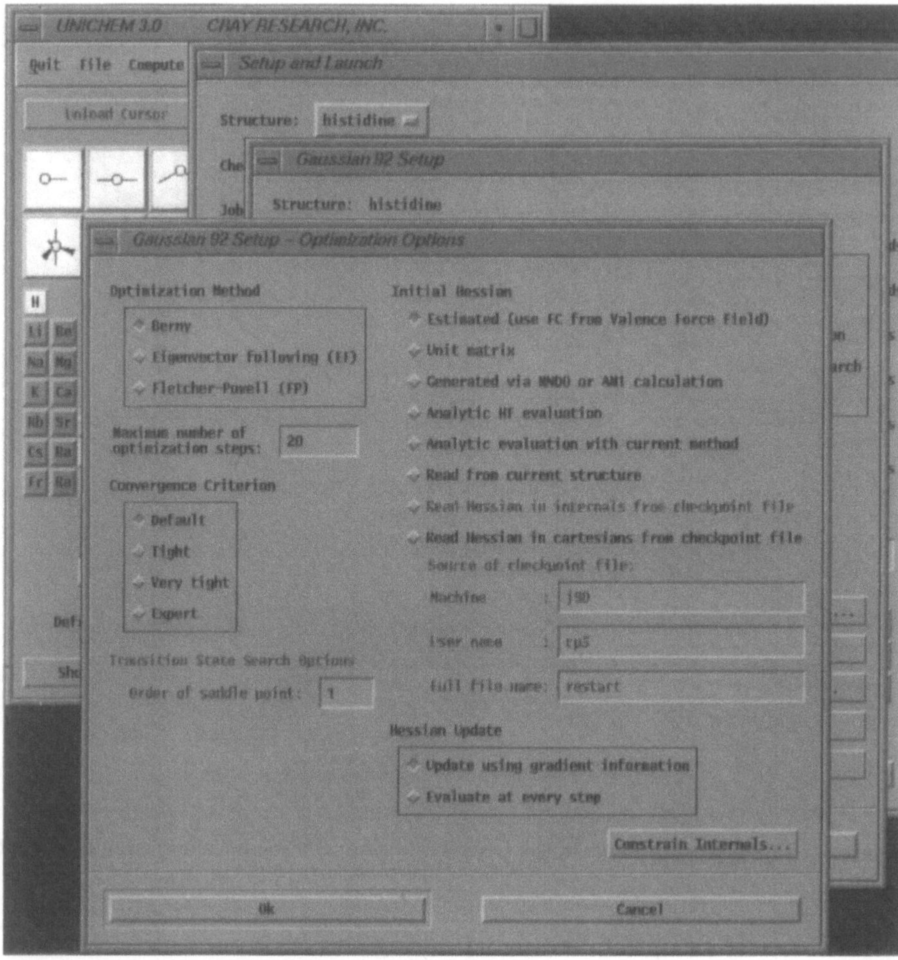

Figure 6. The *Optimization Options* window in the computational chemistry UNICHEM environment.

which involves geometry optimization.

Having completely specified the quantum chemical model and approximation techniques to be used the user returns to the *Setup and Launch* window shown in Figure 4 and clicks on the *Launch* button to initiate execution. By clicking on *Compute* in the main window and selecting *Monitor*, the user can then monitor the progress of his/her job(s).

When the computation is completed the user clicks on *Display* in the main window the view the results. He/she will be interested in the energetics, the optimized geometry and other properties. He/she may wish to view the total electron density using the graphical capabilities of the Sili-

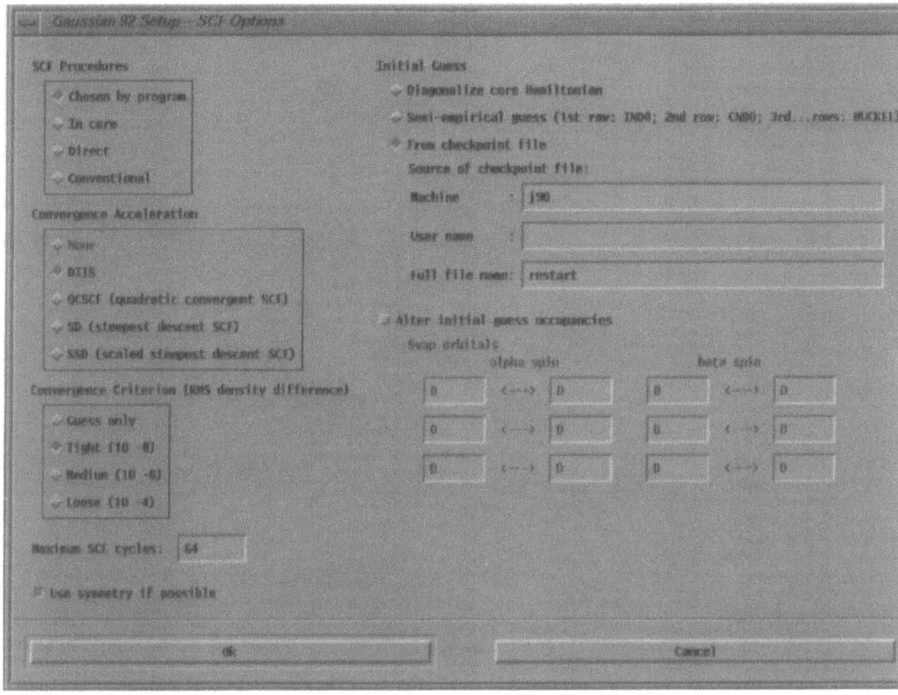

Figure 7. The *SCF Options* window in the UNICHEM computational chemistry environment.

con Graphics workstation. In Figure 11, the total electron density for the histidine molecule is shown with a threshold of 1.5 $e/\mathring{A}^3$; in Figure 12 a threshold of 0.001 $e/\mathring{A}^3$ is used.

3.3. APPROPRIATE APPROXIMATIONS

In 1959, Mulliken and Roothaan wrote, in an article entitled *"Broken bottlenecks and the future of molecular quantum mechanics"* published in the Proceedings of the U.S. National Academy of Science[23]

> *Looking toward the future, it seems certain that colossal rewards lie ahead from large-scale quantum-mechanical calculations on the structure of matter*

Commenting on this prediction in 1982, Schaefer writes[8]

> *We are confident that by the year 2000, essentially all fields of chemistry will acknowledge the accuracy of Mulliken and Roothaan's prophecy.*

Today, in 1996, we still have not reached the stage at which molecular electronic structure calculations can be carried out completely auto-

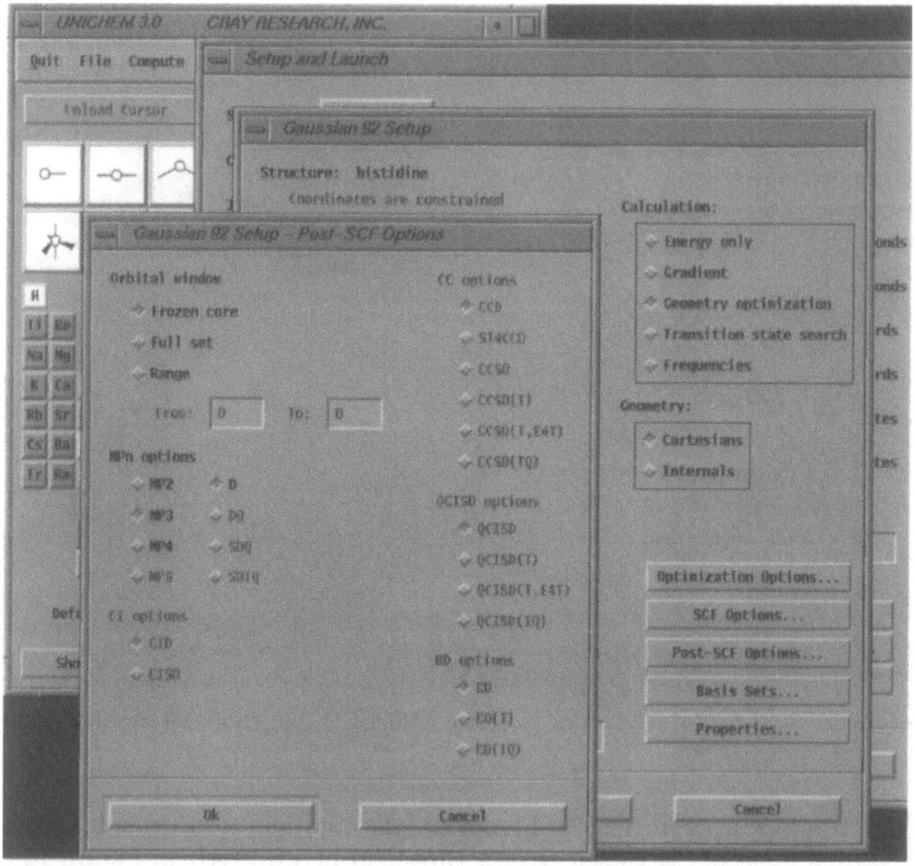

Figure 8. The ***Post-SCF Options*** **in the UNICHEM computational chemistry environment.**

matically. Graphical user interfaces, such as UNICHEM, make quantum chemical packages, such as GAUSSIAN, more accessible, particularly to the non-specialist. But the need to make meaningful calculations which remain computationally tractable whilst providing useful information about the target system requires some user intelligence. The user must select appropriate approximations for the problem in hand.

Many electronic structure calculations are carried out to determine properties which depend directly on the total energy of the system:- geometry optimization, conformational analysis, vibrational frequencies and spectra, etc. The accuracy to which the electronic energy is required in a molecular structure calculation is dependent upon the particular application. For example, bonding interactions typically involve energy changes of $\sim 10^5$ μhartree and, therefore, an accuracy of ~ 1000 μhartree is often

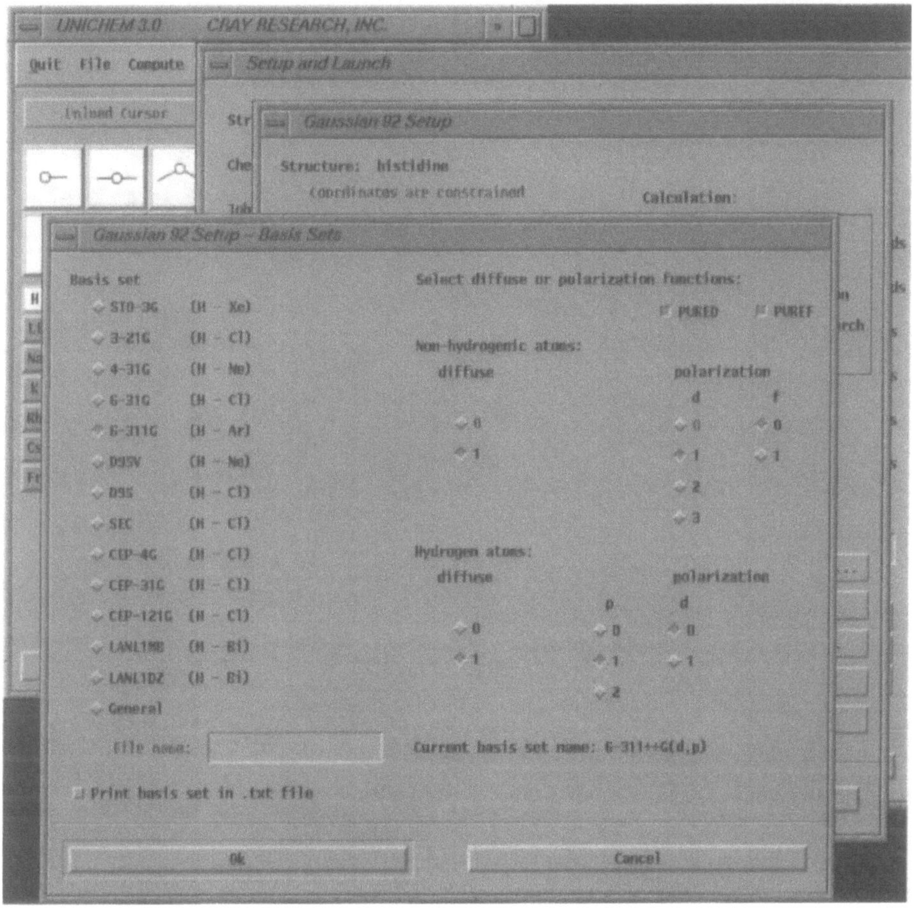

Figure 9. The Basis Sets window in the UNICHEM computational chemistry environment.

regarded as "chemical accuracy". Non-bonded interactions, such as van der Waals interactions, involve minima ranging from $\sim 50 - 1000$ μhartree and thus an accuracy of ~ 1 μhartree is desirable, although yet higher accuracy may be required in studies of many-body effects.

The theoretical model invoked in the study of a particular problem in computational molecular science and the approximations made in its practical realization must be appropriate. For example, one of the key choices to be made in almost all *ab initio* molecular electronic structure studies is the selection of a basis set (or sets) to be employed in the parametrization of the orbitals. Davidson and Feller point out that[45]

> *Occasionally, for reasons of economy or computational feasibility, a basis set is selected for which the computed results are nearly without significance.*

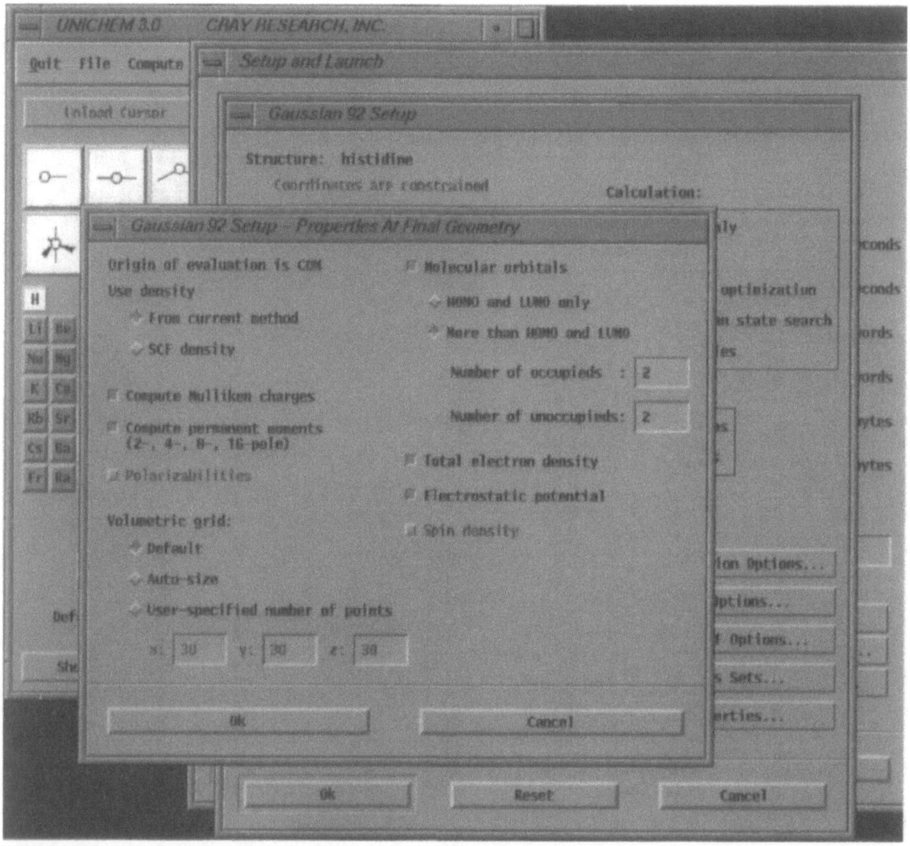

Figure 10. The *Properties* window in the UNICHEM computational chemistry environment.

(The problem of basis set construction will be addressed in greater detail in the following article[46].) For problem solving applications an appropriate approximation is almost invariably one whose strengths and weaknesses have been well documented. A particular theoretical model and approximation scheme must be calibrated by applying it to a range of systems and then a pragmatic approach is adopted in applications. For example, self-consistent field calculations using double-zeta basis sets often yield bond lengths for molecules containing first row atoms that are within 0.01 Å of experimental values. This degree of agreement often deteriorates when the basis set is extended; for example, on adding polarization functions, and is only recovered when correlation effects are accounted for. A pragmatic approach would say that if one requires bond lengths which are within

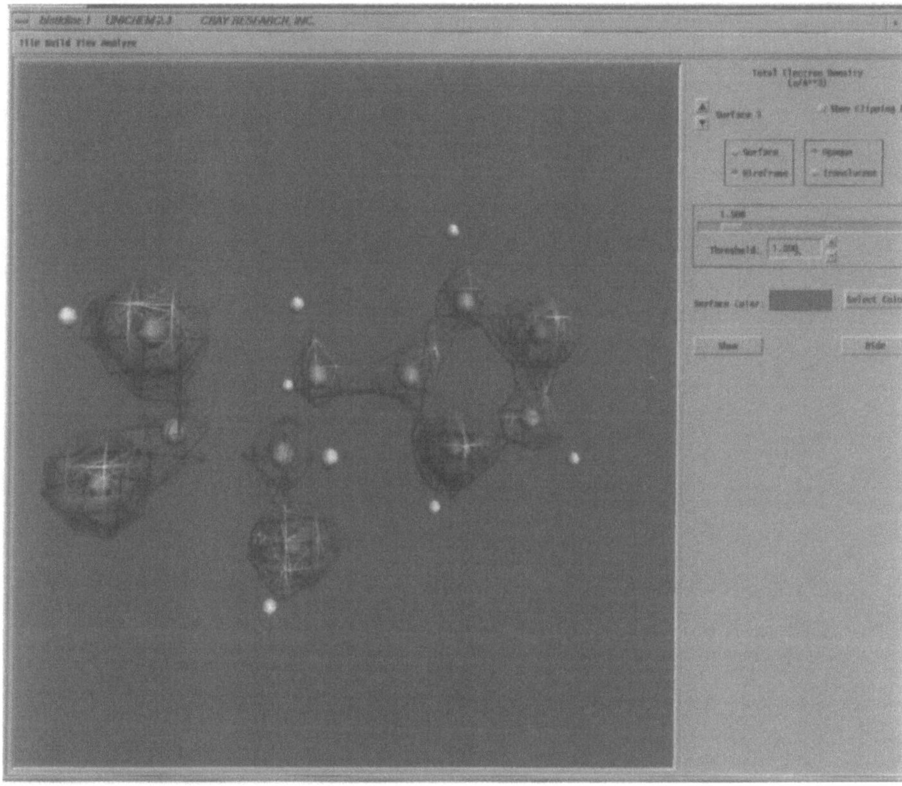

Figure 11. The total electron density with a threshold of 1.5 $e/\AA^3$ generated for the histidine molecule in the UNICHEM computational chemistry environment.

~ 0.01 Å of the experimental values from a computationally tractable approach then the matrix Hartree-Fock model with double-zeta basis sets should be employed even though the apparent accuracy of the results clearly rest on some cancellation of basis set truncation effects and correlation effects.

3.4. PROSPECTS

The utility of *ab initio* quantum mechanical studies of molecules is now firmly established. Their potential is enormous as, for example, K.G. Wilson has emphasized[2]:-

> *The number of unexplored forms of matter which can fit into a small box one centimetre on a side is*
>
> $$92^{(10^{23})}$$

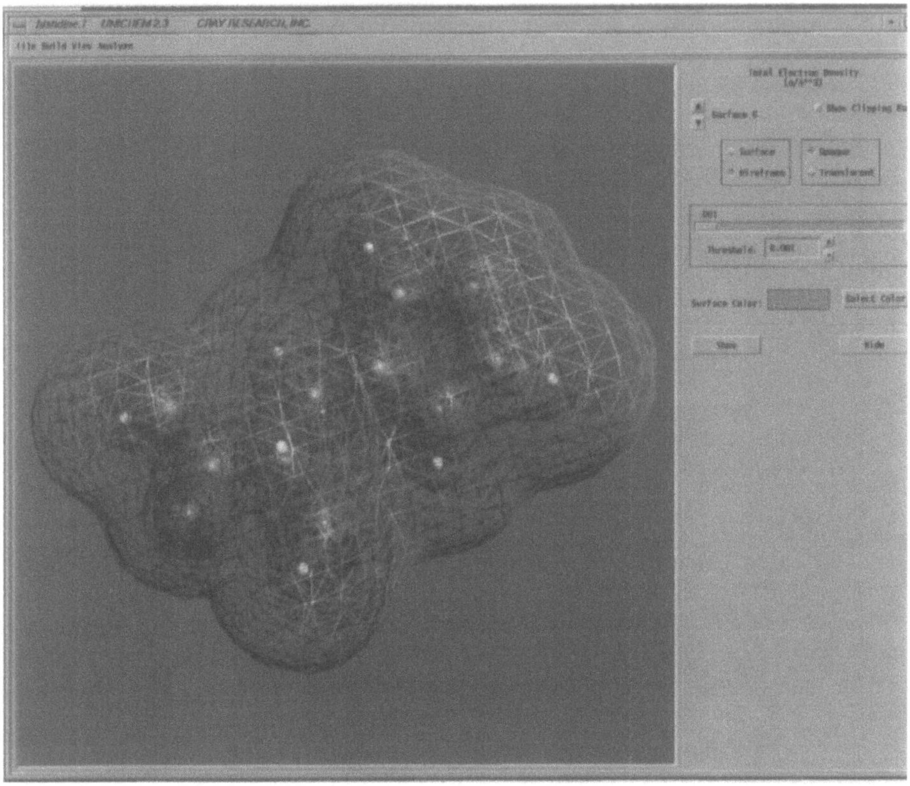

Figure 12. The total electron density with a threshold of 0.001 $e/\AA^3$ generated for the histidine molecule in the UNICHEM computational chemistry environment.

These unexplored forms of matter contain innumerable surprises, as we know from the many extraordinary compounds that are among the ten million already studied. Furthermore there is as yet no known maximum size either of a molecule, or crystalline unit cell, i.e. no maximum above which matter is guaranteed to be repetitive. Hence even centimetre^3 *size chunks might contain new surprises. Furthermore screening of the presently unexplored compounds is likely to lead to improved substitutes for materials of vital importance to the world economy. Areas where improvements would be welcome include:*

1. *energy generation (for example, photosynthesis)*
2. *energy storage (batteries)*
3. *data processing (silicon and magnetic materials)*
4. *structural materials*
5. *telecommunications (optical fibre and optoelectronics)*
6. *drugs (including e.g. drugs to cure addictions)*
7. *catalysts*
8. *and many more*

...

On the whole the vast majority of computational chemists have been content to use general purpose computers in their work designing software to exploit the capabilities of commercially available machines. The view has been expressed that they should be involved in the construction of the machines that they use. Thus, in 1977, Ostlund wrote[47]:-

> *"Computational chemistry is too established and important a field to leave its only significant piece of apparatus to disinterested computer scientists. Chemists in other fields has commonly built very sophisticated instruments - molecular beam machines, electron scattering spectrometers, ion cyclotron resonance spectrometers, etc. ... If computation chemists are to take full advantage of the current revolution in microelectronics, which offers great prospects for cost-effective high-performance computation, it will be necessary to become involved in the design and building of highly parallel architectures specific to specific chemical applications."*

To some extent, some computational chemists are already taking this approach and designing their own computer systems by buying commercially available components and thus tailoring their complete system to their specific need.

In conclusion, let us turn the more general question of the purpose of *ab initio* molecular electronic structure studies. Many contemporary researchers regard the UNICHEM/GAUSSIAN system and packages like it as black boxes, rather like an infra-red spectrometer, from which they demand the highest precision possible. If their problem is to assign the lines in the spectrum of an interstellar molecule then, quite correctly, they demand the highest precision possible. It is salutary, however, to recall the opinions of some of the pioneers in our field, for example, C.A. Coulson[48] wrote

> *"... remarks upon the futility of obtaining accurate numbers, whether by computation or by experiment, unless these numbers can provide us with simple and useful chemical concepts; otherwise, one might as well be interested in a telephone directory."*

whilst E.P. Wigner was rather more explicit in his comments on the role of computation in the physical sciences[49]

"If I had a great calculating machine, I would perhaps apply it to the Schrödinger equation of each metal and obtain its cohesive energy, its lattice constant, etc. It is not clear, however, that I would gain a great deal by this. Presumably, all the results would agree with the experimental values and not much would be learned from the calculation. What would be preferable, instead, would be a vivid picture of the behaviour of the wave function, a simple description of the essence of metallic cohesion, and a understanding of the causes of its variation from element to element. Hence, the task which is before us is not a purely scientific one, it is partly pedagogic. Nor can its solution be unique: the same wave function can be depicted in a variety of ways (just as the cubic close pack lattice can), the same energy can be decomposed in a variety of ways into different basic constituents. Hence the value of any contribution will depend on the taste of the reader. In fact, from the point of view of the present article, the principle purpose of accurate calculation is to assure us that nothing truly significant has been overlooked."

This view, which was not uncommon in an era when calculations were necessarily of a more qualitative nature, would appear to be diametrically opposed to the philosophy adopted in most contemporary studies[50]

"Nowadays, in quantum chemistry ... quantitative accuracy can frequently be achieved in molecular electronic structure studies and, to some extent, the simple concepts that emerged from the pioneering work of the first half of the century are often put to one side. Much will be gained, however, if the computational chemist can devise theories and associated computational algorithms that can not only afford high accuracy but also insight into the problem at hand."

Acknowledgment

The author is grateful to Dr. C. Plant, Rutherford Appleton Laboratory, for his help in preparing the Figures.

References

1. P.A.M. Dirac, Proc. Roy. Soc. **A123**, 714 (1929)
2. K.C. Wilson, Nucl. Phys. B (Proc. Suppl.) **17**, 82 (1990)
3. E.A. Hylleraas, Z. Physik **48**, 469 (1928)
4. H.M. James and A.S. Coolidge, J. Chem. Phys. **1**, 825 (1933)
5. C.A. Coulson, Proc. Camb. phil. Soc. **34**, 204 (1938)

6. D.R. Hartree, W. Hartree and B. Swirles, Phil. Trans. Roy. Soc. (London) **A238**, 229 (1939)
7. B. Swirles, Proc. Roy. Soc. **A152**, 625 (1935); *ibid* **A157**, 680 (1935)
8. H.F. Schaefer III, *Quantum Chemistry. The development of ab initio methods in molecular electronic structure theory, Clarendon Press*, Oxford (1984)
9. D.R. Hartree and A. Porter, Mem. Manchr. lit. phil. Soc. **79**, 51 (1935)
10. D.R. Hartree, Calculating Instruments and Machines, Cambridge University Press (1950)
11. R. McWeeny, in Methods in Computational Chemistry **1**, *vii-ix* (1987)
12. G.G. Hall, Proc. Roy. Soc. (London) **A205**, 541 (1951)
13. C.C.J. Roothaan, Rev. Mod. Phys. **23**, 69 (1951)
14. R. McWeeny, *Dissertation* (University of Oxford, 1948)
15. R. McWeeny, Nature **166**, 21 (1950)
16. S.F. Boys, Proc. Roy. Soc. (London) **A200**, 542 (1950)
17. I. Shavitt, Israel J. Chem. **33**, 357 (1993)
18. S.F. Boys, Proc. Roy. Soc. (London) **A201**, 125 (1950)
19. J. Miller, R.H. Friedman, R.P. Hurst and F.A. Matsen, J. Chem. Phys. **27**, 1385 (1957)
20. R.K. Nesbet, Proc. Roy. Soc. (London) **A230**, 312 (1955)
21. J.A. Pople and R.K. Nesbet, J. Chem. Phys. **22**, 571 (1954)
22. S.F. Boys, G.B. Cook, C.M. Reeves and I. Shavitt, Nature (London) **178**, 1207 (1956)
23. R.S. Mulliken and C.C.J. Roothaan, Proc. U.S. Natl. Acad. Sci. **45**, 394 (1959)
24. M.P. Barnett, Rev. Mod. Phys. **35**, 571 (1963)
25. E. Clementi and D.R. Davis, J. Comput. Phys. **1**, 223 (1966)
26. O. Sinanoğlu, Adv. Chem. Phys. **6**, 315 (1964); *ibid*, **14**, 237 (1969)
27. K.J. Miller and K. Ruedenberg, J. Chem. Phys. **48**, 3415 (1968); D.M. Silver, E.L. Mehler and K. Ruedenberg, *ibid*, **52**, 1174, 1181
28. C.F. Bender and E.R. Davidson, Phys. Rev. **183**, 23 (1969)
29. J. Čižek, Adv. Chem. Phys. **14**, 35 (1969)
30. H.P. Kelly, Adv. Chem Phys. **14**, 129 (1969)
31. J. Gerratt and I.M. Mills, J. Chem. Phys. **49**, 1719 (1968)
32. P. Pulay, Molec. Phys. **17**, 197 (1969)
33. J. Goldstone, Proc. Roy. Soc. (London) A239, 267 (1957)
34. J. Schulman and D.N. Kaufman, J. Chem. Phys. **53**, 477 (1970)
35. U. Kaldor, Phys. Rev. A **7**, 427 (1973)
36. S. Wilson and D.M. Silver, Phys. Rev. A **14**, 1949 (1976)
37. C. Møller and M.S. Plesset, Phys. Rev. **46**, 618 (1934)
38. J. Pople, R. Krishnan, H.B. Schlegel and J.S. Binkley, Intern. J. Quantum Chem. **14**, 545 (1978)
39. M.F. Guest and S. Wilson, in *Supercomputers in Chemistry*, edited by P. Lykos and I. Shavitt, ACS Symposium Series **173**, American Chemical Society, Washington D.C. (1981)
40. J. Almlöf, K. Faegri and K. Korsell, J. Comput. Chem. **3**, 385 (1982)
41. Part III
42. Part IV
43. M.J. Frisch, G.W. Trucks, H.B. Schlegel, P.M.W. Gill, B.G. Johnson, M.A. Robb, J.R. Cheeseman, T. Keith, G.A. Petersson, J.A. Montgomery, K. Raghavachari, M.A. Al-Laham, V.G. Zakrzewski, J.V. Ortiz, J.B. Foresman, J. Cioslowski, B.B. Stefanov, A. Nanayakkara, M. Challacombe, C.Y. Peng, P.Y. Ayala, W. Chen, M.W. Wong, J.L. Andres, E.S. Replogle, R. Gomperts, R.L. Martin, D.J. Fox, J.S. Binkley, D.J. Defrees, J. Baker, J.P. Stewart, M. Head-Gordon, C. Gonzalaz and J.A. Pople, GAUSSIAN94, Gaussian Inc., Pittsburgh PA, 1995
44. UNICHEM was originally developed by Cray Research Inc. Details can be obtained

from Oxford Molecular Ltd., The Medawar Centre, Oxford Science Park, Oxford OX4 4GA, England.
45. E.R. Davidson and D. Feller, Chem. Rev. **86**, 681 (1986)
46. Part II
47. N.S. Ostlund, in *Personal Computers in Chemistry*, American Chemical Society, Washington D.C. (1977)
48. C.A. Coulson, quoted in Rev. Mod. Phys. **45**, 22 (1973)
49. E.P. Wigner, in *Proc. Internat. Confer Theoret. Phys.*, Science Council of Japan (1953)
50. S. Wilson, Methods in Computational Chemistry **3**, 1 (1989)

PRACTICAL AB INITIO METHODS FOR MOLECULAR ELECTRONIC STRUCTURE STUDIES. II. FINITE BASIS SETS AND THE ALGEBRAIC APPROXIMATION.

S. WILSON
Rutherford Appleton Laboratory
Chilton, Oxfordshire OX11 0QX, England

Abstract. The use of finite analytic basis sets in molecular electronic structure calculations is ubiquitous. In this article, a review of the fundamentals of the algebraic aprroximation is given together with a survey of some of the more promising recent developments in the systematic approximation of molecular electronic structure by means of finite basis set expansions.

1. Introduction

Finite difference methods have been employed in atomic structure calculations since the pioneering work of Hartree and his collaborators[1] in the 1930s. The spherical symmetry of the atom allows the wave function to be factorized into an angular part which can be handled analytically and a one-dimensional radial part which is solved numerically using finite differences. Since 1983, the finite difference technique has been applied to diatomic molecules[2]; the use of elliptical coordinates facilitates the analytic treatment of the dependence on the azimuthal angle following by a computationally demanding but nevertheless tractable two-dimensional finite difference algorithm. Finite difference, and, more recently, finite element[3], calculations have provide important atomic and molecular benchmark calculations. However, for calculations which go beyond the Hartree-Fock model and take account of electron correlation effects and for polyatomic molecules the use of finite analytic basis sets is almost obligatory.

In 1990, whilst comparing the then present state of *ab initio* molecular electronic structure calculations with computational lattice gauge theory, K.R. Wilson[4] wrote:-

S. Wilson and G. H. F. Diercksen (eds.), Problem Solving in Computational Molecular Science, 109–158.

Computational quantum chemistry as a discipline is fifty years AHEAD of lattice gauge theory and there are lessons about computing power, algorithm development and interactions with experiment in this history.

and, furthermore, he recognized the importance of finite analytic basis sets in molecular studies as opposed to grids

The most important algorithmic advance was the introduction of systematic algorithms using analytic basis functions in place of numerical grids, first proposed in the early 1950's [5] [6] .

The selection of the basis sets employed puts the "art" into applied quantum chemistry. As the present author observed in 1982 [7].

The vast majority of contemporary molecular calculations adopt a pragmatic approach; no error bounds are determined and the accuracy of a calculation is assessed by comparison with quantities derived from experiment. ... this procedure enables the quality of the basis set to be established and an empirical range of validity ... obtained which can then be used to make educated guesses of the accuracy of a computed molecular property. The art of selecting a basis set a basis set is based on previous experience in treating similar systems using basis sets of comparable quality ... There can be little doubt that the error attributable to the truncation of the basis set is the largest source of error in most accurate calculations ...

Writing in 1986, Davidson and Feller [8] underline the fact that in contemporary quantum chemical studies the choice of basis set is an integral part of the assumed model and a knowledge of the capabilities of that basis set is vital:-

Intelligent reading of the literature requires at least a cursory knowledge of the limitations of other basis sets. Information concerning the likely accuracy of a specific basis set for a particular property is essential in order to judge the adequacy of the computational method and, hence, the soundness of the results. Occasionally, for reasons of economy or computational feasibility, a basis set is selected for which the computed results are nearly without significance.

In 1985, Huzinaga [9] emphasized the

quasi-empirical

element associated with basis set selection in the vast majority of contemporary electronic structure studies. A similar point of view has been expressed by Davidson and Feller [8] who comment that

the smaller the basis set the more ab initio calculations assume an empirical flavour.

Clementi et al[10], writing in 1990, complained that as a result of the failure to address the problem of basis set truncation, the

> *literature ... was flooded with computations which essentially compared experiments with basis set superposition error data*.

The importance of the basis set truncation error was emphasized again in 1992 by Taylor [11] who wrote:-

> *The ultimate accuracy of any calculation is determined by the one-particle basis set. This is one of the most obvious, yet one of the most ignored, observations about quantum chemical calculations. For example, it is in general just not possible to get the right answer for the right reasons using a DZP basis. This is not an argument against using such sets, but their limitations must be kept constantly in mind.*

In using small basis sets one is often exploiting a fortuitous cancellation of errors. For example, for a molecule containing C, N, O and H, the use of the basis set designated 3-21G in a self-consistent field calculation will often lead to a cancellation of the basis set truncation error and the energy associated with electron correlation in calculations of optimal bond lengths. As Taylor[11] points out this is not an argument against using such sets. One has to take a pragmatic approach. If a particular combination of basis set and theory will yield the required accuracy for a particular class of molecules then, in routine quantum chemical studies, one should go ahead and use them. But one should be aware of what one is doing and recognize that is the same combination of basis set and theory may not deliver a comparable accuracy for some other property or, indeed, for some other class of molecules.

In this article, an review of the fundamentals of the algebraic approximation is given together with a survey of some of the more promising recent developments in the systematic approximation of molecular electronic structure by means of finite basis set expansions. First, in the following section, we provide a brief description of the fully numerical, finite difference technique in atomic and molecular applications. The use of finite basis sets and the implementation of the algebraic approximation is considered in general terms in section 3, whilst in section 4 the nature of this approximation is described in some detail for the case of one-electron systems. Nowadays, the basis sets used in molecular studies are comprised almost exclusively of Gaussian functions and the most commonly used functions of this type are discussed in section 5. The construction of basis sets of Gaussian functions in surveyed in section 6, where not only are matrix Hartree-Fock and electron correlation energy calculations considered but *(i)* the use of systematic sequences of even-tempered basis sets is described, *(ii)* recent work on distributed Gaussian basis sets is surveyed, *(iii)* the use of field-dependent basis sets is outlined, and *(iv)* the basis set problem in relativistic electronic structure studies is briefly discussed. In the final section, section 7, a summary is given and the prospects for the future are indicated.

2. Numerical methods

As has been mention above, for atoms the Hartree-Fock equations can be separated by using spherical polar coordinates. The angular equations can be solved analytically and the radial equation can solved by one-dimensional finite difference methods. This has been the "traditional" approach to the atomic structure problem since the pioneering work of Hartree and his coworkers[1] in the 1930s and is still in use today in, for example, the relativistic atomic structure package authored by Grant and his coworkers[12]. However, it has been known since the work of Kelly[13] in the 1960s that the numerical approach can run into difficulties in calculations which go beyond the Hartree-Fock approximation and take account of electron correlation. For example, the sum-over-states expressions which arise in the many-body perturbation treatment of correlation effects needs to be handled with care in fully numerical calculation[14]. In contrast, finite basis set methods yield not only an approximation to the occupied orbitals but also a representation of the spectrum which can be directly employed in the sum-over-states perturbation theory expressions[15] [16].

Over the past decade or so, two-dimensional finite difference Hartree-Fock calculations for diatomic molecules have been shown to be computationally feasible and to give very accurate total energies and occupied orbital energies[2] [17] [18] [19]. The finite difference approach has been refined over recent years both in terms of the accuracy achieved and in terms of efficiency of the computational implementation [19] and has been shown to be competitive with the alternative fully numerical approach, the finite element technique[3] [20]. Finite difference algorithms have been designed to take advantage of the vector processing capabilities of modern high performance computers [21].

It has been stressed [19] that, whilst finite difference and finite element techniques are often described as basis set independent methods,

> *they are in fact grid dependent and ... inadequate grid size can easily spoil the results, ensuring that the Hartree-Fock limit values are never reached.*

These fully numerical approaches to the Hartree-Fock problem for diatomic molecules provide invaluable benchmarks against which the solutions of the matrix Hartree-Fock equations, which are obtained when the algebraic approximation is invoked, can be assessed.

For diatomic molecules, the use of prolate spheroidal coordinates leads to the Hartree-Fock equations in the form of Poisson-like two-dimensional partial differential equations for the molecular orbitals and potentials. Upon high order discretization the corresponding sets of linear equations are solved by the successive overrelaxation method which proved to be both stable and efficient[19] [21]. A detailed description of the finite difference approach to the Hartree-Fock problem for diatomic molecules may be found elsewhere [17] [21] [22]. The treatment of the Hartree-Fock equations involves the solution of Poisson-like, two-dimensional partial differential equations for the molecular orbitals and potentials, which are written in prolate spheroidal coordinates (ξ, η, φ) , $\xi = (r_a + r_b)/R$; $\eta = (r_a - r_b)/R$; where a and b label the nuclei.

The finite difference equations are usually defined on a rectangular region

$$[1, \xi_\infty] \times [-1, 1] \tag{1}$$

with ξ_∞ sufficiently large. The coordinate system actually employed is defined through the transformation

$$\mu = \cosh^{-1} \xi, \quad (0 \leq \mu \leq \infty) \tag{2}$$

and

$$\nu = \cos^{-1} \eta, \quad (0 \leq \nu \leq \pi) \tag{3}$$

so that

$$r_a = \frac{1}{2} R(\cosh \mu + \cos \nu) \tag{4}$$

and

$$r_b = \frac{1}{2} R(\cosh \mu - \cos \nu) \tag{5}$$

In these coordinates the grid points are distributed uniformly according to

$$\mu_i = \mu_{i-1} + \delta_\mu, \; i = 1, 2, ..., n_\mu, \quad \mu_0 = 0 \tag{6}$$

with a step given by $\delta_\mu = R_\infty / n_\mu$, where R_∞ is an approximation to infinity and n_μ is the number of points, and

$$\nu_j = \nu_{j-1} + \delta_\nu, \; j = 1, 2, ..., n_\nu, \quad \nu_0 = 0 \tag{7}$$

with a step given by $\delta_\nu = 2\pi / n_\nu$ where n_ν is the number of points. These grids are designated

$$[n_\nu \times n_\mu; R_\infty] \tag{8}$$

For diatomic molecular systems containing light atoms a uniform distribution of grid points has been shown to afford an adequate description of the spatial behaviour of the molecular orbitals and potentials and to be computationally tractable (see, for example, the work of Kobus *et al*[23] [24] [25]). However, such a uniform grid is no longer be appropriate for diatomic systems containing heavy atoms for which the highly contracted core orbitals and also the valence orbitals in the region close to the nuclei require a very fine grid to properly describe their nodal structure whereas a much more sparse grid is required to adequately describe their tails[26].

3. Finite basis sets and the algebraic approximation

3.1. MATRIX HARTREE-FOCK THEORY AND BEYOND

For arbitrary molecular systems the lack of a suitable coordinate system has necessitated the use of basis set expansion techniques in which each of the one-electron solutions of the Hartree-Fock equations is expanded in terms of some basis set

$$\varphi = \sum_{p=1}^{N} c_p \chi_p \tag{9}$$

where the χ_p are global basis functions (almost invariably in contemporary molecular electronic structure studies these basis functions are taken to be Gaussian-type functions) and the coefficients, c_p, are determined by solution of the matrix Hartree-Fock equations. A set of algebraic equations for the expansion coefficients is obtained. This is the algebraic approximation[27] [28]. The solution of the matrix Hartree-Fock equations yields not only the occupied orbitals which are required to describe the particular electronic state being considered but also orbitals which are not occupied and can be used to develop a description of electron correlation effects via, for example, the many-body perturbation theory expansion[29].

3.2. COMPUTATIONAL CONSIDERATIONS

Algorithms for the solution of the matrix Hartree-Fock equations can be effectively formulated in terms of Basis Linear Algebra Subroutines (BLAS) and, in particular, matrix multiplication, and can, therefore, be determined with great efficiency on vector processing computers and parallel vector processing computers often by employed routines, often supplied by the computer manufacturer, and very carefully tailored to the machine being used[30] [31]. For example, the fourth-order many-body perturbation theory energy component corresponding to diagrams which contain triply excited intermediate states is now recognized as an important fragment of high precision correlation treatments, yet its evaluation leads to an algorithm which depends on the seventh power of the number of basis functions[32].

However, it has been demonstrated that this component can be evaluated very efficiently on parallel vector processing computers such as the CRAY C-90 or the NEC SX-3/44 [33] by constructing the algorithm on matrix multiplication kernels.

3.3. GAUSSIAN BASIS FUNCTIONS

Although during the 1950s and early 1960s exponential functions, or Slater-type functions, were regards as the basis functions of choice, nowadays it is widely recognized that Gaussian basis functions are particularly attractive in both non-relativistic and relativistic molecular electronic structure studies. The following advantages of the algebraic approximation and its implementation with basis sets of Gaussian functions are worth recalling:

(i) Multi-centre, two-electron integrals can be evaluated easily and accurately.

(ii) Because Gaussian functions do not introduce a cusp, they do not have to be atom-centred.

(iii) The Gaussian function is a "natural" choice of basis function if the point nucleus is abandoned in favour of a more realistic finite nucleus [34] [35] [36] [37] such as the "Fermi nucleus" which has the form of a Fermi distribution

$$\rho(r) = \frac{\rho_0}{1 + e^{(r-c)/a}} \tag{10}$$

where c is the *"half-density radius"* and the parameter a is related to the nuclear *"skin thickness"*. The use of finite nuclear models is not restricted to relativistic electronic structure studies. Indeed, there may be some advantage in using such models in non-relativistic studies since it is known that the convergence of finite basis set expansions is dominated by the ability of the functions to describe cusps [38] [39] [40].

(iv) Near linear dependence for the large and flexible basis sets demanded for high precision calculations is less severe in calculation using Gaussian-type functions than for calculations employing exponential-type functions.

(v) Relativistic calculations demand the use of matched basis sets for the large and small components of the spinors, a condition which is readily implemented for Gaussian functions[37].

(vi) The use of the algebraic approximation facilities the introduction of the Breit interaction into the relativistic self-consistent field procedure[41] [42] [43] [44].

(vi) The use of the algebraic approximation provides a representation of the whole of the spectrum: the bound states and the continuum in the non-relativistic formulation, the bound states, the positive energy continuum and the negative energy continuum in the relativistic formulation[45] [42] [43] [44].

4. The algebraic approximation for one-electron systems

In this section some fundamental aspects of the algebraic approximation are illustrated by describing a practical application to some one-electron hydrogenic systems. First, the primitive s-type Gaussian functions employed in the calculations are defined. Then, the matrix representation of the Schrödinger equation obtained by invoking the algebraic approximation with the finite analytical basis set of Gaussian functions is presented and its approximate eigenvalues discussed. Finally, attention is turned to the approximate eigenfunctions obtained within the algebraic approximation.

4.1. PRIMITIVE S-TYPE GAUSSIAN FUNCTIONS

A primitive s-type Gaussian function has the form

$$\chi_p = \left(\frac{2\zeta}{\pi}\right)^{\frac{3}{4}} \exp(-\zeta_p r^2) \tag{11}$$

where ζ_p is the screening parameter or orbital exponent. This function is sometimes called a spherical Gaussian orbital. It is plotted in Figure 1 as a function of r for four different choice of ζ_p. The radius of each function at its half height, $\frac{1}{2}\left(\frac{2\zeta}{\pi}\right)^{\frac{3}{4}}$, is given by

$$r_{\frac{1}{2}} = \sqrt{\frac{1}{\zeta}\ln 2} \tag{12}$$

so that for the most diffuse function with $\zeta = \frac{1}{2}: r_{\frac{1}{2}} \simeq 1.1774$, for $\zeta = 1: r_{\frac{1}{2}} \simeq 0.8326$, for $\zeta = 2: r_{\frac{1}{2}} \simeq 0.5887$, and for the most contracted function with $\zeta = 4: r_{\frac{1}{2}} \simeq 0.4163$.

4.2. THE MATRIX SCHRÖDINGER EQUATION FOR HYDROGENIC SYSTEMS

Using s-type Gaussian functions of the type given in (11) the hamiltonian matrix for a one-electron hydrogenic system, with nuclear charge Z, can be constructed and diagonalized. The matrix eigenvalue equation takes the form

$$\mathbf{hc} = \varepsilon \mathbf{sc} \tag{13}$$

where ε is the eigenvalue and the matrix elements of the hamiltonian and overlap matrix are

$$h_{pq} = \left\langle \chi_p \left| \hat{h} \right| \chi_q \right\rangle; \quad \hat{h} = -\frac{1}{2}\nabla^2 - \frac{Z}{r} \tag{14}$$

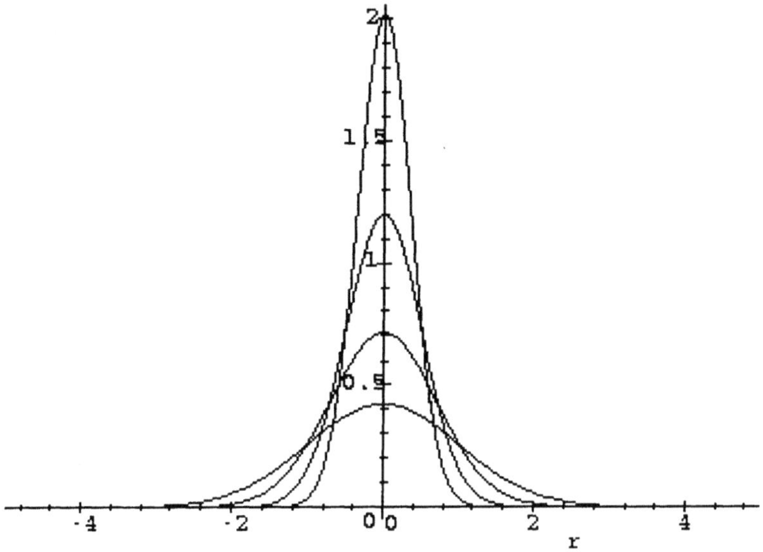

Figure 1. Normalized s-type Gaussian functions with exponents, ζ, of $\frac{1}{2}$, 1, 2 and 4.

and

$$s_{pq} = \langle \chi_p \mid \chi_q \rangle \quad (15)$$

respectively. $\mathbf{c}$ is a vector whose elements are the expansion coefficients, c_p, defined in equation(9).

Even tempered basis sets, in which the exponents are taken to form a geometric series

$$\zeta_k = \alpha\beta^k, \; k = 1, 2, ..., N, \; \alpha > 0, \; \beta > 1, \quad (16)$$

afford a systematic technique for generating large basis sets. Systematic sequences of even-tempered basis sets will be discussed in more detail in section 6 but for the purposes of the present discussion they provide a mechanism for generating basis sets capable of supporting increasing levels of accuracy.

The particular sequence of basis sets employed here was originally developed by Schmidt and Ruedenberg[46] for the ground state of the hydrogen atom. The eigenfunctions are expressed as a linear combination of the Gaussian basis functions. The energy spectrum obtained in this way for the Ga^{30+} ion using an even-tempered basis set of 10 Gaussian functions is shown in Table 1. The column headed ϵ_{calc} gives the approximate energy eigenvalues, in Hartree, resulting from the matrix diagonalization.

This problem can, of course, be solved exactly and the column headed ϵ_{exact} gives the corresponding exact energies also in Hartree. The column headed δ gives the difference between the calculated and exact energies in μhartree. Examination of Table 1 reveals that this basis set will support approximations for the ground state and the lowest seven excited states of s-symmetry. The approximations for the bound states are not at all accurate since the basis set is one that was original designed for the hydrogen atom and does not contain functions which are sufficiently contracted for an accurate description of the Ga^{30+} ion. Indeed, the ground state energy is in error by almost 40 Hartree! In addition to the eight bound states, the basis set of ten functions supports two virtual states with positive energies providing a crude representation of the continuum.

Table 1

Spectrum of the non-relativistic hydrogen-like Ga^{30+} supported by an even-tempered basis set of 10 s-type Gaussian functions.

i	$\epsilon_{calc}/Hartree$	$\epsilon_{exact}/Hartree$	$\delta/\mu Hartree$
0	-440.8148663983	-480.5000000000	39685133.6017
1	-114.8125794821	-120.1250000000	5312420.5179
2	-51.7662050327	-53.3888888889	1622683.8562
3	-29.1512534000	-30.0312500000	879996.6000
4	-18.2614471024	-19.2200000000	958552.8976
5	-12.1341895943	-13.3472222222	1213032.6279
6	-9.0282341149	-9.8061224490	777888.3341
7	-6.4133209341	-7.5078125000	1094491.5659
8	96.8817504395	-5.9320987654	
9	556.2420165559	-4.8050000000	

Now consider a calculation for the Ga^{30+} ion using a basis set of 15 s-type Gaussian functions. (It should be noted that as the basis set is increased in size the values of α and β employed in generating the even-tempered exponents decrease so that the smallest (largest) exponent, corresponding to the most diffuse (contracted) basis function, is smaller (larger) than the corresponding exponent for the case of 10 functions. This will be discussed in more detail in section 6.) The lower part of the energy spectrum obtained for an even-tempered basis set of 15 primitive s-type Gaussian functions is shown in Table 2. There is a significant improvement in the accuracy with which the bound states are described. The error in the ground state energy is reduced to less than 2 Hartree. Nine bound states are supported and the remaining six virtual states provide a representation of the continuum.

Table 2
Lower part of the spectrum of the non-relativistic hydrogen-like Ga^{30+} supported by an even-tempered basis set of 15 s-type Gaussian functions.

i	$\epsilon_{calc}/Hartree$	$\epsilon_{exact}/Hartree$	$\delta/\mu Hartree$
0	−478.7605823172	−480.5000000000	1739417.6828
1	−119.9057931579	−120.1250000000	219206.8421
2	−53.3228585632	−53.3888888889	66030.3257
3	−29.9886039937	−30.0312500000	42646.0063
4	−19.1243166693	−19.2200000000	95683.3307
5	−13.1170348492	−13.3472222222	230187.3730
6	−9.5416283341	−9.8061224490	264494.1149
7	−7.3351179422	−7.5078125000	172694.5578
8	−5.2407022929	−5.9320987654	691396.4725
9	31.8866487955	−4.8050000000	
10	208.3215487505	−3.9710743802	
11	737.7079302722	−3.3368055556	

The representation of the spectrum supported by an even-tempered basis set consisting of 30 s-type Gaussian functions is displayed in Table 3. In this case eleven bound states and 19 virtual states are supported and the accuracy of the ground state energy and the first four excited states is reduced to the *"sub-milliHartree"* level. It should be remembered that the sequence of even-tempered basis sets that are being used was designed for the hydrogen atom ground state and not Ga^{30+}!

Table 3
Lower part of the spectrum of the non-relativistic hydrogen-like Ga^{30+} supported by an even-tempered basis set of 30 s-type Gaussian functions.

i	$\epsilon_{calc}/Hartree$	$\epsilon_{exact}/Hartree$	$\delta/\mu Hartree$
0	−480.4996996123	−480.5000000000	300.3877
1	−120.1249621027	−120.1250000000	37.8973
2	−53.3888719392	−53.3888888889	16.9497
3	−30.0312286984	−30.0312500000	21.3016
4	−19.2196150368	−19.2200000000	384.9632
5	−13.3435793919	−13.3472222222	3642.8303
6	−9.7875870850	−9.8061224490	18535.3640
7	−7.4500013296	−7.5078125000	57811.1704
8	−5.8446164838	−5.9320987654	87482.2817
9	−4.7524735163	−4.8050000000	52526.4837
10	−3.7308585610	−3.9710743802	240215.8192
11	7.1121381803	−3.3368055556	

Increasing the size of the basis set still further, in Table 4 the lower part spectrum for the Ga^{30+} ion supported by a basis set of 50 s-type Gaussian functions is displayed. The accuracy of the four lowest energy level is now beyond the *"sub-μHartree"* level and eight states are described with *"sub-millihartree"* accuracy.

Table 4
Lower part of the spectrum of the non-relativistic hydrogen-like Ga^{30+} supported by an even-tempered basis set of 50 s-type Gaussian functions.

i	$\epsilon_{calc}/Hartree$	$\epsilon_{exact}/Hartree$	$\delta/\mu Hartree$
0	−480.4999999695	−480.5000000000	0.0305
1	−120.1249999960	−120.1250000000	0.0040
2	−53.3888888821	−53.3888888889	0.0068
3	−30.0312499417	−30.0312500000	0.0583
4	−19.2199980597	−19.2200000000	1.9403
5	−13.3471996909	−13.3472222222	22.5313
6	−9.8060118995	−9.8061224490	110.5495
7	−7.5073389405	−7.5078125000	473.5595
8	−5.9294257055	−5.9320987654	2673.0599
9	−4.7879751996	−4.8050000000	17024.8004
10	−3.8932797788	−3.9710743802	77794.6014
11	−3.1114635335	−3.3368055556	225342.0220

Finally, in Table 5, the lower part of the Ga^{30+} spectrum supported by 100 s-type Gaussian function is shown. Of course, a basis set of this size would not normally be used to describe hydrogenic systems, but it is instructive to examine the behaviour of the algebraic approximation in the limit of a very large basis set. For the case of 100 basis functions, the six lowest levels of s symmetry are described to *"sub-nanoHartree"* level. Two further states are described to a *"sub-μHartree"* accuracy and a total of 11 states are given to a *"sub-milliHartree"* level.

Table 5
Lower part of the spectrum of the non-relativistic hydrogen-like Ga^{30+} supported by an even-tempered basis set of 100 s-type Gaussian functions

i	$\epsilon_{calc}/Hartree$	$\epsilon_{exact}/Hartree$	δ
0	−480.5000000000	−480.5000000000	0.0000
1	−120.1250000000	−120.1250000000	0.0000
2	−53.3888888889	−53.3888888889	0.0000
3	−30.0312500000	−30.0312500000	0.0000
4	−19.2200000000	−19.2200000000	0.0000
5	−13.3472222211	13.3472222222	0.0008
6	−9.8061224458	−9.8061224490	0.0031
7	−7.5078123536	−7.5078125000	0.1464
8	−5.9320962888	−5.9320987654	2.4766
9	−4.8049760464	−4.8050000000	23.9536
10	−3.9708879394	−3.9710743802	186.4408
11	−3.3355481532	−3.3368055556	1257.4024

Now using a basis set of 100 functions to describe a hydrogenic atom is not particularly efficient. A very much more compact representation of the occupied orbital could have been obtained by optimizing the exponents for the problem in hand and/or using contracted Gaussian basis functions, i.e. fixed linear combinations of primitive Gaussian functions. However, the use of a systematic sequence of even-tempered Gaussian function for this simple systems serves to emphasize three important points:-

i) exponent optimization can be avoided (This is a non-linear optimization which can be computationally demanding and, indeed, prohibitive for larger basis sets),

ii) a representation of the complete spectrum is obtained supporting not only an accurate representation of the lower bound states but also of the continuum (This is required in any calculation which goes beyond an independent electron model and takes account of electron correlation effects),

iii) the method can be systematically extended to achieve higher and higher accuracy until computational linear dependence becomes significant and the associated numerically problems associated with the use of a finite word length render the calculation inaccurate..

Near computational linear dependence in a basis set is usually measure by examining the eigenvalues of the overlap matrix. The lowest eigenvalues of the overlap matrices corresponding to the one-electron problems considered in Table 1-5 are shown in Table 6. The increasing near computational linear dependence with increasing basis set size is evident. The accuracy of calculation that can be supported by a given basis set is determined by the word length of the particular computing machine employed and the details of the matrix diagonalization algorithm employed.

Table 6
Lowest eigenvalues of the overlap matrix for the even-tempered basis sets of s-type Gaussian functions employed in the calculations for Ga^{30+}.

No. of basis functions	ϵ_0
10	0.0038837066850944
15	0.0006151487544774
30	0.0000117636996674
50	0.0000002731836860
100	0.0000000003502244

4.3. THE APPROXIMATE EIGENFUNCTION

Having consider the energy spectrum for the hydrogenic systems supported by basis sets of Gaussian functions of increasing size, the ground state eigenfunction and its approximation by a linear combination of s-type Gaussians will be briefly considered.

It is important to recognize that the approximation of the ground state wave function by a combination of Gaussians yields the best representation "in the mean" and not from "point to point". This is illustrated in

Figure 2 where the orbital amplitude difference function for the hydrogen atom is shown for a basis set of 30 Gaussian functions [taken from the work of Kobus *et al*[47]]. The exact ground state wave function for the hydrogenic atom is, of course, $\sqrt{Z^3/\pi}\exp(-Zr)$, where Z is the nuclear charge. The orbital amplitude difference function, that is, the difference between the exact wave function and its approximation by a linear combination of Gaussian functions can therefore be determined exactly. From Figure 2 the oscillatory nature of this function is immediately apparent. The oscillations of largest amplitude occur very close to the origin and are associated with the cusp at this point. The increasing amplitude of the oscillation with r are associated with the singularity which is known to exist at $r \to \infty$ [47]. It is clear that it is entirely possible to obtain an approximation to a function as a linear combination of Gaussian functions which is quite good "in the mean" but is in fact rather poor at certain points. This can be crucially important in studies of, for example, the electric field gradient at the nucleus. (Although such difficulties in this particular case would be alleviated by using a finite nuclear model.)

5. Gaussian basis functions

5.1. SPHERICAL HARMONIC AND CARTESIAN GAUSSIAN BASIS FUNCTIONS

Two types of Gaussian basis functions are in widespread use - spherical harmonic Gaussian basis functions and Cartesian Gaussian basis functions.

Spherical harmonic Gaussian-type functions are usually are written in the form

$$\chi_{k\ell m} = R_\ell(r)\, r^\ell\, Y_\ell^m(\theta,\varphi) \tag{17}$$

where the $Y_\ell^m(\theta,\varphi)$ are normalized spherical harmonics and the radial functions, $R_\ell(r)$, have the form

$$R_\ell(r) = \{(\zeta)^{2\ell+3}\, 2^{4\ell+7}/\pi(2\ell+1)!!^2\}^{\frac{1}{4}}\, \exp(-\zeta\, r^2) \tag{18}$$

with the exponent, $\zeta > 0$.

Cartesian Gaussian-type functions are usually written in the form

$$N_{\ell mn}(\zeta) x^\ell y^m z^n \exp(-\zeta r^2) \tag{19}$$

with ℓ, m and n being integers greater than or equal to zero. The normalization factor is given by

$$\left[\frac{1}{N_{\ell mn}(\zeta)}\right]^2 = \zeta^{-\frac{2(\ell+m+n)+3}{2}} \left(\frac{\pi}{2}\right)^{\frac{3}{2}} (2\ell-1)!!\,(2m-1)!!\,(2n-1)!! \tag{20}$$

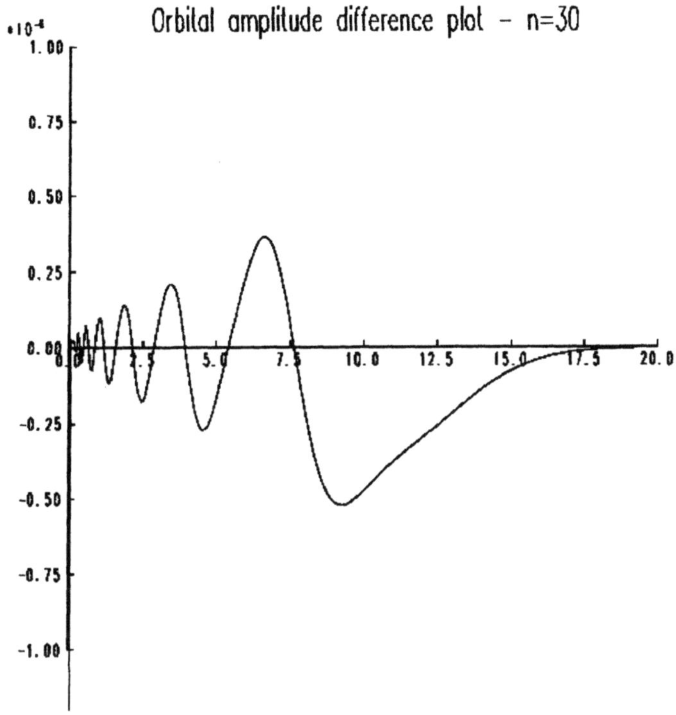

Figure 2. Orbital amplitude difference function for the ground state of the hydrogen atom described by a linear combination of thirty s-type Gaussian functions taken from a systematically generated sequence of even-tempered basis sets.

s-type Gaussian basis functions correspond to $\ell = 0$ in the case of spherical harmonic Gaussian functions, equation(17), and to $\ell = m = n = 0$ in the case of Cartesian Gaussian functions, equation(19). Primitive s-type Gaussian functions have identical form for both spherical harmonic and Cartesian Gaussian basis sets. Typical forms have already been displayed in Figure 1.

5.2. P-TYPE GAUSSIAN BASIS FUNCTIONS

Primitive p-type Gaussian functions also have identical form for both spherical harmonic and Cartesian Gaussian functions.

The unnormalized p_x-type Gaussian function takes the form

$$x \exp(-\zeta r^2) \tag{21}$$

In Figure 3, this function is plotted in the xy-plane for the case $\zeta = 1$. The origin is at the point (x,y)=(0,0). p-type Gaussian functions can be used to

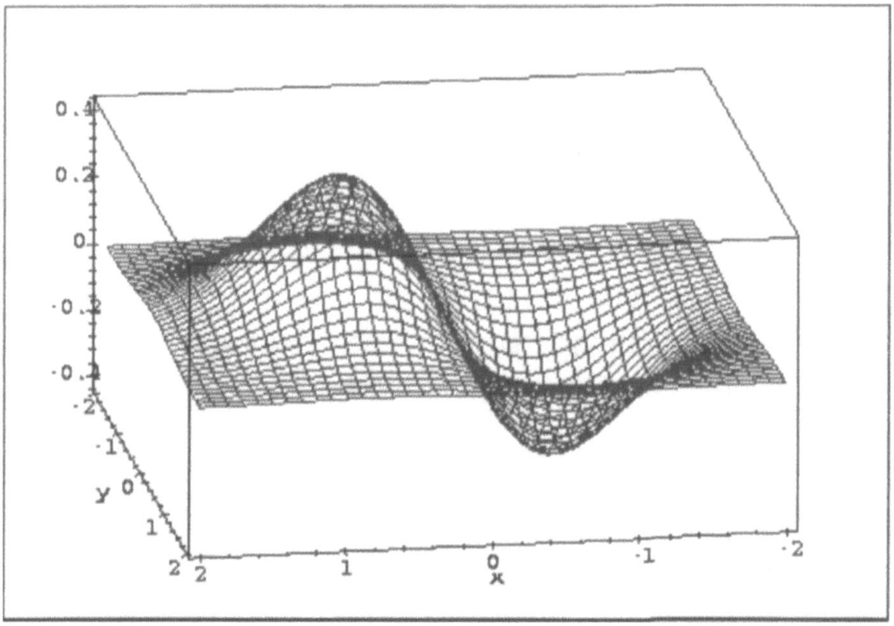

Figure 3. An unnormalized p_x Gaussian function plotted in the xy-plane.

approximate atomic functions of p-symmetry or to describe the polarization of s-type functions in a molecular environment or in a field.

In Figure 4 normalized p_x-type functions are plotted along the x-axis for different choices of ζ; in particular, for $\zeta = \frac{1}{2}$, 1, 2 and 4. The function has positive (negative) amplitude for positive (negative) values of x and a node at the origin. The larger the value of ζ the larger the amplitude of the p_x-type function and the smaller the value of x at which it occurs.

5.3. D-TYPE CARTESIAN GAUSSIAN BASIS FUNCTIONS

There are six Cartesian d-type Gaussian functions for which

$$\ell + m + n = 2 \tag{22}$$

in expression(19). Explicitly, they have the forms

$$x^2 \exp(-\zeta r^2) \tag{23}$$

$$xy \exp(-\zeta r^2) \tag{24}$$

$$y^2 \exp(-\zeta r^2) \tag{25}$$

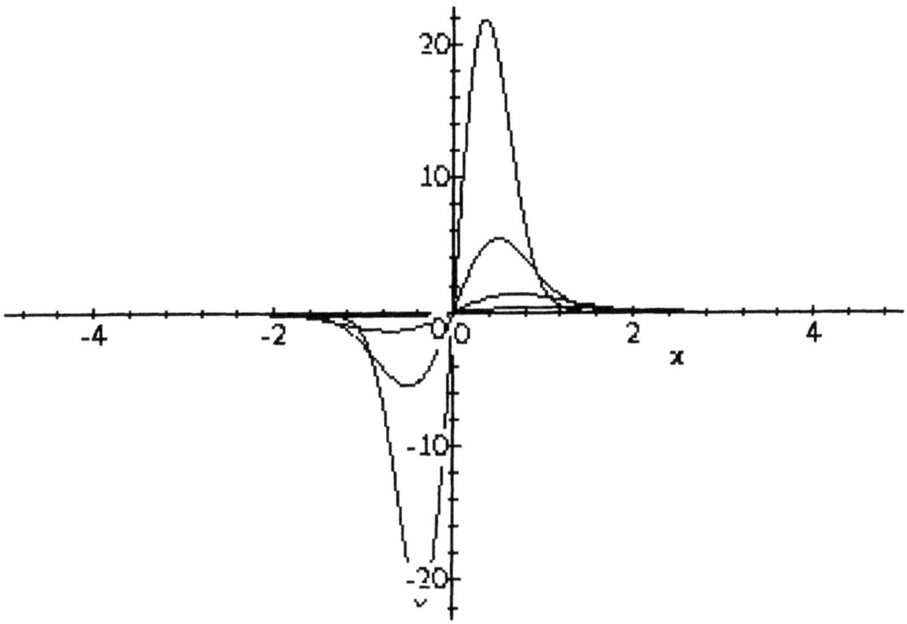

Figure 4. Normalized p-type Gaussian functions with exponents, ζ, of $\frac{1}{2}$, 1, 2 and 4.

$$xz \exp(-\zeta\, r^2) \tag{26}$$

$$yz \exp(-\zeta\, r^2) \tag{27}$$

and

$$z^2\ \exp(-\zeta\, r^2) \tag{28}$$

A plot of the first of the Cartesian d-type functions (d_{x^2}) on the x-axis is provided in Figure 5.
A plot of the Cartesian d_{xy} function on the xy plane is given in Figure 6.

5.4. F-TYPE CARTESIAN GAUSSIAN BASIS FUNCTIONS

There are ten f-type Cartesian Gaussian functions for which

$$\ell + m + n = 3 \tag{29}$$

in expression(19). The f-type Cartesian Gaussian functions have the form

$$x^3\ \exp(-\zeta\, r^2) \tag{30}$$

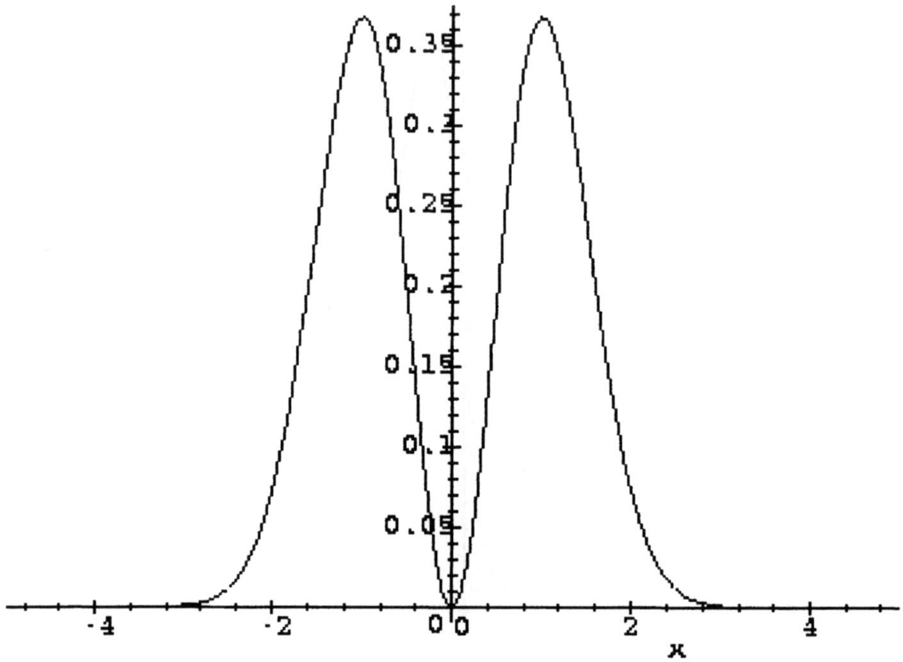

Figure 5. An unnormalized d_{x^2} cartesian Gaussian function on the x-axis.

$$x^2\ y\ \exp(-\zeta\ r^2) \tag{31}$$

$$x^2\ z\ \exp(-\zeta\ r^2) \tag{32}$$

$$x\ y^2\ \exp(-\zeta\ r^2) \tag{33}$$

$$x\ y\ z\ \exp(-\zeta\ r^2) \tag{34}$$

$$x\ z^2\ \exp(-\zeta\ r^2) \tag{35}$$

$$y^3\ \exp(-\zeta\ r^2) \tag{36}$$

$$y^2\ z\ \exp(-\zeta\ r^2) \tag{37}$$

$$y^2\ z\ \exp(-\zeta\ r^2) \tag{38}$$

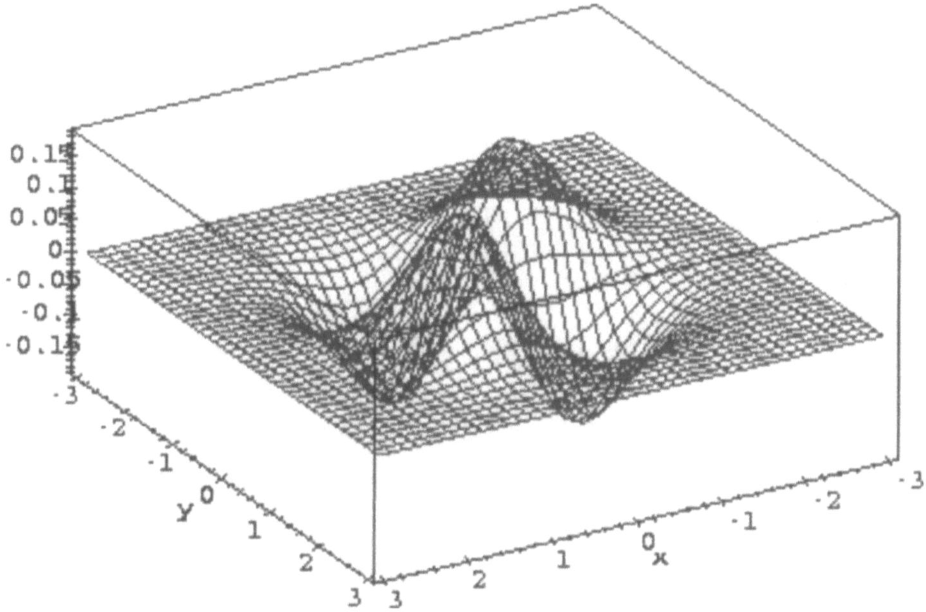

Figure 6. An unnormalized d_{xy} Gaussian function plotted in the xy-plane.

and

$$z^3 \exp(-\zeta\, r^2) \tag{39}$$

A plot showing the first of these Cartesian f-type functions is shown in Figure 7. This function is also displayed in Figure 8 where it is plotted in turn on the x axis, the y axis and the z axis.

5.5. D-TYPE SPHERICAL HARMONIC GAUSSIAN BASIS FUNCTIONS

There are five d-type spherical harmonic Gaussian functions which can be obtained by orthogonalization of the d-type Cartesian Gaussian functions. Specifically, the d-type spherical harmonic Gaussian functions have the form

$$(3z^2 - r^2) \exp(-\zeta\, r^2) \tag{40}$$

$$(x^2 - y^2) \exp(-\zeta\, r^2) \tag{41}$$

$$xy \exp(-\zeta\, r^2) \tag{42}$$

$$xz \exp(-\zeta\, r^2) \tag{43}$$

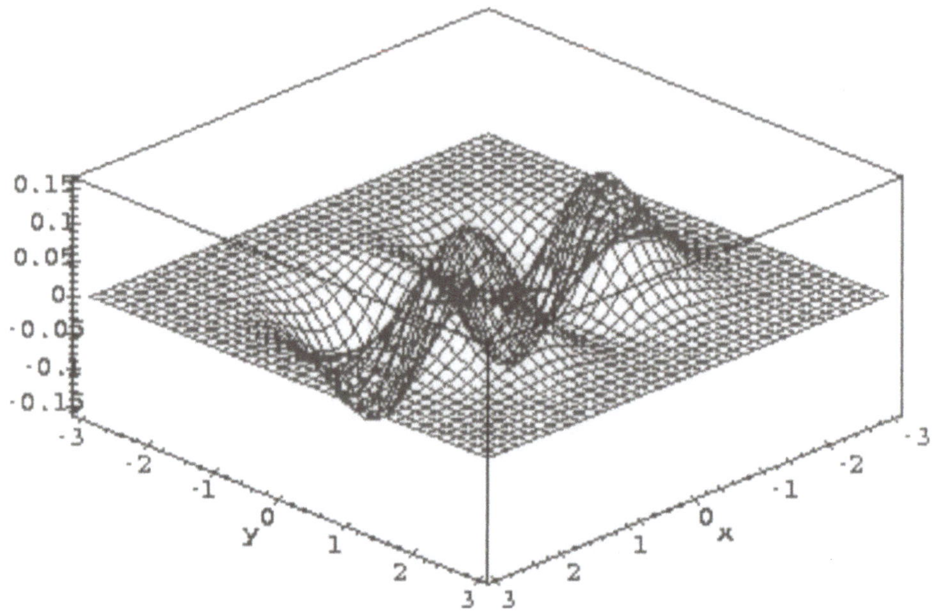

Figure 7. The f-type cartesian Gaussian function $x^2y\exp(-\zeta r^2)$ plotted in the xy-plane.

$$yz\exp(-\zeta\ r^2) \tag{44}$$

A sixth spherical harmonic Gaussian function can be formed by taking linear combinations of the d-type Cartesian Gaussian functions. This function has the form

$$(x^2+y^2+z^2)\exp(-\zeta\ r^2) = r^2\ \exp(-\zeta\ r^2) \tag{45}$$

and can be seen to be a function with s-symmetry. In accurate calculations demanding large basis sets numerical problems associated with the linear dependence of the s-type functions and that linear combination of Cartesian d-type which has s-symmetry can arise. It is for this reason that Cartesian Gaussian functions are not suitable for accurate calculations.

5.6. F-TYPE SPHERICAL HARMONIC GAUSSIAN BASIS FUNCTIONS

Cartesian Gaussian-type functions of f-type can be orthogonalized to give seven functions corresponding to spherical harmonic Gaussian-type functions of f-type. Specifically, the f-type spherical harmonic Gaussian-type functions have the form

$$(5x^2-3r^2)x\ \exp(-\zeta\ r^2) \tag{46}$$

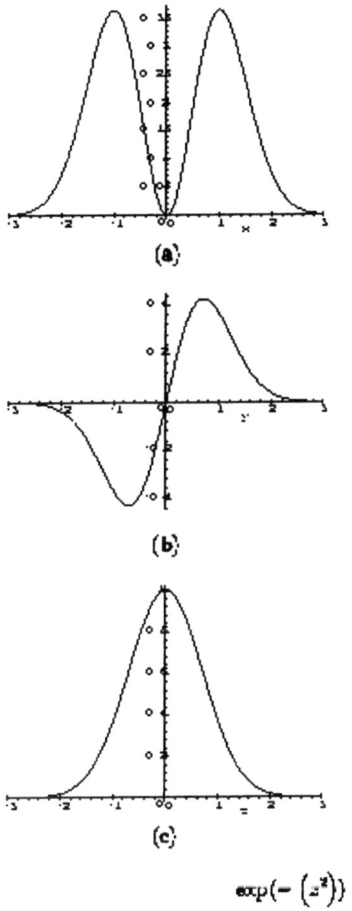

Figure 8. The f-type cartesian Gaussian function plotted on *(a)* the x axis, *(b)* the y axis, *(c)* the z axis.

$$(5y^2 - 3r^2)y \ \exp(-\zeta \ r^2) \tag{47}$$

$$(5z^2 - 3r^2)z \ \exp(-\zeta \ r^2) \tag{48}$$

$$(y^2 - z^2)x \ \exp(-\zeta \ r^2) \tag{49}$$

$$(z^2 - x^2)y \ \exp(-\zeta \ r^2) \tag{50}$$

$$(x^2 - y^2)z \ \exp(-\zeta \ r^2) \tag{51}$$

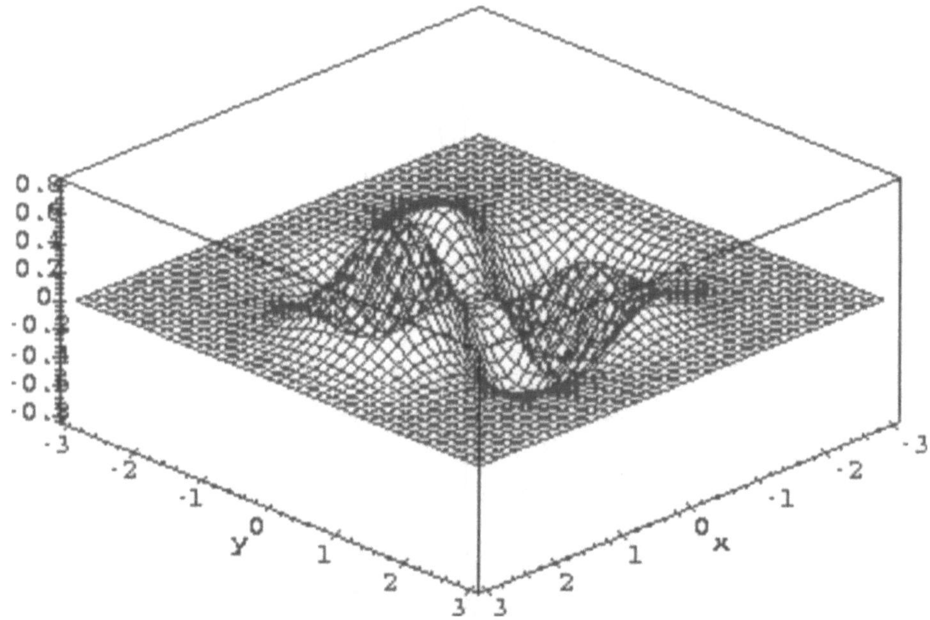

Figure 9. The f-type spherical harmonic Gaussian function plotted in the xy-plane.

and

$$xyz \exp(-\zeta\ r^2) \tag{52}$$

¿From the ten Cartesian Gaussian functions of f-type three linear combinations can be taken which correspond to p-type spherical harmonic Gaussian functions. Specifically, the p-type spherical harmonic Gaussian functions have the form

$$x\ r^2\ \exp(-\zeta\ r^2) \tag{53}$$

$$y\ r^2\ \exp(-\zeta\ r^2) \tag{54}$$

$$z\ r^2\ \exp(-\zeta\ r^2) \tag{55}$$

In Figure 9, the first of the f-type spherical harmonic Gaussian function given above is plotted in the xy-plane. This function is also plotted in Figure 10 where it is shown on the x, y and z axes.

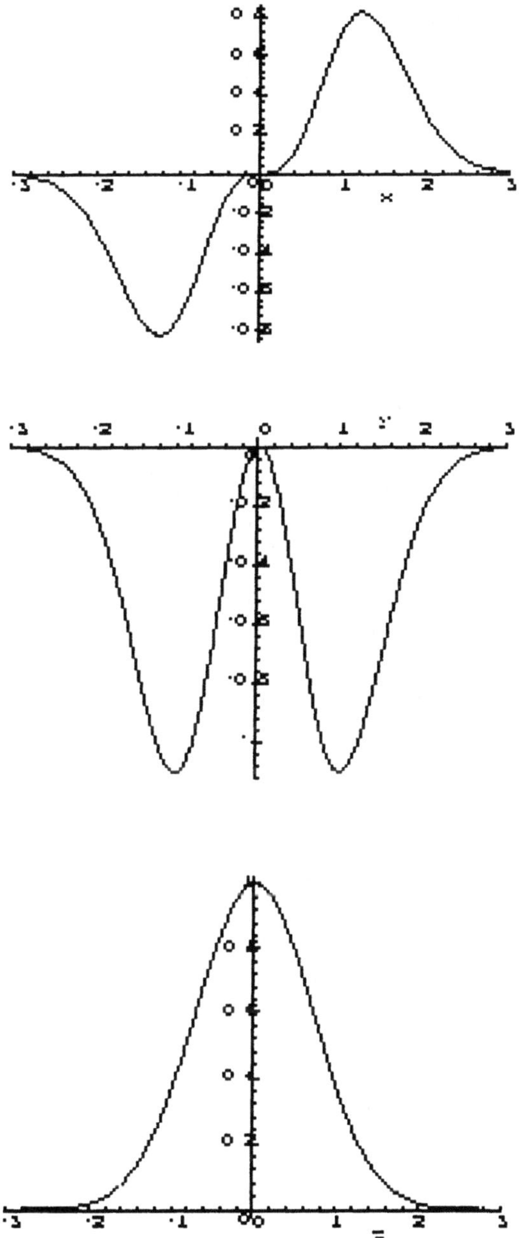

Figure 10. The f-type spherical harmonic Gaussian function plotted on *(a)* the x axis, *(b)* the y axis, *(c)* the z axis.

6. Gaussian basis sets

The selection of the basis set $[\chi_p, p = 1, ..., N]$ for a particular molecular electronic structure calculation puts the "art" into quantum chemistry. Most often the basis set is constructed from functions which have been optimized for an atomic system supplemented with so-called polarization functions to describe the distortion of the atomic orbitals in a molecular environment. In contemporary quantum chemical studies the choice of basis set is an integral part of the assumed model and a knowledge of the capabilities of that basis set is vital. Davidson and Feller point out that[8]

> *Occasionally, for reasons of economy or computational feasibility, a basis set is selected for which the computed results are nearly without significance.*

In this section, the construction of basis sets is considered first for the Hartree-Fock model and then for studies which take account of electron correlation effects. Some of the more frequently used prescriptions for constructing molecular basis sets are briefly reviewed.

The construction of systematic sequences of even-tempered basis sets and universal basis sets has been shown to provide an avenue for the progressive refinement of electronic structure studies both within the Hartree-Fock approximation and beyond. However, high precision calculations are known to require the use of higher harmonics when a basis set is constructed exclusively from atom-centred subsets. Recent work has shown that there is some advantage in using basis subsets centred on off-atom positions and the concept of the distributed basis set has been shown to be useful. Particular care is required in the study of atoms and molecules in electric (and magnetic) fields and the use field dependent basis set is known to provide an efficient procedure which avoids the need to supplement the basis set used to describe the atom or molecule in zero field with a large and flexible set of polarization functions to describe the effects of the applied field. Electric field variant basis sets have also be used to described the effect of the field resulting from the other nuclei on the basis set associated with a particular nucleus. Finally, the problem of basis set construction for heavy elements and in particular for relativistic electronic structure calculations on molecules containing heavy elements is briefly addressed. This topic is discussed more fully in Part IV[48].

6.1. MATRIX HARTREE-FOCK CALCULATIONS

The accuracy of the total electronic energy obtained in matrix Hartree-Fock calculations is often employed as a measure of the quality of the basis set employed in such calculations. Using the ground state of the carbon dioxide molecule as a prototype, the results of calculations carried out using a basis set which is capable of supporting an accuracy approaching the Hartree-Fock limit can be used to access the accuracy achieved in matrix Hartree-Fock calculations carried out using some of the "standard" basis sets which are commonly employed in molecular electronic structure studies. The GAUSSIAN[49] program in the UNICHEM[50] environment has been used by Moncrieff and Wilson[51] to explore the accuracy with which the total energy is determine by exploring a range of the pre-define basis sets available in these programs. The results are collected in Table 7.

The basis sets consisting of a minimum set of Slater-Type Orbitals each approximated by a linear combination of three primitive Gaussian functions and, following Pople and his coworkers[52] [53] designated "STO-3G", were the smallest employed in Table 7. For first row atoms these basis sets consist of two contractions of 3 s-type Gaussian primitive and a single contraction of 3 p-type Gaussian primitives obtained by least-squares fitting of Slater-type functions and with the exponents of the valence shell s- and p-type primitive functions being constrained to be equal. A small "split-valence" basis set, designated "3-21G" by Pople and his coworkers[54] [55] [56], was employed. This basis sets contains the same number of Gaussian primitive functions as the STO-3G sets, the core orbital being a contraction of 3 primitive functions and the valence shell being described by two basis functions; one a contraction of two primitive functions and the second an uncontracted function. Again, the exponents of the valence shell primitive s- and p-type functions being constrained to be equal. In the "4-31G" "split-valence" basis set, the core function is a contraction of four primitive functions and the two valence functions are a contraction of three primitives and an uncontracted function. Basis sets of the type commonly designated "6-31G", due to Pople and his coworkers[57] [58] [59] [60] [61] were also used. The convention that * and $^+$ denote basis sets which have been augmented with polarization functions and diffuse function, respectively, is widely employed. In Table 7, supplementary polarization functions are defined parenthetically. The "6-311" basis sets from Pople and his coworkers were also used. Two double-zeta basis sets proposed by Dunning[62] were included in Table 7. The basis sets designated "D95V" is described as a valence double-zeta set whilst that designated "D95" is a full double zeta set. The "correlation-consistent" Gaussian basis sets developed by Dunning and his co-workers[63]$^-$[67] have been specifically designed to treat correlation effects. They are a sequence of contracted basis sets in which the original prescription of Raffenetti[68] using the occupied atomic self-consistent field orbitals is employed together with supplementary uncontracted primitives and polarization functions. The correlation consistent basis sets employed in the matrix Hartree-Fock calculations carried out in the work of Moncrieff and Wilson[51] are those designated *cc*-pVDZ, *cc*-pVTZ, *cc*-pVQZ. In Table 7, E is the total electronic energy given by the matrix Hartree-Fock model for the carbon dioxide ground state and ε_{BSTE} is the basis set truncation error - the difference between the energy obtained with a given basis set and that obtained with the basis set designed by Moncrieff and Wilson[51] to approach the Hartree-Fock limit for the energy.

The Moncrieff-Wilson basis set for the CO_2 ground state was developed from a molecular basis set set for the CO ground state[23] which had been shown to yield an accuracy approaching the sub-μhartree level for the energy in comparisons with the results of finite difference calculations. The basis set contains functions of s, p, d and f symmetry centred on each of the atoms in the molecule together with functions of s, p and d symmetry at the two bond centres. The basis set is written as $28s14p14d14f$: $24s10p11d$ *bc*, where in the notation A : B *bc*, A denotes the atom-centred basis subsets and B denotes the bond-centred subsets.

Table 7
The accuracy of matrix Hartree-Fock calculations for the ground state of the carbon dioxide molecule using certain standard basis sets of Cartesian Gaussian-type functions.

Basis set	$-E/hartree$	$\varepsilon_{BSTE}/hartree$
STO-$3G$	−185.064696	2.6607
3-21G	−186.561175	1.1642
4-31G	−187.327956	0.3975
6-31G	−187.514949	0.2105
6-31$G(d)$	−187.632831	0.0926
6-31$G(2d)$	−187.644118	0.0813
6-31$G(3d)$	−187.649951	0.0755
6-31$G(3df)$	−187.658687	0.0667
6-31 + $G(3df)$	−187.664125	0.0613
6-311G	−187.566207	0.1592
6-311 + $G(3df)$	−187.708745	0.0167
$D95V$	−187.552962	0.1725
$D95$	−187.553282	0.1721
cc-$pVDZ$	−187.651108	0.0743
aug-cc-$pVDZ$	−187.662824	0.0626
cc-$pVTZ$	−187.707256	0.0182
aug-cc-$pVTZ$	−187.709507	0.0159

It is also of interest to compare the energy errors documented in Table 7 with the correlation energies estimated by means of many-body perturbation theory using some of the "standard" basis sets. (All orbitals are taken to be "active" in the correlation energy calculations reported in the remainder of this subsection and Cartesian Gaussian basis functions are employed.) For the $STO-3G$ basis set, a basis set not designed for correlation studies, the second order energy is $E_2 \simeq -0.1995$ Hartree, which is just 7.5% of the corresponding basis set truncation error (2.6607 Hartree). On the other hand, for the basis set $6-31G(2d)$ the basis set truncation error, ε_{BSTE}, is 0.0813 Hartree and the coefficients in the many-body perturbation expansion through fourth order are $E_2 \simeq -0.5438$ Hartree, $E_3 \simeq +0.0144$ Hartree and $E_4 \simeq -0.0404$ Hartree. It can seen, therefore, that ε_{BSTE} has a magnitude of $\sim 15\%$ of E_2 but is larger than the higher order energy coefficients E_3 is $\sim 17.7\%$ of ε_{BSTE} and E_4 is $\sim 49.7\%$ of

6.2. ELECTRON CORRELATION ENERGY CALCULATIONS

It has been recognized for many years that the calculation of correlation energies of high accuracy demands the use of large and flexible basis sets. This requirement, in turn, demands not only the availability of powerful computing machines but also appropriate theories, together with associated algorithms, which are capable of effectively exploiting their capabilities. It is perhaps worth noting that for the large basis sets which are required for high precision calculations full configuration interaction calculations would appear impossible. (Weyl's number for a fourteen electron singlet systems described by 500 basis functions is $\sim 2.8 \times 10^{29}$.) Similarly, quadratic configuration interaction or coupled cluster calculations with large and flexible basis sets would appear intractable.

On the other hand, low order many-body perturbation theory[73] [74] [29] [75] [76] using the large, flexible basis sets capable of supporting high precision in electron correlation energy studies can be carried out with a standard commercially available code implemented on typical contemporary computing machines. The many-body perturbation theory lends itself to effective implementation on parallel processing computers[77] [78] which have been recognized to show considerable potential when effectively implemented in appropriate hardware and exploited through software. It is known that by using localized orbitals, both occupied and unoccupied, the many-body perturbation theoretic approach can provide a useful technique for large molecules which can be further improved by exploiting techniques such as Feenberg scaling[79]$^-$[92]. The use of superfluid reference functions opens the route to treatments of dissociative processes[93] [94]. The relativistic extension of the present approach via the Furry picture bound state picture of quantum electrodynamics is well established[95]-[98].

In Table 8, the results of matrix Hartree-Fock calculations for the ground state of the nitrogen molecule obtained by employing various basis sets which have been designed to describe electron correlation effects are compared with the results obtained in the near Hartree-Fock limit study by Moncrieff and Wilson[99] (see also [100]). The second-order correlation energy components corresponding to the basis sets cited in Table 8 are given in Table 9 where a comparison with the near second-order limit results of Moncrieff and Wilson[99] is made. In Tables 10 and 11 the deviations between the energies obtained with a given basis set and the limiting values of Moncrieff and Wilson are given for the matrix Hartree-Fock energies and the second-order correlation energy components, respectively.

Table 8

Comparison of the matrix Hartree-Fock energy for the nitrogen molecule ground state obtained by using various published basis sets which were designed to describe electron correlation effects with the results obtained in the near Hartree-Fock limit study by Moncrieff and Wilson[99]

Basis set	N	M	E_{HF}
cc-pVDZ [a]	28	78	-108.954 553 2
aug-cc-pVDZ [b]	46	96	-108.961 045 0
cc-pVTZ [c]	60	120	-108.984 093 4
aug-cc-pVTZ [d]	92	152	-108.985 317 4
cc-pVQZ [e]	110	184	-108.991 735 3
aug-cc-pVQZ [f]	160	234	-108.992 205 2
cc-pV5Z [g]	182	276	-108.993 419 8
aug-cc-pV5Z [h]	254	348	-108.993 610 5
Partridge 1 [i]	82	82	-108.909 689 8
Partridge 2 [i]	98	98	-108.910 752 8
Partridge 3 [i]	114	114	-108.910 642 7
aug-DZ-ANO [j]	46	354	-108.980 259 8
aug-TZ-ANO [k]	92	560	-108.990 606 5
ANO [l]	110	574	-108.992 222 3
Moncrieff & Wilson 1[m]			-108.9938245
Moncrieff & Wilson 2[n]			-108.9937920

[a] correlation consistent-polarized Valence Double Zeta basis set due to Dunning[63]
[b] augmented correlation consistent-polarized Valence Double Zeta basis set due to Dunning et al[63] [64]
[c] correlation consistent-polarized Valence Triple Zeta basis set due to Dunning[63]
[d] augmented correlation consistent-polarized Valence Triple Zeta basis set due to Dunning et al[63] [64]
[e] correlation consistent-polarized Valence Quadruple Zeta basis set due to Dunning[63]
[f] augmented correlation consistent-polarized Valence Quadruple Zeta basis set due to Dunning et al[63] [64]
[g] correlation consistent-polarized Valence Quintuple Zeta basis set due to Dunning[63]
[h] augmented correlation consistent-polarized Valence Quintuple Zeta basis set due to Dunning et al[63] [64]
[i] Basis set due to Partridge[69]
[j] augmented Double Zeta Atomic Natural Orbital basis set due to Widmark et al[70] [71]
[k] augmented Triple Zeta Atomic Natural Orbital basis set due to Widmark et al[70] [71]
[l] Atomic Natural Orbital basis set due to Almlof and Taylor[72]
[m] From the work of Moncrieff and Wilson[99] using *spdf* : *spdf bc* basis set with n=15
[n] From the work of Moncrieff and Wilson[99] using *spdfgh* : *spdfgh bc* basis set with $n = 10$

Table 9

Comparison of the second-order correlation energies for the nitrogen molecule ground state obtained by using various published basis sets with were designed to describe electron correlation effects with the results obtained in the near second-order energy limit by Moncrieff and Wilson[99].[a]

Basis set	N	M	E_2
cc-pVDZ [a]	28	78	-0.309 596 8
aug-*cc*-pVDZ [b]	46	96	-0.320 927 6
cc-pVTZ [c]	60	120	-0.398 573 9
aug-*cc*-pVTZ [d]	92	152	-0.408 650 7
cc-pVQZ [e]	110	184	-0.455 105 8
aug-*cc*-pVQZ [f]	160	234	-0.458 837 8
cc-pV5Z [g]	182	276	-0.477 699 3
aug-*cc*-pV5Z [h]	254	348	-0.479 908 5
Partridge 1 [i]	82	82	-0.350 614 4
Partridge 2 [i]	98	98	-0.351 115 1
Partridge 3 [i]	114	114	-0.351 442 0
aug-DZ-ANO [j]	46	354	-0.329 320 0
aug-TZ-ANO [k]	92	560	-0.408 052 0
ANO [l]	110	574	-0.440 545 5
Moncrieff & Wilson 1[m]			-0.5201846
Moncrieff & Wilson 2[n]			-0.5304337

[a] See notes in Table 8

Table 10

Deviation of the ground state matrix Hartree-Fock energies for the nitrogen molecule obtained using various standard basis sets of Gaussian functions from the "near Hartree-Fock limit" results of Moncrieff and Wilson[99].[a]

Basis set	$\delta(E_{HF})$
cc-pVDZ [a]	39 271.3
aug-*cc*-pVDZ [b]	32 779.5
cc-pVTZ [c]	9 731.1
aug-*cc*-pVTZ [d]	8 507.1
cc-pVQZ [e]	2 089.2
aug-*cc*-pVQZ [f]	1 619.3
cc-pV5Z [g]	404.7
aug-*cc*-pV5Z [h]	214.0
Partridge 1 [i]	909 689.8
Partridge 2 [i]	83 071.7
Partridge 3 [i]	83 181.8
aug-DZ-ANO [j]	13 564.7
aug-TZ-ANO [k]	3 218.0
ANO [l]	1 602.2
Moncrieff & Wilson 1[m]	0.0
Moncrieff & Wilson 2[n]	32.5

[a]See notes in Table 8

Table 11
Deviation of the ground state second order correlation energies for the nitrogen molecule obtained using various standard basis sets of Gaussian functions from the "near second-order limit" results of Moncrieff and Wilson[99].[a]

Basis set	$\delta(E_2)$
cc-pVDZ [a]	220836.9
aug-cc-pVDZ [b]	209506.1
cc-pVTZ [c]	131859.8
aug-cc-pVTZ [d]	121783.0
cc-pVQZ [e]	75327.9
aug-cc-pVQZ [f]	71595.9
cc-pV5Z [g]	52734.4
aug-cc-pV5Z [h]	50525.2
Partridge 1 [i]	179819.3
Partridge 2 [i]	179318.6
Partridge 3 [i]	178991.7
aug-DZ-ANO [j]	201113.7
aug-TZ-ANO [k]	122381.7
ANO [l]	89888.2
Moncrieff & Wilson 1[m]	10249.1
Moncrieff & Wilson 2[n]	0.0

[a]See notes in Table 8

6.3. SYSTEMATIC SEQUENCES OF EVEN-TEMPERED GAUSSIAN BASIS SETS

The controlled reduction of basis set truncation errors is an essential prerequisite of a quantitative quantum chemistry and recent years have witnessed a growing interest in the systematic implementation of the algebraic approximation [16] [101], *i.e.* finite basis set expansions, in both non-relativistic and relativistic studies.

Basis set truncation effects are widely recognized as the main source of error in contemporary molecular electronic structure calculations. Over thirty years ago in 1963, Schwartz [102] emphasized that

> ... *an orderly plan of attack is... essential for any mathematical analysis of the convergence rate. The old habit of picking the "best" (chosen by art) choice of a fixed number of terms is to be discarded if one wants to see how the problem converges.*

He continued

> *One must choose a set of basis functions and then gradually add more ... in some systematic manner.*

Ruedenberg and his co-workers [103] pointed out that even-tempered basis sets[1] afforded a systematic method for generating atomic orbital basis sets [104]–[108]. Feller and Ruedenberg [109] and Schmidt and Ruedenberg [46] showed that sequences of such basis sets can be constructed that systematically approach a complete set and thus approach the Hartree-Fock limit.

There is a theorem in mathematics which states that the monomials $\left\{x^{\lambda_n}\right\}_{n=1}^{\infty}$, where $\lambda_n > -\frac{1}{2}$ $(n = 1, 2, 3, ...)$ are fixed distinct real numbers, form a complete set in $L^2\left[0,1\right]$ if and only if

$$\sum_{n=1}^{\infty} \frac{1+2\lambda_n}{1+\lambda_n^2} = \infty \tag{56}$$

is valid. This is the Müntz-Szász theorem [114], [120].

In molecular structure calculations the basis sets most frequently employed consists of Gaussian-type functions which at their simplest have the form

$$\chi \sim \exp(-\zeta_k r^2) \tag{57}$$

in which ζ_k is an orbital exponent. Now Klahn [120] has presented a generalized Müntz-Szász theorem which essentially states that a set of Gaussian-type functions is asymptotically complete in $L\left(R^+\right)$, if and only if

$$\Gamma(N) = \sum_{k=1}^{N} \frac{\zeta_k}{1+\zeta_k^2} \tag{58}$$

and

$$\lim_{N\to\infty} \Gamma(N) = \infty \tag{59}$$

holds.

Now it has been observed that exponents which have been carefully optimized for atoms often form a good approximation to a geometric progression

$$\zeta_k = \alpha\beta^k,\ \beta > 1,\ k = 1, 2, ..., N \tag{60}$$

Conversely, if it is assumed that the exponents do form a geometric progression and the parameters α and β are optimized for atoms then there is found to be little lost in accuracy. Basis sets developed in this way are termed even-tempered basis sets (for a discussion see [16]) and open up the possibility of constructing the large and flexible basis sets that are inevitably required for calculations of high precision.

[1]Even-tempered Gaussian basis sets were first suggested by R. McWeeny in his *D. Phil.* thesis (Oxford, 1948).

As the number of basis functions, N, is increased we require that our basis set approach a complete set. The generalized Müntz-Szász theorem can be used to show that this is not the case if α and β are held fixed as N is increased. One possible choice which does lead to a complete set as $N \to \infty$ is

$$\begin{aligned} \alpha &\to 0 \\ \beta &\to 1 \\ \beta^N &\to \infty \end{aligned} \tag{61}$$

and these limits can be guaranteed by generating successive basis sets according to the following empirical recursions [16]

$$\alpha_N = \left[\frac{\beta_N - 1}{\beta_{N-1} - 1}\right]^a \alpha_{N-1}, \; a > 0 \tag{62}$$

$$\ln \beta_N = \left[\frac{N}{N-1}\right]^b \ln \beta_{N-1}, \; -1 < b < 0 \tag{63}$$

This approach is illustrated in Figure 11 where a systematic sequence of even-tempered basis sets of Gaussian-type functions is used to calculate the Hartree-Fock energy for the ground state of the F^- ion[121]. In this Figure, the notation $A[B]$ indicates that system A is studied using a sequence of basis sets originally designed for B. $[B']$ indicates that a diffuse function is added to basis sets $[B]$. $[B'']$ indicates that two diffuse function of each symmetry type are added to the basis set $[B]$. It can be seen that provided one diffuse function of each symmetry type is added to the basis set designed for the neutral species, the anion can be accurately described within the Hartree-Fock model within the algebraic approximation.

6.4. DISTRIBUTED GAUSSIAN BASIS SETS

Molecular basis sets are almost invariably constructed from atomic basis sets. In 1960, Nesbet [122] pointed out that molecular basis sets containing only basis functions types necessary to reach the atomic Hartree-Fock limit, the so-called isotropic basis set, cannot possibly account for polarization in molecular binding. Two approaches to the problem of designing molecular basis sets can be identified:-

a) the addition of atom-centred polarization functions to the atomic basis sets

b) the addition of off-centre functions of the same symmetry as the atomic Hartree-Fock basis set.

In the early sixties, Reeves, Harrison and Fletcher [123]–[125] established important elements of the current 'mainstream' usage of Gaussian basis sets. They concluded that [126]

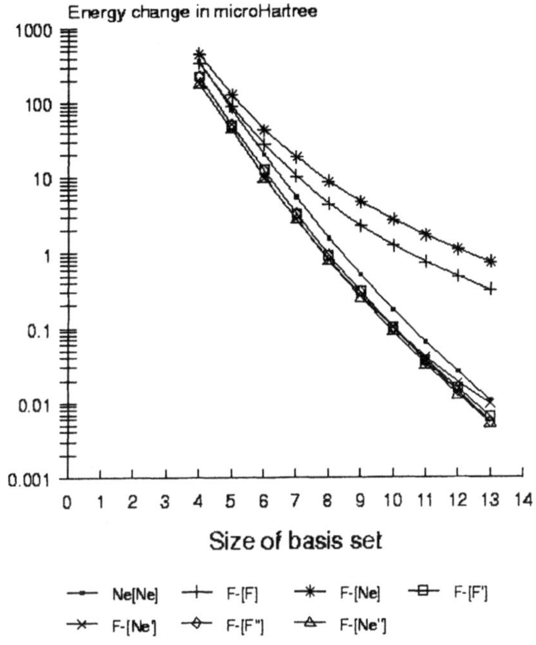

Figure 11. Convergence of the matrix Hartree-Fock energy for the F^- ion with increasing size of basis set. See text for details.

> *no significant advantage was gained by letting the basis functions float away from the nuclear centres, that orbital exponents should be distributed approximately in geometric progression, and that it is not advantageous to optimize separate sets of Gaussians for the description of individual shells.*

These conclusions support approach *a)* and form the foundation of the majority of current mainstream practice (see the review by Shavitt [126] for a masterly overview). However, these conclusions were drawn from calculations using basis sets consisting of a small number of functions ($\sim 10 - 10^2$). Today, the so-called 'direct' algorithms[127], which avoid the storage of the two-electron integrals over the basis functions, have enabled calculations using in excess of 10^3 to be carried out on a workstation and calculations using basis sets of a size approaching 10^4 to be seriously contemplated. It is, therefore, timely to reconsider approaches *a)* and *b)* in the light of these developments. We note that these approaches are not mutually exclusive in that hybrid prescriptions for basis set design can be adopted.

A promising development is the distributed basis set[128] [129] in which the exponents and the distribution of the Gaussian functions are generated according to empirical prescriptions. It should be noted that Gaussian functions are particularly well suited for such a procedure since, unlike exponential basis functions, they do not introduce a cusp. This approach is distinct from the use Gaussian lobe functions[130] and the Floating Spherical Gaussian Orbital (F.S.G.O.) model[131]–[133]in that there is no attempt to mimic higher harmonics directly and non-linear optimization is avoided. This allows the use of large basis sets resulting in high precision. In this section, we describe two distributed Gaussian basis sets which have been discussed recently.

A particularly simple distributed basis set is the Gaussian Cell model, introduced by Murrell and his coworkers in 1974[134] [135], in which the functions are arranged on a regular lattice, with one function centred on each lattice point and all functions having the same exponent. Not surprisingly, this simple model does not achieve high accuracy. However, recent work on one-electron diatomic and triatomic systems have demonstrated how, with simple refinements[136]–[138] it can be very significantly enhanced and accuracies approaching 1 μHartree can be achieved.

An alternative method for generating a distributed basis set is based on a Laplace transform. The 'natural' basis functions for diatomic molecules are elliptical functions[141]-[150]. A Laplace transform[151], [152] can be used to obtain an integral representation of an elliptical function in terms of Gaussian functions. This Laplace transform provides a prescription for the generation of a systematic sequence of distributed basis sets of Gaussian functions from a sequence of atomic even-tempered sets[140].

Use of the Laplace transform for the elliptical function provides an exact representation of χ by means of a continuous distribution of Gaussian functions over the variables η_a and η_b:

$$\begin{aligned}\chi\left(r_a, r_b; \zeta_a, \zeta_b, R\right) &= \frac{1}{\sqrt{2\pi R^3}}\frac{\zeta_a\zeta_b}{4\pi}\int_0^\infty d\eta_a\int_0^\infty d\eta_b\,\eta_a^{-\frac{3}{2}}\eta_b^{-\frac{3}{2}} \\ &\left(-\frac{1}{4}\left[\frac{\zeta_a^2}{\eta_a}+\frac{\zeta_b^2}{\eta_b}\right]\right)\exp(-\{\eta_a r_a^2+\eta_b r_b^2\}) \quad (64)\end{aligned}$$

Equation (64) can be converted into a finite expansion in terms of Gaussian-type functions by performing the double integration with any one of a number of numerical quadrature schemes. In the expansion

$$\chi\left(r_a, r_b; \zeta_a, \zeta_b, R\right) \approx \sum_{p=1}^{N}\sum_{q=1}^{N} c_{ap}\, c_{bp}\exp(-\eta_p r_a^2)\exp(-\eta_q r_b^2) \quad (65)$$

the parameters η_p and c_p are specified by the chosen quadrature. Using the Gaussian product theorem[153], (65) may be written

$$\chi\left(r_a, r_b; \zeta_a, \zeta_b, R\right) \approx \sum_{p=1}^{N}\sum_{q=1}^{N} c_{ap}c_{bp}\gamma_{pq}\exp(-\eta_{pq} r_{pq}^2) \quad (66)$$

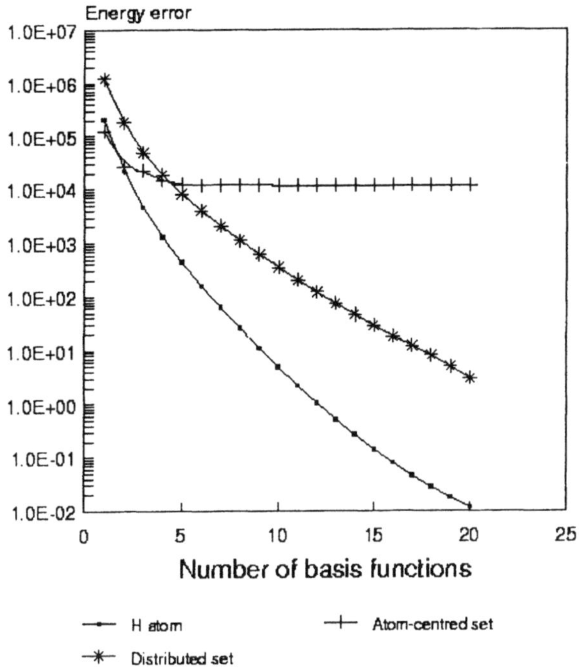

Figure 12. Convergence of the calculated energy of the hydrogen atom and the hydrogen molecular ion observed with a systematic sequence of even-tempered basis sets of 1s Gaussian functions. For the molecular system the basis sets are either atom-centred or distributed along the line passing through the nucleui according to the Laplace transform of an elliptical function.

where the coefficients γ_{pq} are given by

$$\gamma_{pq} = \exp\left(-\frac{\eta_p\eta_q}{\eta_p+\eta_q}R^2\right) \tag{67}$$

and the screening parameters are

$$\eta_{pq} = \eta_p + \eta_q \tag{68}$$

The coordinate r_{pq} is defined with respect to the point (x_{pq}, y_{pq}, z_{pq}) where

$$x_{pq} = \frac{\eta_p x_a + \eta_q x_b}{\eta_p + \eta_q} \tag{69}$$

with similar relations for y_{pq} and z_{pq}.

In Figure 12 the convergence of the calculated energy of the hydrogen atom and the hydrogen molecular ion observed with a systematic sequence

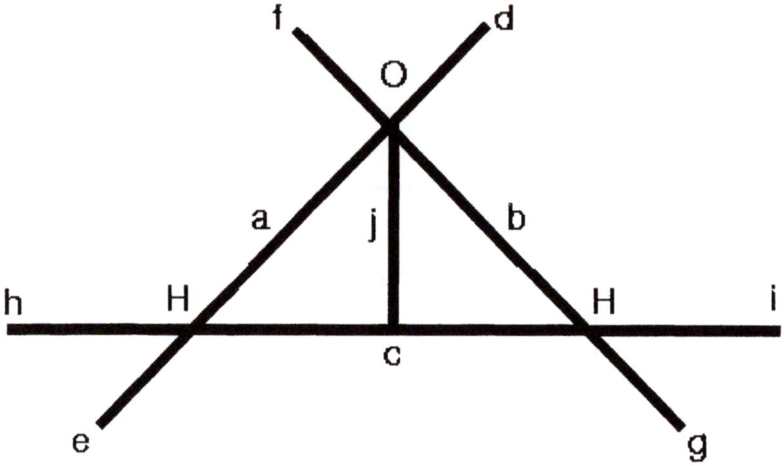

Figure 13. Distributed Gaussian basis set for the water molecule employed in the recent study by Moncrieff and Wilson [see text for details].

of even-tempered basis sets of s-type Gaussian functions is displayed. For the molecular system the basis sets are either atom-centred or distributed along the internuclear axis according to the Laplace transform of an elliptical function. The atom-centered basis set leads to an incorrect asymptotic limit for the H_2^+ ion whereas the distributed set converges to the correct limit although not as rapidly as for the hydrogen atom case.

The use of basis functions centred on points other than the atomic nuclei is not new. In particular, the utility of bond centre functions in matrix Hartree-Fock calculations has been known for many years[154]-[165]. In Figure 13, the distribution of basis subsets employed in a recent study by Moncrieff and Wilson [166] is shown. The points a, b and c are the mid-points of the OH_1, OH_2 and H_1H_2 vectors. The points $d - i$ lie beyond the atoms on these vectors, with $d - g$ being on the OH vectors and h and i lying on the vector HH. The point j lies at the mid-point of the line joining the point c to the oxygen nucleus. The basis subset centred on the O nucleus contains both s and p functions; all other subsets contain only s functions. In Figure 14A, the components of the matrix Hartree-Fock energy, with respect to atom-centred sets, when calculated with the sequence (i) O, H_1, H_2; (ii) a, b; (iii) c; (iv) d, e, f, g; (v) h, i; (designated sequence A) whilst in Figure 14B the sequence (i) O, H_1, H_2; (ii) a, b; (iii) c; (iv) j (designated sequence B) is considered. The latter sequence leads to a slightly lower energy. In both sequences the bulk of the energy recovered is associated with the basis subsets placed at the points a, b and c.

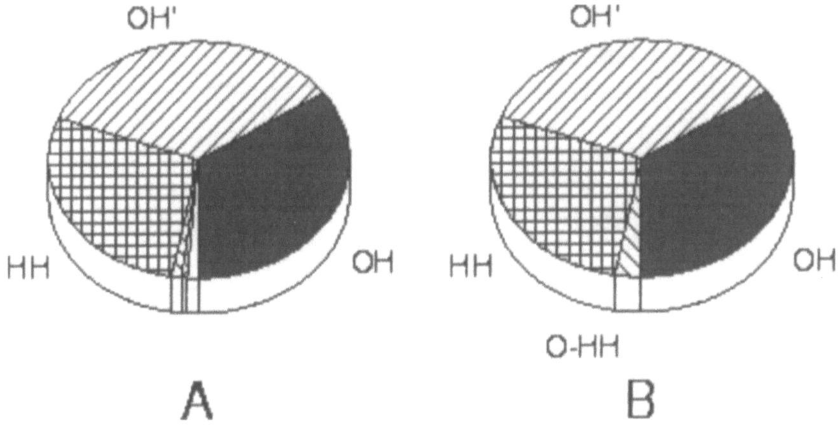

Figure 14. Components of the matrix Hartree-Fock energy when calculated with distributed Gaussian basis sets as follows: A- *(i)* O,H_1,H_2, *(ii)* a,b, *(iii)* c, *(iv)* d,e,f,g, *(v)* h,i; B- *(i)* O,H_1,H_2, *(ii)* a,b, *(iii)* c, *(iv)* j.

The components of the second-order many-body perturbation theory associated with the sequences A and B of distributed basis sets are shown in the pie charts in Figure 15. Again the sequence B leads to a slightly lower energy. However, in contrast to the situation for the matrix Hartree-Fock energy, a significant fraction of the correlation energy is associated in centres other than the mid-points of the OH_1, OH_2 and H_1H_2 vectors. In the case of sequence A, the subsets located at points d, e, f and g deliver a substantial fraction of the correlation energy whereas in sequence B, the subsets located at the centre j recover a sizeable fraction of the second-order energy. The reader is referred to the work of Moncrieff and Wilson[166] for further details.

In Figure 16 a comparison of second order correlation energy components reported in the work of Moncrieff and Wilson are compared with those reported by other workers. The approach used in generating each of the energies given in Figure 16 are:-

a) *spd*; *spd*; *spd* - functions of s, p and d symmetry on the O, H atom centres and the OH bond centre;

b) *spd*; *spd*; *spd*; *spd* - as in a) together with a set of functions of s, p and d symmetry at the HH mid-point;

c) *spdf*; *spd* - functions of s, p, d and f symmetry on the O atom centre and functions of s, p and d symmetry on the H atom centres;

d) *spdf*; *spd*; *spd* - as in c) together with a set of functions of s, p and d symmetry at the OH mid-point;

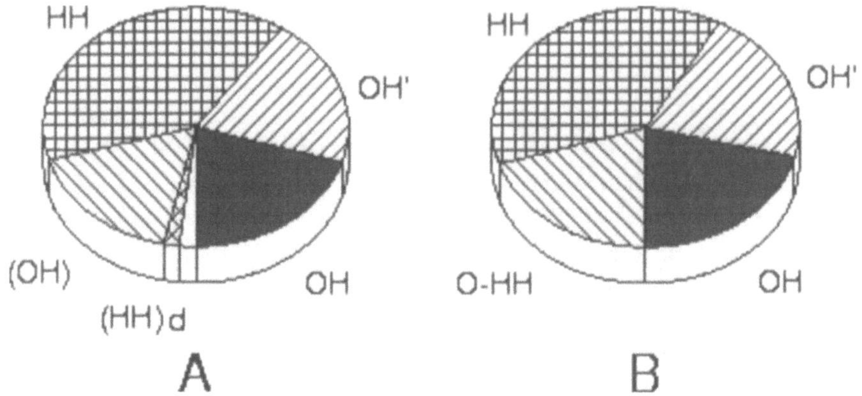

Figure 15. Components of the second-order correlation energy when calculated with distributed Gaussian basis sets as follows: A- *(i)* O,H_1,H_2, *(ii)* a,b, *(iii)* c, *(iv)* d,e,f,g, *(v)* h,i; B- *(i)* O,H_1,H_2, *(ii)* a,b, *(iii)* c, *(iv)* j.

e) - as in d) together with a set of functions of s, p and d symmetry at the HH mid-point;

f) *spdfg*; *spdf* - atom centred sets of s, p, d and f symmetry on the H atoms and s, p, d, f and g on the O atom;

g) *spdfg*; *spdf* - a somewhat larger number of functions for each symmetry type included in f)

h) *spdfg*; *spdf*; *spdf* - a set of s, p, d and f functions at both of the OH bond centres to f);

i) *spdfg*; *spdf*; *spdf*; *spdf*- a set of s, p, d and f functions at HH bond centre to h);

j)-r) represent extrapolated second order correlation energies based on the energies defined above (details of the extrapolation procedures can be found in the work of Moncrieff and Wilson[166];

s) valence double zeta polarization set[167];

t) correlation consistent set cc-pVQZ[168];

u) correlation consistent set cc-pV5Z[168];

v) augmented correlation consistent set aug-cc-pVTZ[168];

w) augmented correlation consistent set aug-cc-pVQZ[168];

x) a basis set due to Chakravorty and Davidson[169];

y) a basis set due to Chakravorty and Davidson[169];

z) a basis set due to Clementi et al[170];

A) a calculation due to Bukowski et al[171] which incorporates factors depending on the interelectronic distance in the sum-over-states perturbation expression;

B) a "standard" second order calculation due to Klopper[172];

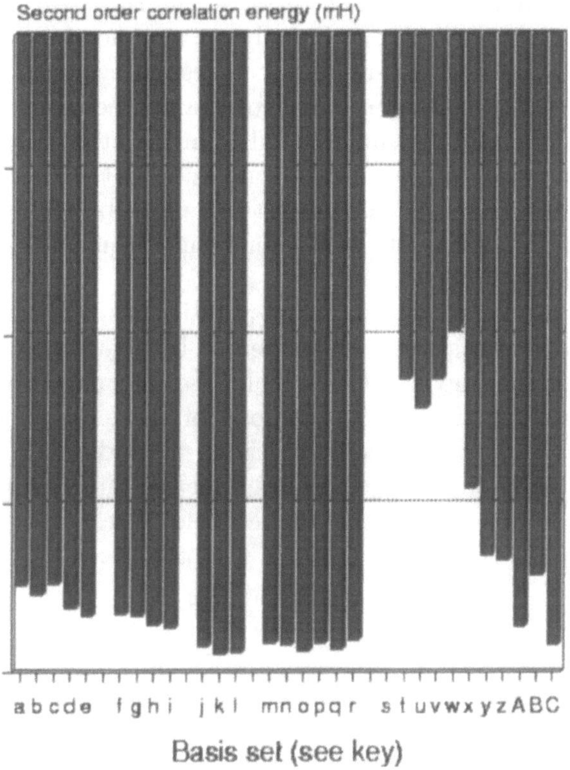

Figure 16. Comparison of second-order correlation energy components obtained in the present work with those reported previously:- a) *spd*; *spd*; *spd*, b) *spd*; *spd*; *spd*; *spd*, c) *spdf*, *spd*, d) *spdf*; *spd*; *spd*, e) *spdf*; *spd*; *spd*; *spd*, f) s4, g) s8, h) s12, i) s16, j) $E_2(\infty)_{[n=15,\ell=3]}$, k) $E_2'(\infty)_{[n=15,\ell=3]}$, l) $E_2''(\infty)_{[n=15,\ell=3]}$, m) $E_2(\infty)_{[n=10,\ell=3]}$, n) $E_2(\infty)_{[n=10,\ell=4]}$, o) $E_2'(\infty)_{[n=10,\ell=3]}$, p) $E_2'(\infty)_{[n=10,\ell=4]}$, q) $E_2''(\infty)_{[n=10,\ell=3]}$, r) $E_2''(\infty)_{[n=10,\ell=4]}$, s) Olsen *et al* valence double zeta polarization[167], t) Feller cc-pVQZ[168], u) Feller cc-pV5Z[168], v) Feller aug-cc-pVTZ[168], w) Feller aug-cc-pVQZ[168], x) Chakravorty and Davidson[169], y) Chakravorty and Davidson[169], z) Clementi et al[170], A) Bukowski et al[171], B) Klopper MP2[172], C) Klopper MP2-R12[172]

C) a calculation due to Klopper[172] which incorporates factors depending on the interelectronic distance in the sum-over-states perturbation expression.

6.5. ELECTRIC FIELD VARIANT BASIS SETS

Molecular basis sets are most often constructed from functions which have been designed for atomic systems augmented by so-called polarization functions capable of describing distortions of the atomic orbitals in a molecular environment. Whereas the description of the Hartree-Fock ground state of an atom requires only a limited number of symmetry types in the basis set, such a limitation does not apply for molecular systems. The Hartree-Fock ground state of the nitrogen atom, for example, can be described in terms of functions of s and p symmetry only. On the other hand, the description of the ground state of the nitrogen molecule in the Hartree-Fock approximation formally requires all symmetry type to be included. In Figure 12, an incorrect description of the H_2^+ ion results from the use of atom-centred functions of s symmetry only.

The calculation of atomic polarizabilities has traditionally necessitated the use of atomic basis sets supplemented by polarization fields to describe the effects of the applied electric field. An alternative approach is to use a basis set of functions which respond of the external field, such as the electric field variant basis sets described by Sadlej[173] [174]. Such basis sets are no larger than those employed for the unperturbed atom but each basis function depends on the applied field. These functions are obtained by replacing the coordinate vector $\mathbf{r}$ with the vector

$$\mathbf{r}' = \mathbf{r} - \frac{\lambda F}{\zeta^2} \tag{70}$$

where ζ is the exponent, F is the applied field strength and λ is a parameter which is taken to be the same for all functions in the set. The electric field variant basis set generated from the even-tempered basis set of s-type Gaussian functions shown in Figure 1 is displayed in Figure 17. This Figure demonstrates that in basis sets of the electric field variant type the most contracted functions, i.e. those with the largest exponents, are displaced least by the applied field whilst those which the exponents are smaller, i.e. more diffuse functions, the displacement is greater. Variational minimization of the second order energy can be used to optimize the value of.

Now when an atom is considered in a molecular environment, a field is induced at a given nucleus by the other nuclei in the systems. Wilson and Moncrieff[128] [129] used electric field variant basis sets to describe the distortion of an atom in a molecular environment in a number of simple molecular systems. The electric field variant basis sets can be regarded as a special type of distributed basis set.

6.6. BASIS SETS FOR RELATIVISTIC ELECTRONIC STRUCTURE CALCULATIONS

The relativistic electronic structure problem contains many subtleties which are not present in the non-relativistic problem. We therefore prefer the use of four-component theories which can then be systematically refined and

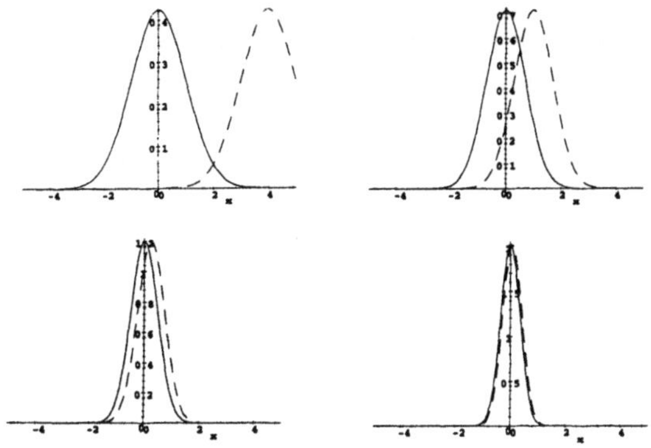

Figure 17. Electric field variant basis set of s-type Gaussian functions corresponding to the even-tempered set displayed in Figure 1.

approximated. Such an approach holds open, for example, to possibility of taking electron correlation treatments beyond the "no-virtual-pair" approximation since it afford a representation of the negative energy sea. On the other hand, this approach allows for the development of well-defined approximation schemes as basis sets are systematically extended.

Four-component theories require the use of matched basis sets for the large and small components of the relativistic spinors. In this way, a clean separation of the positive and negative energy branches of the relativistic spectrum is maintained and the non-relativistic limit is recovered when the speed of light is scaled. If the large component basis set contains functions of s-type then the small component basis set contains functions of p-type, if the large component set contains functions of p-type then the small component basis set contains a linear combination of functions of s- and d-type, etc.

Basis sets capable of supporting high precision in non-relativistic studies can be easily modified to serve as a basis for relativistic molecular structure calculations. Essentially, additional contracted basis functions are required in the relativistic case to describe the contraction of the single particle state functions in the core region whilst additional diffuse functions are needed to account for the expansion of the valence single particle state functions. In relativistic electronic structure calculations within the algebraic approximation matched basis sets must be employed for the large and small components of the wave function. The large and small components of the relativistic wave function must be approximated in terms of basis functions which satisfy the condition[96]

$$\chi_{i\kappa}^{S} \propto \left(\frac{d}{dr} + \frac{\kappa}{r} \right) \chi_{i\kappa}^{L} \tag{71}$$

There is a one-to-one correspondence between the large and small component basis functions. If the large component basis set contains s functions then the small component basis sets must contain p functions with the same exponents; if the large component basis set contains p functions then the small component basis set must contain functions which are a linear combination of s and d functions; in general, if the large component basis functions have the form

$$\chi^{L}_{im\kappa} = N_{i\kappa} r^{\ell} \exp(-\zeta_i r^2) \Omega^{\mathrm{sgn}(\kappa)}_{jm}(\theta, \varphi) \tag{72}$$

then the small component basis functions have the form

$$\chi^{S}_{im\kappa} = \imath N_{i\kappa} \left((\ell + \kappa + 1) - 2\zeta r^2 \right) r^{\ell - 1} \exp(-\zeta_i r^2) \Omega^{-\mathrm{sgn}(\kappa)}_{jm}(\theta, \varphi) \tag{73}$$

where κ is the usual angular quantum number, $2\jmath = 2|\kappa| - 1$, $-\jmath \leq m \leq \jmath$. The restriction of the large components basis set to functions of low symmetry when using distributed basis sets imposes corresponding restrictions on the small component basis set.

The use of distributed basis sets lowers the number of symmetry types required to approximate the large component of the spinor and as a consequence the number of symmetry types arising in the parametrization of the small component. The algebraic complexity of molecular integral evaluation over Gaussian-type functions increases markedly with ℓ. For basis set containing only functions of s- and p-type it has long been recognized that the integral evaluation can be made very efficient by explicitly coding "special formulae" which are much simpler than the general case. Efficient integral evaluation is an essential ingredient of the so-called "direct" algorithms in relativistic electronic structure calculations.

7. Summary and prospects

A review of the fundamentals of the algebraic approximation has been given in this article together with a survey of some of the more promising recent developments in the systematic approximation of molecular structure by means of finite basis set expansions. Gaussian basis sets are employed in almost all contemporary molecular electronic structure calculations and the advantages of this approach have been summarized. The systematic implementation of finite basis set approximations has been shown to afford high precision.

Central to the efficient implementation of the algebraic approximation is the effective exploitation of high performance computers. Vector processors, massively parallel processors, parallel vector processors, workstation clusters, heterogeneous computer networks all offer potential computing devices for relativistic molecule electronic structure calculations. We are currently exploring the use of heterogeneous computer network, performing each part of the calculation on the most appropriate architecture.

About ten years ago, Davidson and Feller wrote:

> *It may be possible in the future to have programs make an informed decision about the choice of basis set based on the results of thousands of previous calculations which are accessible in a data base, but that time has not yet come. The responsibility still rests with the program user.*

However, it is now apparent that the use of high performance computing machines together with fast algorithms for integral evaluation described in the following article will render the systematic implementation of the algebraic approximation described in the present work feasible in "routine" molecular electronic structure calculations in the near future.

Acknowledgment

I would like to acknowledge numerous conversations about finite basis set methods, the algebraic approximation and its accurate implementation with Dr. B.H. Wells, Dr. J.W. Thompson and Dr. D. Moncrieff.

References

1. D.R. Hartree, W. Hartree and B. Swirles, Phil. Trans. Roy. Soc. (London) **A238**, 229 (1939)
2. L. Laaksonen, P. Pyykkö and D. Sundholm, Intern. J. Quantum Chem. **23** 309, 319 (1983)
3. D. Heinemann, D. Kolb and B. Fricke, Chem. Phys. Lett. **137**, 180 (1987); *ibid.* **145**, 125 (1988); Phys. Rev. A **38**, 4994 (1988)
4. K.G. Wilson, Nucl. Phys. B (Proc. Suppl.) **17**, 82 (1990)
5. G.G. Hall, Proc. Roy. Soc. A205, 541 (1951)
6. C.C.J. Roothaan, Rev. Mod. Phys. **23**, 69 (1951)
7. S. Wilson, in *Methods in Computational Molecular Physics*, NATO ASI Series **113**, p. 71, edited by G.H.F. Diercksen and S. Wilson, Reidel, Dordrecht (1983)
8. E.R. Davidson and D. Feller, Chem. Rev. **86**, 681 (1986)
9. S. Huzinaga, Comput. Phys. Rept. **2**, 279 (1985)
10. E. Clementi, S. Chakravorty, G. Corongiu and V. Sonnad, in *Modern Techniques in Computational Chemistry*, edited by E. Clementi, ESCOM, Leiden (1990)
11. P.R. Taylor, in *European Summer School in Quantum Chemistry*, Lecture Notes in Chemistry **58**, p. 399, edited by B.O. Roos, Springer-Verlag, Berlin (1992)
12. I.P. Grant, in *Relativistic effects in atoms and molecules*, Methods in Computational Chemistry **2**, edited by S. Wilson, Plenum Press, New York (1988)
13. H.P. Kelly, Phys. Rev. **131**, 684 (1963); *ibid.* A **134**, 1450 (1964); *ibid.* B **136**, 896 (1964); Adv. Theoret. Phys. **2**, 75 (1968); *Adv. Chem. Phys.* **14**, 129 (1969)
14. P.J. Rossky and M. Karplus, J. Chem. Phys. **67**, 5419 (1977)
15. H.M. Quiney, I.P. Grant and S. Wilson, J. Phys. B: At. Mol. Phys. **18**, 577 (1985)
16. S. Wilson, Adv. Chem. Phys. **67**, 439 (1987)

17. L. Laaksonen, P. Pyykkö and D. Sundholm, Comput. Phys. Rept. **4**, 313 (1986)
18. P. Pyykkö, in *Numerical Determination of the Electronic Structure of Atoms, Diatomic and Polyatomic Molecules*, edited by M. Defranceschi and J. Delhalle, Kluwer, Dordrecht (1989)
19. J. Kobus, Chem. Phys. Lett. **202**, 7 (1993)
20. D. Heinemann, A. Rosen and B. Fricke, Phys. Scripta **42**, 692 (1990)
21. J. Kobus, Comput. Phys. Commun. **78**, 247 (1994)
22. J. Kobus, L. Laaksonen and D. Sundholm, Comput. Phys. Commun. **98**, 346 (1996)
23. J. Kobus, D. Moncrieff and S. Wilson, J. Phys. B: At. Mol. Opt. Phys. **27**, 5139 (1994)
24. J. Kobus, D. Moncrieff and S. Wilson, J. Phys. B: At. Mol. Opt. Phys. **27**, 2867 (1994)
25. D. Moncrieff, J. Kobus and S. Wilson, J. Phys. B: At. Mol. Opt. Phys. **28**, 4555 (1995)
26. J. Kobus, D. Moncrieff and S. Wilson, Mol. Phys. **86**, 1315 (1995)
27. I.G. Kaplan, Symmetry of Many-Electron Systems, *p.* 269, Academic Press, London (1975)
28. S. Wilson and D.M. Silver, Phys. Rev. A **14**, 1949 (1976)
29. S. Wilson, *Electron correlation in molecules*, Clarendon Press, Oxford (1984)
30. M.F. Guest and S. Wilson, in *Supercomputers in Chemistry*, edited by P. Lykos and I. Shavitt, ACS Symposium Series **173**, American Chemical Society, Washington D.C. (1981)
31. V.R. Saunders and M.F. Guest, Comput. Phys. Commun. **26**, 389 (1982)
32. S. Wilson and V.R. Saunders, Comput. Phys. Commun. **19**, 293 (1980)
33. D. Moncrieff, V.R. Saunders and S. Wilson, in *Proceedings Supercomputing Symposium '92 "Grand Challenges in Supercomputing"*, edited by C. Krupat, Environment Canada, Canadian Meteorological Centre, Dorval, Quebec (1992)
34. A. Bohr and V.F. Weisskopf, Phys. Rev. **77**, 94 (1950)
35. Y. Ishikawa, R. Barretty and R.C. Binning, Chem. Phys. Letts. **121**, 130 (1985)
36. Y. Ishikawa and H.M. Quiney, Intern. J. Quantum Chem. Symp. **21**, 523 (1987)
37. L. Laaksonen, I.P. Grant and S. Wilson, J. Phys. B: At. Mol. Phys. **21**, 1969 (1988)
38. B. Klahn and J.D. Morgan III, J. Chem. Phys. **81**, 410 (1984)
39. R.N. Hill, J. Chem. Phys. **83**, 1173 (1985)
40. J. Morgan III, in *Numerical Determination of the Electronic Structure of Atoms, Diatomic and Polyatomic Molecules*, edited by M. Defranceschi and J. Delhalle, NATO-ASI Series C **271**, Kluwer, Dordrecht (1989)
41. H.M. Quiney, I.P. Grant and S. Wilson, J. Phys. B: At. Mol. Phys. **20**, 1413 (1987)
42. H.M. Quiney, I.P. Grant and S. Wilson, in *Numerical Determination of the Electronic Structure of Atoms, Diatomic and Polyatomic Molecules*, edited by M. Defranceschi and J. Delhalle, NATO-ASI Series C **271**, Kluwer, Dordrecht (1989)
43. H.M. Quiney, I.P. Grant and S. Wilson, in *Many-Body Methods in Quantum Chemistry*, Lecture Notes in Chemistry **52**, edited by U. Kaldor, Springer, Berlin (1989)
44. H.M. Quiney, I.P. Grant and S. Wilson, J. Phys. B: At. Mol. Opt. Phys. **22**, L15 (1989)
45. H.M. Quiney, I.P. Grant and S. Wilson, Physica Scripta **36**, 460 (1987)
46. M.W. Schmidt and K. Ruedenberg, J. Chem. Phys. **71**, 3951 (1979)
47. J. Kobus, D. Moncrieff and S. Wilson, Molec. Phys. **86**, 1315 (1995)
48. Part IV.
49. M.J. Frisch, G.W. Trucks, H.B. Schlegel, P.M.W. Gill, B.G. Johnson, M.A. Robb, J.R. Cheeseman, T. Keith, G.A. Petersson, J.A. Montgomery, K. Raghavachari, M.A. Al-Laham, V.G. Zakrzewski, J.V. Ortiz, J.B. Foresman, J. Cioslowski, B.B. Stefanov, A. Nanayakkara, M. Challacombe, C.Y. Peng, P.Y. Ayala, W. Chen, M.W. Wong, J.L. Andres, E.S. Replogle, R. Gomperts, R.L. Martin, D.J. Fox, J.S. Binkley, D.J. Defrees, J. Baker, J.P. Stewart, M. Head-Gordon, C. Gonzalaz and J.A. Pople, GAUSSIAN94, Gaussian Inc., Pittsburgh PA, 1995

50. UNICHEM was originally developed by Cray Research Inc. Details can be obtained from Oxford Molecular Ltd., The Medawar Centre, Oxford Science Park, Oxford OX4 4GA, England.
51. D. Moncrieff and S. Wilson, J. Phys. B: At. Mol. Opt. Phys. **28**, 4007 (1995)
52. W.J. Hehre, R.F. Stewart and J.A. Pople, J. Chem. Phys. **51**, 2657 (1969)
53. J.B. Collins, P. v. R. Schleyer, J.S. Binkley and J.A. Pople, J. Chem. Phys. **64**, 5142 (1976)
54. J.S. Binkley, J.A. Pople and W.J. Hehre, J. Amer. Chem. Soc. **102**, 939 (1980)
55. M.S. Gordon, J.S. Binkley, J.A. Pople, W.J. Pietro and W.J. Hehre, J. Amer. Chem. Soc. **104**, 2797 (1982)
56. W.J. Pietro, M.M. Francl, W.J. Hehre, D.J. Defrees, J.A. Pople and J.S. Binkley, J. Amer. Chem. Soc. **104**, 5039 (1982)
57. R. Ditchfield, W. Hehre and J.A. Pople, J. Chem. Phys. **54**, 724 (1971)
58. W. Hehre, R. Ditchfield and J.A. Pople, J. Chem. Phys. **56**, 2257 (1972)
59. P.C. Hariharan and J.A. Pople, Molec. Phys. **27**, 209 (1974)
60. M.S. Gordon Chem. Phys. Lett. **76**, 163 (1980)
61. P.C. Hariharan and J.A. Pople, Theo. Chim. Acta. **28**, 213 (1973)
62. T.H. Dunning, Jr., and P.J. Hay, in *Modern Theoretical Chemistry*, edited by H.F. Schaefer III, Plenum, New York, 1976
63. T.H. Dunning, Jr., J. Chem. Phys. **90**, 1007 (1989)
64. R.A. Kendall, T.H. Dunning, Jr. and R.J. Harrison, J. Chem. Phys. **96**, 6796 (1992)
65. D.E. Woon and T.H. Dunning, Jr., J. Chem. Phys. **98**, 1358 (1993)
66. D.E. Woon and T.H. Dunning, Jr., J. Chem. Phys. **100**, 2975 (1994)
67. D.E. Woon and T.H. Dunning, Jr., J. Chem. Phys. **103**, 4572 (1995)
68. R.C. Raffenetti, J. Chem. Phys. **58**, 4452 (1973)
69. Partridge H 1989 J. Chem. Phys. **90** 1043
70. Widmark PO, Malmqvist PA and Roos BO 1990 Theoret. chim. Acta. **77** 291
71. Widmark PO, Persson BJ and Roos BO 1991 Theoret. chim. Acta. **79** 419
72. Almlöf J and Taylor PR 1987 J. Chem. Phys. **86** 4070
73. N.H. March, W.H. Young and S. Sampanthar, *The many-body problem in quantum mechanics*, Cambridge University Press (1967)
74. I. Lindgren and J. Morrison, *Atomic many-body theory*, (Springer-Verlag, Berlin, 1982)
75. F.E. Harris, H.J. Monkhorst and D.L. Freeman, *Algebraic and diagrammatic methods in many-fermion theory*, (Oxford, University Press, New York, 1992)
76. S. Wilson, Comput. Phys. Repts. **2**, 389 (1985)
77. D. Moncrieff, D.J. Baker and S. Wilson 1989 Comput. Phys. Commun. **55** 31
78. Márquez AM and Dupuis M 1995 J. Comput. Chem. **16** 395
79. Feenberg E 1956 Phys. Rev. **103** 1116
80. Wilson S, Silver DM and Farrell RA 1977 Proc. Roy. Soc. (London) **A 356** 363
81. Wilson S 1979 J. Phys. B: At. Mol. Phys. **12** L135
82. Wilson S 1981 J. Phys. B: At. Mol. Phys. **14** L31
83. Wilson S 1981 Theoret. chim. Acta **59** 71
84. Sadlej AJ and Wilson S 1981 Molec. Phys. **44** 229
85. Silver DM and Wilson S 1982 Proc. Roy. Soc. (London) A **383** 477
86. Wilson S, Kellö V, Noga J and Urban M 1983 Molec. Phys. **50** 1323
87. Dietz K, Schmidt C, Warken M and Hess BA 1993 J. Phys. B: At. Mol. Opt. Phys. **26** 1885
88. Dietz K, Schmidt C, Warken M and Hess BA 1993 J. Phys. B: At. Mol. Opt. Phys. **26** 1897
89. Dietz K, Schmidt C, Warken M and Hess BA 1993 Chem. Phys. Letts. 207 281
90. Schmidt C, Warken M and Handy NC 1993 Chem. Phys. Lett. **211** 272
91. Dietz K, Schmidt C, Warken M and Hess BA 1994 Chem. Phys. Lett. **220** 397
92. Dietz K, Schmidt C, Warken M and Hess BA 1994 J. Chem. Phys. **100** 7421

93. T.E. Sorensen, W.B. England, D.M. Silver and E.O. Steinborn, Theoret. chim. Acta. **84**, 1 (1992)
94. T.E. Sorensen, W.B. England and D.M. Silver, J. Phys. B: At. Mol. Opt. Phys. **22**, L539
95. Wilson S 1988 Meth. Comput. Chem. **2** 73
96. Quiney HM, Grant IP and Wilson S 1989 in *Many body methods in Quantum Chemistry*, edited by U. Kaldor, Lecture Notes in Chemistry **52**, 331 (Springer-Verlag, Berlin)
97. Wilson S, Grant IP and Gyorffy BL 1991 editors, *The Effects of Relativity in Atoms, Molecules and the Solid State*, (Plenum, New York)
98. Wilson S 1992 in *Methods in Computational Molecular Physics*, edited by S. Wilson and G.H.F. Diercksen, NATO ASI Series B293, *p.* 253 (Plenum, New York)
99. D. Moncrieff and S. Wilson, J. Phys. B: At. Mol. Opt. Phys. **29**, 2425 (1996)
100. D.Moncrieff & S.Wilson, J.Phys.B: At. Mol. & Opt. Phys. **26**, 1605 (1993); Chem. Phys. Lett. **209**, 423 (1993)
101. S. Wilson, in *Methods in Computational Molecular Physics*, ed. G.H.F. Diercksen and S. Wilson, Reidel, Dordrecht (1983)
102. C. Schwartz, Meth. Comput. Phys. **2**, 241 (1962)
103. K. Ruedenberg, R.C. Raffenetti and R. Bardo, in *Energy structure and reactivity, Proc. 2nd. Boulder Research Confer. in Theoret. Chem.*, Wiley, New York (1973)
104. R.C. Raffenetti, J. Chem. Phys. **58**, 4452 (1973)
105. R.C. Raffenetti, J. Chem. Phys. **59**, 5936 (1973)
106. R.C. Raffenetti and K. Ruedenberg, J. Chem. Phys. **59**, 5978 (1973)
107. R. Bardo and K. Ruedenberg, J. Chem. Phys. **59**, 5956, 5966 (1973)
108. R. Bardo and K. Ruedenberg, J. Chem. Phys. **60**, 918 (1974)
109. D.F. Feller and K. Ruedenberg, Theoret. chim. Acta. **52**, 231 (1979)
110. S. Wilson, Theoret. chim. Acta **57**, 53 (1980); S. Wilson, Theoret. chim. Acta **58**, 31 (1980)
111. D.M. Silver, S. Wilson and W.C. Nieuwpoort, Intern. J. Quantum Chem. **14**, 635 (1978); D.M. Silver and W.C. Nieuwpoort, Chem. Phys. Lett. **57**, 421 (1978); D.M. Silver and S. Wilson, J. Chem. Phys. **69**, 3787 (1978); S. Wilson and D.M. Silver, Chem. Phys. Lett. **63**, 367 (1979); S. Wilson and D.M. Silver, J. Chem. Phys. **72**, 2159 (1980); S. Wilson, in *Specialist Periodical Report: Theoretical Chemistry* **4,** 1 (Royal Society of Chemistry, London, 1981); S. Wilson and D.M. Silver, J. Chem. Phys. **77**, 3674 (1982); S. Wilson, Molec. Phys. **43**, 1489 (1983); D. Moncrieff and S. Wilson, J. Phys. B: At. Mol. Opt. Phys. **27**, 1 (1994)
112. E. Clementi and G. Corongiu, Chem. Phys. Lett. **90**, 359 (1982); *Geometric Basis Sets for Molecular Computations, IBM Tech. Rept. POK-11 (1982);* E. Clementi, S.J. Chakravorty, G. Corongiu and V. Sonnad, in *Modern Techniques in Computational Chemistry*, ed. E. Clementi, ESCOM, Leiden (1990)
113. S. Huzinaga, Comput. Phys. Repts. **2** (1985) 281
114. H. Muntz, *Festschrift*, edited by H.A. Schwartz, p.303, Springer-Verlag, Berlin (1914)
115. O. Szász, Math. Ann. **77** (1926)
116. R. Courant and D. Hilbert, *Methoden der Mathamatischen Physik*, Vol. 1, p.86 (Springer, 1931)
117. R.E.A. Paley and N. Wiener, American Mathematical Society Colloquium Publications, Vol. XIX, pp. 32-36 (New York, 1934)
118. A. Schönhage, *Approximationstheorie*, pp. 49-52, (de Gruyter, Berlin, 1971)
119. J.R. Higgins, *Completeness and Basis Properties of Sets of Special Functions*, pp. 95-97, (Cambridge University Press, Cambridge, 1977)
120. B. Klahn, J. Chem. Phys. **83**, 5748 (1985)
121. A.S. Shalabi and S. Wilson, J. Molec. Struct. (THEOCHEM) **341**, 165 (1995)
122. R.K. Nesbet, Rev. Mod. Phys. **32**, 272 (1960)
123. C.M. Reeves, J. Chem. Phys. **39**, 1 (1963)

124. C.M. Reeves and M.C. Harrison, J. Chem. Phys. **39**, 11 (1963)
125. R. Fletcher and C.M. Reeves, Comput. J. **6**, 287 (1963)
126. I. Shavitt, Israel J. Chem. **33**, 357 (1993)
127. J. Almlof, K. Faegri and K. Korsell, J. Comput. Chem. **3**, 385 (1982)
128. S. Wilson and D. Moncrieff, Molec. Phys. **80**, 461 (1993)
129. D. Moncrieff and S. Wilson, Molec. Phys. **82**, 523 (1994)
130. H. Preuss, Z. Naturforsch. **11a**, 823 (1956); H. Preuss, Z. Naturforsch. **19a**, 1335 (1964); *ibid.* **20a**, 17, 21 (1965); J.L. Whitten, J. Chem. Phys. **39**, 349 (1963); F. Driessler and R. Ahlrichs, Chem. Phys. Lett. **23**, 571 (1973); V.R. Saunders, in *Methods in Computational Molecular Physics*, edited by G.H.F. Diercksen and S. Wilson, Reidel, Dordrecht
131. A.A. Frost, J. Chem. Phys. **47**, 3707 (1967)
132. A.A. Frost, J. Chem. Phys. **47**, 3714 (1967)
133. B. Ford, G.G. Hall and J.C. Packer, Intern. J. Quantum Chem. **4**, 533 (1970)
134. L.M. Haines, J.N. Murrell, B.J. Ralston, and D.J. Woodnutt, J. Chem. Soc. Faraday Transactions II **70**, 1794 (1974)
135. B.J. Ralston, D. Phil. Dissertation, University of Sussex, 1974
136. B.J. Ralston and S. Wilson, J. Molec. Struct. THEOCHEM **341**, 115 (1995)
137. S. Wilson, J. Molec. Struct. THEOCHEM **357**, 37 (1995)
138. S. Wilson, Intern. J. Quant. Chem. **60**, 47 (1996)
139. S. Hackel, D. Heinemann, D. Kolb and B. Fricke, Chem. Phys. Lett. **206** (1993) 91
140. S. Wilson, J. Phys. B: At. Mol. Opt. Phys. **28**, L495 (1995)
141. F.E. Harris, J. Chem. Phys. **32** (1960) 3
142. D.D. Ebbing, J. Chem. Phys. **36** (1962) 1361
143. F.E. Harris and H.S. Taylor, J. Chem. Phys. **38** (1963) 2591
144. H.S. Taylor, J. Chem. Phys. **39** (1963) 3375
145. H.S. Taylor and F.E. Harris, Molec. Phys. **6** (1963) 183
146. S. Wilson, J. Phys. B: At. Mol. Phys. **16** (1983) L429
147. D.M. Bishop and L.M. Cheung, J. Chem. Phys. **78** (1982) 1396
148. B.H. Wells & S. Wilson, J. Phys. B: At. Mol. Phys. **19** (1985) 17
149. J.W. Thompson and S. Wilson, J. Phys. B: At. Mol.Opt. Phys. **23** (1990) 2205
150. J.W. Thompson and S. Wilson, J. Phys. B: At. Mol.Opt. Phys. **25** (1992) 621
151. D.M. Silver, Molec. Phys. **22** (1971) 1069
152. W. Kutzelnigg, Intern. J. Quantum Chem. **51** (1994) 447
153. I. Shavitt, in *Methods in Computational Physics* **2** (1963) 1
154. H. Preuss, Z. Naturforsch **19**a, 1335 (1964); ibid **20**a, 17, 21 (1965)
155. S. Rothenberg and H.F. Schaefer III, J. Chem. Phys. **54**, 2765 (1971)
156. W. Butscher, S. Shih, R.J. Buenker and S.D. Peyerimhoff, Chem. Phys. Lett. **52**, 457 (1977)
157. D. Neisius and G. Verhaegen, Chem. Phys. Lett. **66**, 358 (1979)
158. L. Lawlor, K. Vasudevan and F. Grein, Chem. Phys. Lett. **75**, 79 (1980)
159. P. Mach and O. Kysel, J. Comput. Chem. **6**, 312 (1985)
160. J.M.L. Martin, J.P. Francois and R. Gijbels, J. Comput. Chem. **10**, 152 (1989)
161. J.M.L. Martin, J.P. Francois and R. Gijbels, J. Comput. Chem. **10**, 875 (1989)
162. J.M.L. Martin, J.P. Francois and R. Gijbels, Theoret. chim. Acta **76**, 195 (1989)
163. F.M. Tao and Y.K. Pan, J. Chem. Phys. **97**, 4989 (1992)
164. F.M. Tao and Y.K. Pan, Molec. Phys. **81**, 507 (1994)
165. P. Čársky and M. Urban, *Ab initio Calculations. Methods and Applications in Chemistry*, Lecture Notes in Chemistry **16** (Springer-Verlag, Berlin)
166. D. Moncrieff and S. Wilson, J. Phys. B: At. Mol. Phys. **29** (1996) 6009
167. J. Olsen, P. Jorgensen, H. Koch, A. Balkova and R.J. Bartlett, 1996 J. Chem. Phys. **104** 8007
168. D. Feller 1992 J. Chem. Phys. **96** 6104

169. S.J. Chakravorty and E.R. Davidson, 1993 J. Phys. Chem. **97** 6374
170. E. Clementi, G. Corongiu and O.G. Stradella, 1990 Modern Techniques in Computational Chemistry, MOTECC-90, ESCOM Science Publishers, Leiden, pp. 343
171. R. Bukowski, B. Jeriorski, S. Rybak, and K. Szalewicz, 1995 J. Chem. Phys. **102** 888
172. W. Klopper 1995 J. Chem. Phys. **102** 6168
173. A.J. Sadlej 1977 Chem. Phys. Lett. **47**, 50
174. S. Wilson and A.J. Sadlej 1981 Theo. chim. Acta. **60**, 19

PRACTICAL *AB INITIO* METHODS FOR MOLECULAR ELECTRONIC STRUCTURE STUDIES. III. MOLECULAR INTEGRALS OVER GAUSSIAN-TYPE FUNCTIONS

S. WILSON

Rutherford Appleton Laboratory
Chilton, Oxfordshire OX11 0QX, England

Abstract. The evaluation of molecular integrals over Gaussian-type functions lies at the heart of practical realization of contemporary quantum chemical theories. The aims of the present article are two-fold. First, to provide an introduction to the problem of molecular integral evaluation for the quantum chemistry package user; to give a elementary, yet fairly thorough, introduction to the steps, both analytical and computational, involved in the evaluation of the simplest of integrals. Second, to provide a brief overview of developments that have taken place in the field of molecular integral evaluation since the 1983 review by Saunders.

1. Introduction

A major advance in the realization of a computationally tractable approach to the molecular electronic structure problem was made by Hall[1] and, independently, by Roothaan[2] when they formulated the Hartree-Fock equations using finite basis set expansions thereby obtaining a set of algebraic equations for the expansion coefficients which were well suited to automatic computation. The use of Gaussian basis functions was first suggested by McWeeny[3] [4] but actually implemented by Boys[5]. (An historical review has been given by Shavitt[6].) The construction of Gaussian basis sets has been considered in the preceding article [7].

The accurate and efficient computation of molecular integrals constitutes a major portion of any large *ab initio* molecular electronic structure calculation. Formally, the number of two-electron integrals arising from a

S. Wilson and G. H. F. Diercksen (eds.), Problem Solving in Computational Molecular Science, 159–184.

basis set of N functions increases as

$$\frac{1}{8}N(N+1)\left(N^2+N+2\right)\sim\frac{N^4}{8} \tag{1}$$

So for a basis set of just 10 functions there are, neglecting any symmetry, $\sim$ 1540 two-electron integrals; for a basis set of 100 functions there are 12, 753, 775 integrals; for a basis set of 1000 basis functions there are $\sim 1.3\times10^{11}$ integrals. In practice, the number of integrals which contribute significantly to the energy may be much smaller scaling as $\sim N^2$, particularly for extended molecules.

For "traditional" algorithms, in which integrals are evaluated once, stored on disk and processed repeated during the self-consistent-field and any subsequent correlation treatment, the computational requirement are not only the central processor time for integral evaluation but also the disk required for their storage. For a machine with a 64 bit/8 byte word length, a basis set of 100 functions requires 102 Megabytes of disk store; a basis set of 1000 functions necessitates 1040 Gigabytes.

The recognition in the early eighties that central processor power was increasing more rapidly than the speed of disk access led to the introduction of "direct" algorithms in which integrals are evaluated as required. Almlöf *et al*[8] demonstrated that the direct self-consistent field approach was far more efficient that had been previously thought since each Fock matrix contribution is a product of a density matrix element and a two-electron integral. In each iteration the Fock matrix is known and can be used to effectively eliminate larger numbers of electron repulsion integrals. The elimination of the need to stored and repeated read integrals from disk opened up the possibility of using significantly larger basis sets than had been previously possible. However, the introduction of direct algorithms for molecular electronic structure studies fuelled the need for more efficient integral evaluation techniques since, for example, integrals are evaluated in each iteration of the self-consistent field procedure.

The need to efficiently evaluate the two electron integrals required for energy derivatives has become increasingly important as geometry optimization as become an important ingredient of many *ab initio* electronic structure studies. The development of *ab initio* four-component relativistic molecular electronic structure algorithms also provides an impetus for more efficient algorithms for electron repulsion integrals.

In view of the pivotal rôle played by the two-electron integral evaluation stage of any *ab initio* electronic structure calculation the user of any quantum chemical program should have some understanding of the mechanics of molecular integral evaluation. The aims of the present article are two-fold. First, to provide an introduction to the problem of molecular integral evaluation for the quantum chemistry package user; to give a elementary, yet fairly thorough, introduction to the steps, both analytical and computational, involved in the evaluation of the simplest of integrals. Second, to provide a brief overview of developments that have taken place in the field of molecular integral evaluation since the review by Saunders[9] published in the Proceedings of the NATO Advanced Study Institute *Methods in Computational Molecular Physics* in 1983 [See also the review published by Saunders[10] in 1975].

2. Integrals over s-type primitive Gaussian basis functions

2.1. INTEGRALS OVER S-TYPE GAUSSIAN FUNCTIONS

Consider four s-type Gaussian functions centred on points A, B, C and D with exponents ζ_{aA}, ζ_{bB}, ζ_{cC} and ζ_{dD}. The electron repulsion integral may be written

$$\begin{aligned}\mathcal{I}_{abcd;ABCD} &= \int d\tau_1 \int d\tau_2 \exp(-\zeta_{aA} r_{1A}^2) \exp(-\zeta_{cC} r_{2C}^2) \frac{1}{r_{12}} \\ &\quad \exp(-\zeta_{bB} r_{1B}^2) \exp(-\zeta_{dD} r_{2D}^2) \end{aligned} \tag{2}$$

where the electrons are labelled 1 and 2, and the indices a, b, c and d distinguish the different functions centred on the points A, B, C and D, respectively. The Cartesian co-ordinates of the two electrons are

$$\mathbf{r}_1 = (x_1, y_1, z_1) \tag{3}$$

and

$$\mathbf{r}_2 = (x_2, y_2, z_2) \tag{4}$$

The Cartesian co-ordinates of the four points upon which the Gaussian functions are centred are written

$$\mathbf{A} = (A_x, A_y, A_z) \tag{5}$$

$$\mathbf{B} = (B_x, B_y, B_z) \tag{6}$$

$$\mathbf{C} = (C_x, C_y, C_z) \tag{7}$$

and

$$\mathbf{D} = (D_x, D_y, D_z) \tag{8}$$

Then

$$\mathbf{r}_{1A} = \mathbf{r}_1 - \mathbf{A} \tag{9}$$

$$\mathbf{r}_{1B} = \mathbf{r}_1 - \mathbf{B} \tag{10}$$

$$\mathbf{r}_{2C} = \mathbf{r}_2 - \mathbf{C} \tag{11}$$

$$\mathbf{r}_{2D} = \mathbf{r}_2 - \mathbf{D} \tag{12}$$

and

$$r_{1A}^2 = (x_1 - A_x)^2 + (y_1 - A_y)^2 + (z_1 - A_z)^2 \tag{13}$$

$$r_{1B}^2 = (x_1 - B_x)^2 + (y_1 - B_y)^2 + (z_1 - B_z)^2 \tag{14}$$

$$r_{2C}^2 = (x_2 - C_x)^2 + (y_2 - C_y)^2 + (z_2 - C_z)^2 \tag{15}$$

$$r_{2D}^2 = (x_2 - D_x)^2 + (y_2 - D_y)^2 + (z_2 - D_z)^2 \tag{16}$$

The infinitesimal volume elements for electrons 1 and 2 are

$$d\boldsymbol{\tau}_1 = dx_1 dy_1 dz_1 \tag{17}$$

and

$$d\boldsymbol{\tau}_2 = dx_2 dy_2 dz_2 \tag{18}$$

respectively.

2.2. THE GAUSSIAN PRODUCT THEOREM

The Gaussian product theorem allows the product of Gaussian functions associated with electron 1 to be written as a single Gaussian function with a centre lying on the line segment AB:-

$$\exp(-\zeta_{aA} r_{1A}^2) \exp(-\zeta_{bB} r_{1B}^2) = \Lambda_p \exp(-\zeta_{pP} r_{1P}^2) \tag{19}$$

where

$$\zeta_{pP} = \zeta_{aA} + \zeta_{bB} \tag{20}$$

$$\mathbf{P} = \frac{\zeta_{aA}\mathbf{A} + \zeta_{bB}\mathbf{B}}{\zeta_{pP}} \tag{21}$$

so that

$$P_\lambda = \frac{\zeta_{aA} A_\lambda + \zeta_{bB} B_\lambda}{\zeta_{pP}}, \quad \lambda = x, y, z \tag{22}$$

and

$$\Lambda_p = \exp\left(-\frac{\zeta_{aA}\zeta_{bB}}{\zeta_{pP}} (\mathbf{A} - \mathbf{B})^2\right) \tag{23}$$

where

$$(\mathbf{A} - \mathbf{B})^2 = (A_x - B_x)^2 + (A_y - B_y)^2 + (A_z - B_z)^2 \tag{24}$$

The radial overlap Λ_p decreases as a Gaussian with increasing separation of the centres A and B.

Similarly, for electron 2 the Gaussian product theorem takes the form

$$\exp(-\zeta_{cC} r_{2C}^2) \exp(-\zeta_{dD} r_{2D}^2) = \Lambda_q \exp(-\zeta_{qQ} r_{2Q}^2) \tag{25}$$

where

$$\zeta_{qQ} = \zeta_{cC} + \zeta_{dD} \tag{26}$$

$$\mathbf{Q} = \frac{\zeta_{cC}\mathbf{C} + \zeta_{dD}\mathbf{D}}{\zeta_{qQ}} \tag{27}$$

so that

$$Q_\lambda = \frac{\zeta_{cC}C_\lambda + \zeta_{dD}D_\lambda}{\zeta_{qQ}}, \ \lambda = x, y, z \tag{28}$$

and

$$\Lambda_q = \exp\left(-\frac{\zeta_{cC}\zeta_{dD}}{\zeta_{qQ}}(\mathbf{C}-\mathbf{D})^2\right) \tag{29}$$

where

$$(\mathbf{C}-\mathbf{D})^2 = (C_x - D_x)^2 + (C_y - D_y)^2 + (C_z - D_z)^2 \tag{30}$$

The disposition of the points **A**, **B**, **C**, **D**, **P** and **Q** for a typical two-electron, four-centre integral is illustrated in Figure 1.

The four centre, two-electron integral has now been reduced to a two-centre, two-electron integral which may be written

$$\begin{aligned} \mathcal{I}_{abcd;ABCD} &= \mathcal{I}_{pq;PQ} \\ &= \Lambda_p\Lambda_q \int d\boldsymbol{\tau}_1 \int d\boldsymbol{\tau}_2 \exp(-\zeta_{pP}r_{1P}^2)\frac{1}{r_{12}}\exp(-\zeta_{qQ}r_{2Q}^2) \end{aligned} \tag{31}$$

This completes the first of two major steps in the evaluation of the two-electron integral over s-type Gaussian functions. This first step is common to all derivations. There is some choice in the second step.

2.3. LAPLACE TRANSFORM OF $\frac{1}{R_{12}}$

To make further progress, introduce the Laplace transform[1]

$$\frac{1}{r_{12}} = \frac{1}{\sqrt{\pi}}\int_0^\infty d\eta \exp(-\eta r_{12}^2)\,\eta^{-\frac{1}{2}} \tag{32}$$

[1]In this article the "traditional" derivation of the basic two-electron repulsion integral is given[10]. Nowadays, the Laplace transformation

$$\frac{1}{r_{12}} = \frac{2}{\sqrt{\pi}}\int_0^\infty d\eta \exp(-\eta^2 r_{12}^2)$$

is used more frequently. The interested reader may work through the derivation for this case.

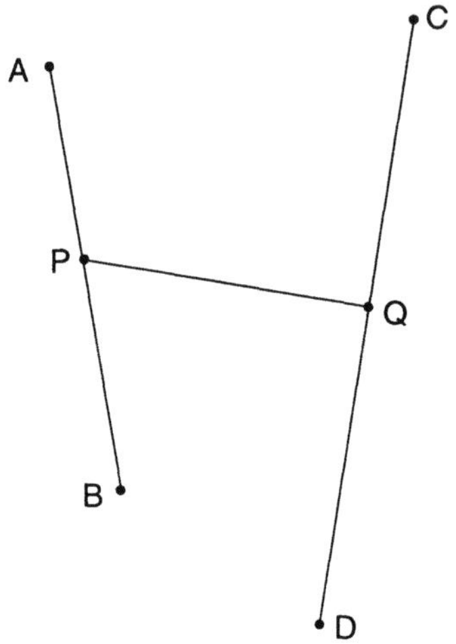

Figure 1. The disposition of the points A, B, C, D, P and Q for a typical two-electron, four-centre integral

Substituting (32) in (31) yields

$$\begin{aligned}\mathcal{I}_{pq;PQ} &= \frac{\Lambda_p\Lambda_q}{\sqrt{\pi}}\int_0^\infty d\eta\int d\boldsymbol{\tau}_1\int d\boldsymbol{\tau}_2\ \exp(-\zeta_{pP}r_{1P}^2)\exp(-\zeta_{qQ}r_{2Q}^2)\\ &\quad \exp(-\eta r_{12}^2)\eta^{-\frac{1}{2}} \qquad (33)\\ &= \frac{\Lambda_p\Lambda_q}{\sqrt{\pi}}\int_0^\infty d\eta\int d\boldsymbol{\tau}_1\int d\boldsymbol{\tau}_2\ \exp(-\zeta_{pP}r_{1P}^2-\zeta_{qQ}r_{2Q}^2\\ &\quad -\eta r_{12}^2)\eta^{-\frac{1}{2}} \qquad (34)\end{aligned}$$

where the square of the interelectronic distance is

$$r_{12}^2 = (x_1-x_2)^2+(y_1-y_2)^2+(z_1-z_2)^2 \qquad (35)$$

The argument of the exponential factor in (34) may be written

$$\begin{aligned}-\zeta_{pP}r_{1P}^2-\zeta_{qQ}r_{2Q}^2-\eta r_{12}^2 &= -\zeta_{pP}\left((x_1-P_x)^2+(y_1-P_y)^2+(z_1-P_z)^2\right)\\ &\quad -\zeta_{qQ}\left((x_2-Q_x)^2+(y_2-Q_y)^2+(z_2-Q_z)^2\right)\\ &\quad -\eta\left((x_1-x_2)^2+(y_1-y_2)^2+(z_1-z_2)^2\right)\end{aligned}$$

Gathering together the x-dependent components, and separately the y- and z-components allows the integral (34) to be written in the form

$$\mathcal{I}_{pq;PQ} = \frac{\Lambda_p \Lambda_q}{\sqrt{\pi}} \int_0^\infty d\eta \mathcal{I}_{pq;PQ;x} \mathcal{I}_{pq;PQ;y} \mathcal{I}_{pq;PQ;z} \eta^{-\frac{1}{2}} \tag{36}$$

where

$$\begin{aligned} \mathcal{I}_{pq;PQ;x} &= \int_{-\infty}^{\infty} dx_1 \int_{-\infty}^{\infty} dx_2 \exp(-\zeta_{pP}(x_1 - P_x)^2 - \zeta_{qQ}(x_2 - Q_x)^2 \\ &\quad -\eta(x_1 - x_2)^2) \end{aligned} \tag{37}$$

and similarly for $\mathcal{I}_{pq;PQ;y}$ and $\mathcal{I}_{pq;PQ;z}$:

$$\begin{aligned} \mathcal{I}_{pq;PQ;\lambda} &= \int_{-\infty}^{\infty} d\lambda_1 \int_{-\infty}^{\infty} d\lambda_2 \exp(-\zeta_{pP}(\lambda_1 - P_\lambda)^2 - \zeta_{qQ}(\lambda_2 - Q_\lambda)^2 \\ &\quad -\eta(\lambda_1 - \lambda_2)^2), \\ \lambda &= x, y, z \end{aligned} \tag{38}$$

The major steps in the evaluation of the electron repulsion integral are now completed. All that remains is to evaluate the integrals $\mathcal{I}_{pq;PQ;\lambda}$ and then integrate over η. This will be done in the following sections thereby completing the evaluation.

2.4. EVALUATION OF THE INTEGRAL (36)

To evaluate the integral $\mathcal{I}_{pq;PQ;x}$ make the substitutions

$$\mu = x_1 - P_x \tag{39}$$

$$\nu = x_2 - Q_x \tag{40}$$

so that

$$\begin{aligned} (x_1 - x_2)^2 &= (\mu - \nu + P_x - Q_x)^2 \\ &= (\mu - \nu + \overline{PQ}_x)^2 \end{aligned} \tag{41}$$

where

$$\overline{PQ}_x = P_x - Q_x \tag{42}$$

giving for the argument of the exponential factor in (37)

$$\begin{aligned} -\zeta_{pP}\mu^2 - \zeta_{qQ}\nu^2 - \eta(\mu - \nu + \overline{PQ}_x)^2 &= -\eta\overline{PQ}_x^2 - (\zeta_{pP} + \eta)\,\mu^2 \\ &\quad - (\zeta_{qQ} + \eta)\,\nu^2 - 2\eta\overline{PQ}_x\mu \\ &\quad +2\eta\overline{PQ}_x\nu + 2\eta\mu\nu \end{aligned} \tag{43}$$

which, together with $d\mu d\nu = dx_1 dx_2$, gives

$$\mathcal{I}_{pq;PQ;x} = \exp(-\eta \overline{PQ}_x^2) \int_{-\infty}^{\infty} d\mu \{ \exp\left(-\left(\zeta_{pP} + \eta\right) \mu^2 - 2\eta \overline{PQ}_x \mu \right)$$
$$\int_{-\infty}^{\infty} d\nu \exp\left(-\left(\zeta_{qQ} + \eta\right) \nu^2 + 2\eta \overline{PQ}_x \nu + 2\eta\mu\nu \right) \} \tag{44}$$

The evaluation of the integral over ν will be considered in the next section (2.5) and the integration over μ will be carried out in section 2.6. The reader may wish to pass over these details on first reading and proceed directly to section 2.7 where the evaluation of the two-electron integral is completed by carrying out the integration over η.

2.5. INTEGRATION OVER ν

The integral over ν may be written

$$\mathcal{J}_\nu = \int_{-\infty}^{\infty} d\nu \exp\left(-\left(\zeta_{qQ} + \eta\right) \nu^2 + 2\eta \left(\overline{PQ}_x + \mu \right) \nu \right) \tag{45}$$

The required integral has the form

$$\mathcal{J} = \int_{-\infty}^{\infty} dx \exp(-\alpha x^2 + \beta x) \tag{46}$$

and can be evaluated by the making the expansion

$$\exp(\beta x) = 1 + \beta x + \left(\frac{1}{2}\beta^2\right) x^2 + \left(\frac{1}{6}\beta^3\right) x^3 + \left(\frac{1}{24}\beta^4\right) x^4 + O\left(x^5\right) \tag{47}$$

integrating term by term and rearranging, giving the result

$$\begin{aligned} \int_{-\infty}^{\infty} dx \exp(-\alpha x^2 + \beta x) &= \exp\left(\frac{\beta^2}{4\alpha}\right) \int_{-\infty}^{\infty} dx \exp(-\alpha x^2) \\ &= \sqrt{\frac{\pi}{\alpha}} \exp\left(\frac{\beta^2}{4\alpha}\right) \end{aligned} \tag{48}$$

Putting $\alpha = (\zeta_{qQ} + \eta)$ and $\beta = 2\eta \left(\overline{PQ}_x + \mu \right)$ in (45), integration over ν leads to the result

$$\mathcal{J}_\nu = \sqrt{\frac{\pi}{(\zeta_{qQ} + \eta)}} \exp\left(\frac{\eta^2 \left(\overline{PQ}_x + \mu \right)^2}{(\zeta_{qQ} + \eta)} \right) \tag{49}$$

which can be rearranged to give

$$\mathcal{J}_\nu = \sqrt{\frac{\pi}{(\zeta_{qQ}+\eta)}} \exp\left(\frac{\eta^2\overline{PQ}_x^2}{\zeta_{qQ}+\eta}\right) \exp\left(\frac{\eta^2}{\zeta_{qQ}+\eta}\mu^2\right)$$
$$\exp\left(\frac{2\eta^2\overline{PQ}_x}{\zeta_{qQ}+\eta}\mu\right) \tag{50}$$

Substitution into (44) gives

$$\mathcal{I}_{pq;PQ;x} = \exp(-\eta\overline{PQ}_x^2) \exp\left(\frac{\eta^2\overline{PQ}_x^2}{\zeta_{qQ}+\eta}\right) \sqrt{\frac{\pi}{(\zeta_{qQ}+\eta)}}$$
$$\times \int_{-\infty}^{\infty} d\mu \exp\left(-(\zeta_{pP}+\eta)\mu^2 - 2\eta\overline{PQ}_x\mu\right) \exp\left(\frac{\eta^2}{\zeta_{qQ}+\eta}\mu^2\right)$$
$$\exp\left(\frac{2\eta^2\overline{PQ}_x}{\zeta_{qQ}+\eta}\mu\right) \tag{51}$$

and the integration over ν is completed.

2.6. INTEGRATION OVER μ

The argument of the combined exponential factors in the integrand of (51) is

$$-(\zeta_{pP}+\eta)\mu^2 - 2\eta\overline{PQ}_x\mu + \frac{\eta^2}{\zeta_{qQ}+\eta}\mu^2 + \frac{2\eta^2\overline{PQ}_x}{\zeta_{qQ}+\eta}\mu \tag{52}$$

Collecting terms in μ^2 and μ gives

$$\left[-(\zeta_{pP}+\eta) + \frac{\eta^2}{\zeta_{qQ}+\eta}\right]\mu^2 - \left[2\eta\overline{PQ}_x - \frac{2\eta^2\overline{PQ}_x}{\zeta_{qQ}+\eta}\right]\mu \tag{53}$$

The coefficient of μ^2 may then be rearranged

$$\left[\frac{-(\zeta_{pP}+\eta)(\zeta_{qQ}+\eta)+\eta^2}{(\zeta_{qQ}+\eta)}\right] = \left[\frac{-(\zeta_{pP}\zeta_{qQ} + (\zeta_{pP}+\zeta_{qQ})\eta)}{(\zeta_{qQ}+\eta)}\right] \tag{54}$$

and the coefficient of μ is

$$-\left[2\eta\overline{PQ}_x - \frac{2\eta^2\overline{PQ}_x}{(\zeta_{qQ}+\eta)}\right] = \left[-2\frac{\left(\eta\overline{PQ}_x(\zeta_{qQ}+\eta)\right) - \eta^2\overline{PQ}_x}{(\zeta_{qQ}+\eta)}\right] \tag{55}$$
$$= \left[-2\frac{\eta\overline{PQ}_x\zeta_{qQ}}{(\zeta_{qQ}+\eta)}\right]$$

so that (50) becomes

$$\mathcal{I}_{pq;PQ;x} = \exp(-\eta QP_x^2)\exp\left(\frac{\eta^2 QP_x^2}{\zeta_{qQ}+\eta}\right)\sqrt{\frac{\pi}{(\zeta_{qQ}+\eta)}} \times \int_{-\infty}^{\infty} d\mu \exp\left(\frac{-\left(\zeta_{pP}\zeta_{qQ}+(\zeta_{pP}+\zeta_{qQ})\,\eta\right)}{(\zeta_{qQ}+\eta)}\mu^2 - 2\frac{\eta\overline{PQ}_x\zeta_{qQ}}{(\zeta_{qQ}+\eta)}\mu\right) \quad (56)$$

Using (48) again with $\alpha = \left(\zeta_{pP}\zeta_{qQ}+(\zeta_{pP}+\zeta_{qQ})\,\eta\right)/(\zeta_{qQ}+\eta)$ and $\beta = -2\eta\overline{PQ}_x\zeta_{qQ}/(\zeta_{qQ}+\eta)$ yields the result

$$\mathcal{I}_{pq;PQ;x} = \exp(-\eta\overline{PQ}_x^2)\exp\left(\frac{\eta^2\overline{PQ}_x^2}{\zeta_{qQ}+\eta}\right)\sqrt{\frac{\pi}{(\zeta_{qQ}+\eta)}} \sqrt{\frac{\pi\,(\zeta_{qQ}+\eta)}{\left(\zeta_{pP}\zeta_{qQ}+(\zeta_{pP}+\zeta_{qQ})\,\eta\right)}}\exp\left(\frac{\left(2\eta\overline{PQ}_x\zeta_{qQ}\right)^2}{(\zeta_{qQ}+\eta)^2} \times \frac{(\zeta_{qQ}+\eta)}{4\left(\zeta_{pP}\zeta_{qQ}+(\zeta_{pP}+\zeta_{qQ})\,\eta\right)}\right) \quad (57)$$

Rearranging

$$\mathcal{I}_{pq;PQ;x} = \pi\frac{1}{\sqrt{\left(\zeta_{pP}\zeta_{qQ}+(\zeta_{pP}+\zeta_{qQ})\,\eta\right)}} \exp\left(\begin{array}{c} -\eta\overline{PQ}_x^2 + \frac{\eta^2\overline{PQ}_x^2}{\zeta_{qQ}+\eta} \\ +\frac{\left(\eta^2\overline{PQ}_x^2\zeta_{qQ}^2\right)}{(\zeta_{qQ}+\eta)\left(\zeta_{pP}\zeta_{qQ}+(\zeta_{pP}+\zeta_{qQ})\eta\right)}\end{array}\right) \quad (58)$$

Combining the first two terms in the argument of the exponential gives

$$-\eta\overline{PQ}_x^2 + \frac{\eta^2\overline{PQ}_x^2}{\zeta_{qQ}+\eta} = \frac{-\eta\overline{PQ}_x^2(\zeta_{qQ}+\eta)+\eta^2\overline{PQ}_x^2}{\zeta_{qQ}+\eta} = \frac{-\eta\overline{PQ}_x^2\zeta_{qQ}}{\zeta_{qQ}+\eta} \quad (59)$$

so that the integral becomes

$$\mathcal{I}_{pq;PQ;x} = \pi\frac{1}{\sqrt{\left(\zeta_{pP}\zeta_{qQ}+(\zeta_{pP}+\zeta_{qQ})\,\eta\right)}}$$

$$\exp(\frac{-\eta \overline{PQ}_x^2 \zeta_{qQ}}{\zeta_{qQ}+\eta} + \frac{\left(\eta^2 \overline{PQ}_x^2 \zeta_{qQ}^2\right)}{\left(\zeta_{qQ}+\eta\right)\left(\zeta_{pP}\zeta_{qQ}+\left(\zeta_{pP}+\zeta_{qQ}\right)\eta\right)} \tag{60}$$

$$+\frac{\left(\eta^2 \overline{PQ}_x^2 \zeta_{qQ}^2\right)}{\left(\zeta_{qQ}+\eta\right)\left(\zeta_{pP}\zeta_{qQ}+\left(\zeta_{pP}+\zeta_{qQ}\right)\eta\right)}) \tag{61}$$

Combining the two terms in the argument of the exponential function gives

$$\frac{-\eta \overline{PQ}_x^2 \zeta_{qQ}\left(\zeta_{pP}\zeta_{qQ}+\left(\zeta_{pP}+\zeta_{qQ}\right)\eta\right)+\eta^2 \overline{PQ}_x^2 \zeta_{qQ}^2}{\left(\zeta_{qQ}+\eta\right)\left(\zeta_{pP}\zeta_{qQ}+\left(\zeta_{pP}+\zeta_{qQ}\right)\eta\right)} \tag{62}$$

and then

$$\frac{-\eta \overline{PQ}_x^2 \zeta_{pP}\zeta_{qQ}^2-\eta^2 \overline{PQ}_x^2 \zeta_{pP}\zeta_{qQ}-\eta^2 \overline{PQ}_x^2 \zeta_{qQ}^2+\eta^2 \overline{PQ}_x^2 \zeta_{qQ}^2}{\left(\zeta_{qQ}+\eta\right)\left(\zeta_{pP}\zeta_{qQ}+\left(\zeta_{pP}+\zeta_{qQ}\right)\eta\right)} \tag{63}$$

or

$$\frac{-\eta \overline{PQ}_x^2 \zeta_{pP}\zeta_{qQ}\left(\zeta_{qQ}+\eta\right)\eta}{\left(\zeta_{qQ}+\eta\right)\left(\zeta_{pP}\zeta_{qQ}+\left(\zeta_{pP}+\zeta_{qQ}\right)\eta\right)} \tag{64}$$

Cancelling the factors of $\left(\zeta_{qQ}+\eta\right)\eta$ gives

$$\frac{-\overline{PQ}_x^2 \zeta_{pP}\zeta_{qQ}\eta}{\left(\zeta_{pP}\zeta_{qQ}+\left(\zeta_{pP}+\zeta_{qQ}\right)\eta\right)} \tag{65}$$

so that the integral becomes

$$\begin{aligned} \mathcal{I}_{pq;PQ;x} &= \pi\left(\zeta_{pP}\zeta_{qQ}+\left(\zeta_{pP}+\zeta_{qQ}\right)\eta\right)^{-\frac{1}{2}} \\ &\quad \exp\left(-\frac{\overline{PQ}_x^2 \zeta_{pP}\zeta_{qQ}\eta}{\left(\zeta_{pP}\zeta_{qQ}+\left(\zeta_{pP}+\zeta_{qQ}\right)\eta\right)}\right) \end{aligned} \tag{66}$$

Defining

$$\rho = \frac{\zeta_{pP}\zeta_{qQ}}{\zeta_{pP}+\zeta_{qQ}} \tag{67}$$

The denominator in the argument of the exponential function becomes

$$\zeta_{pP}\zeta_{qQ}+\left(\zeta_{pP}+\zeta_{qQ}\right)\eta = \left(\zeta_{pP}+\zeta_{qQ}\right)\left(\frac{\zeta_{pP}\zeta_{qQ}}{\zeta_{pP}+\zeta_{qQ}}+\eta\right) \tag{68}$$

$$= \left(\zeta_{pP}+\zeta_{qQ}\right)\left(\rho+\eta\right) \tag{69}$$

so that the total argument is

$$\frac{\overline{PQ}_x^2 \zeta_{pP}\zeta_{qQ}\eta}{\left(\zeta_{pP}+\zeta_{qQ}\right)\left(\rho+\eta\right)} \tag{70}$$

which can be written as

$$\frac{\overline{PQ}_x^2 \rho\eta}{\rho+\eta} \tag{71}$$

The required integral $\mathcal{I}_{pq;PQ;x}$ thus assumes the form

$$\mathcal{I}_{pq;PQ;x} = \pi \left(\zeta_{pP} + \zeta_{qQ}\right)^{-\frac{1}{2}} (\rho+\eta)^{-\frac{1}{2}} \exp\left(-\frac{\overline{PQ}_x^2 \rho\eta}{\rho+\eta}\right) \tag{72}$$

with $\mathcal{I}_{pq;PQ;y}$ and $\mathcal{I}_{pq;PQ;z}$ being given similarly by

$$\begin{aligned} \mathcal{I}_{pq;PQ;\lambda} &= \pi \left(\zeta_{pP} + \zeta_{qQ}\right)^{-\frac{1}{2}} (\rho+\eta)^{-\frac{1}{2}} \exp\left(-\frac{\overline{PQ}_\lambda^2 \rho\eta}{\rho+\eta}\right), \\ \lambda &= x, y, z \end{aligned} \tag{73}$$

The integration over μ is thus completed.

2.7. INTEGRATION OVER η

Recall that the total integral required is

$$\mathcal{I}_{abcd;ABCD} = \mathcal{I}_{pq;PQ} = \frac{\Lambda_p \Lambda_q}{\sqrt{\pi}} \int_0^\infty d\eta \mathcal{I}_{pq;PQ;x} \mathcal{I}_{pq;PQ;y} \mathcal{I}_{pq;PQ;z} \eta^{-\frac{1}{2}} \tag{74}$$

which on substituting (73) becomes

$$\begin{aligned} \mathcal{I}_{pq;PQ} &= \Lambda_p \Lambda_q \pi^{\frac{5}{2}} \left(\zeta_{pP} + \zeta_{qQ}\right)^{-\frac{3}{2}} \int_0^\infty d\eta \eta^{-\frac{1}{2}} (\rho+\eta)^{-\frac{3}{2}} \\ &\times \exp\left(-\frac{\overline{PQ}_x^2 \rho\eta}{\rho+\eta}\right) \exp\left(-\frac{\overline{PQ}_y^2 \rho\eta}{\rho+\eta}\right) \exp\left(-\frac{\overline{PQ}_z^2 \rho\eta}{\rho+\eta}\right) \end{aligned} \tag{75}$$

or combining the exponential factors

$$\begin{aligned} \mathcal{I}_{pq;PQ} &= \Lambda_p \Lambda_q \pi^{\frac{5}{2}} \left(\zeta_{pP} + \zeta_{qQ}\right)^{-\frac{3}{2}} \int_0^\infty d\eta \eta^{-\frac{1}{2}} (\rho+\eta)^{-\frac{3}{2}} \\ &\exp\left(-\frac{\overline{PQ}^2 \rho\eta}{\rho+\eta}\right) \end{aligned} \tag{76}$$

where

$$\overline{PQ}^2 = \overline{PQ}_x^2 + \overline{PQ}_y^2 + \overline{PQ}_z^2 \tag{77}$$

Now make the substitution

$$\sigma^2 = \frac{\eta}{\rho+\eta} \tag{78}$$

for which

$$d\sigma.2\sigma = d\eta \frac{\rho}{(\rho+\eta)^2} \tag{79}$$

so that

$$\begin{aligned} d\sigma.\frac{2}{\rho} &= d\eta \frac{1}{\sigma}\frac{1}{(\rho+\eta)^2} \\ &= d\eta \frac{(\rho+\eta)^{\frac{1}{2}}}{\eta^{\frac{1}{2}}}\frac{1}{(\rho+\eta)^2} \\ &= d\eta\, \eta^{-\frac{1}{2}}(\rho+\eta)^{-\frac{3}{2}} \end{aligned} \tag{80}$$

The integration limits for η are $(0,\infty)$ and, therefore, those for σ are $(0,1)$. Equation (76) therefore becomes

$$\mathcal{I}_{pq;PQ} = 2\Lambda_p\Lambda_q\pi^{\frac{5}{2}}(\zeta_{pP}+\zeta_{qQ})^{-\frac{3}{2}}\frac{1}{\rho}\int_0^1 d\sigma \exp\left(-\rho\overline{PQ}^2\sigma\right) \tag{81}$$

Then, rearranging

$$\begin{aligned} (\zeta_{pP}+\zeta_{qQ})^{-\frac{3}{2}}\frac{1}{\rho} &= \frac{1}{(\zeta_{pP}+\zeta_{qQ})(\zeta_{pP}+\zeta_{qQ})^{\frac{1}{2}}}\frac{\zeta_{pP}+\zeta_{qQ}}{\zeta_{pP}\zeta_{qQ}} \\ &= \frac{1}{\zeta_{pP}\zeta_{qQ}(\zeta_{pP}+\zeta_{qQ})^{\frac{1}{2}}} \end{aligned} \tag{82}$$

enables the integral to be written

$$\mathcal{I}_{pq;PQ} = 2\Lambda_p\Lambda_q\pi^{\frac{5}{2}}(\zeta_{pP}\zeta_{qQ}(\zeta_{pP}+\zeta_{qQ}))^{-\frac{1}{2}}\int_0^1 d\sigma \exp\left(-\rho\overline{PQ}^2\sigma\right) \tag{83}$$

which after carrying out the integration over σ gives the result

$$\mathcal{I}_{pq;PQ} = 2\Lambda_p\Lambda_q\pi^{\frac{5}{2}}(\zeta_{pP}\zeta_{qQ}(\zeta_{pP}+\zeta_{qQ}))^{-\frac{1}{2}}F_0(\rho\overline{PQ}^2) \tag{84}$$

which, on substitution for ρ, gives the final result

$$\begin{aligned} \mathcal{I}_{abcd;ABCD} &= 2\Lambda_p\Lambda_q\pi^{\frac{5}{2}}(\zeta_{pP}\zeta_{qQ}(\zeta_{pP}+\zeta_{qQ}))^{-\frac{1}{2}} \\ &\quad \times F_0\left(\frac{\zeta_{pP}\zeta_{qQ}}{\zeta_{pP}+\zeta_{qQ}}\overline{PQ}^2\right) \end{aligned} \tag{85}$$

where $F_0(x)$ is closely related to the error function, erf

$$F_0(x) = \frac{1}{2}\sqrt{\frac{\pi}{x}}\,\mathrm{erf}\,(\sqrt{x}) \tag{86}$$

The functions $F_0(x)$ and $\mathrm{erf}(\sqrt{x})$ cannot be evaluated analytically and, as will be described below, usually found by tabular interpolation.

3. Integrals over contracted s-type Gaussian functions

3.1. CONTRACTED GAUSSIAN FUNCTIONS

Frequently, contracted Gaussian basis functions are used in molecular electronic structure calculations; that is, fixed linear combinations of primitive Gaussian functions:

$$\phi = \sum_a D_{aA} \exp(-\zeta_{aA} r_{1A}^2). \tag{87}$$

The evaluation of integrals over contracted Gaussian-type functions demands the evaluation of quantities such as

$$[\mathbf{0}]^{(0)} = D_{aA} D_{bB} D_{cC} D_{dD} \mathcal{I}_{abcd;ABCD} \tag{88}$$

which can be written more explicitly as

$$\begin{aligned}[\mathbf{0}]^{(0)} &= D_{aA} D_{bB} D_{cC} D_{dD} \int d\boldsymbol{\tau}_1 \int d\boldsymbol{\tau}_2 \exp(-\zeta_{aA} r_{1A}^2) \exp(-\zeta_{cC} r_{2C}^2) \frac{1}{r_{12}} \\ &\quad \exp(-\zeta_{bB} r_{1B}^2) \exp(-\zeta_{dD} r_{2D}^2)\end{aligned} \tag{89}$$

The coefficients D_{pP} are termed contraction coefficients.

3.2. THE ALGORITHM OF GILL *ET AL*

Following Gill et al[22] an efficient algorithm for the evaluation of the $[0]^{(0)}$ can be written down as follows:-

Define the following quantities associated with the shells aA and bB

$$\sigma_{pP} = \frac{1}{\zeta_{aA} + \zeta_{bB}} \tag{90}$$

$$\mathbf{P} = (\zeta_{aA}\mathbf{A} + \zeta_{bB}\mathbf{B})\, \sigma_{pP} \tag{91}$$

$$U_{pP} = (\pi\sigma_{pP})^{3/2} D_{aA} D_{bB} \exp\left(-\zeta_{aA}\zeta_{bB}\sigma_{pP} |\mathbf{A} - \mathbf{B}|^2\right) \tag{92}$$

and the following quantities associated with the shells cC and dD

$$\sigma_{qQ} = \frac{1}{\zeta_{cC} + \zeta_{dD}} \tag{93}$$

$$\mathbf{Q} = (\zeta_{cC}\mathbf{C} + \zeta_{dD}\mathbf{D})\, \sigma_{qQ} \tag{94}$$

$$U_{qQ} = (\pi\sigma_{qQ})^{3/2} D_{cC} D_{dDB} \exp\left(-\zeta_{cCA}\zeta_{dD}\sigma_{qQ} \left|\mathbf{C} - \mathbf{D}\right|^2\right) \qquad (95)$$

These shell-pair quantities can be evaluated in a preliminary loop running only over pairs of shells.

The main loop runs over all possible quartets of shells. Within this loop the following parameters are defined

$$R^2 = \left|\mathbf{P} - \mathbf{Q}\right|^2 \qquad (96)$$

$$\vartheta^2 = \frac{1}{\sigma_{pP} + \sigma_{qQ}} \qquad (97)$$

$$T = \vartheta^2 R^2 \qquad (98)$$

and

$$U = U_{pP} U_{qQ} \qquad (99)$$

It has been shown above that the integral $[0]^{(0)}$ may be written

$$[0]^{(0)} = U(2\vartheta^2)^{\frac{1}{2}} \left(\frac{2}{\pi}\right)^{\frac{1}{2}} \int_0^1 \exp(-Tu^2)du \qquad (100)$$

In most calculations, this expression is generalize to

$$[0]^{(m)} = U(2\vartheta^2)^{\frac{1}{2}} \left(\frac{2}{\pi}\right)^{\frac{1}{2}} \int_0^1 u^{2m} \exp(-Tu^2)du \qquad (101)$$

where

$$0 \leq m \leq L \qquad (102)$$

for some given L. The use of these more general functions will become apparent in the following section where integrals over functions of higher angular symmetries are discussed.

Following Gill et al[22], consider the evaluation of each of the parameters in the shell-quartet loop. R^2 is the most expensive to compute; direct evaluation demanding 8 floating point operations. Precomputing the quantities

$$\omega = \zeta_{aA} \left|\mathbf{A} - \mathbf{B}\right| / (\zeta_{aA} + \zeta_{bB}) \qquad (103)$$

$$\mu = (\mathbf{A} - \mathbf{B}) \cdot (\mathbf{B} - \mathbf{Q}) / \left|\mathbf{A} - \mathbf{B}\right| \qquad (104)$$

and

$$\nu = \left|\mathbf{B} - \mathbf{Q}\right|^2 - \mu^2 \qquad (105)$$

allows R^2 to be evaluated as

$$R^2 = (\omega + \mu)^2 + \nu \qquad (106)$$

requiring only 3 floating point operations.

The determination of ϑ^2 can be achieved in just 2 floating point operations and furthermore it can be scaled without additional computation

$$\frac{\vartheta^2}{2\Delta} = \frac{1}{2\Delta} \frac{1}{\sigma_{pP} + \sigma_{qQ}} \tag{107}$$

where 2Δ is an interpolation interval to be defined below. Similarly, T can be scaled without incurring any additional computation,

$$\frac{T}{2\Delta} = \frac{\vartheta^2}{2\Delta} R^2 \tag{108}$$

For $T > T_0$, the asymptotic formula

$$\left(\frac{2}{\pi}\right)^{\frac{1}{2}} \int_0^1 u^{2m} \exp(-Tu^2)\, du \sim \frac{(2m-1)!!}{(2T)^{m+\frac{1}{2}}} \tag{109}$$

gives immediately the simple result

$$[0]^{(m)} \sim (2m-1)!! \frac{U}{R^{2m+1}} \tag{110}$$

For $T \leq T_0$ Gill et al write (101) in the form

$$[0]^{(m)} = (2m-1)!! U \left(\frac{\vartheta^2}{2\Delta}\right)^{m+\frac{1}{2}} G_m(T) \tag{111}$$

where

$$G_m(T) = \frac{(4\Delta)^{m+\frac{1}{2}}}{(2m-1)!!} \left(\frac{2}{\pi}\right)^{\frac{1}{2}} \int_0^1 u^{2m} \exp(-Tu^2)\, du \tag{112}$$

They compute $G_L(T)$ using an efficient interpolation scheme and obtain the remaining $G_m(T)$, $(0 < m < L)$ from the downward recursion

$$G_m(T) = \left(\frac{2}{\pi}\right)^{\frac{1}{2}} \frac{(4\Delta)^{m+\frac{1}{2}}}{(2m-1)!!} \exp(-T) + \left(\frac{T}{2\Delta}\right) G_{m+1}(T) \tag{113}$$

The interpolation is carried out on a grid defined by

$$X_j = (2j+1)\,\Delta, \quad j = 0, 1, \ldots \tag{114}$$

so that the index j is given by

$$j = \mathrm{INT}\left(\frac{x}{2\Delta}\right) \tag{115}$$

Gill et al[22] found a cubic Chebyshev interpolation formula to be "a compromise between expense and accuracy".

$$f(x) \approx f_0 + \frac{x}{2\Delta}\left(f_1 + \frac{x}{2\Delta}\left(f_2 + \frac{x}{2\Delta}f_3\right)\right) \tag{116}$$

The optimized algorithm developed by Gill et al[22] for generating the $[\mathbf{0}]^{(m)}$ integrals is summarized in Figure 2.

Step		No. of flops
(1)	$R^2 = (\omega+\mu)^2 + \nu$	3
(2)	$\frac{\vartheta^2}{2\Delta} = \frac{1}{2\Delta}\frac{1}{\sigma_{pP}+\sigma_{qQ}}$	2
(3)	$\frac{T}{2\Delta} = \frac{\vartheta^2}{2\Delta}R^2$	1
(4)	**IF** $\left[\frac{T}{2\Delta}\right] < \left[\frac{T_{critical}}{2\Delta}\right]$ **THEN**	1
(5)	$J = \mathrm{INT}\left[\frac{T}{2\Delta}\right]$	1
(6)	$G_L(T) = g_0(j,L) + \frac{T}{2\Delta}\left[g_1(j,L) + \frac{T}{2\Delta}\left[g_2(j,L) + \frac{T}{2\Delta}g_3(j,L)\right]\right]$	6
(7)	$\exp(-T) = e_0(j) + \frac{T}{2\Delta}\left[e_1(j) + \frac{T}{2\Delta}\left[e_2(j) + \frac{T}{2\Delta}e_3(j)\right]\right]$	6
(8)	$G_m(T) = \sqrt{\frac{2}{\pi}}\frac{(4\Delta)^{m+\frac{1}{2}}}{(2m+1)!!}\exp(-T) + \left[\frac{T}{2\Delta}\right]G_{m+1}(T)$	3L
(9a)	$[0]^{(m)} = (2m-1)!!U_{pP}U_{qQ}\left[\frac{\vartheta^2}{2\Delta}\right]^{m+\frac{1}{2}}G_m(T)$	2(L+1)+$\surd$
	ELSE	
(9b)	$[0]^{(m)} = (2m-1)!!U_{pP}U_{qQ}\frac{1}{R^{2m+1}}$	(L+1)+$\surd$
	END IF	

Figure 2. The algorithm of Gill *et al*

4. Integrals over Gaussian functions of p-symmetry and higher

The integrals $[\mathbf{0}]^{(0)}$ may be written

$$[\mathbf{0}]^{(0)} = U(2\vartheta^2)^{\frac{1}{2}} \left(\frac{2}{\pi}\right)^{\frac{1}{2}} \int_0^1 \exp(-Tu^2)du \tag{117}$$

are electron repulsion integrals for s-type Gaussians. The integrals $[\mathbf{0}]^{(m)}$

$$[\mathbf{0}]^{(m)} = U(2\vartheta^2)^{\frac{1}{2}} \left(\frac{2}{\pi}\right)^{\frac{1}{2}} \int_0^1 u^{2m} \exp(-Tu^2)du \tag{118}$$

where

$$0 \leq m \leq L \tag{119}$$

are intermediates which arise in the evaluation of integrals involving functions of p-symmetry and higher. There has been considerable progress since the review of Saunders[9] in obtaining recurrence relations which allow the efficient evaluation of integrals involving functions of higher symmetries, in the determination of the integrals which arise when energy gradients are evaluated and which show promise in the relativistic molecular integral problem.

4.1. THE OBARA - SAIKA SCHEME

In 1986, Obara and Saika[15] presented a recurrence relation for uncontracted Cartesian Gaussian functions which was found to be very suitable for efficient implementation on vector processing computers. The Obara-Saika scheme was found to be more economic than either the McMurchie-Davidson[11] or the Rys quadrature[12] [13] [14] approaches, and, indeed, more efficient than any of the hybrid schemes[9] suggested in the early '80s. Schlegel[16] had earlier considered a very similar scheme. Indeed, the Obara-Saika scheme has its origins in the seminal work of Boys[5].

The Obara-Saika scheme is based on a recurrence relation for incrementing angular momenta. Specifically, the Obara-Saika scheme exploits the following recurrence relation

$$\begin{aligned}
[(\mathbf{a}+\mathbf{1}_\mu)\mathbf{b} \mid \mathbf{cd}]^{(m)} &= (P_\mu - A_\mu)[\mathbf{ab} \mid \mathbf{cd}]^{(m)} + (W_\mu - P_\mu)[\mathbf{ab} \mid \mathbf{cd}]^{(m+1)} \\
&+ \frac{a_i}{2\zeta}\left([(\mathbf{a}-\mathbf{1}_\mu)b \mid \mathbf{cd}]^{(m)} - \frac{\eta}{\zeta+\eta}[(\mathbf{a}-\mathbf{1}_\mu)b \mid \mathbf{cd}]^{(m+1)}\right) \\
&+ \frac{b_i}{2\zeta}\left([\mathbf{a}(\mathbf{b}-\mathbf{1}_\mu) \mid \mathbf{cd}]^{(m)} - \frac{\eta}{\zeta+\eta}[\mathbf{a}(\mathbf{b}-\mathbf{1}_\mu) \mid \mathbf{cd}]^{(m+1)}\right) \\
&+ \frac{c_i}{2(\zeta+\eta)}\left([\mathbf{ab} \mid (\mathbf{c}-\mathbf{1}_\mu)d]^{(m+1)} + [\mathbf{ab} \mid \mathbf{c}(\mathbf{d}-\mathbf{1}_\mu)]^{(m+1)}\right)
\end{aligned} \tag{120}$$

where $\mathbf{a} = (a_x, a_y, a_z)$ and a_μ, $\mu = x, y, z$ are the powers of x, y and z in the Cartesian Gaussian basis function centred on $\mathbf{A}$:

$$x_A^{a_x} y_A^{a_y} z_A^{a_z} \exp(-\zeta r_A^2) \tag{121}$$

$\mathbf{b}$, $\mathbf{c}$ and $\mathbf{d}$ are similarly defined for centres B, C and D, respectively. $\mathbf{1}_\mu = (\delta_{x\mu}, \delta_{y\mu}, \delta_{z\mu})$.

4.2. THE HEAD-GORDON - POPLE SCHEME

Head-Gordon and Pople[17] tailored the method of Obara and Saika for contracted Gaussian functions and simplified the recurrence relations. In particular, they developed the recurrence

$$\begin{aligned} (\mathbf{a}\,(\mathbf{b}+\mathbf{1}_\mu)| &= ((\mathbf{a}+\mathbf{1}_\mu)\,\mathbf{b}| + (A_\mu - B_\mu)\,(\mathbf{ab}| \\ \mu &= x, y, z \end{aligned} \tag{122}$$

The recurrence relations obtained by Head-Gordon and Pople do not involve exponents or different values of m.

As an example of the use of the recurrence relations of Head-Gordon and Pople in the evaluation of integrals involving functions of higher angular momentum, consider the integrals $(\mathbf{fp} \mid \mathbf{gd})$. For the bra we have from (122)

$$(\mathbf{fp}| = (\mathbf{gs}| + (A_\mu - B_\mu)\,(\mathbf{fs}| \tag{123}$$

whilst for the ket

$$|\mathbf{gd}) = |\mathbf{hp}) + (C_\mu - D_\mu)\,|\mathbf{gp}) \tag{124}$$

which can be further simplified by using 122 again to get

$$|\mathbf{hp}) = |\mathbf{is}) + (C_\mu - D_\mu)\,|\mathbf{hs}) \tag{125}$$

and

$$|\mathbf{gp}) = |\mathbf{hs}) + (C_\mu - D_\mu)\,|\mathbf{gs}) \tag{126}$$

Hence for the integral block $(\mathbf{fp} \mid \mathbf{gd})$, use of (123) gives

$$(\mathbf{fp} \mid \mathbf{gd}) = (\mathbf{gs} \mid \mathbf{gd}) + (A_\mu - B_\mu)\,(\mathbf{fs} \mid \mathbf{gd}) \tag{127}$$

with the first integral on the right-hand-side being given (by using (124))

$$(\mathbf{gs} \mid \mathbf{gd}) = (\mathbf{gs} \mid \mathbf{hp}) + (C_\mu - D_\mu)\,(\mathbf{gs} \mid \mathbf{gp})\,. \tag{128}$$

The first integral on the right-hand-side of expression (128) can be further simplified by using (125) to give

$$(\mathbf{gs} \mid \mathbf{hp}) = (\mathbf{gs} \mid \mathbf{is}) + (C_\mu - D_\mu)\,(\mathbf{gs} \mid \mathbf{hs}) \tag{129}$$

Similarly, the second integral on the right-hand-side of (128) can be simplified by using (126) giving

$$(\mathbf{gs} \mid \mathbf{gp}) = (\mathbf{gs} \mid \mathbf{hs}) + (C_\mu - D_\mu)\,(\mathbf{gs} \mid \mathbf{gs}) \tag{130}$$

The second integral on the right-hand-side of (127) may be simplified by exploiting (124)

$$(\mathbf{fs} \mid \mathbf{gd}) = (\mathbf{fs} \mid \mathbf{hp}) + (C_\mu - D_\mu)\,(\mathbf{fs} \mid \mathbf{gp}) \tag{131}$$

which, in turn, can be further reduced by means of (125)

$$(\mathbf{fs} \mid \mathbf{hp}) = (\mathbf{fs} \mid \mathbf{is}) + (C_\mu - D_\mu)\,(\mathbf{fs} \mid \mathbf{hs}) \tag{132}$$

and (126)

$$(\mathbf{fs} \mid \mathbf{gp}) = (\mathbf{fs} \mid \mathbf{hs}) + (C_\mu - D_\mu)\,(\mathbf{fs} \mid \mathbf{gs}) \tag{133}$$

The integral block $(\mathbf{fp} \mid \mathbf{gd})$ is generated from the blocks $(*\mathbf{s} \mid *\mathbf{s})$ or $(*\mathbf{0} \mid *\mathbf{0})$.

Further progress on the problem of integral evaluation has been made by Gill *et al*[18] [19] [22] [23], Lindh *et al*[25] [26] [27], and Hamilton and Schaefer[20] [21].

5. Systematic approximation of molecular integrals

The large number of two-electron integrals which arise in most quantum chemical calculations has motivated many studies of reliable schemes for approximating them. Indeed, the whole of semi-empirical quantum chemistry can be regarded as the development of schemes for approximating the two-electron integrals which arise in *ab initio* molecular electronic structure calculations.

5.1. THE TWO-ELECTRON INTEGRAL SUPERMATRIX

The two-electron integrals can be arranged as a symmetric, positive definite supermatrix with rows and columns labelled by the index

$$(ij),\ j = 1, 2, \ldots, i;\ i = 1, 2, \ldots, n \tag{134}$$

so that the supermatrix takes the form

$$\begin{pmatrix} & 11 & 21 & 22 & 31 & 32 & 33 & \dots \\ 11 & [1111] & & & & & & \\ 21 & [2111] & [2121] & & & & & \\ 22 & [2211] & [2221] & [2222] & & & & \\ 31 & [3111] & [3121] & [3122] & [3131] & & & \\ 32 & [3211] & [3221] & [3222] & [3231] & [3232] & & \\ 33 & [3311] & [3321] & [3322] & [3331] & [3332] & [3333] & \\ \dots & & & & & & & \end{pmatrix} \quad (135)$$

A unique index can be assigned to each row/column by means of the formula

$$(ij) = [i(i-1)]/2 + j \quad (136)$$

Systematic approximation of the molecular integrals arising in a particular calculation should address the problem of approximating the whole of this supermatrix rather than the individual elements.

5.2. APPLICATION OF THE SCHWARTZ INEQUALITY

Application of the Schwartz inequality to the two-electron repulsion integrals gives[30]

$$[pq \mid rs] \leq \sqrt{[pq \mid pq]}.\sqrt{[rs \mid rs]} \quad (137)$$

The two-centre integrals $[pq \mid pq]$ and $[rs \mid rs]$ can be evaluated rapidly and the above inequality might facilitate the elimination of large numbers of electron repulsion integrals without sacrificing accuracy of calculated properties. Introducing a threshold τ_1, quantum chemical algorithms can be constructed in which all integrals for which

$$J_{pqrs} < \tau_1 \quad (138)$$

where

$$J_{pqrs} = \sqrt{[pq \mid pq]}.\sqrt{[rs \mid rs]} \quad (139)$$

are neglected for a suitably chosen τ_1. A further refinement might be to define a second threshold τ_2 so that integrals for which

$$\tau_1 < J_{pqrs} < \tau_2 \quad (140)$$

can be approximated and only integrals for which

$$\tau_2 < J_{pqrs} \quad (141)$$

are explicitly evaluated. In the direct self-consistent field procedure[8] and in other algorithms which are both direct and iterative the thresholds might be changed from iteration to iteration and reduced as convergence is approached.

For large, that is extended, systems, the total number of two-electron integrals which have a magnitude less than some chosen tolerance may be very large and the cumulative effect of neglecting these integrals may be quite significant. Indeed, the Schwartz inequality discussed above can result in an unacceptable loss of accuracy for very large molecular systems even though the threshold may be set quite tightly.

5.3. FAST GAUSSIAN METHODS

Much recent work in *ab initio* quantum chemistry has been directed towards the development of "fast Gaussian" methods which scale linearly $[O(n)]$ or as $O(n \log n)$ with the number of electrons[31]-[39] . Indeed, in studies of the many-body problem across a range of applications for both the classical and the quantum formulation[40]-[49], a variety of "fast hierarchical" methods have been introduced which significantly reduce the complexity of the N-body problem by subdividing space into a fixed hierarchy of cells and exploiting a tree-like data structure. In "many-body tree" approaches the root (parent) problem is recursively subdivided into smaller cells (children). The complexity of the algorithm is reduced by approximating the distribution of particles within a cell as a series (multipole) expansion which converges rapidly in the far field. Unlike the classical many-body problem the quantum mechanical formulation involves continuous distributions. Molecular integrals that involve charge distributions which do not penetrate may be handled via a multipole approach whilst integrals with overlapping charge distributions are near-field and must be evaluated. The efficiency of these hierarchical multipole methods in quantum chemical application is critically dependent on the effective partition of near- and far-field interactions. This partition is complicated by the fact that it depends not only on the extent and separation of the charge distributions but also on the tolerance imposed and the order of multipole expansion employed.

Carlson and Rushbrooke [50] introduced a (real arithmetic) two-centre multipole expansion in 1950. The expansion expresses the electron-electron interaction as a sum of products of three terms - the first depending on the coordinates of electron 1 with respect to centre **P** defined in Figure 1, the second depending on the distance $\mathbf{P} - \mathbf{Q}$, and the third depending on the coordinates of electron 2 with respect to centre **Q**. Explicitly, the multipole expansion may be written

$$\frac{1}{|\mathbf{r}-\mathbf{r}'|} = \sum_{\ell_p}^{\infty}(-1)^{\ell_p}\sum_{\ell_q}^{\infty}\sum_{m_p=-\ell_p}^{\ell_p}\sum_{m_q=-\ell_q}^{\ell_q}$$
$$\Big(C_{\ell_p}^{m_p}[\mathbf{r}-\mathbf{P}]\ \mathcal{C}_{\ell_p+\ell_q}^{m_p+m_q}[\mathbf{P}-\mathbf{Q}]\ C_{\ell_q}^{m_q}[\mathbf{r}'-\mathbf{Q}]$$
$$+S_{\ell_p}^{m_p}[\mathbf{r}-\mathbf{P}]\ \mathcal{S}_{\ell_p+\ell_q}^{m_p+m_q}[\mathbf{P}-\mathbf{Q}]\ S_{\ell_q}^{m_q}[\mathbf{r}'-\mathbf{Q}]$$
$$-S_{\ell_p}^{m_p}[\mathbf{r}-\mathbf{P}]\ \mathcal{C}_{\ell_p+\ell_q}^{m_p+m_q}[\mathbf{P}-\mathbf{Q}]\ S_{\ell_q}^{m_q}[\mathbf{r}'-\mathbf{Q}]$$
$$+C_{\ell_p}^{m_p}[\mathbf{r}-\mathbf{P}]\ \mathcal{S}_{\ell_p+\ell_q}^{m_p+m_q}[\mathbf{P}-\mathbf{Q}]\ C_{\ell_q}^{m_q}[\mathbf{r}'-\mathbf{Q}]\Big) \quad (142)$$

where

$$C_\ell^m[\mathbf{r}] = |\mathbf{r}|^\ell(\ell+m)!\,P_\ell^m(\cos\theta_{\mathbf{r}})\ \cos(m\phi_{\mathbf{r}}) \quad (143)$$

and

$$S_\ell^m[\mathbf{r}] = |\mathbf{r}|^\ell(\ell+m)!\,P_\ell^m(\cos\theta_{\mathbf{r}})\ \sin(m\phi_{\mathbf{r}}) \quad (144)$$

are multipole tensors, and

$$\mathcal{C}_\ell^m[\mathbf{R}] = |\mathbf{R}|^{-(\ell+1)}\frac{1}{(\ell-m)!}P_\ell^m(\cos\theta_{\mathbf{R}})\cos(m\phi_{\mathbf{R}}) \quad (145)$$

and

$$\mathcal{S}_\ell^m[\mathbf{R}] = |\mathbf{R}|^{-(\ell+1)}\frac{1}{(\ell-m)!}P_\ell^m(\cos\theta_{\mathbf{R}})\sin(m\phi_{\mathbf{R}}) \quad (146)$$

Computations are simplified by the following relations between the multipole tensors for positive and negative values of m:-

$$C_\ell^{-m}[\mathbf{r}] = (-1)^m C_\ell^m[\mathbf{r}] \quad (147)$$

$$S_\ell^{-m}[\mathbf{r}] = (-1)^{(m+1)} S_\ell^m[\mathbf{r}] \quad (148)$$

$$\mathcal{C}_\ell^{-m}[\mathbf{r}] = (-1)^m \mathcal{C}_\ell^m[\mathbf{r}] \quad (149)$$

$$\mathcal{S}_\ell^{-m}[\mathbf{r}] = (-1)^{(m+1)} \mathbf{S}_\ell^m[\mathbf{r}] \quad (150)$$

The reader is referred elsewhere for further details[31]-[39].

5.4. CHOLESKY DECOMPOSITION

It has been shown in the previous section how the evaluation of integrals can be simplified in studies of extended molecules where there is no significant overlap between the two charge distributions involved for a significant fraction of the integrals. Accurate studies of small molecules necessitate the use of large basis sets for which the corresponding charge distributions do overlap significantly and the methods described above are not applicable. However, computational linear dependence amongst the charge distributions involved in the two-electron integral supermatrix can be exploited by means of a Cholesky decomposition. A brief outline of this approach is given below. Further details can be found elsewhere[51]-[55].

The two-electron supermatrix, $\mathbf{V} = V_{ij;kl}$, is a symmetric, positive definite matrix may be written in the form

$$\mathbf{V} = \mathbf{L}\mathbf{L}^{\dagger} \tag{151}$$

where $\mathbf{L}$ is a lower triangular matrix and $\mathbf{L}^{\dagger}$ is its transpose. Explicitly, this matrix product may be written

$$V_{(ij),(kl)} = \sum_{(pq)} L_{(ij),(pq)} L_{(ij),(pq)} \tag{152}$$

where the index (pq) runs over all possible charge distributions. However, in the presence of computational linear dependence, an approximation to $\mathbf{V}$ may be written

$$V_{(ij),(kl)} = \sum_{(pq)}^{\nu} L_{(ij),(pq)} L_{(ij),(pq)} \tag{153}$$

where the effective numerical rank of the two-electron integral matrix, ν, is considerably less than the total number of charge distributions.

A very stable algorithm for the construction of $\mathbf{L}$ is known which most importantly does not require the construction of the full $\mathbf{V}$ supermatrix. For $(ij) = 1, 2, ..., [n(n+1)]/2$

$$L_{(ij),(ij)} = \sqrt{V_{(ij),(ij)} - \sum_{(kl)=1}^{(ij)-1} L_{(ij),(kl)}} \tag{154}$$

and

$$\begin{aligned} L_{(pq),(kl)} &= \frac{\left[V_{(mn),(ij)} - \sum_{(kl)=1}^{(ij)-1} L_{(pq),(kl)} L_{(ij),(kl)}\right]}{L_{(ij),(ij)}}, \\ (pq) &= [j(j+1)]/2, \; ..., \; [n(n+1)]/2 \end{aligned} \tag{155}$$

The summations are omitted when the upper index is zero.

6. Summary and concluding remarks

The evaluation of molecular integrals over Gaussian-type functions lies at the heart of the practical realization of contemporary quantum chemical theories. In this article, an elementary, yet fairly thorough, introduction to the steps, both analytical and computational, involved in the evaluation of the simplest of two-electron, four-centre integrals has been given.

Secondly, a brief overview of developments that have taken place in the field of molecular integral evaluation since the review by Saunders[9] at the NATO Advanced Study Institute on *Methods in Computational Molecular Physics* held in 1983 has been given. The increasing use of 'direct' methods during the 1980s first in self-consistent field calculations and then in some correlation energy studies fuelled the need for more efficient integral evaluation algorithms. On the other hand, the availability of highly efficient 'direct' algorithms has spurred both the study of large (extended) molecules and the introduction of high precision basis sets for small molecule studies. The advances reported by Obara and Saika, by Schlegel, by Head-Gordon and Pople, and by others for evaluating integrals involving higher harmonics has been outlined and the application of these ideas in gradient studies and in relativistic electronic structure calculations has been indicated.

The systematic approximation of the large number of two-electrons which can arise in molecular electronic structure studies has assumed increasing importance in recent years fuelled by the need to treat larger (more extended) molecular systems by *ab initio* techniques and the requirement of higher precision. An overview of the multipole expansion and recent progress on its use in the approximation two-electron integrals which involve charge distributions which do not overlap has been given. The use of the Cholesky decomposition in exploiting the computational linear dependence which arising in the two-electron integrals supermatrix to reduce the number of integrals which have to be evaluated has also been described. Undoubtedly, the future will see the development of large basis sets capable of supporting high precision calculations for extended molecules. Such a development will require the synthesis of the multipole expansion technique and Cholesky decomposition algorithm.

References

1. G.G. Hall, *Proc. Roy. Soc.* **A 205**, 541 (1951)
2. C.C.J. Roothaan, *Rev. Mod. Phys.* **23**, 69 (1951)
3. R. McWeeny, *Dissertation,* University of Oxford (1948)
4. R. McWeeny, *Nature* **166**, 21 (1950)
5. S.F. Boys, *Proc. Roy. Soc.* **A 200**, 542 (1950)
6. I. Shavitt, *Israel J. Chem.* (1993)
7. Part II
8. J. Almlöf, K. Faegri, Jr., and K. Korsell, *J. Comput. Chem.* **3**, 385 (1985)
9. V.R. Saunders, in *Methods in Computational Molecular Physics*, edited by G.H.F. Diercksen and S. Wilson (Reidel, Dordrecht, 1983)
10. V.R. Saunders, in *Computational Techniques in Quantum Chemistry and Molecular Physics*, edited by G.H.F. Diercksen, B.T. Sutcliffe and A. Veillard (Reidel,

Dordrecht, 1975)
11. L.E. McMurchie and E.R. Davidson, *J. Comput. Phys.* **26**, 218 (1978)
12. H.F. King and M. Dupuis, *J. Comput. Phys.* **21**, 144 (1976)
13. M. Dupuis, J. Rys and H.F. King, *J. Chem. Phys.* **65**, 111 (1976)
14. J. Rys, M Dupuis and H.F. King, *J. Comput. Chem.* **4**, 154 (1983)
15. S. Obara and A. Saika, *J. Chem. Phys.* **84**, 3963 (1985)
16. H.B. Schlegel, *J. Chem. Phys.* **77**, 3676 (1982)
17. M. Head-Gordon and J.A. Pople, *J. Chem. Phys.* **89**, 5777 (1988)
18. P.M.W. Gill, M. Head-Gordon and J.A. Pople, *Intern. J. Quantum Chem. Symp.* **23**, 269 (1989)
19. P.M.W. Gill, M. Head-Gordon and J.A. Pople, *J. Phys. Chem.* **94**, 5564 (1990)
20. T.P. Hamilton and H.F. Schaefer III, *Chem. Phys.* **150**, 163 (1991)
21. T.P. Hamilton and H.F. Schaefer III, *Can J. Chem.* **70**, 416 (1992)
22. P.M.W. Gill, B.G. Johnson and J.A. Pople, *Intern. J. Quantum Chem.* **40**, 745 (1991)
23. P.M.W. Gill and J.A. Pople, *Intern. J. Quantum Chem.* **40**, 753 (1991)
24. I. Panas, *Chem. Phys. Letts.* **184**, 86 (1991)
25. U. Ryu, Y.S. Lee and R. Lindh, *Chem. Phys. Letts.* **185**, 562 (1991)
26. R. Lindh, U. Ryu and B. Liu, *J. Chem. Phys.* **95**, 5889 (1991)
27. R. Lindh, *Theoret. chim. Acta* **85**, 423 (1993)
28. S. Ten-no and S. Iwata, *Chem. Phys. Letts.* **211**, 259 (1993)
29. B.G. Johnson, P.M.W. Gill and J.A. Pople, *Chem. Phys. Letts.* **206**, 229 (1993)
30. M. Häser and R. Ahlrichs, *J. Comput. Chem.* **10**, 104 (1989)
31. C.A. White, B. Johnson, P.M.W. Gill and M. Head-Gordon, *Chem. Phys. Letts.* **230**, 8 (1994)
32. C.A. White, B. Johnson, P.M.W. Gill and M. Head-Gordon, *Chem. Phys. Letts.* **253**, 268 (1996)
33. R. Kutteh and J.B. Nicholas, *Chem. Phys. Letts.* **238**, 173 (1996)
34. M. Challacombe, E. Schwegler and J. Almlof, in *Computational Chemistry:Review of Current Trends*, edited by J. Leczszynski, pp. 53-107, World Scientific, Singapore (1996)
35. E. Schwegler and M. Challacombe, *J. Chem. Phys.* **105**, 2726 (1996)
36. M. Challacombe, E. Schwegler and J. Almlof, *J. Chem. Phys.* **104**, 4685 (1996)
37. D.L. Strout and G.E. Scuseria, *J. Chem. Phys.* **102**, 8448 (1995)
38. J.C. Burrant, G.E. Scuseria and M.J. Frisch, *J. Chem. Phys.* **105**, 8969 (1996)
39. M.C. Strain, G.E. Scuseria and M.J. Frisch, *Science* **271**, 51 (1996)
40. J. Barnes and P. Hut, *Nature* **324**, 446 (1986)
41. L. Hernquist, *Ap. J. Suppl.* **64**, 715 (1987)
42. J. Barnes and P. Hut, *Ap. J. Suppl.* **70**, 389 (1989)
43. S. Pfalzner and P. Gibbon, *Comput. Phys. Commun.* **79**, 24 (1994)
44. J.K. Salmon, *Intern. J. Super Appl.* **8**, 129 (1994)
45. M.S. Warren and J.K. Salmon, *Comput. Phys. Commun.* **87**, 266 (1995)
46. L Greengard and V. Rokhlin, *J. Comput. Phys.* **73**, 325 (1987)
47. K.E. Schmidt and M.A. Lee, *J. Stat. Phys.* **63**, 1223 (1991)
48. H.G. Petersen, D. Soelvason, J.W. Perram and E.R. Smith, *J. Chem. Phys.* **101**, 8870 (1994)
49. H.G. Petersen, E.R. Smith and D. Soelvason, *Proc. Roy. Soc.* **A 448**, 401 (1994)
50. B.C. Carlson and G.S. Rushbrooke, Proc. Cambridge Philos. Soc. **46**, 626 (1950)
51. N. Beebe and J. Linderberg, *Intern. J. Quantum Chem.* **12**, 683 (1977)
52. I. Røeggen and E. Wisløff-Nielsen, *Chem. Phys. Letts.* **132**, 154 (1986)
53. S. Wilson, *Meth. Comput. Chem.* **1**, 251(1987)
54. S. Wilson, *Comput. Phys. Commun.* **58**, 71 (1990)
55. S. Wilson, in *Methods in Computational Molecular Physics*, edited by S. Wilson and G.H.F. Diercksen, (Plenum, New York, 1992)

PRACTICAL *AB INITIO* METHODS FOR MOLECULAR ELECTRONIC STRUCTURE STUDIES. IV. RELATIVISTIC MANY-BODY PERTURBATION THEORY

S. WILSON
Rutherford Appleton Laboratory
Chilton, Oxfordshire, England

Abstract. The generalization of the familiar non-relativistic diagrammatic many-body perturbation theory of atomic and molecular electronic structure to the relativistic case is surveyed. It is noted that the treatment of relativistic and quantum electrodynamic effects in atoms and molecules requires the reintroduction of physics that was discarded in early studies of the many-body problem by Goldstone and others in which interactions were taken to be instantaneous and special relativity was ignored. Recent progress is described emphasizing three aspects: *i)* the representation of the positive and negative energy branches of the Dirac spectrum afforded by the algebraic approximation, and *ii)* the non-additivity of relativistic and correlation effects in calculations using the Dirac-Coulomb hamiltonian and the Dirac-Breit hamiltonian. Future directions of research are briefly described emphasizing particularly, *i)* improvements in the accuracy of matrix Dirac-Hartree-Fock-Breit calculations and in sum-over-states perturbative electron correlation calculations for molecules, and *ii)* the use of Bardeen-Cooper-Schrieffer reference models in calculations based on the diagrammatic many-body perturbation theory to afford a description of bond breaking processes.

1. Introduction

For more than twenty five years, the many-body perturbation theory[1] [2], when formulated in the algebraic approximation[3] [4] [5], has formed the cornerstone for practical treatment of the correlation problems in molecules. Since the early 1980s, we have witnessed a growing interest in the effects of relativity on the electronic structure of atoms and molecules. Over the past

S. Wilson and G. H. F. Diercksen (eds.), Problem Solving in Computational Molecular Science, 185–213.

decade the theoretical and computational machinery has been put in place for a relativistic many-body perturbation theory of atomic and molecular electronic structure[6] [7] [8] [9] [10] [11] [12].

Pyykkö[13] has recently surveyed the influence of relativity on periodic trends and provided a concise summary of previous reviews of relativistic electronic structure theory. This development can be seen, on the one hand, as a result of a growing awareness of the importance of relativity in accounting for the properties of heavy atoms and molecules containing them. The inadequacy of physical models which either neglect relativity or which treat it as a small perturbation has fuelled this development. For heavy atoms, the mean speed of electrons close to the nucleus is a substantial fraction of the speed of light, so that a fully relativistic electronic structure theory is required, both at the independent particle level and beyond. The treatment of relativity and the introduction of quantum electrodynamic effects is essential since these may be more important than electron correlation in heavy elements, and there is no evidence that these effects are simply additive. On the other hand, the implementation of relativistic electronic structure theories is dependent on technological developments, which have resulted in computing machines powerful enough to make calculations on heavy atoms, and molecules containing them, meaningful. Vector processing and, more recently, parallel processing techniques are playing a vital role in rendering the algorithms which arise in relativistic electronic structure studies tractable.

This article is a supplement to the report presented by the author at the 1988 Symposium on *Many-body methods in quantum chemistry* held at Tel Aviv University[9]. In that report, we suggested that

> "*If one were to hazard a guess as to where the major new developments in quantum chemistry will be in the next fifteen years, the proper treatment of relativity and the introduction of quantum electrodynamic effects seems to us to be likely candidates*"

and

> "*...firm foundations for this project have already been laid down*".

This article may be considered a "half term" report.

In section 2, a very brief overview of the salient features of the relativistic many-body perturbation theory is given giving particular emphasis to those features which differ from the familiar non-relativistic formulation and to its relation with quantum electrodynamics. Two aspects of the relativistic many-body perturbation theory are considered in more detail in section 3 and 4, respectively: in section 3, the representation of the Dirac spectrum in the algebraic approximation is discussed; in section 4, we consider the non-additivity of relativistic and electron correlation effects, and demonstrate the use of the Dirac–Hartree-Fock-Coulomb-Breit reference hamiltonian. In the final section, we give a summary and consider the impact of recent developments in non-relativistic methodology on the relativistic many-body perturbation theory of atomic and molecular electronic structure: i) the "elimination" of basis set truncation errors in molecular calculations; ii) the use of many-body perturbation theory to describe chemically bonded systems using Bardeen-Cooper-Schrieffer reference models.

2. Relativistic many-body perturbation theory

The theoretical description of any many body system is usually approached in two distinct stages. First, the solution of some independent particle model yielding a set of quasi-particles, or dressed particles, which are then used to formulate a systematic scheme for describing the corrections to the model. Perturbation theory, when developed with respect to a suitable reference model, affords the most systematic approach to the correlation problem which today, because it is non-iterative and, therefore, computationally very efficient, forms the basis of the most widely used approaches in contemporary electronic structure calculations, particularly when developed with respect to a Møller-Plesset zero-order hamiltonian.

In Figure 1, the non-relativistic and relativistic spectra for an electron in a typical mean field potential are shown schematically. The most significant distinguishing feature is the presence of the negative energy branch for $\varepsilon < -2c^2$ in the latter. According to Dirac [14] the levels in this lower continuum are always filled, subject to the Pauli principle, in the true vacuum thereby preventing the radiative decay of bound, positive-energy states. The negative energy "sea" is not observable as an entity since only differences between states can be detected.

The negative energy continuum and the occupied positive energy states are taken to be below the Fermi level within the particle-hole formalism in which the events which occur in a relativistic many-body system may be described. In contrast to the non-relativistic particle-hole formalism, the relativistic extension involves no restriction on particle number. Electrons can be excited from the negative energy continuum into an unoccupied level in the positive energy branch thereby producing a vacancy in the lower continuum which can be related to a positron state. Only the total charge is conserved and not the individual numbers of electrons and positrons. Even hydrogenic systems are implicitly infinitely many-bodied theories, making second quantized formulations mandatory, and, although more elegant formulations [15] have followed Dirac's original theory, the physical content of these theories is essentially unchanged.

For low energy processes electron-positron pairs will appear as interme-

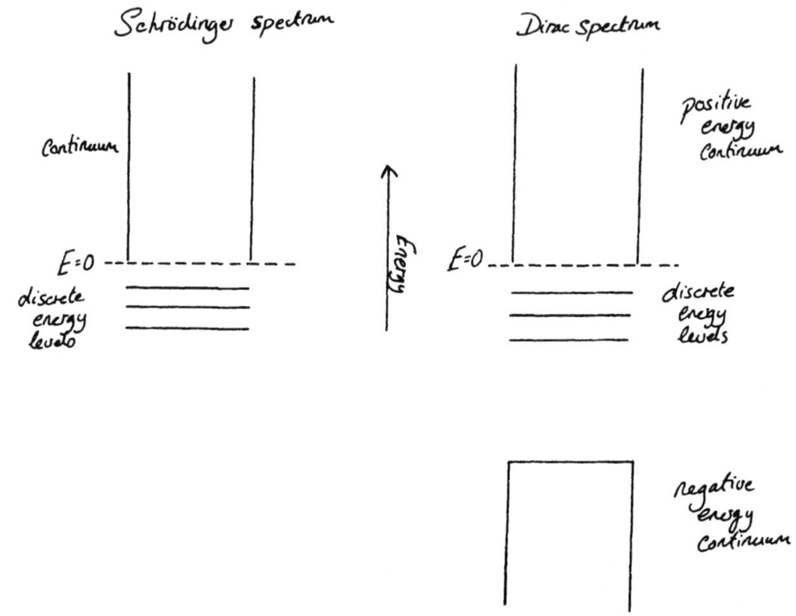

Figure 1. Comparison of the Dirac and Schrödinger spectra for atoms and molecules

diate states in perturbation summations. At higher energies (~1MeV) real electron-positron pairs may be produced. Dirac comments [16]

> *"A quantum electrodynamics which demand conservation of the number of charged particles is ... out of touch with physical reality." ... "It seems that the classical concepts of an electron is no longer a useful model in physics, except possibly for elementary theories that are restricted to low-energy phenomena."*

It is, therefore, valid to ask to what extent can chemistry or, more precisely, molecular electronic structure, be described by *"elementary"* theory?

In his derivation of the linked diagram theorem of many-body perturbation theory in 1957, Goldstone [17] made use of the formalism of quantum electrodynamics. developed in the late 1940's by Tomonaga [18] [19], Schwinger [20] [21] and Feynman [22]. In particular, he used a time-dependent formulation, even for a time-independent perturbation.

Goldstone used a second quantized particle-hole formalism based on an arbitrary choice of vacuum state. The interaction representation, which is intermediate between the Schrödinger and Heisenberg pictures, was employed and the energy was evaluated by the Gell-Mann-Low formalism [23] with hamiltonian

$$H = H_0 + e^{-\alpha t} H_1 \tag{1}$$

where α is a factor which is set to 0 at the end of the discussion. The model state at t =-∞ is allowed to evolve to the fully interacting state at $t = 0$

$$|\psi(0)\rangle = U_\alpha(0, -\infty)\, |\psi_0(-\infty)\rangle \tag{2}$$

and Wick's theorem [24] is employed to evaluate the time-ordered products of creation and annihilation operators arising in the evolution operator, $U_\alpha(0, -\infty)$.

A more symmetrical treatment was presented by Sucher [25] which allows the interacting state to evolve to $t = +\infty$

$$|\psi(\infty)\rangle = S_\alpha(\infty, -\infty)\, |\psi_0(-\infty)\rangle \tag{3}$$

This S matrix approach leads to the modified Gell-Mann-Low formula for the energy shift

$$\Delta E = \lim_{\alpha \to 0} \frac{\imath\alpha}{2} \frac{\partial}{\partial\lambda} \ln \langle S_\alpha \rangle \tag{4}$$

where λ, which is set to 1 at the end of the derivation, is a factor multiplying H_1. The S matrix approach [26] has the advantage that it can be represented in terms of fully covariant Feynman diagrams.

In his work, Goldstone[17] introduced the graphical techniques into many-body physics making use of Feynman-like diagrams. However, interactions were taken to be instantaneous and the effects of relativity were ignored. The growing interest in the treatment of relativistic and quantum electrodynamic effects in atoms and molecules is necessitating the reintroduction of physics that has been known for over forty years. The appropriate machinery is Furry's bound interaction picture of quantum electrodynamics [27]. (For more recent discussions see Quiney *et al* [9] [28].)

The Furry bound state interaction picture of quantum electrodynamics relies on an expansion of the second-quantized electron field operator in terms of single-particle solutions of the Dirac equation for a static external field. This external field may be thought of as some mean atomic or molecular potential, whose single-particle spectrum can be divided into positive- and negative- energy branches. This can always be done for the usual elements of the Periodic Table, although problems arise for superheavy atomic nuclei.

Investigation of the connection between the Many-Body Perturbation Theory (*M.B.P.T.*) approach and the Furry representation of *Q.E.D.* has been shown to allow a precise definition of *Q.E.D.* effects [29]. Every

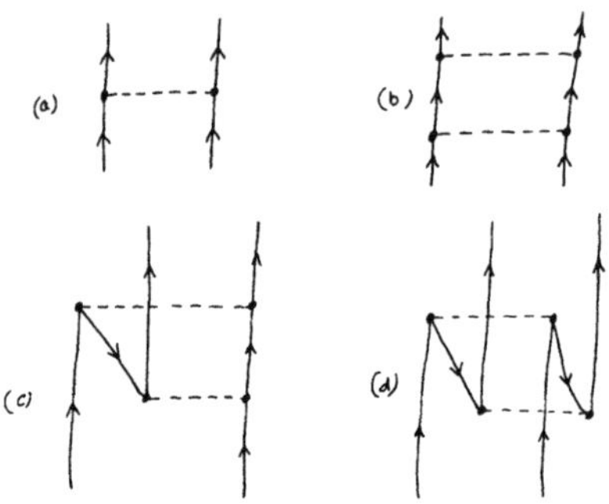

Figure 2. Examples of MBPT, *(a)* and *(b)*, and QED, *(c)* and *(d)*, diagrams

M.B.P.T. diagram has a corresponding Feynman diagram, but there are Feynman diagrams that have no *M.B.P.T.* counterpart.

In Figure 2, examples of *M.B.P.T.* and *Q.E.D.* diagrams are given. Sapirstein has observed [29]

> *"The interface of these two apparently different, but actually intimately related approaches contains a great deal of beautiful physics".*

Q.E.D. provides a framework for describing the role of the negative energy states in the Dirac theory and the divergences which arise in studies of electrodynamic interactions [30] [31] [32]. In the following sections, it will be indicated how the use of the algebraic approximation to generate a discrete representation of the Dirac spectrum has opened the way for the transcription of the rules of *Q.E.D.* into practical algorithms for the study of many-electron systems.

We now have a well defined prescription for the calculation of the properties of atoms and molecules within a relativistic formulation. As in the non relativistic case, the relativistic many-body perturbation theory becomes increasingly complicated in higher orders and in practice it is possible to take the expansion to about fourth order with the size of basis set that is required for calculations of useful accuracy. Sapirstein [29] has recently re-iterated the view that

> *"All-order methods allow for certain subsets of MBPT expressions of higher order to be calculated, but in general other subsets will remain uncalculated."*

3. Representation of the Dirac spectrum in the algebraic approximation

It is well known that by invoking the algebraic approximation for non-relativistic many-body problems not only is an approximate expansion for the occupied orbitals obtained in the independent particle model but also a representation of the virtual spectrum is obtained which forms an essential ingredient of, for example, a many-body perturbation theory treatment of electron correlation effects [4] [33] [34]. The virtual spectrum affords an effective quadrature over the intermediate scattering states which arise in second- and higher- orders. In the case of the relativistic many-body problem, the algebraic approximation also provides a representation of the whole spectrum; the positive energy branch, consisting of both occupied states and virtual states, and the negative energy branch.

Unless carefully implemented the representation of the Dirac spectrum obtained within the algebraic approximation may exhibit undesirable properties which are not encountered in non-relativistic studies. In particular, an inappropriate choice of basis set may obliterate the separation of the spectrum into positive and negative energy branches. So-called *"intruder states"* may arise, which are impossible to classify as being of either *"positive"* or *"negative"* energy character. The Furry bound state interaction picture of quantum electrodynamics is thereby undermined.

It is now firmly established that the problems associated with what has been termed the "finite basis set disease", may be avoided if the basis functions reflect the four-component structure of a relativistic spinor. The four component basis functions must themselves be solutions of a Dirac equation. This imposes a well defined relation between the scalar basis functions employed for the large and small components. A firm foundation established for the *ab initio* treatment of the electronic structure of atoms and of molecules containing heavy atoms and the interactions between them [9] [35] [36] [37] [38] [39] [40] [41]. The application of the algebraic approximation has made possible a unified approach to the study of relativistic, many-body and quantum electrodynamic effects. The basis functions must be matched in pairs and contain information about the relativistic kinetic energy operator $\boldsymbol{\sigma}.\mathbf{p}$, as well as being sufficiently flexible to represent the nuclear and the screening potentials. There is thus a well-defined relation between the basis functions employed for the large and small components which is analogous to the imposition of boundary conditions in finite difference calculations. Figure 3 shows a schematic representation of the Dirac spectrum and of the Schrödinger spectrum generated in the algebraic approximation.

At this point the suitable of Gaussian-type basis sets for relativistic atomic and molecular electronic structure studies is worth emphasizing. The use of Gaussian basis functions in molecular electronic structure cal-

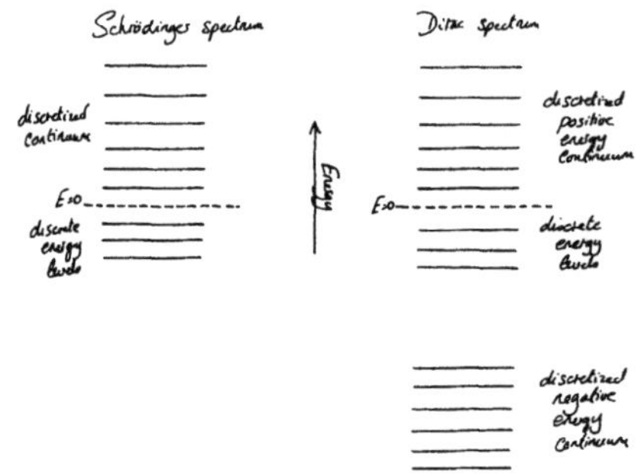

Figure 3. Schematic representation of the Dirac spectrum and of the Schrödinger spectrum generated in the algebraic approximation.

culations was firstly suggested by McWeeny [42] [43] and independently by Boys [44]. However, the supposedly "more physical" Slater (exponential) Type Orbitals (STO) remained the basis function of choice because they "correctly" describe not only the cusp associated with the nucleus on which they are centred but also afford a suitable representation of the long range behaviour. Shavitt [45] records that

> *"Boys did not show much enthusiasm for the use of Gaussian basis sets, and concentrated much of his attention in the mid- and late fifties on devising various schemes for the evaluation of STO multicentre integrals".*

The use of Gaussian basis functions in relativistic molecular electronic structure studies is particularly attractive and the following advantages, some of which are shared by non-relativistic studies, are worth noting:

(i) The multi centre, two-electron integrals can be evaluated easily because of the Gaussian Product theorem. This is the original reason for the introduction of Gaussian-type basis functions in non-relativistic molecular electronic structure calculations. It becomes even more important in relativistic calculations because of the larger number of integrals arising.

(ii) The integrals can be evaluated accurately. This is particularly important when calculating small energy differences such as, for example, in studies of van der Waals interactions between polyatomic molecules. In contrast, the evaluation of multi-centre, two-electron integrals over exponential-type functions involves numerical integration and is usually accompanied by some loss of accuracy. The absolute precision requirements of relativistic calculations are comparable with those of non-relativistic studies. However, because in relativistic calculations the interest is often in

molecules containing heavy atoms, which have comparatively large total energies, the relative precision requirements are somewhat higher than in non-relativistic calculations.

(iii) Because Gaussian-type functions do not introduce a cusp, they do not have to be atom-centred. The use of off-centre Gaussian-type basis function can be an effective way of introducing the effects of higher harmonics on the centre. Off-centre exponential-type functions introduce an unphysical cusp into the approximation of the single particle function. (For recent work on off-centre Gaussian-type functions see the work of Moncrieff and Wilson [46] [47])

(iv) The Gaussian function is a natural choice of basis function if the point nucleus is abandoned in favour of a more realistic finite nucleus [41] [48] [49] [50] [51] such as the "Fermi nucleus" which has the form of a Fermi distribution

$$\rho(r) = \frac{\rho_0}{1 + e^{(r-c)/a}} \tag{5}$$

where c is the *"half-density radius"* and the parameter a is related to the nuclear *"skin thickness"*. The use of finite nuclear models is not restricted to relativistic electronic structure studies. Indeed, there may be some advantage in using such models in non-relativistic studies since it is known that the convergence of finite basis set expansions is dominated by the ability of the functions to describe cusps [52] [53].

(v) Near linear dependence problems for the large and flexible basis sets demanded for high precision calculations are less severe in calculation using Gaussian-type functions than they are for calculations employing exponential-type functions.

(vi) Relativistic calculations demand the use of matched basis sets for the large and small components of the spinors. If Gaussian-type basis sets are employed then a $1s$ function in the large component basis set requires the use of a $2p$ function in the small component basis set. However, if exponential-type basis sets are employed then the use of a 1s function in the basis set for the large component requires that a corresponding $1p$ function be included in the small component basis set. The exponential-type $1p$ function

$$\chi_{1p}(r,\theta,\varphi) = \exp(-\zeta r)\ Y_{10}(\theta,\varphi) \equiv \exp(-\zeta r)\ \cos\theta \tag{6}$$

is multi valued at the origin. Moreover, integral programs for non-relativistic calculations using exponential-type basis sets do not usually encompass $1p$ basis functions.

(vii) The use of the algebraic approximation facilities the introduction of the Breit interaction into the self-consistent field procedure since once the integral over the Breit operator have been evaluated for the chosen basis

set they can be incorporated in the construction of the Dirac-Fock matrix in a manner analogous to the two-electron Coulomb integrals This is not a specific argument in favour of Gaussian-type basis functions. However, it should be contrasted with the situation in finite difference calculations which, for reasons of computational economy, have invariably included the Breit interaction as a first order perturbation after the self-consistent field iterations have converged.

(viii) The use of the algebraic approximation provides a representation of the whole of the relativistic spectrum: the bound states, the positive energy continuum and the negative energy continuum. Atomic finite difference calculations only routinely afford an approximation to the occupied orbitals. As we have pointed out previously [9]

> *"This is rather unsatisfactory from a formal point of view, since the considerable technical effort required to develop a relativistic theory are largely wasted if we use a manifestly incomplete basis of positive-energy Dirac spinors. If we wish to clarify the importance of magnetic effects in the many-body problem by using some satisfactory modification of the Breit operator or to explore the origin of quantum electrodynamic effects in many-electron systems, then we must be prepared to expand Dirac wave functions in a complete basis including both negative- and positive-energy contributions,..."*

3.1. A MODEL PROBLEM

Utilization of the representation of the Dirac spectrum afford by the algebraic approximation will be considered for two cases: a model problem, which provides some useful physical insights, and the independent particle model obtained within the Dirac-Hartree-Fock ansätz.

Consider a simple model problem consisting of a hydrogenic atom with nuclear charge Z perturbed by an additional charge Z'. Aspects of this model have been studied within the framework of non-relativistic quantum mechanics by a number of authors [54] [55] [56] [57]. In this case the exact energies are known not only in order of the perturbation series but also for each diagrammatic term.

When formulated within the framework of relativistic quantum mechanics, the model can be solved exactly in each order of the perturbation expansion and affords a sensitive test of the implementation of the algebraic approximation [6].

This zero-order problem may be solved exactly to give a ground state energy (the Sommerfeld eigenvalue) of

$$\epsilon_0 = -c^2 \left[1 - \left(1 - \frac{Z^2}{c^2} \right)^{\frac{1}{2}} \right] \tag{7}$$

which may be expanded in powers of $1/c^2$

$$\epsilon_0 = -\frac{1}{2} Z^2 - \frac{1}{8} \frac{Z^4}{c^2} - \frac{1}{16} \frac{Z^6}{c^4} - \ldots \tag{8}$$

The perturbed problem has a ground state the eigenvalue given by

$$\epsilon^{(\lambda)} = -c^2 \left[1 - \left(1 - \frac{(Z + \lambda Z')^2}{c^2} \right)^{\frac{1}{2}} \right] \tag{9}$$

It should be noted that in the relativistic case the exact energy is not given by summing the perturbation expansion through second-order as it is in the non-relativistic case; third- and higher-order energy coefficients are non-zero.

The energy coefficients in the perturbation series can be obtained by repeated differentiation of the perturbed energy expression with respect to Z'. For the ground state the first four terms in the perturbation series are readily shown [6] to be

$$\epsilon_1 = -Z \left(1 - \frac{Z^2}{c^2} \right)^{-\frac{1}{2}} (Z') \tag{10}$$

$$\epsilon_2 = -\frac{1}{2} \left(1 - \frac{Z^2}{c^2} \right)^{-\frac{3}{2}} (Z')^2 \tag{11}$$

$$\epsilon_3 = -\frac{1}{2} \frac{Z}{c^2} \left(1 - \frac{Z^2}{c^2} \right)^{-\frac{5}{2}} (Z')^3 \tag{12}$$

$$\epsilon_4 = -\frac{1}{8} \frac{1}{c^2} \left(1 - \frac{Z^2}{c^2} \right)^{-\frac{7}{2}} \left(1 + 2\frac{Z^2}{c^2} \left(2 + 3\frac{Z^2}{c^2} \right) \right) (Z')^4 \tag{13}$$

The terms in the perturbation series for the energy for an arbitrary state have been given elsewhere [6].

The diagrammatic representation of the perturbation series is given in Figure 4 for the non-relativistic case. In the relativistic case, the energy components corresponding to each of the diagrams are : *(i)* in first order

$$\mathcal{E}_1^{(\mu)} = \langle \Phi^{(\mu)} \mid H_1 \mid \Phi^{(\mu)} \rangle \tag{14}$$

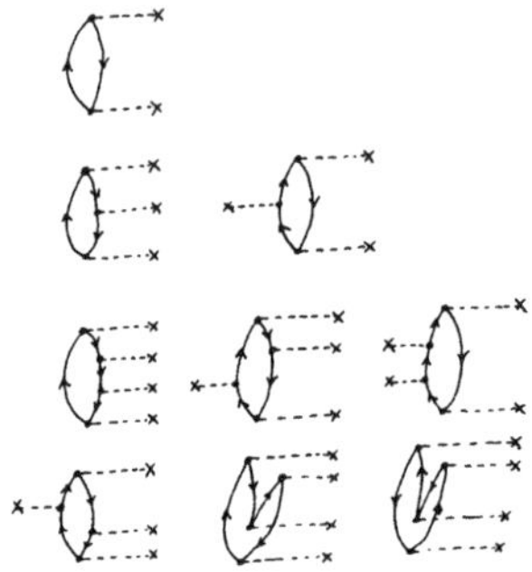

Figure 4. Diagrammatic representation of the perturbation series for the hydrogenic model problem in the non-relativistic case.

(ii) in second order

$$\mathcal{E}_2^{(\mu)} = \sum_{a\neq\mu} \frac{\langle\Phi^{(\mu)} \mid H_1 \mid \Phi^{(a)}\rangle\langle\Phi^{(a)} \mid H_1 \mid \Phi^{(\mu)}\rangle}{\mathcal{E}_0^{(\mu)} - \mathcal{E}_0^{(a)}} \tag{15}$$

(iii) in third order there are two terms distinguished by the nature of the central interaction involving either a particle line

$$\mathcal{E}_3^{(\mu)}(p) = \sum_{a,b\neq\mu} \frac{\langle\Phi^{(\mu)} \mid H_1 \mid \Phi^{(a)}\rangle\langle\Phi^{(a)} \mid H_1 \mid \Phi^{(b)}\rangle\langle\Phi^{(b)} \mid H_1 \mid \Phi^{(\mu)}\rangle}{\left(\mathcal{E}_0^{(\mu)} - \mathcal{E}_0^{(a)}\right)\left(\mathcal{E}_0^{(\mu)} - \mathcal{E}_0^{(b)}\right)} \tag{16}$$

or a hole line

$$\mathcal{E}_3^{(\mu)}(h) = -\sum_{a\neq\mu} \frac{\langle\Phi^{(\mu)} \mid H_1 \mid \Phi^{(a)}\rangle\langle\Phi^{(\mu)} \mid H_1 \mid \Phi^{(\mu)}\rangle\langle\Phi^{(a)} \mid H_1 \mid \Phi^{(\mu)}\rangle}{\left(\mathcal{E}_0^{(\mu)} - \mathcal{E}_0^{(a)}\right)^2} \tag{17}$$

(iv) in fourth order there are six terms distinguished by the nature of the two central interaction lines involving two particle lines

$$\mathcal{E}_4^{(\mu)}(pp) = \sum_{a,b,c\neq\mu} \frac{\langle\Phi^{(\mu)} \mid H_1 \mid \Phi^{(a)}\rangle\langle\Phi^{(a)} \mid H_1 \mid \Phi^{(b)}\rangle\langle\Phi^{(b)} \mid H_1 \mid \Phi^{(c)}\rangle}{\left(\mathcal{E}_0^{(\mu)} - \mathcal{E}_0^{(a)}\right)\left(\mathcal{E}_0^{(\mu)} - \mathcal{E}_0^{(b)}\right)\left(\mathcal{E}_0^{(\mu)} - \mathcal{E}_0^{(c)}\right)}$$
$$\times\langle\Phi^{(c)} \mid H_1 \mid \Phi^{(\mu)}\rangle \tag{18}$$

a particle (hole) line and a hole (particle) line

$$\mathcal{E}_4^{(\mu)}(ph) = \mathcal{E}_4^{(\mu)}(hp) = -\sum_{a,b\neq\mu} \frac{\langle\Phi^{(\mu)} \mid H_1 \mid \Phi^{(a)}\rangle\langle\Phi^{(\mu)} \mid H_1 \mid \Phi^{(\mu)}\rangle}{\left(\mathcal{E}_0^{(\mu)} - \mathcal{E}_0^{(a)}\right)^2\left(\mathcal{E}_0^{(\mu)} - \mathcal{E}_0^{(b)}\right)}$$

$$\times\langle \mathbf{\Phi}^{(a)} \mid H_1 \mid \mathbf{\Phi}^{(b)}\rangle\langle \mathbf{\Phi}^{(b)} \mid H_1 \mid \mathbf{\Phi}^{(\mu)}\rangle \tag{19}$$

or two hole lines

$$\mathcal{E}_4^{(\mu)}(hh) = -\sum_{a\neq\mu} \frac{\langle \mathbf{\Phi}^{(\mu)} \mid H_1 \mid \mathbf{\Phi}^{(a)}\rangle \left(\langle \mathbf{\Phi}^{(\mu)} \mid H_1 \mid \mathbf{\Phi}^{(\mu)}\rangle\right)^2 \langle \mathbf{\Phi}^{(a)} \mid H_1 \mid \mathbf{\Phi}^{(\mu)}\rangle}{\left(\mathcal{E}_0^{(\mu)} - \mathcal{E}_0^{(a)}\right)^3} \tag{20}$$

There are two fourth order renormalization terms which can be added to give

$$\mathcal{E}_4^{(\mu)}(R_1 + R_2) = -\mathcal{E}_2^{(\mu)} \sum_{a\neq\mu} \frac{\langle \mathbf{\Phi}^{(\mu)} \mid H_1 \mid \mathbf{\Phi}^{(a)}\rangle\langle \mathbf{\Phi}^{(a)} \mid H_1 \mid \mathbf{\Phi}^{(\mu)}\rangle}{\left(\mathcal{E}_0^{(\mu)} - \mathcal{E}_0^{(a)}\right)^2} \tag{21}$$

In the above expressions, $\mathbf{\Phi}^{(i)}$ denotes the relativistic 4-spinor

$$\mathbf{\Phi}^{(i)} = \begin{pmatrix} \mathbf{\Phi}_L^{(i)} \\ \mathbf{\Phi}_S^{(i)} \end{pmatrix} \tag{22}$$

with large and small components given by 2-spinors

$$\mathbf{\Phi}_L^{(i)} \tag{23}$$

and

$$\mathbf{\Phi}_S^{(i)} \tag{24}$$

respectively. It should be noted that the summations in these equations include the negative energy states.

Table 1 demonstrates the convergence of the calculated relativistic energy values with increasing size of basis set for a systematically constructed sequence of even-tempered basis sets.

Table 1.
Energy coefficients in the diagrammatic relativistic perturbation theory for the hydrogenic model problem described in the text with Z=92, Z'=1 and κ=+1 using a systematic sequence of exponential-type, even-tempered basis sets[†]

N	ϵ_1	ϵ_2	ϵ_3	ϵ_4
11	−3.326 024 54	−0.360 572 72	−0.003 605 26	−0.000 050 01
12	−3.326 024 56	−0.360 495 90	−0.003 606 45	−0.000 050 66
13	−3.326 024 54	−0.360 638 62	−0.003 616 74	−0.000 050 91
14	−3.326 024 28	−0.360 583 92	−0.003 615 26	−0.000 051 12
15	−3.326 024 55	−0.360 647 51	−0.003 619 72	−0.000 051 21
16	−3.326 024 55	−0.360 614 68	−0.003 618 20	−0.000 051 27
Exact	−3.326 024 54	−0.306 651 50	−0.003 619 65	−0.000 051 28

[†] taken from the work of Quiney *et al*[9].
Atomic units are used throughout.

3.2. THE DIRAC-HARTREE-FOCK METHOD WITHIN THE ALGEBRAIC APPROXIMATION

The Dirac-Hartree-Fock self-consistent field method was introduced by Swirles [58] [59] and developed for atoms by Grant [28] [60], by Lindgren and Rosén [62], and by Desclaux [63] using finite difference methods. Roothaan's students Synek [64] and, in particular, Kim [65] demonstrated that introduction of the algebraic approximation into self-consistent field calculations may lead to catastrophic failure within the relativistic framework. For almost twenty years the successful implementation of the a relativistic quantum chemistry has had to await a successful resolution of these difficulties. However, as described in the previous section these difficulties are now fully resolved.

The Dirac-Hartree-Fock iterative process can be interpreted as a method of seeking cancellations of certain one- and two-body diagrams [8] [51]. The self-consistent field procedure can be regarded as a sequence of rotations of the trial orbital basis into the final Dirac-Hartree-Fock orbital set, *each set in this sequence forming a basis for the Furry bound-state interaction picture of quantum electrodynamics.* The self-consistent field potential involves contributions from the negative energy states of the *unscreened* spectrum so that the Dirac-Hartree-Fock method defines a *stationary* point in the space of possible configuration, rather that a variational *minimum*, as is the case in non-relativistic theory.

In contrast to the non-relativistic Hartree-Fock method, the energy in the Dirac-Hartree-Fock does not necessarily converge from above. This can be easily demonstrated by expanding the trial solution in a complete converged Dirac-Hartree-Fock orbital set. The negative energy components of the expansion are eliminated more rapidly than the positive energy excited orbitals and the late stages of the iterative sequence are similar to those of a non-relativistic Hartree-Fock problem.

Table 2
Relativistic matrix Dirac-Hartree-Fock calculations for the ground state of the argon atom from a sequence of even-tempered basis sets†

N	α	β	E/hartree	$\Delta E/\mu$hartree
9	0.500	1.500	−528.643 948 19	–
10	0.484	1.516	−528.683 364 15	39415.96
11	0.470	1.486	−528.684 311 73	947.58
12	0.458	1.462	−528.684 289 73	−22.00
13	0.447	1.440	−528.684 430 68	140.95
14	0.437	1.421	−528.684 432 46	1.78
15	0.429	1.404	−528.684 448 56	16.10
16	0.421	1.389	−528.684 449 90	1.34
17	0.413	1.376	−528.684 450 46	5.6
finite difference			−528.684 450 09	

† based on H.M.Quiney, I.P.Grant & S.Wilson, in *Many-Body Methods in Quantum Chemistry* edited by U.Kaldor, Lecture Notes in Chemistry **52**, 307 (1989)

The results of Diraac-Hartree-Fock calculations performed with a systematic sequence of even-tempered basis sets for the ground state of the argon atom are displayed in Table 2. These results demonstrate the convergence of the calculated total electronic energy with basis set size, N. For the largest basis set considered, $N = 17$, the difference between the total energy values determined with finite basis set expansions and finite difference techniques is a fraction of a μHartree.

4. Non-additivity of relativistic and correlation effects

Within the framework of relativistic quantum mechanics, the total electronic energy of an atom or molecule may be written as

$$E = E_{reference,non-relativistic} + \Delta E_{correlation} + \Delta E_{relativistic} + \Delta E_{non-additive} \tag{25}$$

where $E_{reference,non-relativistic}$ is most often taken to be the Hartree-Fock energy. The correlation energy is defined as

$$\Delta E_{correlation} = E_{exact,non-relativistic} - E_{reference,non-relativistic} \tag{26}$$

The relativistic energy is often defined as the difference in energy between the non-relativistic and the relativistic reference systems

$$\Delta E_{relativistic} = E_{reference,relativistic} - E_{reference,non-relativistic} \tag{27}$$

The non-additive component of the energy is often heuristically neglected. Formally, it may be written

$$\begin{aligned}\Delta E_{non-additive} = E_{exact,relativistic} - E_{reference,non-relativistic} \\ -\Delta E_{correlation} - \Delta E_{relativistic}\end{aligned} \tag{28}$$

or

$$\begin{aligned}\Delta E_{non-additive} = \left(E_{exact,relativistic} - E_{exact,non-relativistic}\right) \\ - \left(E_{reference,relativistic} - E_{reference,non-relativistic}\right)\end{aligned} \tag{29}$$

Correlation energies and relativistic energies cannot, of course, be observed. They are theoretical concepts arising from the models that we have employed in describing many-body systems.

4.1. A MODEL PROBLEM

Let us return to the model problem of a hydrogenic atom with nuclear charge Z perturbed by increasing this charge to $Z + Z'$. Let us examine the non-additivity in this model. We shall restrict our attention to the ground state for which the energy of the unperturbed or reference system within the framework of non-relativistic quantum mechanics is

$$E_{reference,non-relativistic} = -\frac{1}{2}Z^2 \tag{30}$$

whilst for the perturbed problem we have

$$E_{exact,non-relativistic} = -\frac{1}{2}\left(Z + \lambda Z'\right)^2 \tag{31}$$

For the corresponding relativistic problem, the ground state energy for the reference problem is

$$E_{reference,relativistic} = -c^2\left[1 - \left(1 - \frac{Z^2}{c^2}\right)^{\frac{1}{2}}\right] \tag{32}$$

and for the perturbed problem we have

$$E_{exact,relativistic} = -c^2 \left[1 - \left(1 - \frac{(Z + \lambda Z')^2}{c^2} \right)^{\frac{1}{2}} \right] \tag{33}$$

The total energy for the perturbed system $E_{perturbed,relativistic}$ may be put in the form

$$\begin{aligned} E \equiv E_{exact,relativistic} &= E_{reference,non-relativistic} \\ &+ \Delta E_{relativistic} + \Delta E_{perturbative} + \Delta E_{non-additive} \end{aligned} \tag{34}$$

where

$$\Delta E_{relativistic} = E_{reference,relativistic} - E_{reference,non-relativistic} \tag{35}$$

is the "relativistic correction" and

$$\Delta E_{perturbative} = E_{exact,non-relativistic} - E_{reference,non-relativistic} \tag{36}$$

is the correction to the energy within the framework of non-relativistic quantum mechanics; that is, the term analogous to the correlation energy for a true many-body problem. The non-additive energy component is given by

$$\begin{aligned} \Delta E_{non-additive} &= (E_{exact,relativistic} - E_{reference,non-relativistic}) \\ &- (E_{reference,relativistic} - E_{reference,non-relativistic}) \end{aligned} \tag{37}$$

The exact energy may be written as a double perturbation series

$$E \equiv E_{exact,relativistic} = \left(\sum_{i,j=0}^{\infty} \lambda^i \mu^j \epsilon_{ij} \right)_{\lambda=1,\mu=1} \tag{38}$$

where $\epsilon_{00} \equiv E_{reference,non-relativistic}$, the first index, i, labels the perturbative change in the nuclear charge and the second index, j, labels the relativistic effects arranged in powers of $1/c^2$. The correction to the energy within the framework of non-relativistic quantum mechanics may be written

$$\Delta E_{perturbative} = \left(\sum_{i=1}^{\infty} \lambda^i \epsilon_{i0} \right)_{\lambda=1} = \epsilon_{10} + \epsilon_{20} \tag{39}$$

and the relativistic energy may be written as

$$\Delta E_{relativistic} = \left(\sum_{j=1}^{\infty} \mu^j \epsilon_{0j} \right)_{\mu=1} \tag{40}$$

The energy component termed "non-additive" may be written as the double summation

$$\Delta E_{non-additive} = \left(\sum_{i=1}^{\infty} \sum_{j=1}^{\infty} \lambda^i \mu^j \epsilon_{ij} \right)_{\lambda=1,\mu=1} \tag{41}$$

Explicit expressions for the energy coefficients, ϵ_{ij}, may be obtained by expanding the expressions (10-13) in powers of $1/c^2$. From the first order energy coefficient (10) this gives

$$\lambda\epsilon_1 = \lambda \left(-ZZ' - \mu \left(\frac{1}{2} \frac{Z^3}{c^2} Z' \right) - \mu^2 \left(\frac{3}{8} \frac{Z^5}{c^4} Z' \right) - \ldots \right) = \lambda\epsilon_{10} + \lambda\mu\epsilon_{11} + \lambda\mu^2\epsilon_{12} + \ldots \tag{42}$$

In second order, the energy coefficient (11) leads to

$$\begin{aligned} \lambda^2\epsilon_2 &= \lambda^2 \left(-\frac{1}{2} (Z')^2 - \mu \left(\frac{3}{4} \frac{Z^2}{c^2} (Z')^2 \right) - \mu^2 \left(\frac{15}{16} \frac{Z^4}{c^4} (Z')^2 \right) - \ldots \right) \\ &= \lambda^2\epsilon_{20} + \lambda^2\mu\epsilon_{21} + \lambda^2\mu^2\epsilon_{22} + \ldots \end{aligned} \tag{43}$$

The third order energy coefficient (12) gives

$$\lambda^3\epsilon_3 = \lambda^3 \left(\mu \left(-\frac{1}{2} \frac{Z}{c^2} (Z')^3 \right) - \mu^2 \left(\frac{5}{4} \frac{Z^3}{c^4} (Z')^3 \right) - \ldots \right) = \lambda^3\mu\epsilon_{31} + \lambda^3\mu^2\epsilon_{32} + \ldots \tag{44}$$

whilst from the fourth order energy coefficient (13) the expansion

$$\lambda^4\epsilon_4 = \lambda^4 \left(\mu \left(-\frac{1}{8} \frac{1}{c^2} (Z')^4 \right) - \mu^2 \left(\frac{15}{16} \frac{Z^2}{c^4} (Z')^4 \right) - \ldots \right) = \lambda^4\mu\epsilon_{41} + \lambda^4\mu^2\epsilon_{42} + \ldots \tag{45}$$

is obtained.

The analytic expressions for energy coefficients arising in the double perturbation series (38) are collected in Table 3 for low orders. In Table 4, the numerical values of some of the low order energy coefficients are given in Hartree atomic units for the case $Z = 100$, $Z' = 1$. The unperturbed problem is therefore Fm^{99+} and the perturbed problem is Md^{100+}. It can clearly be seen that the non-additive terms ϵ_{ij}, $i > 0$, $j > 0$ are by no means negligible. The reference energy, ϵ_{00}, is changed by some 13.3% on adding the first order relativistic term, ϵ_{01}, a further 3.5% on adding the second order relativistic term, ϵ_{02}, and by 1.2% on adding ϵ_{03}. The first order correlation correction, ϵ_{10}, is changed by some 26.6% by the first order relativistic term, ϵ_{11}, and a further 10.6% by the second order relativistic term, ϵ_{12}. The second order correlation energy, ϵ_{20}, is changed by 79.9% on adding the first order relativistic terms, ϵ_{21}. This simple model serves to demonstrate the inadequacy of approaches which assume relativistic and correlation effects are simply additive.

Table 3
Energy coefficients in the double perturbation expansion for the hydrogenic model problem

$j\backslash i$	0	1	2	3	4	5
0	$-\frac{1}{2}Z^2$	$-ZZ'$	$-\frac{1}{2}Z'^2$	0	0	0
1	$-\frac{1}{8}\frac{Z^4}{c^2}$	$-\frac{1}{2}\frac{Z^3}{c^2}Z'$	$-\frac{3}{4}\frac{Z^2}{c^2}(Z')^2$	$-\frac{1}{2}\frac{Z}{c^2}(Z')^3$	$-\frac{1}{8}\frac{1}{c^2}(Z')^4$	...
2	$-\frac{1}{16}\frac{Z^6}{c^4}$	$-\frac{3}{8}\frac{Z^5}{c^4}Z'$	$-\frac{15}{16}\frac{Z^4}{c^4}(Z')^2$	$-\frac{5}{4}\frac{Z^3}{c^4}(Z')^3$	$-\frac{15}{16}\frac{Z^2}{c^4}(Z')^4$	...
3	$-\frac{5}{128}\frac{Z^8}{c^6}$	$-\frac{5}{16}\frac{Z^7}{c^6}Z'$	...	...	...	
4	$-\frac{7}{256}\frac{Z^{10}}{c^8}$	$-\frac{35}{128}\frac{Z^9}{c^8}Z'$	...			
5	...	...				

Table 4
Values of $\left(\lambda^i\mu^j\epsilon_{pq}\right)_{\lambda=1,\mu=1}$ for $Z = 100$, $Z' = 1$in Hartree.

$j\backslash i$	0	1	2	3	4
0	−5 000.000 000	−100.000 000	−0.500 000	0	...
1	−665.991 699	−26.639 664	−0.399 595	...	
2	−177.418 106	−10.645 083	...		
3	−59.079 483	...			
4	...				

4.2. NON-ADDITIVITY OF CORRELATION AND RELATIVISTIC EFFECTS IN MANY-ELECTRON SYSTEMS

Let us now return to many-electron systems but before addressing the non-additivity of electron correlation effects and relativistic effects we need to concern ourselves with the nature of the electron-electron interaction.

The Breit interaction is an integral part of the total covariant Coulomb interaction, and is essential for maintaining gauge invariance. In his original application to the helium atom, Breit encountered difficulties which were attributed to unphysical contributions from the negative energy states, and, following analysis by Bethe and Salpeter[66], it was concluded that erroneous results were inevitable if the Breit interaction were to be used other than in first order perturbation theory.

The commonly quoted form of the Breit operator is the frequency independent part:-

$$B_{12} = -\frac{1}{2r_{12}}\left\{\boldsymbol{\alpha}_1.\boldsymbol{\alpha}_2 + \frac{[(\boldsymbol{\alpha}_1.\mathbf{r}_{12})(\boldsymbol{\alpha}_2.\mathbf{r}_{12})]}{r_{12}}\right\} \qquad (46)$$

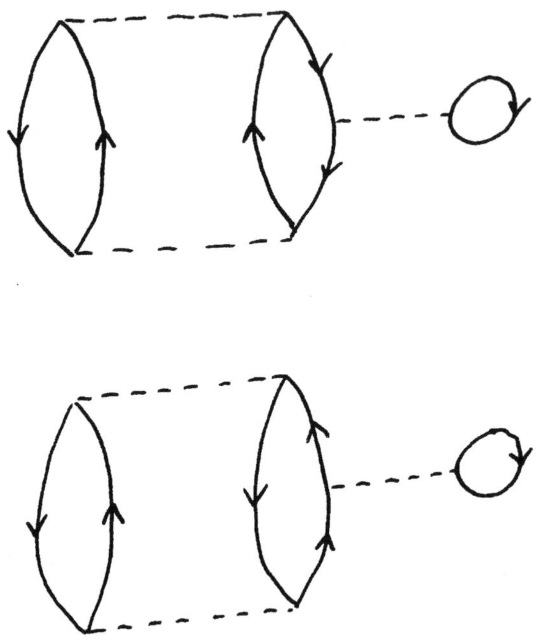

Figure 5. Some typical bubble diagrams

It has been established [40], that this operator can be included in self-consistent field calculation. It has the same general form as the Coulomb operator and may be conveniently incorporated within the algebraic approximation.

It should be noted that finite difference atomic structure calculations have traditionally included the Breit interaction by means of perturbation theory, but that this is for computational rather than theoretical reasons. This perturbative approach leads to a double perturbation expansion when we go beyond the independent electron model and consider the effects of electron correlation. "Bubble" diagrams, such as those shown in Figure 5, arise. By employing a single covariant Coulomb operator we can ensure that all such diagrams vanish identically, leading to a relativistic many-body perturbation theory which is very similar to the non-relativistic case.

In Table 5 the results of various calculations using finite basis sets for the argon atom ground state are present. The finite basis set Hartree-Fock energy is compared with the corresponding finite difference value in the first two rows. The finite basis set Dirac-Hartree-Fock-Coulomb energy is compared with the corresponding finite difference value in the following two rows. In both cases the degree of agreement between total energies obtained from the finite basis set and finite difference calculations is at least an order of magnitude less the energy changes disucssed below. In the finite basis set approximation it is straightforward to calculate the Dirac-Hartree-Fock-Breit energy in which the frequency independent part of the Breit operator is incorporated in the calculation self-consistently. This is compared with a calculation in which the first order Breit energy is calculated using the Dirac-Hartree-Fock-Coulomb as a reference. It can be seen that the difference between the total energies obtained by including the Breit interaction as a first order perturbation and including it self-consistently is very small for this particular system.

The lower part of Table 5 records the results of three electron correlation energy calculations - in each case second order many-body perturbation theory is used to describe correlation effects. In the relativistic calculations the no-virtual pair approximation is employed. The first of the correlated calculations is a standard Hartree-Fock plus second-order perturbation theory study, designated HF+MBPT2. The second calculation is relativistic, uses a Dirac-Hartree-Fock-Coulomb reference function and is designated DHFC+MBPT2. The third calculation is also relativistic but uses a Dirac-Hartree-Fock-Breit reference function and is designated DHFB+MBPT2.

Table 5
Relativistic finite basis set calculations for the argon ground state

	$E/Hartree$
Hartree-Fock (HF) energy†	−526.817 48
(Finite difference HF energy)‡	−526.817 51
Dirac-Hartree-Fock-Coulomb (DHFC) energy*	−528.684 45
(Finite difference DHFC energy)*	−528.684 45
Dirac-Hartree-Fock-Breit (DHFB) energy*	−528.552 12
First order Breit energy*	+0.132 37
Dirac-Hartree-Fock-Cooulomb energy plus first order Breit energy*	−528.552 09
Non-relativistic correlation energy (HF/MBPT-2)*	−0.638 68
Total non-relativistic energy (HF+MBPT-2)*	−527.456 16
Relativistic correlation energy (DHFC/MBPT-2)*	−0.639 42
Total relativistic energy (DHFC+MBPT-2)*	−529.323 87
Relativistic correlation energy (DHFB/MBPT-2)*	−0.646 20
Total relativistic energy (DHFB+MBPT-2)*	−529.198 32

† B.H. Wells & S. Wilson, J. Phys. B: At. Mol. Phys. **19**, 2411 (1986)
‡ C. Froese Fischer, *The Hartree-Fock Method for Atoms. A numerical approach*, (New York, Interscience, 1977)
* H.M. Quiney, I.P. Grant & S. Wilson, J. Phys. B: At. Mol. Phys. **23**, L271 (1990)

From the results for the ground state of the argon atom presented in Table 6 the non-additive component of the energy can be determined. The details of this analysis are presented in Table 6 for both the Dirac-Hartree-Fock-Coulomb ($DHFC$) and the Dirac-Hartree-Fock-Breit ($DHFB$) reference functions. In this Table, $\Delta E_{correlation}$, denotes the second-order correlation energy determined with respect to a Hartree-Fock reference function in a non-relativistic formulation. $\Delta E_{relativistic}$ denotes the difference between the total energy calculated within either the $DHFC$ model or the $DHFB$ model and the non-relativistic Hartree-Fock reference. It is important to note that $\Delta E_{relativistic}$ is approaching three times the correlation energy correction, $\Delta E_{correlation}$. The non-additive component of the energy is the difference between the $MBPT2$ energy component calculated in either the $DHFC$ or the $DHFB$ model and the corresponding non-relativistic energy component. For the argon atom ground state the non-additive component is a small percentage of the correlation energy ($\sim 1\%$ for the $DHFB$ case). It is an order of magnitude larger for the $DHFB$ case than it is for the $DHFC$ case.

Table 6
Non-additivity of relativistic and correlation energies in the argon atom ground state†

	$\underline{DHFC}$	$\underline{DHFB}$
$\Delta E_{correlation}$	−638.68	−638.68
$\Delta E_{relativistic}$	−1866.97	−1734.64
$\Delta E_{non-additive}$	−0.74	−7.52

† Energies are given in milliHartree

5. Prospects

5.1. BASIS SET TRUNCATION ERRORS

In non-relativistic molecular structure calculations the use of basis set expansion techniques, the algebraic approximation, is almost mandatory[33] [34]. Finite difference techniques, which are often used in atomic structure calculations [67], and finite element methods have been exploited in Hartree-Fock calculations for diatomic molecules[68] [69] [70] [71] [72], for which a suitable coordinate system is available, but for arbitrary polyatomic molecules and in calculations which take account of correlation effects the algebraic approximation afford the most general approach even though such calculations may possess quite severe basis set truncation errors.

Over the past three years, it has been clearly demonstrated that Hartree-Fock calculations for diatomic molecules containing first-row atoms carried out with finite basis sets expansions [74] [75] [76] can lead to total energies approaching a sub-μhartree level of accuracy which had previously only been realized in fully numerical calculations using finite difference [68] [70] or finite element [72] techniques.

In a pioneering series of finite basis set calculations on diatomic systems Wahl[78], Huo [79]and Cade, Sales and Wahl [77] established that it is important : to start with atomic basis sets which yield near Hartree-Fock atomic energies; to add to these atomic sets two sets of d-type functions and one set of f-type function to describe the polarization of the atom in the molecular environment; it is usually more efficient to extend the basis set rather than exhaustively optimize the exponents. A series of matrix Hartree-Fock calculations [74] [75] for the ground state of the nitrogen molecule, at its equilibrium nuclear configuration, using systematically constructed even-tempered basis sets of Gaussian-type functions including both d-type and f-type functions demonstrated that such calculations *cannot* yield total Hartree-Fock energies which are comparable in accuracy with those given by fully numerical finite difference/finite element approaches.

However, by adding functions at the bond centre to describe the build up of charge in the bond region improved accuracy was obtained in the finite basis set calculations and a total energy within ~ 2 μhartree of the fully numerical value obtained. Specifically, finite basis set calculations gave a total electronic energy of -108.9938234 hartree [75], finite difference calculations gave a total electronic energy of -108.9938257 hartree[70], and finite element calculations gave a total electronic energy of -108.993826 hartree [71] [72]. Calculations [76] for the ground state of the carbon monoxide molecule have demonstrated that this level of accuracy can be maintained over that part of the molecular potential energy curve which is adequately described by the Hartree-Fock ansatz.

The idea of using of bond functions in molecular structure calculations is an old one [80]. Indeed, a number of authors have advocated the use of basis functions located at the bond centre in both self-consistent field and in calculations taking account of correlation effects[81] [82] [83] [84] [85]. Such functions were introduced to provide the same effects as the addition of a set of polarization functions but at a considerably lower cost. Davidson and Feller [86] point out that

> *For first-row diatomics the introduction of a single (s,p) set at the centre of the bond provides* 90% *of the energy lowering obtained with a single set of d functions at the nuclear centres.*

In 1980, Čársky and Urban noted that general rules for selecting exponents and positions of the bond functions are still lacking [87].

There is no unique prescription for the construction of basis sets capable of yielding sub-μhartree accuracy. Recent matrix Hartree-Fock calculations [47] [46] for small diatomic systems described by orbitals of σ-symmetry have shown that such systems can be described to a sub-μhartree level of accuracy using basis sets containing only functions of s type provided that the basis functions are distributed along the internuclear axis according to empirical rules inspired by simple chemical considerations regarding the changes to the electron distribution during bonding. For the ground state of the hydrogen molecule at its equilibrium geometry a total electronic energy of $-1.133\ 629\ 29$ hartree using a basis set of 150 s-functions. This should be compared with an energy of $-1.133\ 629\ 37$ hartree obtained by using atom-centred basis sets containing functions of s, p, d, and f symmetry and the exact energy which is $-1.133\ 629\ 57$ hartree.

Two-dimensional finite difference Hartree-Fock calculations for diatomic molecules are computationally feasible and give very accurate total energies and occupied orbital energies [68] [69] [70]. The finite difference approach has been refined over recent years both in terms of the accuracy achieved and in terms of efficiency of the computational implementation [70] and has been shown to be competitive with the alternative fully numerical approach, the finite element technique[72]. Algorithms have been designed to take advantage of the vector processing capabilities of modern high performance computers [88]. It has been stressed [70] that, whilst finite difference and finite element techniques are often described as basis set independent methods,

> *they are in fact grid dependent and ... inadequate grid size can easily spoil the results, ensuring that the Hartree-Fock limit values are never reached.*

Table 7
Some calculated energies for the ground state of the HCl molecule with an internuclear separation of 2.40 bohr.†

	$E/Hartree$
$E(MHF)$	−460.060 54
$E(NHF)$	−460.113 05
$E(MDHFC)$	−461.525 38
ε_{BSTE}	0.054 52
$\varepsilon_{Rel.}$	−1.464 56
$\varepsilon_{Non-Rel.\ MBPT-2}$	−0.304 06

MHF : Matrix Hartree-Fock; NHF : Numerical Hartree-Fock;
$MDHFC$: Matrix Dirac-Hartree-Fock-Coulomb;
$BSTE$: Basis Set Truncation Error
† L. Laaksonen, I.P. Grant & S. Wilson, J. Phys. B: At. Mol. Opt. Phys. **21**, 1969 (1988)

The results of some prototype calcualtions for the ground state of the hydrogen chloride molecule are presented in Table 7. This molecule, which is isoelectronic with the argon atom considered in Table 5, has an experimental equilibrium internuclear distance of 2.40 bohr. $E(MHF)$ is the energy obtained by using the matrix Hartree-Fock model with a standard basis set defined in the work of Laaksonen, Grant and Wilson. $E(NHF)$ is the energy obtained by using the finite difference Hartree-Fock technique. $E(MDHFC)$ is the energy obtained by using the matrix Dirac-Hartree-Fock-Coulomb model with the same standard basis set. ε_{BSTE} is the basis set truncation error determined from $E\,(MHF)$ and $E(NHF)$. $\varepsilon_{Rel.}$ is the relativistic correction determined as the difference between $E(MDHFB)$ and $E(MHF)$. $\varepsilon_{Non-Rel.\ MBPT-2}$ is the non-relativistic second order many-body perturbation theory. For the ground state of HCl it can be seen that $\varepsilon_{BSTE} < \varepsilon_{Non-Rel.\ MBPT-2} < \varepsilon_{Rel.}$

5.2. BARDEEN-COOPER-SCHRIEFFER REFERENCE MODELS

Sorenson *et al*[89]-[94] identify two general classes of single-particle models for non-relativistic many-body perturbation theory studies:- *(i) normal* models, which include the Hartree-Fock theory, in which the quasi-particle interactions are localized, and *(ii) superfluid* models, which have bound states connecting two particles or two holes and in which particle-hole levels become fractionally occupied. In case *(ii)* the model physical vacuum is a Bardeen-Cooper-Schrieffer function and the mean field is a sum of the Hartree (direct), Fock (exchange) and a pairing potential arising from the pairing interactions which describe the bound state.

To set up the Bardeen-Cooper-Schrieffer function the single particle states are divided into core levels which are designated C, paired levels which are designated P, and unoccupied or virtual levels, which are designated V. The subset P is defined by

$$P = \bigcup_I \{h_I \cup p_I\} \tag{47}$$

where h_I and p_I label holes and particles mixed by the starting model for bond I. The Bardeen-Cooper-Schrieffer unperturbed function is written

$$|-\rangle = \left\{\prod_{i\in C} \hat{\alpha}_i \hat{\alpha}_{\bar{i}}\right\} \prod_I \left\{\prod_{i\in h_I} \left(\frac{\hat{\alpha}_i \hat{\alpha}_{\bar{i}}}{v_i}\right) \prod_{a\in p_I} \left(\frac{\hat{\alpha}_a \hat{\alpha}_{\bar{a}}}{v_a}\right)\right\} \tag{48}$$

where barred (unbarred) subsets count spin-down (spin-up). The first factor is written in terms of the creation and annihilation operators $\hat{a}_k^\dagger$ and $\hat{a}_k$, respectively, and describes the model core. The second factor describes the model pair operators and is written in terms of the second quantized solutions of the pairing interaction which is obtained by a special Bogoliubov-Valatin transformation

$$\hat{\alpha}_k = u_k \hat{a}_k - v_k \hat{a}_{\bar{k}}^\dagger \tag{49}$$

$$\hat{\alpha}_{\bar{k}} = u_k \hat{a}_{\bar{k}} + v_k \hat{a}_{\bar{k}}^\dagger \tag{50}$$

$k > 0$ and $k = -k$ are paired. The use of such a reference function has enabled the application of the many-body perturbation theory to a variety of chemically bonded systems including F_2 with its six lone pairs and the triply bonded N_2 molecule, both of which are stringent tests. Satisfactory results were obtained by employing third-order theory.

The construction of a suitable unperturbed Hamiltonian operator is described in the original publications by Sorenson *et al*[89]-[94]. The diagrammatic terms in the many-body perturbation theory with respect to the Bardeen-Cooper-Schrieffer function are in one-to-one correspondence with those which arise in the familiar particle-hole theory based on a single determinantal unperturbed function and the computational demands of the perturbation expansions based on the normal and on the superfluid models are essentially the same. Indeed, the diagrammatic terms arising from both models can be evaluated by means of the same code.

Now the diagrammatic perturbation theory is formulated using second quantization methods. The relativistic electronic structure problems is infinitely many-bodied and therefore the sue of second quantization techniques is mandatory. Just as the non-relativistic formal formalism can be readily transferred to the relativistic case for the normal class of single particle starting model, it would appear that for the superfluid class a similar transfer is possible. Work on this problem is in hand and will be reported in due course.

This concludes our "half term" report. A *"full term"* report will be produced in due course.

This article is based on a lecture given at the 8th International Congress of Quantum Chemistry Satellite Meeting on *Electron correlation in atoms and molecules: New methods and applications*, Smolenice, Slovakia, 14-18 June, 1994.

References

1. N.H. March, W.H. Young and S. Sampanthar, *The Many-Body Problem in Quantum Mechanics*, Cambridge University Press (1967)
2. S. Wilson, *Electron correlation in molecules*, Clarendon Press, Oxford (1984)
3. U. Kaldor, Phys. Rev. **A7**, 427 (1973)
4. S. Wilson and D.M. Silver, Phys. Rev. A14, 1969 (1976)
5. I.G. Kaplan, *Symmetry of Many-Electron Systems*, translated by J. Gerratt, Academic Press, New York & London (1975) [p. 269, the term "algebraic approximation" is employed to describe the use of finite analytic basis sets in molecular electronic structure calculations]
6. H.M. Quiney, I.P. Grant and S. Wilson, J. Phys. B: At. Mol. Phys. **18**, 2805 (1985)
7. H.M. Quiney, in *Relativistic Effects in Atoms and Molecules*, Meth. Comput. Chem. **2** 227 (1988)
8. S. Wilson, in *Relativistic Effects in Atoms and Molecules*, Meth. Comput. Chem. **2** 73 (1988)
9. H.M. Quiney, I.P. Grant and S. Wilson, in *Many-body methods in quantum chemistry*, Lecture Notes in Chemistry **52**, edited by U. Kaldor, Springer-Verlag, Berlin (1989)
10. H.M. Quiney, I.P. Grant and S. Wilson, J. Phys. B: At. Mol. Phys. **23**, L271 (1990)
11. S. Wilson, in *The Effects of Relativity in Atoms, Molecules and the Solid State*, edited by S. Wilson, I.P. Grant and B.L. Gyorffy, Plenum Press, New York (1991)
12. S. Wilson, in *Methods in Computational Molecular Physics*, NATO ASI Series B, Volume 293, edited by S. Wilson and G.H.F. Diercksen, p.253, Plenum Press, New York (1992)
13. P. Pyykkö, in *The Effects of Relativity in Atoms, Molecules and the Solid State*, edited by S. Wilson, I.P. Grant and B.L. Gyorffy, Plenum Press, New York (1991)
14. P.A.M. Dirac, Proc. Roy. Soc. London A **133**, 60 (1931)

15. W.H. Furry and J.R. Oppenheimer, Phys. Rev. **45**, 245 (1934)
16. P.A.M. Dirac, *Principles of Quantum Mechanics (Fourth edition)*, p. *v*, Clarendon Press, Oxford (1957)
17. J. Goldstone, Proc. Roy. Soc. (London) **A239**, 267 (1957)
18. S. Tomonaga, Progr. Theoret. Phys. **1**, 27 (1946)
19. S. Tomonaga, Phys. Rev. **74**, 224 (1948)
20. J. Schwinger, Phys. Rev. **74**, 1439 (1948)
21. J. Schwinger, Phys. Rev. **75**, 651 (1949)
22. R.P. Feynman, Phys. Rev. **76**, 749 (1949)
23. M. Gell-Mann and F. Low, Phys. Rev. **84**, 350 (1951)
24. G.C. Wick, Phys. Rev. **80**, 268 (1951)
25. J. Sucher, Phys. Rev. **107**, 1448 (1957)
26. P.J. Mohr, Phys. Rev. **A32**, 1949 (1985)
27. W.H. Furry, Phys. Rev. **81**, 115 (1951)
28. I.P. Grant and H.M. Quiney, Adv. At. Molec. Phys. **23**, 37 (1988)
29. J. Sapirstein, Physica Scripta **T46**, 52 (1993)
30. S. Schweber, *An Introduction to Relativistic Quantum Field Theory*, Harper and Row, New York (1961)
31. F. Mandl and G. Shaw, *Quantum Field Theory*, Wiley, New York (1984)
32. C. Itzykson and J.-B. Zuber, *Quantum Field Theory*, McGraw-Hill, New York (1980)
33. S. Wilson, Adv. Chem. Phys. **67**, 439 (1987)
34. S. Wilson, *Basis sets for electronic structure calculations*, Plenum Press, New York (in preparation)
35. Y.S. Lee and A.D. McLean, J. Chem. Phys. **76**, 735 (1982)
36. K.G. Dyall, I.P. Grant and S. Wilson, J. Phys. B: At. Mol. Phys. **17**, 493 (1984)
37. K.G. Dyall, I.P. Grant and S. Wilson, J. Phys. B: At. Mol. Phys. **17**, L45 (1984)
38. K.G. Dyall, I.P. Grant and S. Wilson, J. Phys. B: At. Mol. Phys. **17**, 1201 (1984)
39. R.E. Stanton and S. Havriliak, J. Chem. Phys. **81**, 1910 (1984)
40. H.M. Quiney, I.P. Grant and S. Wilson, J. Phys. B: At. Mol. Phys. **20**, 1413 (1987)
41. H.M. Quiney, I.P. Grant and S. Wilson, Physica Scripta **36**, 460 (1987)
42. R. McWeeny, *Dissertation*, Oxford, 1948
43. R. McWeeny, Nature **166**, 21 (1950)
44. S.F. Boys, Proc. Roy. Soc. (London) **A200**, 542 (1950)
45. I. Shavitt, Israel J. Chem. **33**, 357 (1993)
46. D. Moncrieff and S. Wilson, Molec. Phys. **82**, 523 (1994)
47. S. Wilson and D. Moncrieff, Molec. Phys. **80**, 461 (1993)
48. A. Bohr and V.F. Weisskopf, Phys. Rev. **77**, 94 (1950)
49. Y. Ishikawa, R. Barretty and R.C. Binning, Chem. Phys. Letts. **121**, 130 (1985)
50. Y. Ishikawa and H.M. Quiney, Intern. J. Quantum Chem. Symp. **21**, 523 (1987)
51. L. Laaksonen, I.P. Grant and S. Wilson, J. Phys. B: At. Mol. Phys. **21**, 1969 (1988)
52. B. Klahn and J. Morgan III, J. Chem. Phys. **81**, 410 (1984)
53. J. Morgan III, in *Numerical Determination of the Electronic Structure of Atoms, Diatomic and Polyatomic Molecules*, edited by M. Defranceschi and Joseph Delhalle, p. 49, NATO ASI Series C Volume 271, Kluwer, Dordrecht (1989)
54. R.E. Trees, Phys. Rev. **102**, 1553 (1956)
55. B.F. Gray, J. Chem. Phys. **36**, 1801 (1962)
56. P.J. Rossky and M. Karplus, J. Chem. Phys. **67**, 5419 (1977)
57. H.M. Quiney, I.P. Grant and S. Wilson, J. Phys. B: At. Mol. Phys. **18**, 577 (1985)
58. B. Swirles, Proc. Roy. Soc. (London) **A152**, 625 (1935)
59. B. Swirles, Proc. Roy. Soc. (London) **A157**, 680 (1935)
60. I.P. Grant, Proc. Roy. Soc. (London) **A262**, 555 (1961)
61. I.P. Grant, in *Relativistic effects in atoms and molecules*, Meth. Comput. Chem. **2**, 1, Plenum Press, New York (1988)
62. I. Lindgren and A. Rosén, Case Studies in Atomic Physics **4**, 97 (1974)
63. J.-P. Desclaux, in *Relativistic effects in atoms, molecules and solids*, Proceedings of

NATO Advanced Study Institute, edited by G. Malli, p. 115, Plenum Press, New York (1983)
64. M. Synek, Phys. Rev. **136**, 1556 (1964)
65. Y.K. Kim, Phys. Rev. **154**, 17 (1967); *erratum* 159, 190 (1958)
66. H.A. Bethe and E.E. Salpeter, *The Quantum Mechanics of One- and Two-Electron Systems*, Springer-Verlag, Berlin (1957)
67. C. Froese Fisher, *The Hartree-Fock Method for Atoms. A numerical approach*, (New York, Interscience, 1977)
68. L. Laaksonen, P. Pyykkö and D. Sundholm, Comput. Phys. Rept. **4**, 313 (1986)
69. P. Pyykkö, in *Numerical Determination of the Electronic Structure of Atoms, Diatomic and Polyatomic Molecules*, edited by M. Defranceschi and J. Delhalle, NATO ASI Series C271, *p.* 161 (1989)
70. J. Kobus, Chem. Phys. Lett. **202**, 7 (1993)
71. D.Heinemann, B.Fricke & D.Kolb, Phys. Rev. A**38**, 4994 (1988)
72. D. Heinemann, A. Rosen and B. Fricke, Physica Scripta **42**, 692 (1990)
73. H.M. Quiney, I.P. Grant and S. Wilson, J. Phys. B: At. Mol. Phys.
74. D.Moncrieff and S. Wilson, Chem. Phys. Lett. **209**, 423 (1993)
75. D.Moncrieff and S. Wilson, J. Phys. B: At. Mol. Opt. Phys. **26**, 1605 (1993)
76. J. Kobus, D. Moncrieff and S. Wilson, J. Phys. B: At. Mol. Opt. Phys. **27**, 5139 (1994)
77. P.E. Cade, K.D. Sales and A.C. Wahl, J. Chem. Phys. **44**, 1973 (1966)
78. A.C. Wahl, J. Chem. Phys. **41**, 2600 (1964)
79. W.M. Huo, J. Chem. Phys. **43**, 624 (1965)
80. H. Preuss, Z. Naturforsch **19a**, 1335 (1964); ibid **20a**, 17, 21 (1965)
81. S. Rothenberg and H.F. Schaefer III, J. Chem. Phys. **54**, 2765 (1971)
82. W. Butscher, S. Shih, R.J. Buenker and S.D. Peyerimhoff, Chem. Phys. Lett. **52**, 457 (1977)
83. D. Neisius and G. Verhaegen, Chem. Phys. Lett. **66**, 358 (1979)
84. L. Lawlor, K. Vasudevan and F. Grein, Chem. Phys. Lett. **75**, 79 (1980)
85. P. Mach and O. Kysel, J. Comput. Chem. **6**, 312 (1985)
86. E.R. Davidson and D. Feller, Chem. Rev. **86**, 681 (1986)
87. P. Čársky and M. Urban, *Ab initio Calculations. Methods and Applications in Chemistry*, Lecture Notes in Chemistry **16** (Springer-Verlag, Berlin)
88. J. Kobus, Comput. Phys. Commun. **78**, 247 (1994)
89. T.E. Sorensen, *Bardeen-Cooper-Schrieffer-Lipkin-Nogami Theory: A New Method for Electronic Structure Calculations with Applications to Potential Energy Curves for H_2, LiH, FH, F_2 and N_2*, Dissertation, University of Wisconsin-Milwaukee, August, 1989
90. T.E. Sorensen, W.B. England and D.M. Silver, J. Phys. B: At. Mol. Opt. Phys. **22**, L539 (1989)
91. T.E. Sorensen, W.B. England, D.M. Silver and E.O. Steinborn, Theor. chim. Acta. **84**, 1 (1992)
92. T.E. Sorensen, W.B. England and D.M. Silver, Theor. chim. Acta. **84**, 21 (1992)
93. T.E. Sorensen, W.B. England and D.M. Silver, Theor. chim. Acta. **84**, 37 (1992)
94. T.E. Sorensen, W.B. England and D.M. Silver, Intern. J. Quantum Chem: Quantum Chemistry Symposium **27**, 467 (1993)

MODELS FOR SIMULATING MOLECULAR PROPERTIES IN CONDENSED SYSTEMS

MATI KARELSON
Department of Chemistry, University of Tartu
Jakobi St. 2, Tartu EE 2400, Estonia

AND

GEERD H.F. DIERCKSEN
Max-Planck-Institut für Astrophysik
Karl-Schwarzschild-Strasse 1, Garching bei München, D-85700, Deutschland

1. Introduction

The disordered condensed media (liquids, solutions, polymers, glasses etc.) are most widespread objects in chemistry and physics. As a rule, the quantum-chemical methods give a description of isolated individual molecules that corresponds to the infinitely rare gas at $0K$, i.e. to the conditions close to the intergalaxial space. Therefore, it is of large theoretical and practical importance to extend the quantum description of atomic and molecular systems to the disordered condensed media, including pure liquids, solutions, polymeric and glassy materials. The quantum theory of solid matter with ordered crystal lattices has been extensively developed on the basis of use of the translational symmetry in such systems and will not be reviewed in the present lecture notes. The reader is referred to excellent reviews on this subject [1, 2]. However, we will consider the theory which is applicable to low symmetry molecular impurities in regular crystals. In the following, we wish to review theoretical methods used to approach different chemical and physical aspects of the solvation of molecular systems in disordered condensed media.

The complex process of solvation corresponds to a manifold of the structural, geometrical and electronic changes both in the solvent and solute in the course of the dissolution of the latter from infinitely diluted gaseous phase. It is customary to distinguish between the equilibrium and non-

S. Wilson and G. H. F. Diercksen (eds.), Problem Solving in Computational Molecular Science, 215–248.

equilibrium solvation effects. The equilibrium effects reflect the difference in the chemical potential and in the electronic and geometrical structure of the solute molecule in the gas phase and in the solution at the thermodynamic equilibrium. The Gibbs free energy of solvation is accordingly defined as follows

$$\Delta G_{sol} = \Delta G_s - \Delta G_{gas} \tag{1}$$

where both the free energy of the solute in the solution ΔG_s and the free energy of the solute in the gas phase ΔG_{gas} can be rigorously presented by the respective partition functions in these two media

$$\Delta G_s = RT \ln Z_s \tag{2}$$

$$\Delta G_{gas} = RT \ln Z_{gas} \tag{3}$$

It should be emphasised that all components of the partition function

$$Z = Z_{tr}\, Z_{rot}\, Z_{vib}\, Z_{el} \tag{4}$$

i.e. the translational, rotational, vibrational and electronic partition functions, respectively, may be affected by the solvation of the molecule. Therefore, the rigorous *ad hoc* calculation of the solvation phenomena is very complicated and requires the use of precise physical theory for the description of different motions of particles and molecules in different media.

The free energy of solvation, ΔG_{sol}, can be also divided into the respective enthalpic and entropic contributions, ΔH_{sol} and ΔS_{sol}, i.e.

$$\Delta G_{sol} = \Delta H_{sol} - T\,\Delta S_{sol} \tag{5}$$

This separation brings about the differentiation between the enthalpic and entropic solvent effects, each of which should be treated using different, in principle, theoretical methods.

Historically, the enormous complexity of solvational phenomena has given rise to numerous empirical (parametric) and physical models of solvation. The parametric approach is based on the widespread observance of the linear free energy relationships [3, 4]. According to this approach, the solvation is considered as a manifold of physically independent processes which can be described by different empirical solvent effect scales. In most cases, two types of processes are considered, the first of which correspond to the non-specific or macroscopic dielectric polarization effects whereas the second type of solvent effects involves the formation of directed semichemical bonds (e.g., hydrogen bonds) between the solute and solvent molecules in solution. Notably, the solvent scales used in the development of solvation linear free energy relationships (often called the linear solvation energy relationships - $LSER$ [4]) are almost exclusively derived from different sets of empirical (experimental) data, each of which presumably corresponds to a single solvent effect according to the above-given classification. The more notable deficiency of such empirical approach is that no predictions can be made for chemical and physical processes with no experimental information available or solvents with unknown parameters.

A theoretically more justified approach to the solvent effects is based on the construction of certain physical models for the description of different aspects of solvation phenomena. Traditionally, the following solvation processes are considered separately: (i) the cavity formation in the solvent for the solute molecule; (ii) the placement of the solute into the cavity and the accompanying electrostatic and orientational polarization and the dispersion interactions between the solute and solvent molecules; (iii) the formation of hydrogen bonds or complexes between the solute and solvent molecules. The cavity formation effects are usually treated using the statistical mechanical methods whereas the polarization and dispersion effects can be described using the combination of statistical mechanical and quantum mechanical theory. An accurate description of the hydrogen bonding in liquid media requires also the use of advanced quantum mechanical methods.

The theoretical approaches discussed above apply in the cases when the solute is fully equilibrated with the solvent and *vice versa*. However, many physical processes and phenomena (spectroscopic transitions, instantaneous application of external electric or magnetic fields in condensed media, etc.) are characterized by much shorter time scales than the characteristic relaxation time of the solvent. Therefore, in these cases the solvent is in a non-equilibrium state during the observable process and should be treated according to the time-dependent theory.

2. Entropic and cavity formation contributions to the free energy of solvation

According to Ben-Naim's definition of the solvation process [5-7], it consists the transfer of a solute M from a fixed position in the ideal gas phase to a fixed position in solvent S, at constant temperature T, pressure p and the chemical composition. The Gibbs free energy of solvation is then related to the work necessary to carry out this transfer, $W(M|S)$, also called the coupling work of M with S. A simple statistical derivation leads the following expression for the solvation free energy [8]

$$\Delta G_{sol} = W\,(M|S) + RT\,\left(\frac{q_{rot,g}\,q_{vib,g}}{q_{rot,s}\,q_{vib,s}}\right) + p\,\Delta V - RT\,\ln\left(\frac{n_{M,g}\,\Lambda^3_{M,g}}{n_{M,s}\,\Lambda^3_{M,s}}\right) \quad (6)$$

where $q_{rot,g}, q_{vib,g}, q_{rot,s}$ and $q_{vib,s}$ are the microscopic partition functions for rotation and vibration of M, in the gas phase and in the solution, respectively; $n_{M,g}$ and $n_{M,s}$ are the number densities of the solute M in the corresponding phases; and $\Lambda_{M,g}$ and $\Lambda_{M,s}$ are the respective translational momentum partition functions. In Eq. 6, the $p\,\Delta V$ term is usually neglected as being very small ($\sim 10^{-3}$ $kcal/mol$). The sum of two first terms in this equation represent the solvation free energy at equal numerical densities of the solute in the gas phase and in the solution (denoted also as ΔG_{sol}). The coupling work of M and S is usually decomposed as follows:

$$W(M|S) = \Delta G_{el} + \Delta G_{cav} + \Delta G_{disp} + \Delta G_{rep} + \Delta G_{HB} \quad (7)$$

where ΔG_{el} is the solvent electrostatic polarization term, ΔG_{cav} is the term representing the work needed to create the cavity for a solute in the liquid, ΔG_{disp} and ΔG_{rep} are the solute-solvent dispersion and exchange-repulsion interaction energy, respectively, and ΔG_{HB} is the specific hydrogen-bonding energy between the solute and solvent.

The last term in Eq.6 is usually cited as the liberation free energy difference [6, 7] and has essentially an entropic origin. In the case of infinite isotropic solutions of the same numerical density of the solute as in the respective gaseous phase, this term is zero. However, if restrictions apply to the translational motion of the solute molecule in a condensed medium, either due to the specific bonding with the solvent or because of the formation of solvational cages around the solute molecule, the liberation free energy may become significant. There are no limitations to the movement of molecules in the ideal gas. The corresponding additional freedom in motion as compared to the molecules in rigid matrix is called "communal entropy" and may be recovered only partly in liquids and solutions [9]. In many cases, this entropy term is neglected in the treatment of the solvation free energy [8] and for the aqueous solutions of alkanes it has been shown that this term is very small [10]. However, the assumption about ideal mixing in solution also requires the interchangeability of the solvent and solute molecules in terms of their respective volumes, i.e. the partial molar volumes of the components of the mixture should be the same. Obviously, this is not valid for most solutions, particularly for the aqueous solutions, due to the small size of the water molecule as compared to most solutes.

Proceeding from the ratio of volumes occupied by the solute and solvent molecules in solution, Flory and Huggins [11-13] and Hildebrand [14] had developed the expression for the partial molar entropy of chain polymers in nondilute solutions. Recently, an analogous formula has been derived for the calculation of the respective entropic term for the solutions of low molecular weight compounds [10,15-18]

$$\Delta S_{FH} = -R\left[\ln \phi_M + \phi_S\,(1 - \frac{V_M}{V_S})\right] \tag{8}$$

where

$$\phi_M = \rho_{M,S}\, V_M \tag{9}$$

$$\phi_S = 1 - \phi_M \tag{10}$$

V_M is the volume of the large (solute) molecule in the solution, V_S is the volume of the small (solvent) molecule in the solution, and $\rho_{M,S}$ is the number density of M in S. The quantities ϕ_M and ϕ_S are the respective volume fractions of M and S in solution. ΔS_{FH} is usually called the Flory-Huggins approximation to the combinatorial entropy of combining the solute with the solvent [18].

The use of the Flory-Huggins combinatorial entropy for the aqueous solutions of low molecular weight compounds, in particular for their dilute aqueous solutions has been severely criticized [19, 20]. It has been pointed out that the entropy of mixing will include not only the combinatorial entropy, but also that due to the dissimilarity in free volume of the two components and that due to the structural effects. Also, the solution configuration cannot be assumed random during the dissolution of the solute molecule.

Another term in the free energy of solvation having predominantly entropic nature is the free energy of the cavity formation ΔG_{cav} (cf. Eq. 7). The concept of cavity formation in the process of dissolution of a compound in the pure liquid was introduced by Uhlig and Eley [21], following a suggestion by Susskind and Kasarnowski [22]. Several theoretical expressions, based on the shape and size of the solute and various parameters of the solvent have been derived for the calculation of ΔG_{cav}. The simplest approach proceeds from the abstraction of the surface tension on the boundary between the solute cavity and the solvent. In this case, the free energy of cavity formation is taken simply proportional to the surface of the solute cavity, S_M :

$$\Delta G_{cav} = \sigma\, S_M \tag{11}$$

where σ is the surface tension of the solvent (in appropriate units). This formula has been applied in many cases, e.g. for the evaluation of the free energy of transfer of electroneutral solutes between different solvents [8]. It has been also extended to account for the size of the solvent molecule as follows:

$$\Delta G_{cav} = \sigma\, S_M - RT\, \ln(1 - V_S\, n_S) \tag{12}$$

where V_S is the intrinsic volume of a solvent molecule and n_S is the number density of the solvent. It has been suggested to use different surface tension values for different atomic types in the solute molecule [23]. Thus,

$$\Delta G_{cav} = C + \sum_i \sigma_i\, A_i \tag{13}$$

where A_i are the solvent-accessible surface areas of different atoms in the solute molecule and C and σ_i are empirically adjustable parameters. The quality of the description of experimental data has been, however, not significantly improved by the introduction of those additional empirical parameters.

Another theoretical construction for the calculation of the free energy of cavity formation proceeds from the theory of microscopic curved surfaces [24-27]. According to Sinanoglu [27, 28],

$$\Delta G_{cav} = k_s^g\,(V_S|V_M)\,\sigma\,\Sigma \tag{14}$$

where Σ is the area of the cavity and k_s^g is a correction factor, characteristic for a given solvent and depending on the ratio of molecular volumes of the solvent and solute. This factor has been approximated by the following formula

$$k_s^g\,(V_S|V_M) = 1 + \left(\frac{V_S}{V_M}\right)^{2/3} [\,k_s^g\,(1) - 1\,] \tag{15}$$

where k_s^g is estimated from the solubility of a given solute in a given solvent. The main deficiency of this approach is connected with the introduction of additional empirical information, often not readily available.

It has been also suggested to estimate the free energy of cavity formation from the data on isothermal compressibility, β_T, as follows [29-31]

$$\Delta G_{cav} = \frac{V_{cav}}{\beta_T} + C \tag{16}$$

where V_{cav} is the volume of the cavity and C is a constant term. An excellent correlation has been established between the free energy of cavity formation and thermal volume of the cavity, both calculated from the results of the molecular dynamics simulation of liquid water. However, the microscopic isothermal compressibility of water, calculated from the slope of Eq. 16, was found to be about an order higher than the respective experimental value for water ($\beta_T(calc) = 23.5$ vs. $\beta_T(exp) = 3.14$). Therefore, it is not feasible to use directly the macroscopic surface tension or compressibility of the solvent for the respective microscopic model quantities.

A different, basically entropic approach to the calculation of the free energy of cavity formation proceeds from the scaled particle theory (SPT) [32-35]. The free energy of the formation of a spherical cavity in a solvent, ΔG_{cav}, can be calculated proceeding from the SPT as follows

$$\Delta G_{cav} = RT \left\{ 1 - \ln(1-y) + \left(\frac{3y}{1-y}\right)\frac{a_M}{a_S} + \left[\frac{3y}{1-y} + \frac{9}{2}\left(\frac{y}{1-y}\right)^2\right]\left(\frac{a_M}{a_S}\right)^2 \right\} \tag{17}$$

where

$$y = \frac{4\pi \rho a_S^3}{3} \tag{18}$$

is the reduced number density of the solvent. In the last formulae, a_M and a_S denote the intrinsic radii of the solute and solvent molecules, respectively, and ρ is the number density of the solvent. In the case of an ellipsoidal cavity, the SPT cavity formation energy is given by the following equation [36]

$$\Delta G_{cav} = RT \left\{ 1 - \ln(1-y) + \left(\frac{\alpha y}{1-y}\right)\frac{a}{a_S} + \left[\frac{\beta y}{1-y} + \gamma\left(\frac{y}{1-y}\right)^2\right]\left(\frac{a}{a_S}\right)^2 \right\} \tag{19}$$

where α,β,and γ denote the geometrical coefficients and a is the characteristic length of the ellipsoid (the major semi-axis).

Recently, the scaled particle theory has been extended to dilute solutions of arbitrary shaped solutes and has been successfully applied for the calculation of the solvation free energy of hydrocarbons in aqueous solutions [37].

3. Dispersion interaction in condensed media

The dispersion interaction between two atomic or molecular systems has been rather thoroughly studied at different levels of theory [38-41]. The modelling of the dispersion interaction in condensed media is much less elaborated and proceeds either from the discrete molecular description of the liquid or from the continuum dielectric model. According to a contemporary classification [8], the theoretical approaches to the dispersion effect in solutions can be divided into following classes: (i) pair-potential approaches; (ii) reaction field based approaches; and (iii) cavity surface-dispersion energy relationship approaches.

The pair-potential approach is based on the discrete representation of the pairs of solvent and solute molecules or some fragments of them. Traditionally, the dispersion potentials are expressed as truncated asymptotic expansions in powers of $1/r$, the reciprocal of the distance between the interacting entities [8]

$$U_{ms}\,(disp) = \sum_{k=6,8,10} d_{ms}^{(k)}\, r_{ms}^{-k} \tag{20}$$

where the indexes m and s denote the structural entities (atoms, bonds, chemical groups) belonging to the solute and solvent molecules, M and S, respectively. The powers in expansion (Eq. 20) are based on the formal theory of two-body interactions. Only the first term of the expansion ($k = 6$) is often applied in the practical calculations. The expansion coefficient d_{ms}^{6} may be approximated, for instance, by the well-known London formula

$$d_{ms}^{(6)} = -\frac{3}{2}\,\alpha_m\,\alpha_s\,\frac{\bar{I}_m\,\bar{I}_s}{\bar{I}_m + \bar{I}_s} \tag{21}$$

where α_m and α_s are the isotropic polarizabilities for interacting systems and $\bar{I}_m$ and $\bar{I}_s$ are the mean excitation energies of these systems. This approximate formula is, in principle, valid only for interacting atomic systems. In the case of molecular systems, the atomic or group polarizabilities and local excitation energies are no more isotropic and require the use of the respective tensorial quantities. Another question arises in connection with the absence of information about of accurate solute-solvent atom-atom distribution functions in dense media. Those can be calculated either using the molecular dynamics or Monte Carlo computer simulations or from the experimental scattering data on the respective systems. However, almost all these methods give only the averaged distribution functions and lack, therefore, the information about the local anisotropy of the atom-atom distributions.

In the first approximation [42], the dispersion potential is limited to the dipole-dipole term and the mean excitation energies are approximated by the respective ionization potentials for the solute and solvent molecules. Thus, when the solute molecule is surrounded by a small cluster of solvent molecules, the dispersion energy is given by the following formula [42]:

$$U_{MS}\,(disp) = -\frac{x}{4}\,\frac{\bar{I}_m\,\bar{I}_s}{\bar{I}_m + \bar{I}_s}\sum_{u=1}^{B_M}\sum_{v=1}^{B_S}\left\{ r_{uv}^{-6}\, Tr\left[\mathbf{T}_{uv}\,\mathbf{A}_u\,\mathbf{T}_{uv}\,\mathbf{A}_v\right]\right\} \tag{22}$$

where B_M and B_S are the numbers of bonds in M and S respectively, I_m and I_s are the mean excitation energies of these species, T_{uv} is the tensor

$$\mathbf{T}_{uv} = 3\,\frac{\vec{r}_{uv}}{r_{uv}} \otimes \frac{\vec{r}_{uv}}{r_{uv}} - 1 \tag{23}$$

where r_{uv} and $\vec{r}_{uv}$ are the distance and the radius-vector between the bonds u and v, respectively, and $\mathbf{A}_u$ is the polarizability tensor for bond u. The factor x in Eq. 22 is introduced to achieve the agreement between the molecule-molecule pair dispersion potential and a simpler expression derived on the basis of assumption that the dispersion energy between two molecules may be reduced to the sum of independent atom-atom contributions [43, 44]

$$U_{MS}\,(disp) = \sum_m \sum_s d_{ms}^{(6)}\, r_{ms}^{-6} \tag{24}$$

It has been concluded that a simpler atom-atom isotropic pair potential approach with coefficients drawn either from experimental data [44] or computer simulations is superior over the bond partitioning scheme [8, 45].

J. Tomasi and others [8] have developed a scheme which reduces the spatial representation of the dispersion interaction into a surface representation of this interaction, thus introducing the concept of the molecular cavity in the framework of the pair-pair dispersion potential approach. Namely, the average dispersion-repulsion energy of a dense solute-solvent system may be written as follows [8]:

$$\langle E_{disp-rep}\rangle = \int \ldots \int U\,(\Omega)\, g\,(\Omega)\, d\Omega \tag{25}$$

where Ω stands for the set of all coordinates of the molecules involved, $g(\Omega)$ is the distribution function and $U(\Omega)$ is expressed as a sum of two-body dispersion-repulsion potentials. In the case of the fixed geometry of the solute molecule

$$\langle E_{disp-rep}\rangle = n_S \sum_{s\in S} N_S \sum_{m\in M} \sum_k d_{ms}^{(k)} \int r_{ms}^{(k)}\, g_{ms}\,(\vec{r}_{ms}\,)\, dr_{ms}^3 \tag{26}$$

The integrals in the last formula can be limited only to a certain minimum distance between the solute and solvent molecules, for instance, to those corresponding to the van-der-Waals radii of these molecules. In such case, a fixed cavity with the surface Σ_S can be defined for the solute molecule as formed from the space forbidden for the solvent molecules. By introducing the auxiliary vector functions A_{ms}^k such that

$$\vec{\nabla} \cdot A_{ms}^{(k)}\,(r_{ms}) = d_{ms}^{(k)}\, r_{ms}^{-k}\, g_{ms}\,(\vec{r}_{ms}\,) \tag{27}$$

the average dispersion-repulsion energy between the solute and solvent molecules in solution can be written as follows

$$\langle E_{disp-rep}\rangle = n_S \sum_{s\in S} N_S \sum_{m\in M} \sum_{k} \int_{\Sigma_s} A_{ms}^{(k)}\, \vec{n}_\sigma \, d\sigma \tag{28}$$

where $\vec{n}_\sigma$ is the outer normal to the surface Σ_s at the position σ. The integral in the last equation may be calculated numerically using an appropriate partitioning (tessellation) of the surface. A comprehensive overview of different pair-pair dispersion interaction models is given elsewhere [8, 45].

The reaction field approach to the calculation of the solute-solvent dispersion energy in solution is based on the model according to which a quantum mechanically described solute molecule interacts with the medium described as a collection of point-polarizable dipoles. In the direct reaction field (DRF) approach [46-48], the Hamiltonian of the solute molecule is written as

$$\hat{H} = \hat{H}^0 - \frac{1}{2} \sum_{i,j} \sum_{p,q} \mathbf{F}_{ip}^{\dagger}\, \alpha_{pq}\, \mathbf{F}_{jq} \tag{29}$$

where the indexes i and j run over the particles of solute and p and q over the external polarizable dipoles. $\mathbf{F}_{ip}$ is the field generated by particle i at point p. The α_{pq} tensor gives the moment induced at q by a field applied at point p. The Hamiltonian (Eq.29) includes, of course, the electrostatic polarization part in the perturbation term. According to the authors of the DRF method, what is left after subtracting this part is considered as the measure for the dispersion interaction in the dipole-dipole approximation.

Several other dispersion models have been connected with the cavity representation of the reaction field [49-53]. The respective effective Hamiltonian for the solute in dielectric continuum is usually presented by the following equation

$$\hat{H} = \hat{H}^0 + \hat{V}_{el} + \hat{V}_{disp} \tag{30}$$

where the $\hat{V}_{el}$ operator regards to the classical electrostatic polarization term and the $\hat{V}_{disp}$ operator represents the dispersion interaction.

A quantum-mechanical method of calculation of the dispersion energy in solution was developed by Rösch and Zerner [54] on the basis of Abe's theory [55, 56]. According to this method, the dispersion energy, E_{disp} , for a solute molecule in a spherical cavity is given as follows

$$E_{disp} = -\frac{2}{3} \frac{1}{a_S^3\, a_M^3} \sum_{J\neq I} \sum_{K\neq O} \frac{\left(\mu_{IJ}^{M}\right)^2 \left(\mu_{KO}^{S}\right)^2}{E_K^S - E_O^S + E_J^M - E_I^M} \tag{31}$$

where the superscript S refers to the solvent molecule and the superscript M to the solute molecule. Thus, μ_{IJ}^{M} and μ_{KO}^{S} are the transition dipoles between the respective states of the solute (I and J) and the solvent (K and O) molecules. In Eq. 31, E_K^S, E_O^S and E_J^M, E_I^M, denote the energies of the K-th and O-th state of the solvent and of the J-th and I-th state of the solute molecule, respectively. The a_S and a_M are the cavity radii for the solvent and solute molecules in the solution, respectively.

4. Dielectric polarization of condensed media

As discussed above, the solvation free energy, $\triangle G_S$, can divided into the following contributions

$$\triangle G_s = \triangle G_{el} + \triangle G_{cav} + \triangle G_{disp-rep} + \triangle G_{HB} \tag{32}$$

where $\triangle G_{el}$ is the solvent electrostatic polarization term, $\triangle G_{cav}$ is the free energy of the cavity formation, $\triangle G_{disp-rep}$ is the solute-solvent dispersion and exchange-repulsion interaction energy, and $\triangle G_{HB}$ - the specific hydrogen-bonding energy between the solute and solvent molecules. In this lecture, we will concentrate on the methods of calculation of the first term, the solvent electrostatic polarization term.

The solvent electrostatic polarization term can be calculated either using the classical electrostatics or quantum theory. First, we contemplate the classical approach to the electrostatic solvation energy, originated by Kirkwood and Onsager [57, 58] and excellently reviewed by Böttcher and Bordewijk [59]. By assuming the statistical averaging over all solute-solvent intermolecular distances and configurations, the solvent can be represented as a homogeneous polarizable dielectric continuum. The energy of electrostatic interaction between the charge distribution of a solute molecule and the dielectric medium is then given by the following equation [60]

$$E_{el} = \frac{1}{8\pi} \int_V \mathbf{E}_S \cdot \mathbf{E}_0 \, (\epsilon - 1) \, dV \tag{33}$$

where $\mathbf{E}_0$ is the electrostatic field of the charges in the molecule *in vacuo* and $\mathbf{E}_S$ is the modified field in the presence of a dielectric media. Notably, according to Eq. 33, the dielectric constant ϵ of the medium is a function of the space coordinates, i.e. both the interior of the molecule and the surrounding medium are treated by the same equation. However, the integral in the last equation cannot be calculated, in general, analytically and even the numerical integration over V presents a difficult mathematical task. Therefore, the electrostatic equation is usually simplified by the application of the Gauss divergence theorem. By using this theorem, the volume integral (33) is transformed into a surface integral over some boundary

$$E_{el} = \frac{\epsilon - 1}{8\pi} \int_S \Phi_S \cdot \mathbf{E}_0 \vec{n}\, dS \tag{34}$$

where S is the boundary surface, n the unit outward normal vector on S and the reaction potential Φ_S is defined as follows: $\mathbf{E}_S = -grad\,\Phi_S$. Different representations of this integral, corresponding to different molecular shapes, have been applied to find electrostatic solvation energies of compounds in liquids. According to the classical reaction field theory of Kirkwood and Onsager [57-59], the solute molecule is represented by a set of point charges fixed inside of sphere of a radius a_0 and the electrostatic equation (Eq. 34) is solved by applying the appropriate boundary conditions inside and outside the sphere. It is also assumed that the dielectric constant inside the cavity (sphere) is equal to unity (vacuum) and outside the cavity has a constant value, corresponding to the macroscopic dielectric constant of the medium. In that case, the energy of the electrostatic interaction between the solute charge distribution and the surrounding dielectric medium will be given by the following infinite expansion

$$E_{el} = \frac{1}{2} \sum_{i,j} e_i\, e_j \sum_{l=0}^{\infty} \left[\frac{(l+1)(1-\epsilon)}{\epsilon(l+1)+l} \right] \frac{\vec{r}_i^{\,l}\, \vec{r}_j^{\,l}}{a_0^{2l+1}} P_l\,(\cos\Theta_{ij}) \tag{35}$$

where e_i and e_j are the charges inside the sphere at positions $\vec{r}_i$ and $\vec{r}_j$, respectively, and Θ_{ij} is the angle at the centre of the sphere between the vectors $\vec{r}_i$ and $\vec{r}_j$. In this equation, the summation extends over all charged particles (nuclei and electrons) of the solute and P_l are the Legendre polynomials of the l-th order. By expressing the Legendre polynomials as the products of the respective spherical harmonics of order $m(-l \leq m \leq l)$, Eq. 34 can be rewritten as [61]

$$E_{el} = -\frac{1}{2} \sum_{l=0}^{\infty} \sum_{m=-l}^{l} \mathbf{R}_l^m \cdot \mathbf{M}_l^m \tag{36}$$

where

$$\mathbf{R}_l^m = f_l\, \mathbf{M}_l^m \tag{37}$$

and

$$f_l = \frac{(l+1)(\epsilon-1)}{(l+1)\epsilon+l}\, \frac{1}{a_0^{2l+1}} \tag{38}$$

In these equations, $\mathbf{M}_l^m$ and $\mathbf{R}_l^m$ represent the electrical momentum and the respective reaction field components. The first term ($l = 0$) in the infinite expansion (Eq. 35) gives the interaction of the excess (ionic) charge of the solute with the respective reaction field created in the dielectric medium (Born term)

$$E_{Born} = \frac{(1-\epsilon)}{2\epsilon}\,\frac{Q^2}{a_0} \tag{39}$$

where Q is the value of the ionic charge. The next term ($l = 0$) corresponds to the solute dipole interaction with the corresponding reaction field (Onsager dipolar term)

$$E_{Onsager} = \frac{(1-\epsilon)}{(2\epsilon+1)}\,\frac{\vec{\mu}^{\,2}}{a_0^3} \tag{40}$$

where $\vec{\mu}$ is the dipole moment of the solute. In most applications, only these two terms which are the largest by size are considered in the calculation of the electrostatic interaction energy. However, depending on system studied, the interaction of higher electrical moments with the corresponding reaction field may become also significant and the terms corresponding to higher moments of order $2\,l$ (quadrupole, octupole, hexadecapole etc.) should be taken into account.

In many cases, the shape of the solute molecule may be very different from the sphere and therefore, it is necessary to apply the methods of calculation of the electrostatic solvation energy developed for more complex cavities. In the case of an ellipsoidal cavity with main semiaxes a, b, and c, the analytical formulas are still available for the calculation of the charge and dipolar terms of the electrostatic interaction with the reaction field. The charge term is simply

$$E_{Born}^{ell} = \frac{(1-\epsilon)}{2\epsilon}\,\frac{Q^2}{a\,b\,c} \tag{41}$$

whereas in the respective dipolar term [3, 6]

$$E_{Onsager}^{ell} = R\,\vec{\mu} \tag{42}$$

where the reaction field R is presented by the following tensor

$$R = \begin{pmatrix} \frac{3\,A_a\,(1-A_a)(1-\epsilon)\,\mu_a}{a\,b\,c\,[\epsilon+(1-\epsilon)\,A_a]} & 0 & 0 \\ 0 & \frac{3\,A_b\,(1-A_b)(1-\epsilon)\,\mu_b}{a\,b\,c\,[\epsilon+(1-\epsilon)\,A_b]} & 0 \\ 0 & 0 & \frac{3\,A_a\,(1-A_a)(1-\epsilon)\,\mu_a}{a\,b\,c\,[\epsilon+(1-\epsilon)\,A_a]} \end{pmatrix} \tag{43}$$

where A_a, A_b and A_c are the standard ellipsoidal shape factor integrals [59], and μ_a, μ_b and μ_c are the dipole moment components along the main semiaxes of the ellipsoid. Various methods have been developed to define the semiaxes of the ellipsoidal cavity. For instance, these have been taken collinear with the axes of the solute dipole polarizability tensor, and their lengths proportional to the respective eigenvalues [63]. Another definition proceeds from the inertia tensor of the van-der-Waals solid, i.e. of a solid uniform density composed of interlocking van-der-Waals spheres [64]. Also, the ellipsoidal surface has been defined in terms of the best fitting of a given molecular electrostatic isopotential surface [65].

In the case of more complex cavities, e.g. formed by the overlapping van-der-Waals spheres of atoms consisting the solute molecule, the electrostatic energy of interaction between the charge distribution inside the cavity and the respective reaction field can be calculated numerically. According to the classical electrostatics, the potential Φ_S (cf. Eq. 21) can be described at any point of the space in terms of the apparent charge distribution σ on the cavity surface. It is composed from two terms,

$$\Phi_S = \Phi_M + \Phi_\sigma \tag{44}$$

the first of which (Φ_M) corresponds to the electrostatic potential created by the charge distribution of the solute and the second (Φ_σ) is due to the reaction potential by the solvent. The latter is directly connected with the apparent charge distribution on the surface of the cavity as follows:

$$\Phi_\sigma(\vec{r}) = \int_\Sigma \frac{\sigma(\vec{s})}{|\vec{r} - \vec{s}|} d^2\vec{s} \tag{45}$$

where Σ is the cavity surface and s vector defines a point on Σ. The Σ surface is usually divided into appropriate number of triangular tesserea, each of which has an area ΔS_k and contains the charge q_k in some internal point $\vec{s}_k$. Thus, according to this, so-called boundary element method, the reaction potential is found as the following sum over all tesserea

$$\Phi_\sigma(\vec{r}) = \sum_k \frac{q_k}{|\vec{r} - \vec{s}_k|} \tag{46}$$

with

$$q_k = \Delta S_k \, \sigma(\vec{s}_k) \tag{47}$$

In the application of the boundary element method, it is crucial to select appropriate boundary surface for the solute cavity and to proceed as accurate as possible tessellation (triangulation) of this surface. For instance, it has been proposed [66, 67] that in the case of the cavity formation from overlapping van-der-Waals spheres, the atomic van-der-Waals radii should be multiplied by 1.2. Other possibilities of the surface definition include the closed envelope obtained by rolling a spherical probe of adequate diameter on the van-der-Waals surface of the solute molecule [68] and the surface obtained from the positions of the center of such spherical probe around the solute [69].

In the quantum chemical extension of the electrostatic polarization theory of the solvation, the interaction energy between the quantum-mechanically treated solute molecule and the surrounding dielectric continuum is represented by the respective additional terms to the solute Hamiltonian which can be divided into nuclear-nuclear, nuclear-electron and electron-electron contributions [70]. The self-consistent reaction field ($SCRF$) method [71-73] has been used in most cases to find the solutions of the respective Schrödinger equation for a solute molecule in a polarizable dielectric medium. According to this method, the electrostatic interaction energy, E_{el}, between the solute and its reaction field is given as the following difference

$$E_{el} = E_{rf} - E_0 \tag{48}$$

where E_o and E_{rf} are the quantum chemically calculated total energies for the isolated solute molecule and for the solute molecule in the dielectric medium, respectively. If to assume the full relaxation of the polarizable medium in the electrostatic field of the solute molecule, approximated by its dipole, then E_{el} will be directly connected with the expectation value of the total dipole moment of the solute molecule $\vec{\mu}_{tot}$ as follows

$$E_{el} = -\frac{1}{2}\Gamma \langle \vec{\mu}_{tot} \rangle^2 = -\frac{1}{2}\Gamma \left(\langle \Psi | \hat{\mu}_{el} | \Psi \rangle + \vec{\mu}_{nuc} \right)^2 \tag{49}$$

where

$$\hat{\mu}_{el} = \sum_{i=1}^{n} e_i \hat{r}_i \tag{50}$$

and

$$\vec{\mu}_{nuc} = \sum_{k=1}^{N} Z_k \vec{R}_k \tag{51}$$

are the electronic dipole moment operator and the nuclear component of the dipole moment (under the assumption of nuclei with charges Z_k at fixed positions $\vec{R}_k$, respectively. In equation 49, the factor Γ describes the intensity of the reaction field response of the solvent. This factor depends both on the dielectric constant of the solvent and the size and shape of the solute cavity. In the case of the simplest, spherical model for the solute cavity

$$\Gamma = \frac{\epsilon - 1}{(2\epsilon + 1)\, a_0^3} \tag{52}$$

where a_0 is the cavity radius and ϵ is the dielectric constant of medium. In the case of the ellipsoidal cavity with principal axes a, b and c, Γ becomes a tensor as discussed above (cf. Eq. 43), but still the general form of Eq. 49 remains the same. In the case of ionic solutes, the factor Γ has to be modified to maintain the gauge invariancy of Eq. 49. For positive ions (cations) with the ionic charge Q

$$\Gamma_{cation} = \Gamma \frac{n_e}{n_e + Q} \tag{53}$$

and for anions with the negative charge $-Q$

$$\Gamma_{anion} = \Gamma \frac{n_e - Q}{n_e} \tag{54}$$

where n_e is the number of electrons in the solute molecule. Also, the ionic solvation term calculated according to the classical Born formula (Eq. 39) has to be added to the electrostatic solvation energy for ionic solutes

$$E_{el} = -\frac{1}{2}\Gamma \langle \vec{\mu}_{tot} \rangle^2 - \frac{(\epsilon - 1)\, Q^2}{2\epsilon\, a_0} \tag{55}$$

The expectation value of the dipole moment depends, of course, on the molecular wavefunction and therefore, due to the interaction of the solute electron distribution with the solvent, it has to be calculated during each self-consistent field cycle of solving the respective Schrödinger equation. The self-consistent reaction field ($SCRF$) method to find the solution of this equation had been developed by Tapia and Goscinski [71]. Proceeding from the classical expression for the electrostatic solvation energy of a solute molecule in a dielectric medium in the dipole-dipole interaction approximation (cf. Eq. 49, the total energy of the solute is given as follows

$$E = E^0 - \frac{1}{2}\Gamma \left(\langle \Psi|\hat{\mu}|\Psi \rangle \langle \Psi|\hat{\mu}|\Psi \rangle + 2\,\vec{\mu}_{nuc} \langle \Psi|\hat{\mu}|\Psi \rangle + \vec{\mu}^2_{nuc} \right) \tag{56}$$

where $E^0 = \langle\Psi|\hat{H}^0|\Psi\rangle$, $\hat{H}^0$ is the Hamiltonian for the reaction field unperturbed solute molecule and Ψ is the molecular electronic wave function. From Eq. 56, one can construct the variational functional

$$L = E^0 - \frac{1}{2}\Gamma \Big(\langle \Psi|\hat{\mu}|\Psi \rangle \langle \Psi|\hat{\mu}|\Psi \rangle + 2\,\vec{\mu}_{nuc} \langle \Psi|\hat{\mu}|\Psi \rangle + \vec{\mu}^2_{nuc} \Big) - W\, (\langle\Psi|\Psi\rangle - 1) \tag{57}$$

where W is the Lagrange multiplier ensuring the normalization of the variational wave function. Variation of the last equation with respect to the wave function parameters yields

$$\begin{aligned} \delta L &= \delta E^0 - \Gamma\left(\langle\,\delta\Psi|\hat{\mu}|\Psi\,\rangle\,\langle\,\Psi|\hat{\mu}|\Psi\,\rangle + \vec{\mu}_{nuc}\,\langle\,\delta\Psi|\hat{\mu}|\Psi\,\rangle\,\right) - W\,\delta\langle\Psi|\Psi\rangle \\ &= \langle\,\delta\Psi|\hat{H}^0|\Psi\,\rangle - \Gamma\,\langle\,\delta\Psi|\hat{\mu}|\Psi\,\rangle\,\vec{\mu}_{tot} - W\,\langle\,\delta\Psi|\Psi\rangle + c.c. = 0 \qquad (58) \end{aligned}$$

where $\vec{\mu}_{tot} = \vec{\mu}_{nuc} + \langle\Psi|\hat{\mu}_{el}|\Psi\rangle$ is the total variational dipole moment of the solute molecule. The latter has to be found during the $SCRF$ procedure simultaneously with the total energy of the system. From Eq. 58, the following Schrödinger equation is obtained for the electronic state Ψ of the solute molecule

$$\left(\hat{H}^0 - \Gamma\,\langle|\Psi|\hat{\mu}_{tot}|\Psi\rangle\,\hat{\mu}_{el}\right)\,|\Psi\rangle = W\,|\Psi\rangle \qquad (59)$$

W plays the role of the quantum mechanical motif which is directly obtained from the Schrödinger equation as follows

$$\begin{aligned} W &= \langle\,\Psi|\,\hat{H}^0 - \Gamma\,\langle|\Psi|\hat{\mu}_{tot}|\Psi\rangle\,\hat{\mu}_{el}\,|\Psi\rangle \\ &= \langle\,\Psi|\,\hat{H}^0|\Psi\,\rangle - \Gamma\,\langle|\Psi|\hat{\mu}_{tot}|\Psi\rangle\,\langle\,\Psi|\hat{\mu}_{el}\,|\Psi\rangle \qquad (60) \end{aligned}$$

By adding the part for the interaction of the nuclear component of the solute dipole with the total reaction field and assuming

$$\langle\,\Psi|\,\hat{H}^0|\Psi\,\rangle \approx \langle\,\Psi^0|\,\hat{H}^0|\Psi^0\,\rangle \qquad (61)$$

where Ψ^0 is the wavefunction of the solute molecule, unperturbed by the reaction field, one obtains that

$$\begin{aligned} E_{el} &= E_{rf} - E_0 \\ &= \langle\,\Psi|\,\hat{H}^0|\Psi\,\rangle - \Gamma\,\left(\langle|\Psi|\hat{\mu}_{el}|\Psi\rangle + \vec{\mu}\,\right)^2 - \langle\,\Psi^0|\hat{H}^0\,|\Psi^0\rangle \\ &\approx -\Gamma\,\langle\,\vec{\mu}_{tot}\,\rangle^2 \qquad (62) \end{aligned}$$

The comparison of the last equation with the starting equation (Eq. 49) reveals a difference by the factor of two in the final result. Of course, the approximation (Eq. 61) brings up a certain error and it has been therefore proposed [71, 74] and elsewhere to correct the last formula by the addition of the "solvent cost", i.e. the additional work required to reorganize the solvent due to the electrostatic field of the solute

$$E_{el}(corrected) = E_{rf} - E_0 + \frac{1}{2}\Gamma\,\langle\,\vec{\mu}_{tot}\,\rangle^2 \qquad (63)$$

This approach has been described as $ModelA$ in [74] and it seems to be intuitively correct. However, it cannot be proved that the perturbation term

$$\langle \Delta\Psi | \hat{H}^0 | \Delta\Psi \rangle = \langle \Psi | \hat{H}^0 | \Psi \rangle - \langle \Psi^0 | \hat{H}^0 | \Psi^0 \rangle \tag{64}$$

is exactly equal to the total electrostatic solvation energy of the solute (with the opposite sign). Therefore, an alternative derivation of the dipolar reaction field perturbed Schrödinger equation for the solute immersed into a dielectric continuum has been proposed [74, 75] (cf. $ModelB$ in [74]), proceeding from the following variational functional

$$L = E^0 - \frac{1}{4}\Gamma\Big(\langle \Psi|\hat{\mu}_{el}|\Psi \rangle \langle \Psi|\hat{\mu}_{el}|\Psi \rangle + 2\,\vec{\mu}_{nuc} \langle \Psi|\hat{\mu}_{el}|\Psi \rangle + \vec{\mu}^{\,2}_{nuc} \Big) - W\,\left(\langle \Psi|\Psi \rangle - 1 \right) \tag{65}$$

This leads to a Schrödinger equation which has a solution immediately connected to the total electrostatic (dipolar) solvation energy, E_{el},

$$\left(\hat{H}^0 - \frac{1}{2}\Gamma \langle |\Psi|\hat{\mu}_{tot}|\Psi\rangle \, \hat{\mu}_{el} \right) |\Psi\rangle = E^{(e)}_{el} \, |\Psi\rangle \tag{66}$$

$$E_{el} = E^{(e)}_{el} - \frac{1}{2}\Gamma\, \vec{\mu}_{nuc}\vec{\mu}_{tot} - E_0 \tag{67}$$

Expanding the electronic wavefunction as follows [76]

$$|\Psi \rangle = O_s\, A\, [\, \phi_1, ..., \phi_n \,] \tag{68}$$

i.e. describing $|\Psi\rangle$ as a proper spin-projected antisymmetrized product of molecular (or atomic) orbitals, and recalling that the dipole moment operator is an one-electron operator (as are all electric moment operators) the following orbital equations are obtained

$$\mathbf{f}(k)\, \phi_i(k) = \epsilon_i\, \phi_i(k) \tag{69}$$

with

$$\mathbf{f}(k) = \mathbf{f}_0(k) - \Gamma \langle |\Psi|\hat{\mu}_{tot}|\Psi\rangle\, \hat{\mu}_{el}(k) \tag{70}$$

in the case of $ModelA$, and

$$\mathbf{f}(k) = \mathbf{f}_0(k) - \frac{1}{2}\Gamma \langle |\Psi|\hat{\mu}_{tot}|\Psi\rangle\, \hat{\mu}_{el}(k) \tag{71}$$

in the case of $ModelB$, where $\mathbf{f}_0(k)$ is the usual Fock operator for the isolated molecule, ϵ_i is the molecular orbital energy for ϕ_i and $\hat{\mu}_{el}$ is the electronic part of the dipole moment operator. Both equations can be solved iteratively, using the usual SCF procedure and the expectation value of the total dipole moment from the previous SCF cycle.

The initial $SCRF$ approach has been developed further by O. Tapia and others [77-80] into a form of the generalized self-consistent reaction field theory ($GSCRF$). According to this theory, the effective Hamiltonian of the solute in the solvent has the following form

$$\hat{H}_s = \hat{H}_s^0 + \int d\mathbf{r}\boldsymbol{\Omega}_s(\mathbf{r}) \left[V_m^0(\mathbf{r}) + \int d\mathbf{r}' \, G(\mathbf{r},\mathbf{r}') \, \boldsymbol{\Omega}_s(\mathbf{r}') \right] \tag{72}$$

where $\boldsymbol{\Omega}_s(\mathbf{r})$ is the solute charge density operator given by

$$\boldsymbol{\Omega}_s(\mathbf{r}) = -\sum_i \delta(\mathbf{r} - \mathbf{r}_i) + \sum_a Z_a \, \delta(\mathbf{r} - \mathbf{R}_a) \tag{73}$$

where $\mathbf{r}_i$ stands for the i-th electron position vector, $\mathbf{R}_a$ is the position vector of the a-th nucleus with the charge Z_s in the solute and $\delta(\mathbf{r})$ is the Dirac's delta function. The first term in square brackets in Eq. 72, V_m^0, represents the electrostatic potential created by the solvent in the absence of the solute. The second, integral term corresponds to the reaction potential response function of the polarizable solvent. Together these terms produce the reaction field potential applying to the solute molecule in the polarizable dielectric medium. Notably, a principal part of the Hamiltonian (Eq. 72) is the solute charge density which can be represented using different approximations of which the multipolar expansion has been mostly applied for this purpose [78, 80]. By using the distributed multipole model, it is possible to obtain the $GSCRF$ equations for the molecules of a complex shape. However, it has been mentioned that the use of multipole expansions of the solvent electrostatic and reaction potentials in Eq. 73 may cause this Hamiltonian to become unbound [80]. Special damping procedures [81] have been suggested to overcome this difficulty.

The $GSCRF$ total energy of the solute is given by the following equation [78]

$$E_{GSCRF} = \langle \Psi | \hat{H}^0 | \Psi \rangle + \langle \Psi | \int d\mathbf{r}\boldsymbol{\Omega}_s(\mathbf{r}) \, V_m^0(\mathbf{r}) | \Psi \rangle + \frac{1}{2} \langle \Psi | \int d\mathbf{r}\boldsymbol{\Omega}_s(\mathbf{r}) | \Psi \rangle \int d\mathbf{r}' \, G(\mathbf{r},\mathbf{r}') \, \boldsymbol{\Omega}_s(\mathbf{r}') \tag{74}$$

This energy expression can be used to build up the respective variational functional to get the molecular orbitals [77, 79]. A crucial step in the general self-consistent reaction field procedure is the estimation of the solvent charge density needed to obtain the response function $G(\mathbf{r}, \mathbf{r}')$ and the reaction potential. The use of Monte Carlo or molecular dynamics simulations of the system consisting the solute and surrounding solvent molecules has been proposed to find the respective solvent static and polarization densities [78]. The $GSCRF$ methodology has been widely used for the modelling of large biomolecules in aqueous solutions [81-85].

According to the direct reaction field (DRF) method, developed by P. Th. van Duijnen and others [86-89], the classical part of the solute-solvent system (solvent) is treated not as a dielectric continuum but as the distribution of the polarizable point dipoles, interacting with each other. The DRF Hamiltonian of the solute-solvent system is given by the following formula:

$$\hat{H}_{DRF} = \hat{H}^0 - \frac{1}{2}\sum_{i,j}\sum_{p,q} \mathbf{F}_{ip}^{\dagger}\alpha_{pq}\mathbf{F}_{jq} \tag{75}$$

where indexes i an j correspond to the solute particles (electrons and nuclei) and p and q run over the external polarizable points. $\mathbf{F}_{ip}$ is the field of the particle i at the position p, and α_{pq} gives the induced dipole at point q by a field applied at point p. By using Eq. 75, it is possible to get immediately the solution of the respective Schrödinger equation, without the iterative adjustment of the solvent charge distribution and the respective reaction field potential. It has been argued [8] that the use of operator $\hat{H}_{DRF}$ does not correspond to the correct introduction of electrostatic polarization effects that are properly defined only when the average distributions are used. While the $GSCRF$ approach uses the average reaction field model, the DRF method proceeds from the direct reaction field obtained as the linear solute-solvent interaction operator, proportional to the square of the electric field operator. It has been argued that the additional energy contributions can be interpreted as due to the dispersion interaction between the solute and solvent molecules [90]. More recently, the DRF approach has been combined with the continuum approach by dividing the space around the solute into a closer surrounding treated by direct reaction field method and to more distant space represented by the macroscopic dielectric properties of the solvent [91, 92].

The polarizable continuum model (PCM), based on the classical electrostatic boundary element method discussed above (cf. Eq. 44-47), has been proposed by Miertuš, Scrocco and Tomasi [93] and has been extensively used for the calculation of different molecular properties in solution (for a review, cf. [8]). Analogous approaches have also been developed by other groups [94-101]. The PCM model proceeds from the following Schrödinger equation for a solute molecule in the dielectric continuum

$$\left(\hat{H}^0 + \hat{V}_{PCM}\right)\Psi = E\,\Psi \tag{76}$$

where the reaction field potential is expressed as a sum over the small surface elements $\triangle S_k$ on the boundary between the solute and solvent (cf. Eq. 46)

$$\hat{V}_{PCM} = \sum_k \frac{q_k}{|\vec{r} - \vec{s}_k|} \tag{77}$$

with the surface charges defined as

$$q_k = \triangle S_k\,\sigma(\vec{s}_k)$$

The charges on the boundary are found from the electrostatic polarization of the dielectric medium on the surface of the cavity due to the potential derived from the charge distribution of the solute and from other (induced) charges on the surface. The induced surface charge is evaluated iteratively at each step of the SCF procedure to solve the Schrödinger equation (Eq. 76). It has been reported that a simultaneous iteration of the surface charge with the Fock procedure allows to reduce substantially the computation time without the loss in the precision of calculations [102].

Also, it has been shown that the expressions which determine the charges q_k may be given as a set of linear equations. In the matrix form [8]

$$\mathbf{D}\,\mathbf{q}^{0f} = \mathbf{E}_{in} \tag{78}$$

where $\mathbf{D}$ is a square nonsymmetric and nondiagonal matrix with the dimension equal to the number of surface elements, derived from the curvature of the surface and $\mathbf{q}^{of}$ is a column vector, containing the unknown surface charges. $\mathbf{E}_{in}$ is also a column vector collecting the effective components of the solute electric field multiplied by the surface elements

$$E^0_{in,k} = -\triangle S_k\,\vec{\nabla}\,\Phi_{M,in}(\vec{s}_k)\cdot\vec{n}_k \tag{79}$$

The $\mathbf{D}$ matrix depends only on the shape of the cavity and the dielectric constant of the medium. Therefore, when the system of linear equations (Eq. 78) has to be solved several times with different $\mathbf{E}_{in}$, as in the case of the polarizable solute, it may be convenient to work with the single inversion matrix $\mathbf{D}^{-1}$. Several formulations have been proposed to find the respective surface charges [94, 96, 101].

The advantage of the PCM method is in that it is applicable to the solute cavity of practically any shape in the solution. However, it is not clear how precisely should the molecular cavity be defined bearing in mind the classical (quasi-macroscopic) representation of the solvent. It is difficult to perceive that the solvent, e.g. the water molecules, can produce the electrical polarization corresponding to the statistically average distribution in the macroscopic liquid at infinitely small regions on the cavity surface. However, it is conceivable that larger chemical groups in the molecules may possess their own reaction field created by their charge distribution and the reaction fields of other groups in the solute molecule. A multicavity self-consistent reaction field ($MCa\,SCRF$) has been proposed [103] for the description of rotationally flexible molecules in condensed dielectric media. It proceeds from the observation that the interaction of the charge and higher electrical moments of a charge distribution in a spherical cavity with the corresponding reaction fields localized in the center of the cavity does not depend on the position of charge or (point) multipole centers in this cavity [59]. Therefore, it is possible to divide a rotationally flexible solute molecule or a hydrogen-bonded molecular complex between two or more spherical cavities that embed the rotationally separated fragments of the solute or solute and solvent molecules, respectively. Assuming the classical Born-Kirkwood-Onsager charge density expansion (Eq. 35) for each of these fragments, the total energy of the solute in a dielectric medium can be expressed as a sum of terms that correspond to the energies arising from the interaction of the partial charge and the electric moments of a given molecular fragment with the reaction field of its own and the reaction fields of other fragments, as well as from the interaction between the reaction fields of different fragments. The Hartree-Fock-type equations derived from the variational functional for the total energy E can then be solved iteratively using the $SCRF$ procedure.

The $COSMO$ (conductor-like screening model) approach has been developed recently to account for more realistic solute cavity shapes in the solution [104, 105]. This approach proceeds from the assumption about the ideal dielectric screening of the solute by the surrounding conductor-like medium. This corresponds to the infinite value of the dielectric constant of the medium and should be, therefore, primarily applicable for the high dielectric constant solvents (e.g., water)

Finally, considering the electrostatic polarization models of the solute-solvent interaction, we note the virtual charge methods. In the first explicit formulation of this model by Klopman [106], an imaginary particle called solvaton was associated to each atom A of the solute. The charge of the solvaton, Q^s_A, was taken equal in magnitude but opposite in sign to the Mulliken partial charge q_A on the atom A. The interaction energy was then calculated between all partial charges and solvatons but the interactions between the partial charges themselves and solvatons themselves were neglected. This approach has been also incorporated into the SCF procedure [107, 108]. An extension of this model has been used in the $AM1$-SMx

solvational models by Cramer and Truhlar [109-113].

5. Electrostatic interaction and electron correlation in condensed media

In the previous chapter, the quantum-chemical methods for modelling the electrostatic polarization effects in dielectric media were considered at the Hartree-Fock level of theory. It is, of course, of principal importance to study the interplay between the electrostatic interaction in the condensed media and the Coulomb' electron correlation in the solute molecule. The introduction of the electron correlation modifies the charge distribution in the molecule and, correspondingly, the solvent reaction field. On the other hand, the polarization due to the solvent reaction field may modify the electron correlation in the solute molecule. No systematic investigation has been performed, to our knowledge, to separate these effects. However, various approaches have been developed to study the electron correlation effects, in general, in solute molecules embedded into the dielectric media.

First, we consider the combination of the self-consistent reaction field theory with the configuration interaction (CI) formalism [114]. The usual CI wavefunction is constructed as follows

$$\Psi_{CI} = \sum_{a=1}^{N} C_a D_a \tag{80}$$

where C_a are the CI expansion coefficients and D_a are the basis functions (Slater determinants or their linear combinations). In the case of orthogonal basis functions, the normalization condition of the function Ψ_{CI} is given as

$$\sum_{a=1}^{N} C_a^2 = 1 \tag{81}$$

and the coefficients C_a are determined from the following equation

$$\left(\hat{H}^0 + \hat{V}_{rf}\right) \Psi = E\,\Psi \tag{82}$$

where $\hat{H}^0$ is the Hamiltonian for the molecule, unperturbed by the reaction field and $\hat{V}_{rf}$ is the reaction field perturbation. The latter can be presented, for example, in the framework of the boundary element method as follows (cf. Eq. 45)

$$\hat{V}_{rf} = \int d^2\mathbf{r} \frac{\sigma(\mathbf{r}')}{|\mathbf{r} - \mathbf{r}'|} \tag{83}$$

where $\sigma(\mathbf{r}')$ is the charge density on the surface of the cavity.

It is instructive to separate the noninertial and inertial components of the electrostatic polarization of the dielectric medium. The first of them corresponds to the electrostatic polarization of the electron charge distribution in the solvent which is supposedly instantaneous as compared to any electronic or conformational transition of the solute. The second component arises from the orientational polarization of the solvent molecules in the electrostatic field of the solute. The noninertial polarization can be described by the optical dielectric permittivity of the solvent which corresponds to the infinite frequency of external electromagnetic field ($\epsilon_\infty \approx n_D^2$) whereas the inertial polarization represents the slow, orientational part of the total dielectric constant of the solvent, ϵ. In order to separate the noninertial polarization, it is helpful to determine the solute charge density as the sum of the respective nuclear and electronic parts

$$\rho = \rho_n + \rho_e = \sum_A \delta(\mathbf{r} - \mathbf{r}_A)\, Z_A + \sum_{a,b} C_a\, C_b (ab|\mathbf{r}, \mathbf{r}') = \sum_{a,b} C_a\, C_b\, \rho_{ab} \quad (84)$$

where $\delta(\mathbf{r} - \mathbf{r}_A)$ is the Dirac's delta function and $(ab|\mathbf{r}, \mathbf{r}')$ the elements of the single-determinant matrices of interconfigurational transitions. Notably, the values of

$$\rho_{ab} = \delta_{ab}\rho_n - \rho(ab|\mathbf{r}, \mathbf{r}') \quad (85)$$

do not depend on the coefficients C_a and C_b. The noninertial component of the polarization field, Φ_∞, is always in equilibrium and thus it can be represented as follows

$$\Phi_\infty(r) = \sum_{a,b} C_a\, C_b\, \Phi_{ab}^{(\infty)}(\mathbf{r}) \quad (86)$$

where Φ_{ab} is the solution of Eq. 82 for the basic charge distribution ρ_{ab}. Since the latter does not depend on the coefficients C_a and C_b, the values of Φ_{ab} can be determined before the calculation of the wavefunction Ψ_{CI} (Eq. 80). The coefficients C_a can be calculated, as usual, from the matrix equation

$$H|C\rangle = E|C\rangle \quad (87)$$

where C denotes the column-vector of the coefficients and the elements of the matrix H are given as follows [114]

$$H_{ab} = \langle D_a|\hat{H}^0|D_b\rangle + \sum_{a,b} C_a\, C_b \int d^3\mathbf{r}\, \rho_{ab}\, \Phi_{ab}^{(\infty)} + \int d^3\mathbf{r}\, \rho_{ab}\, \Phi_0 \quad (88)$$

where Φ_0 is inertial (nuclear) part of the polarization field. For a given Φ_0, the set of equations (87) can be solved iteratively. Implicitly, the last equations describe electron subsystems of both the solute and solvent, the latter being taken into account as the field of noninertial (electron) polarization of the solvent, Φ^{∞}. The *SCRF CI* methodology has been further elaborated for the description of the molecular electronic spectra in solution. It will be more closely examined in the other lecture series on the calculation of molecular properties and spectra in solution [115].

A multiconfigurational self-consistent reaction field theory (*MCSCRF*) for the description of the electron correlation effects in the molecules in solution has been proposed by K.V. Mikkelsen et al. [116-119]. In their theoretical development, solvent reaction field is described in the framework of Kirkwood's spherical cavity model in the dielectric continuum. The dielectric solvation energy of the solute charge distribution embedded into a spherical cavity is thus presented as the product of the expectation values of the reaction field and the multipole charge moments [116]

$$E_{pol} = -\frac{1}{2}\sum_{lm}\langle\Phi_l^{m\dagger}\rangle\langle M_l^m\rangle \tag{89}$$

where

$$\langle M_l^m\rangle = \sum_k Z_k\, S_l^m(\mathbf{R}_k) - \langle S_l^m\rangle \tag{90}$$

$$\langle\Phi_l^m\rangle = f_l\,\langle M_l^m\rangle \tag{91}$$

and

$$f_l = \frac{(l+1)(\epsilon-1)}{[l+\epsilon(l+1)]\,a_0^{2l+1}} \tag{92}$$

where a_0 is the radius of the spherical solute cavity. The *MC SCRF* solvation energy may be written as [116]

$$E_{sol} = \sum_{lm} g_l(T_{lm})^2 \tag{93}$$

where $g_l = -\frac{1}{2}f_l$ and T_{lm} are the matrix elements obtained as the expectation values of the corresponding nuclear and electronic solvent operators

$$T_{lm} = T_{lm}^n - T_{lm}^e \tag{94}$$

$$T_{lm}^n = \sum_a Z_a\, R^{lm}(\mathbf{R}_a) \tag{95}$$

$$T_{lm}^{e} = R^{lm}(r) = \sum_{p,q} R_{pq}^{lm}\, E_{pq} \tag{96}$$

where R_{lm} are the solvent integrals containing the atomic orbitals and spherical harmonic polynomials used in the multipole expansion of the electrostatic potential, and

$$E_{pq} = \sum_{\sigma} a_{p\sigma}^{+}\, a_{q\sigma} \tag{97}$$

with $a_{p\sigma}^{+}$ and $a_{p\sigma}$ being the creation and annihilation operators for an electron in orbital p with spin σ. The analysis of the relative contributions by different multipole terms into the solvation energy has shown that higher terms in expansion (93) become convergent at about $l = 10$ [116].

A combination of the reaction field approach with the perturbation theory according to the Møller-Plesset scheme has been developed by J-L. Rivail et al. [120, 121]. By starting from the Hamiltonian of the solute molecule as follows

$$\hat{H} = \hat{H}^{0} - M_l^m\, f_l^m \langle\Psi|M_{l'}^{m'}|\Psi\rangle \tag{98}$$

(cf. Eqs. 90-92) where Ψ denotes the exact (correlated) wavefunction, the Hartree-Fock operator may be written as

$$\mathbf{F} = \mathbf{F}^{0} - M_l^m\, f_l^m \langle\Psi|M_{l'}^{m'}|\Psi\rangle \tag{99}$$

where Ψ_0 denotes the electronic wavefunction at the Hartree-Fock level. The Hamiltonian may be then written as a perturbed expression of the Hartree-Fock operator

$$\hat{H} = \mathbf{F} + \left(\hat{H}^{0} - \mathbf{F}^{0}\right) + M_l^m\, f_l^m \left(\langle\Psi_0|M_{l'}^{m'}|\Psi_0\rangle - \langle\Psi|M_{l'}^{m'}|\Psi\rangle\right) \tag{100}$$

The perturbation has two contributions, the standard Møller-Plesset perturbation and the non-linear perturbation due to the solute-solvent interaction. If C_J^i denotes the coefficient of the eigenstate J in the corrections of Ψ to order i, then the perturbation operators $\hat{H}^{(i)}$ of the i-th order are given by the following formulae [121]

$$\hat{H}^{(1)} = \hat{H}^{0} - F^{0} \tag{101}$$

$$\hat{H}^{(2)} = -2\sum_{I\neq 0} C_I^{(1)}\, M_l^m\, f_l^m \langle 0|M_{l'}^{m'}|I\rangle \tag{102}$$

$$\hat{H}^{(3)} = -2\sum_{J\neq 0} C_J^{(2)}\, M_l^m\, f_l^m \langle 0|M_{l'}^{m'}|J\rangle -$$

$$\sum_{J\neq 0}\sum_{P\neq 0} C_J^{(1)} C_P^{(1)} M_l^m f_l^m \langle I|M_{l'}^{m'}|P\rangle \tag{103}$$

The calculation is performed, as usual, by comparing the coefficients of the Schrödinger equation to successive orders.

The first order energy is the same as given by the usual Møller-Plesset treatment ($\langle 0|H|0\rangle$) and the first order electrostatic contribution to the free energy of solvation is identical to the result obtained at the Hartree-Fock level of theory. The second order correction to the free energy is given as follows

$$\triangle G_s^2 = 2\sum_{S\neq 0} C_S^{(2)} \langle 0|M_{l'}^{m'} f_l^m \langle 0|M_{l'}^{m'}|0\rangle -$$

$$\sum_{D\neq 0}\sum_{D'\neq 0} C_D^{(1)} C_{D'}^{(1)} \langle D|M_{l'}^{m'}|D'\rangle f_l^m \langle 0|M_{l'}^{m'}|0\rangle \tag{104}$$

where S stands for the singly excited states, and D and D' for a pair of doubly excited states different by just one orbital. Without excessive difficulty, it is possible to derive the correction terms to the electrostatic free energy of solvation of higher orders [120].

A many-body perturbation theory ($MBPT$) approach has been combined with the polarizable continuum model (PCM) of the electrostatic solvation by F.J. Olivares del Valle et al. [122-124]. The first approximation, called by authors the perturbation theory at energy level (PTE) consists the solution of the PCM problem at the Hartree-Fock level to find the solvent reaction potential and the wavefunction for the calculation of the $MBPT$ correction to the energy. In the second approximation, called the perturbation theory at the density matrix level only (PTD), the calculation of the reaction potential and electrostatic free energy is based on the $MBPT$ corrected wavefunction for the isolated molecule. At the next approximation (perturbation theory at the energy and density matrix level, $PTED$), both the energy and the wave function are solvent reaction field and $MBPT$ corrected.

Recently, the density functional theory (DFT) based methods for the calculation of the electrostatic solvation energy in dielectric media have been also proposed by several groups [125-129].

6. The combined approach to molecular solvation

The enormous complexity of solvation phenomena usually requires the use of several different models to describe the experimental observations in solution as only few chemical or physical processes can be expected to be determined by a single solvation factor. Therefore, for most chemical and physical processes in solutions, a combined theoretical approach should be used.

Historically, the first combined approach is the empirical treatment of the solvent effects on the basis of the linear free energy relationship (*LFER*) methodology (for a comprehensive review, cf. [3]). According to this approach, the experimental data on some phenomenon or property of a chemical or physical system in a series of different solvents (Y_i) are treated using the multiple least-squares method according to the following multilinear expansion

$$Y_i = Y_0 + \sum_k C_k X_{ik} \tag{105}$$

where the summation is performed over different solvent effect scales considered in the *LFER* treatment. In the last equation, X_{ik} are the parameters for a given solvent i and a given solvent scale k, C_k are the expansion coefficients determined by the least-squares procedure and Y_0 is the value of the property Y in the standard solvent 0 (with all $X_{ik} = 0$). Importantly, the parameters X_{ik} are almost always determined from some experimental data which presumably correspond to a single solvation interaction in solution. This limitation restricts the *LFER* approach to molecular structures and solvents with known empirical parameters. An important extension of the *LFER* approach involving the theoretically calculated solvent effect parameters (*TLSER* = *Theoretical Linear Solvation Energy Relationships*) has been recently proposed and efficiently used for the prediction of many chemical and biochemical properties of compounds in different media by G. Famini et al. [130-133]. In most cases, two types of solvent effect scales are considered in the *LFER*, the first of which corresponds to the nonspecific or macroscopic dielectric polarization effects whereas the second type of phenomena involves the formation of directed semichemical bonds (e.g., hydrogen bonds) between the solute and solvent molecules in solution. Also, the terms to account for the cavity formation and the dispersion effect in solution have been proposed in *LSER* treatments (cf. [3]).

Numerous combined theoretical approaches have been proposed for the calculation of ionic and molecular solvation energies. The most straightforward method is the application of some continuum theory (*SCRF*, *PCM* etc.) to the supermolecule complex consisting of the solute molecule and the solvent molecules in its first coordination sphere [134-139]. In this case, the short-range solute-solvent electron correlation, dispersion and exchange-repulsion interactions are taken into account explicitly at the quantum level of theory as the particles (electrons and nuclei) both from the solute and solvent are included explicitly in the respective Schrödinger equation. The long-range electrostatic polarization of the solvent outside the first coordination sphere is treated according to the dielectric theory. Thus, the energy of solvation of a solute molecule calculated according to this method has the following form:

$$E_{sol} = \langle \Psi_{SM} | \hat{H}_{SM}^{(S)} | \Psi_{SM} \rangle - \langle \Psi_M | \hat{H}_M^{(0)} | \Psi_M \rangle - n \langle \Psi_M | \hat{H}_M^{(0)} | \Psi_M \rangle \quad (106)$$

where H_{SM} is the Hamiltonian for the supermolecule (SM) in the solution (in principle, any Hamiltonian which accounts for the solvent dielectric polarization effects, is suitable), and H_M and H_S are the Hamiltonians for the isolated solute (M) and the solvent (S) molecules, respectively. In the last equation, n denotes the number of the solvent molecules included in the supermolecule, Ψ_{SM} is the total wavefunction of the supermolecule immersed into dielectric medium, and Ψ_M and Ψ_S are the wavefunctions for isolated solute and solvent molecules, respectively.

In studying the intermolecular interactions using the supermolecule approach, the "size-extensivity" [140, 141] of the methods applied is of crucial importance. Furthermore, the interaction energies calculated in the supermolecule approach usually suffer from what is called the basis set superposition error ($BSSE$) [141], a spurious energy improvement resulting from the use of truncated basis sets. This error is unavoidable in most practical calculations except for very small systems [140, 141].

The "static" intermolecular interactions which correspond to the fixed geometry of the solute-solvent complex can be also studied by a perturbation" approach [143, 144]. The perturbation theory has some advantages over the supermolecule approach and may therefore be considered conceptually more appropriate for the calculation of intermolecular interaction energies. The interaction energy is calculated directly and it may be separated into components of well defined meaning. But all perturbation theories developed so far for studying (weak) intermolecular interactions have not been, to our knowledge, used for the calculation of solvation phenomena using the supermolecule-continuum approach. However, one should be encouraged to look into this possible extension of the theory which should allow the direct calculation of the solvation energy.

Nevertheless, even at the highest theoretical level accessible for practical calculations, the static approach is strictly valid only for the description of the molecular clusters of fixed geometry. It is well-known that both in the case of strong and weak intermolecular interactions the energy of interaction in the molecular cluster in the gas phase or on the inert-gas matrix is substantially different from the total solvation energy in the condensed phase [145]. A direct solution of a time-dependent Schrödinger equation for the condensed low-order bulk matter, needed to overcome this problem, is premature. Therefore, the molecular dynamics method (MD) [145-149]

based on the computer modelling of a system of particles which interact by the known (model) potential to each other and undergoes the rotational and translational movement in the field caused by this interaction according to the classical (Newtonian) mechanics is widely applied for this purpose. By applying various boundary conditions and performing the calculation of the potential energies and forces for hundreds of thousands configurations obtained by step-by-step time evolution the time-averages such as internal energy (or enthalpy) can be obtained. An alternative is the stochastic Monte Carlo method which is based on the ergodic theorem and provides the ensemble averages calculated from randomly generated and weighted configurations. These methods suffer of several shortcomings, of which the problem of the applicability of the ergodic theorem (i.e. the identity of the time-averaged and ensemble-averaged thermodynamic and dynamic observables), the path sampling and difficulties to obtain precise intermolecular interaction potentials are most serious. Also, the real dense systems, i.e. liquids and solutions are intrinsically quantized systems and therefore a molecular dynamics method should be developed which accounts for the quantum effects in the microscopic system from the first principles.

Finally, numerous computational schemes have been developed to calculate the total molecular solvation energy or free energy using the combination of different theoretical solute-solvent interaction models. From these, we point out the SMx methodology proposed by C.J. Cramer and D.G. Truhlar [109-112] in the framework of the semiempirical level quantum theory ($AM1$ [150]) and which is available as the respective computer code called $AMSOL$ [113]. The original $AM1$-$SM1$ method proceeds from the division of the total molecular solvation energy into the solute-solvent electrostatic and inductive polarization terms, standard-state free energy of cavity creation in the solvent plus the solute-solvent dispersion interaction, and an empirical part of the nuclear motion free energy. The polarization term is presented using the generalized Born formula (as follows):

$$G_p = -\frac{1}{2}\left(1 - \frac{1}{\epsilon}\right) \sum_{k=1}^{N} \sum_{k'=1}^{N} q_k \, q_{k'} \, \gamma_{kk'} \tag{107}$$

where the double summation is performed over all atomic partial charges q_k in the solute molecule, ϵ is the relative dielectric permittivity of the solvent and $\gamma_{kk'}$ - the Coulomb' integral between two centres k and k', parametrized for the interactions with the solvent. The cavity creation plus dispersion term is calculated as

$$G^0_{CD} = \sum_{k=1}^{N} \sigma_k \, A_k \tag{108}$$

where N is the number of atoms in the solute, A_k is the solvent accessible surface area of a given atom and σ_k is the parameter for this atom that is called the accessible surface tension. The latter is obtained from the fit with the experimental data and is, thus, essentially an empirical parameter for a given type of atom. Thus, in essence the SMx methodology represents a semiempirical approach to the calculation of solvent effects.

7. Conclusions

In conclusion, it has to be admitted that in spite of the tremendous effort made in the development of the rigorous theory of the solute-solvent interactions in condensed media, no complete theory is yet available. In most cases, the special approximate theoretical methods have to be used for the solution of specific tasks involving the solvation phenomena.

References

1. Slater, J.C. *Quantum Theory of Molecules and Solids*, Vol. II, McGraw Hill, New York, 1965.
2. André, J.-M.; Delhalle, J.; Brédas, J.-L. *Quantum Chemistry Aided Design of Organic Polymers*, World Scientific Publishing Co., Singapore, 1991.
3. Reichardt, Chr. *Solvents and Solvent Effects in Organic Chemistry*, VCH, Weinheim, 1988.
4. Hammett, L.P. *Physical Organic Chemistry*, McGraw Hill, New York, 1970.
5. Ben-Naim, A. *Water and Aqueous Solutions*, Plenum Press, New York, 1974.
6. Ben-Naim, A. *J. Phys. Chem.* **1978**, 82, 792.
7. Ben-Naim, A. *Solvation Thermodynamics*, Plenum Press, New York, 1987.
8. Tomasi, J.; Persico, M. *Chem. Rev.* **1994**, 94, 2027.
9. Hill, T.H. *An Introduction to Statistical Thermodynamics*, Addison-Wesley, Reading, 1960.
10. Giesen, D.J; Cramer, C.J.; Truhlar, D.G. *J. Phys. Chem.* **1994**, 98, 4141.
11. Huggins, M.L. *J. Chem. Phys.* **1941**, 9, 440.
12. Flory, P.J. *J. Chem. Phys.* **1941**, 9, 660.
13. Hildebrand, J.H. *J. Chem. Phys.* **1947**, 15, 225.
14. Sharp, K.A.; Nicholls, A.; Fine, R.F; Honig, B. *Science* **1991**, 252, 106.
15. Irisa, M.; Nagayama, K; Hirata, F. *Chem. Phys. Lett.* **1993**, 207, 430.
16. Tuñón, I; Silla, E.; Pascual-Ahuir, J.L. *Chem. Phys. Lett.* **1993**, 203, 289.
17. Flory, P.J. *Discuss. Faraday Soc.* **1970**, 49, 7.
18. Sharp, K.A.; Nicholls, A.; Friedman, R.; Honig, B. *Biochemistry* **1991**, 30, 9686.
19. Ben-Naim, A.; Mazo, R.M. *J. Phys. Chem.* **1993**, 93, 10829.
20. Abraham, M.H.; Sakellariou, P. *J. Chem. Soc. Perkin Trans. 2* **1994**, 405.
21. Eley, D.D. *Trans. Faraday Soc.* **1939**, 35, 1281.
22. Susskind, B.; Kasarnowski, I. *Z. Anorg. Allg. Chem.* **1933**, 214, 385.
23. Simonson, T.; Brünger, A.T. *J. Phys. Chem.* **1994**, 98, 4683.
24. Tolman, C.R. *J. Chem. Phys.* **1948**, 16, 758.
25. Kirkwood, J.G.; Buff, F.P. *J. Chem. Phys.* **1949**, 17, 338.
26. Kondo, S. *J. Chem. Phys.* **1956**, 25, 662.
27. Sinanoglu, O. *Chem. Phys. Lett.* **1967**, 1, 283.
28. Sinanoglu, O. in: *Molecular Interactions*, Ratajczak, H.; Orville-Thomas, W.J, Eds.; J. Wiley: New York, 1982., p. 283.

29. Oakenfull, D.G.; Fenwick, D.E. *J. Phys. Chem.* **1974**, 78, 1759.
30. Postma, J.P.M.; Berendsen, H.J.C.; Haak, J.R. *Faraday Symp. Chem. Soc.* **1982**, 17, 55.
31. Gogonea, V.; Osawa, E. *J. Mol. Struct. (Theochem)* **1994**, 311, 305.
32. Reiss, H.; Frisch, H.L.; Lebowitz, J.L. *J. Chem. Phys. 1959*, 31, 369.
33. Frisch, H.L. *Adv. Chem. Phys.* **1963**, 6, 229.
34. Reiss, H. *Adv. Chem. Phys.* **1966**, 9, 1.
35. Pierotti, R.A. *Chem. Rev.* **1976**, 76, 717.
36. Gibbons, R.M. Mol. Phys. 1969, 17, 81.
37. Irisa, M.; Nagayama, K.; Hirata, F. *Chem. Phys. Lett.* **1993**, 207, 430.
38. Hirschfelder, J.D.; Curtiss, C.F.; Bird, R.B. *Molecular Theory of Gases and Liquids*, J. Wiley, New York, 1954.
39. Margenau, H.; Kestner, N.R. *Theory of Intermolecular Forces*, Pergamon, Oxford, 1971.
40. Mahanty, J.; Ninham, W.B. *Dispersion Forces*, Academic Press, New York, 1976.
41. Hobza, P.; Zahradnik, P. *Intermolecular Complexes*, Elsevier, Amsterdam, 1988.
42. Huron, M.J.; Claverie, P. *Chem. Phys. Lett.* **1969**, 4, 429.
43. Kitaigorodsky, A.I. *Tetrahedron* **1961**, 14, 230.
44. Pertsin, A.J.; Kitaigorodsky, A.I. *The Atom-Atom Pair Potential Method*, Springer, Berlin, 1986.
45. Frecer, F.M.; Miertuš, S.; Májeková, M. *J. Mol. Struct. (Theochem)* **1991**, 227, 157.
46. Thole, B.T.; van Duijnen, P.Th. *Theor. Chim. Acta* **1980**, 55, 307.
47. Rullman, J.A.C.; Bellido, M.N.; van Duijnen, P.Th. *J. Mol. Biol.* **1989**, 206, 101.
48. Dijkman, J.P.; van Duijnen, P.Th. *Int. J. Quant. Chem.: Quantum Biol. Symp.* **1991**, 18, 49.
49. Linder, B. *J. Chem. Phys.* **1960**, 33, 668.
50. Linder, B. *J. Chem. Phys.* **1962**, 37, 963.
51. Linder, B. *Adv. Chem. Phys.* **1967**, 12, 225.
52. Rinaldi, D.; Costa-Cabral, B.J.; Rivail, J.-L. *Chem. Phys. Lett.* **1986**, 125, 495.
53. Aguilar, M.A.; Olivares del Valle, F.J. *J. Chem. Phys.* **1989**, 138, 327.
54. Rösch, N.; Zerner, M.C. *J. Phys. Chem.* **1994**, 98, 5817.
55. Abe, T.; Amako, Y.; Nishioka, T.; Azumi, H. Bull. *Chem. Soc. Japan* **1966**, 39, 845.
56. Amos, A.T.; Burrows, B.L. *Adv. Quantum Chem.* **1973**, 7, 289.
57. Kirkwood, J.G. *J. Chem. Phys.* **1939**, 7, 911.
58. Onsager, L. *J. Am. Chem. Soc.* **1936**, 58, 1486.
59. Böttcher, C.J.F. and Bordewijk, P. *Theory of Electric Polarization*, 2nd ed., vol. II, Elsevier Co, Amsterdam, 1978.
60. Abraham, R.J.; Hudson, B.D.; Kermode, M.W.; Mines, J.R. *J. Chem. Soc. Farad. Trans.* **1988**, 84, 1911.
61. Rivail, J.-L.; Rinaldi, D. *Chem. Phys.* **1976**, 18, 233.
62. Karelson, M.M.; Tamm, T.; Katritzky, A.R.; Cato, S.J.; Zerner, M.C. *Tetrahedron Comput. Methodol.* **1989**, 2, 295.
63. Rinaldi, D.; Ruiz-Lopez, M.F.; Rivail, J.-L. *J. Chem. Phys.* **1983**, 78, 834.
64. Rinaldi, D.; Rivail, J.-L.; Rguini, N. *J. Comput. Chem.* **1992**, 13, 675.
65. Rivail, J.-L.; Terryn, B.; Rinaldi, D.; Ruiz-Lopez, M.F. *J. Mol. Struct. (Theochem)* **1988**, 166, 319.
66. Pascual-Ahuir, J.L.; Silla, E. *J. Comput. Chem.* **1990**, 11, 1047.
67. Silla, E.; Tuñón, I.; Pascual-Ahuir, J.L. *J. Comput. Chem.* **1991**, 12, 1077.
68. Richards, F.M. *Annu. Rev. Biophys. Bioeng.* **1977**, 6, 151.
69. Lee, B.; Richards, F.M. *J. Mol. Biol.* **1971**, 55, 379.
70. Hylton, J.; Christoffersen, R.E.; Hall, G.G. *Chem. Phys. Lett.* **1974**, 26, 501.
71. Tapia, O.; Goscinski, O. *Mol. Phys.* **1975**, 29, 1653.
72. Karelson, M.M. *Org. React.* **1980**, 17, 357.

73. Karelson, M.M. *Org. React.* **1980**, 17, 366.
74. Karelson, M.; Zerner, M.C. *J. Phys. Chem.* **1992**, 92, 6949.
75. Karelson, M.M.; Katritzky, A.R.; Zerner, M.C. *Int. J. Quant. Chem.* **1986**, S20, 521.
76. Szabo, A.; Ostlund, N. *Modern Quantum Chemistry*, MacMillan, New York, 1982.
77. Tapia, O. *J. Mol. Struct. (Theochem)* **1991**, 226, 59.
78. Tapia, O.; Colonna, F.; Ángyán, J. *J. Chim. Phys.* **1990**, 87, 875.
79. Ángyán, J. *J. Math. Chem.* **1992**, 10, 93.
80. Jansen, G.; Colonna, F.; Ángyán, J. *Int. J. Quant. Chem.* **1996**, 58, 251.
81. Thole, B.T. *Chem. Phys.* **1981**, 59, 341.
82. Tapia, O.; Johannin, G. *J. Chem. Phys.* **1981**, 75, 3624.
83. Ángyán, J.; Allavena, M.; Picard, M.; Potier, A.; Tapia, O. *J. Chem. Phys.* **1982**, 77, 4723.
84. Tapia, O. in: *Theoretical Models of Chemical Bonding*, Ed. Maksic, Z.B., Springer-Verlag, Berlin; 1990. vol. 4.
85. Ángyán, J.; Chipot, C. *Int. J. Quant. Chem.* **1994**, 52, 17.
86. Thole, B.T.; van Duijnen, P.Th. *Chem. Phys.* **1982**, 71, 211.
87. Thole, B.T.; van Duijnen, P.Th. *Theor. Chim. Acta* **1983**, 63, 209.
88. Thole, B.T.; van Duijnen, P.Th. *Biophys. Chem.* **1983**, 18, 53.
89. van Duijnen, P.Th.; Juffer, A.H; Dijkman, H. *J. Mol. Struct. (Theochem)* **1992**, 260, 195.
90. deVries, A.H.; van Duijnen, P.Th. *Int. J. Quant. Chem.* **1996**, 57, 1067.
91. Dijkman, H.; van Duijnen, P.Th. *Int. J. Quant. Chem. Quantum Biol. Symp.* **1991**, 18, 49.
92. van Duijnen, P.Th.; deVries, A.H. *Int. J. Quant. Chem. Quantum Chem. Symp.* **1995**, 29, 531.
93. Miertuš, S.; Scrocco, E.; Tomasi, J. *Chem. Phys.* **1981**, 55, 117.
94. Hoshi, H.; Sakurai, M.; Inoue, Y. Chûjô, R, *J. Chem. Phys.* **1987**, 87, 1107.
95. Hoshi, H.; Sakurai, M.; Inoue, Y. Chûjô, R, *J. Mol. Struct. (Theochem)* **1988**, 180, 267.
96. Drummond, M.L. *J. Chem. Phys.* **1988**, 88, 5014.
97. Wang, B.; Ford, G.P. *J. Chem. Phys.* **1992**, 97, 4162.
98. Fox, T.; Rösch, N.; Zauhar, R.J. *J. Comput. Chem.* **1993**, 14, 253.
99. Luque, F.J.; Orozco, M.; Bhadane, P.K.; Gadre, S.R. *J. Chem. Phys.* **1994**, 100, 6718.
100. Zauhar, R.J.; Morgan, R.S. *J. Comput. Chem.* **1990**, 11, 603.
101. Pascual-Ahuir, J.L.; Silla, E.; Tomasi, J.; Bonaccorsi, R. *J. Comput. Chem.* **1987**, 8, 778.
102. Zerner, M.C. *private communication.*
103. Karelson, M.; Tamm, T.; Zerner, M.C. *J. Phys Chem.* **1993**, 97, 11901.
104. Klamt, A.; Schüürmann, G. *J. Chem. Soc. Perkin Trans. 2* **1993**, 799.
105. Klamt, A. *J. Phys. Chem.* **1995**, 99, 2224.
106. Klopman, G. *Chem. Phys. Lett.* **1967**, 1, 200.
107. Germer, H.A. *Theor. Chim. Acta* **1974**, 34, 145.
108. Constanciel, R.; Tapia, O. *Theor. Chim. Acta* **1978**, 48, 75.
109. Cramer, C.J.; Truhlar, D.G. *J. Am. Chem. Soc.* **1991**, 113, 8305.
110. Cramer, C.J.; Truhlar, D.G. *J. Am. Chem. Soc.* **1991**, 113, 8552.
111. Cramer, C.J.; Truhlar, D.G. *J. Comput. Chem.* **1992**, 13, 1089.
112. Cramer, C.J.; Truhlar, D.G. *Science* **1992**, 256, 213.
113. Cramer, C.J.; Lynch, G.C.; Hawkins, G.D.; Truhlar, D.G. *AMSOL 4.0. QCPE 606 QCPE Bull.* **1993**, 13, 78.
114. Basilevsky, M.V.; Chudinov, G.E. *Chem. Phys.* **1991**, 157, 245.
115. Karelson, M. in: *Problem Solving in Computational Molecular Science, NATO ASI Series*, Wilson, S. and Diercksen, G.H.F. (Eds.) Kluwer, Dordrecht. 1997.

116. Mikkelsen, K.V.; Ågren, H.; Jensen, H.J.Aa.; Helgaker, T. *J. Chem. Phys.* **1988**, 89, 3086.
117. Ågren, H.; Medina-Llanos, C.; Mikkelsen, K.V.; Jensen, H.J.Aa. *Chem. Phys. Lett.* **1988**, 153, 322.
118. Medina-Llanos, C.; Ågren, H.; Mikkelsen, K.V.; Jensen, H.J.Aa. *J. Chem. Phys.* **1989**, 90, 6422.
119. Mikkelsen, K.V.; Ågren, H. *J. Phys. Chem.* **1990**, 94, 6220.
120. Rivail, J.-L. *C. R. Acad. Sci. Paris* **1990**, 311 II, 307.
121. Rivail, J.-L.; Rinaldi, D.; Ruiz-Lopez, M.F. in: *Theoretical and Computational Models for Organic Chemistry*, Formosinho, S.J.; Csizmadia, I.G. and Arnaut, L.G. (Eds.), Kluwer, Dordrecht. 1991, pp. 77-92.
122. Olivares del Valle, F.J.; Tomasi, J. *Chem. Phys.* **1991**, 150, 139.
123. Olivares del Valle, F.J.; Aguilar, M.A. *J. Comput. Chem.* **1992**, 13, 115.
124. Olivares del Valle, F.J.; Aguilar, M.A. *J. Mol. Struct. (Theochem)* **1993**, 280, 25.
125. Contreras, R.R.; Aizman, A.J. *Int. J. Quant. Chem. Quantum Chem. Symp.* **1991**, 25, 281.
126. Hall, R.J.; Davidson, M.M.; Burton, N.A.; Hillier, I.H. *J. Phys. Chem.* **1995**, 99, 921.
127. Contreras, R.R.; Pérez. P.; Aizman, A.J. *Int. J. Quant. Chem.* **1995**, 56, 433.
128. Mineva, T.; Russo, N.; Toscano, M. *Int. J. Quant. Chem.* **1995**, 56, 663.
129. Tawa, G.J., Martin, R.L.; Pratt, L.R.; Russo, T.V. *J. Phys. Chem.* **1996**, 100, 1515.
130. Famini, G.R.; Kassel, R.J.; King, J.W.; Wilson, L.Y. *Quant. Struct.-Act. Relat.* **1991**, 10, 344.
131. Famini, G.R.; Ashman, W.P.; Mickiewicz, A.P.; Wilson, L.Y. *Quant. Struct.- Act. Relat.* **1992**, 11, 162.
132. Famini, G.R.; Wilson, L.Y. *J. Phys. Org. Chem.* **1992**, 5, 395.
133. Famini, G.R.; Penski, C.A.; Wilson, L.Y. *J. Phys. Org. Chem.* **1992**, 5, 395.
134. Tapia. O.; Lluch, J.M.; Cardenas, R.; Andres, J. *J. Am. Chem. Soc.* **1989**, 111, 829.
135. Freitas, L.C.G.; Longo, R.L.; Simas, A.M. *J. Chem. Soc. Faraday Trans.* **1992**, 88, 189.
136. Rick, S.W.; Berne, B.J. *J. Am. Chem. Soc.* **1994**, 116, 3949.
137. Karelson, M.M. *Org. React.* **1983**, 20, 127.
138. Karelson, M.M.; Zerner, M.C. *J. Am. Chem. Soc.* **1990**, 112, 9405.
139. Diercksen, G.H.F.; Karelson, M.; Tamm, T.; Zerner, M.C. *Int. J. Quant. Chem.* **1994**, S28, 339.
140. Clementi, E.; Corongiu, G. (Eds.) *Computational Chemistry, METECC-95*, STEF, Cagliari, 1995.
141. Boys, S.F.; Bernardi, F. *Mol. Phys.* **1970**, 19, 553.
142. Latajka, Z.; Scheiner, S. *J. Comput. Chem.* **1987**, 8, 674.
143. Arrighini, P. *Intermolecular Forces and Their Evaluation by Perturbation Theory, Lecture Notes in Chemistry*, vol. 25, Springer-Verlag, Berlin, 1981.
144. Jeziorski, B.J.; Mozsynski, R.; Ratkiewicz, A.; Rybak, S.; Szalewicz, K.; Williams, H.L. Chapter 3 in: *Methods and Techniques in Computational Chemistry, METECC-94*, vol. B. STEF, Cagliari, 1994.
145. Stillinger, F.H. *Adv. Chem. Phys.* **1975**, 31, 1.
146. Engström, S.; Jönsson, B.; Impey, R.W. *J. Chem. Phys.* **1984**, 80, 5481.
147. van Gunsteren, W.F. *Current Opinion in Structural Biology* **1993**, 3, 277.
148. Corongiu, C.; Martorana, V. Chapter 3 in: *Methods and Techniques in Computational Chemistry, METECC-94*, vol. C. STEF, Cagliari, 1994.
149. Scott, W.R.P.; van Gunsteren, W.F. Chapter 9 in: *Methods and Techniques in Computational Chemistry, METECC-95*, Ed. by E. Clementi and G. Corongiu, STEF, Cagliari, 1995.
150. Dewar, M.J.S.; Zoebisch, E.G.; Healy, E.F.; Stewart, J.J.P. *J. Am. Chem. Soc.* **1985**, 107, 3902.

ON CALCULATING THE ELECTRONIC SPECTROSCOPY OF VERY LARGE MOLECULES

Michael C. Zerner
Quantum Theory Project
University of Florida
Gainesville , FL 32611

1. Introduction

This chapter will review the many ways in which the excited states of molecules can be calculated. The very fact that there are several ways attests to the observation that none are particularly "right" for all occasions. This is still an area of quantum chemistry of research interest. There is no real "black box" solution that will hold much hope for great accuracy in all situations. Rather, individual cases will need individual attention, and this is especially true for large systems, systems containing transition metal elements, or systems in which the ground state is of open-shell character.

In general, we can characterize the appropriate methods as

1. ΔE Methods, methods in which two separate calculations are performed, and the transition energies obtained from subtraction,
2. Configuration Interaction (CI) methods,
3. Direct methods, in which the transition energies are obtained directly, as in Green's function or propagator calculations, and
4. Hybrid methods, in which the above 3 models are combined.

Generally, ΔE refers to ΔE(SCF), in which two self-consistent field (SCF) calculations are performed, and the transition obtained as the difference in these two results. An example of a hybrid method, would be ΔE(SCF-CI), in which the transition energies are obtained as the difference between two SCF-CI calculations. I include in the CI model any calculation in which the states of interest are obtained by the diagonalization of a matrix containing as its elements $\langle\Psi_I| H |\Psi_J\rangle$. $|\Psi_J\rangle$ might be a spin adopted configuration in the CI model, but it could also be a couple cluster wave function as well. The most common direct method that is used for spectroscopy today is the RPA model [1,2]. These models will be revisited below.

A great deal of these studies, especially as they relate to larger systems, are done using semi-empirical models. As this is, indeed, a method used greatly by us, I will draw many examples from the Intermediate Neglect of Differential Overlap (INDO/s) model that our colleagues have helped us design over the years [3-6]. This model was parametrized directly on the spectroscopy of mid-sized molecules (i.e., benzene, pyridine, etc.) at the CIS (CI - single excitation) level of theory and has a high reliance on parameters derived from atomic

S. Wilson and G. H. F. Diercksen (eds.), Problem Solving in Computational Molecular Science, 249–289.

spectroscopy.

2. The Basic Methods

2.1 ΔE METHODS

Consider a closed-shell ground state for ease of example. An SCF calculation is performed on this state using any number of computer codes available. The wave function is given by

$$\Psi_0 = |\phi_1(1)\overline{\phi_1(2)}...\phi_i(i)\overline{\phi_i(i)}..\phi_n(2n-1)\overline{\phi_n(2n)}| \quad (1)$$

and yields an energy, $E_0(SCF)$. A bar over a molecular orbital (mo) ϕ_i designates beta spin.

The situation is shown in figure 1.

An excitation of an electron from occupied orbital ϕ_i to virtual (unoccupied) orbital ϕ_a yields a wave function given by

$$^{1,3}\Psi_0 = \{|\phi'_1(1)\overline{\phi'_1(2)}...\phi'_i(i)\overline{\phi'_a(i+1)}..\phi'_n(2n-1)\overline{\phi'_n(2n)}| \pm$$

$$|\phi'_1(1)\overline{\phi'_1(2)}...\phi'_a(i)\overline{\phi'_i(i+1)}..\phi'_n(2n-1)\overline{\phi'_n(2n)}|\}/\sqrt{2}$$

$$\Rightarrow \{\phi'_i(i)\phi'_a(i+1)\}\{(\alpha\beta \pm \beta\alpha)/\sqrt{2}\} \quad (2)$$

and an energy designated $^{1,3}E_{ia}(SCF)$. The structure of this wave function forms the basis, in this case, of a Restricted Open-Shell Hartree-Fock [ROHF) calculation, and the primes on the mo's designate that these are not the same orbitals that were used in the ground state calculation [7-9].

This is not the only way one could proceed. One could also allow for different orbitals for different spins (DODS), the most common method, the Unrestricted Hartree-Fock (UHF) model (10). Usually this model can be used only for the states of highest multiplicity for an open-shell structure, and for the above case, only the excited triplet,

$$^{uhf}\Psi_{ia} = |\phi_1^{\alpha}\overline{\phi_1}^{\beta}-\phi_n^{\alpha}\overline{\phi_n}^{\beta}\phi_i^{\alpha}\phi_a^{\alpha}| \quad (3)$$

Such a wave function is not an eigenfunction of spin,

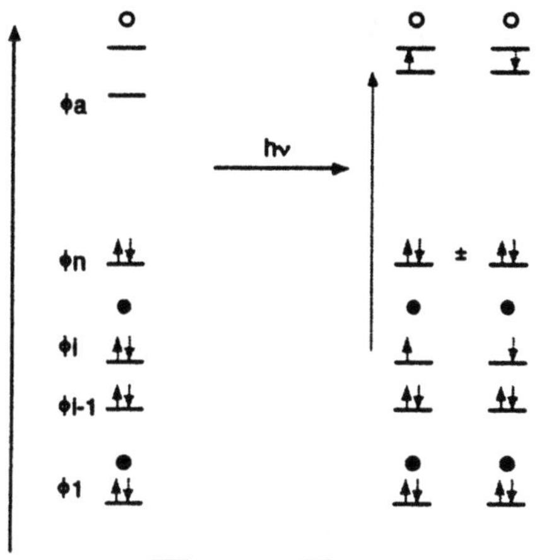

$\{\Psi_0 = |\phi_1(1)\overline{\phi_1(2)}..\phi_l(2l-1)\overline{\phi_l(2l)}..\phi_n(2n-1)\overline{\phi_n(2n)}|$

$\Longrightarrow E_0(SCF)\}$

$\text{---- h}\nu \text{ ---->}$

$^{1,3}\Psi_0 = [\ |\phi_1(1)\overline{\phi_1}(2)..\phi_l(2l-1)\overline{\phi_a}(2l)..\phi_n(2n-1)\overline{\phi_n}(2n)| \pm$

$|\phi_1(1)\overline{\phi_1}(2)..\phi_a(2l-1)\overline{\phi_l}(2l)..\phi_n(2n-1)\overline{\phi_n}(2n)|\ ]/\sqrt{2}$

$\Longrightarrow {}^{1,3}E_{la}$

Figure 1. A pictorial representation of a ground state RHF calculation, leading to an excited state ROHF calculation

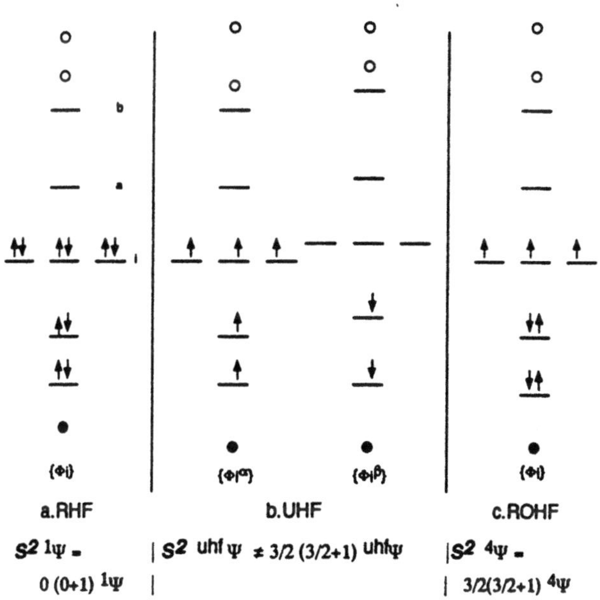

Figure 2. Contrasting a UHF calculation which has different orbitals for α and β spin electrons, and aa ROHF calculation, in which the spatial parts of formally doubly occupied orbitals are restricted to be the same, a s required by spin symmetry and the Hartree-Fock procedure.

$$S^2\,{}^{uhf}\Psi_{ia} \neq s(s+1)\,{}^{uhf}\Psi_{ia} \tag{4}$$

Note the contrast between this, and the ROHF function

$$^{3}\Psi_{ia} = |\phi_1 \bar{\phi}_1 \cdots \phi_n \bar{\phi}_n \phi_i \phi_a| \tag{5a}$$

$$S^2\,{}^{3}\Psi_{ia} = s(s+1)\,{}^{3}\Psi_{ia} = 2\,{}^{3}\Psi_{ia} \tag{5b}$$

which is, by construction an eigenfunction of S^2. The situation is contrasted in figure 2. Note further, in general, that

$$^{uhf}E_{ia}(SCF) \leq {}^{3}E_{ia}(SCF) \tag{6}$$

It is often desirable to create an eigenfunction of S^2, either by projection (8) giving an exact eigenfunction of spin [10a,12,13],

$$O_S\,{}^{uhf}\Psi_{ia} = {}^{2s+1}\Psi_{ia} \tag{7}$$

or by annihilation of the next spin contaminant[10d,14], in the above case, the quintet $^{5}\Psi_{ia}$ (9,10), giving an approximate eigenfunction of S^2

$$\begin{aligned} A_S\,{}^{uhf}\Psi_{ia} &= [S^2 - (s+1)(s+2)]\,{}^{uhf}\Psi_{ia} \\ &= a_1(-2(s+1))\,{}^{2s+1}\Psi_{ia} + a_2(0)\cancel{{}^{2s+3}\Psi_{ia}} + a_3\,(2(s+1))\,{}^{2s+5}\Psi_{ia} + \cdots\cdots\cdots \\ &\Longrightarrow \sim\ {}^{2s+1}\Psi_{ia} \end{aligned} \tag{8}$$

A typical example of spin annihilation is given by a UHF INDO/s calculation of the sextet ground state of T_d $FeCl_4^-$. This is a case of 5 unpaired spins, with expected value of S^2

$$S(S+1) = 5/2\,(7/2) = 8.75$$

The value obtained in the calculation is

$$\langle S^2\rangle_{uhf} = 8.753297$$

after annihilation

$$\langle S^2\rangle_{AA} = 8.750002$$

The annihilation leads to a change in energy given by

$$\Delta E = 0.00019h = 0.12\text{kcal/mol}$$

Encouraged by this, one might try to calculate the excited two electron open-shell singlet states also, and annihilate the triplet. In general, this is not successful. Higher contaminants than the triplet are present.

Projection is a far more complex procedure than is annihilation[13a,13g]. Nevertheless, we have recently examined this(13a,13g,15), with remarkably balanced results obtained for problems in which magnetism is of interest (i.e., ferromagnetism vs. antiferromagnetism).

2.2 CI METHODS

This procedure starts in a very similar manner to the $\Delta E(SCF)$ model, with a wave function describing the excited states of similar form. In our example of a closed-shell ground state,

$$^{1,3}\Psi_0 = \{|\phi_1(1)\overline{\phi_1(2)}...\phi_i(i)\overline{\phi_a(i)}..\phi_n(2n-1)\overline{\phi_n(2n)}| \pm |\phi_1(1)\overline{\phi_1(2)}...\phi_a(i)\overline{\phi_i(i)}..\phi_n(2n-1)\overline{\phi_n(2n)}|\}/\sqrt{2} \quad (9)$$

The orbitals used here are the same as those obtained in the ground state SCF. Because of this, it is relatively straight forward to write the energy of the approximate excited state as

$$^{1,3}\Delta E_{ia} = \varepsilon_a - \varepsilon_i - J_{ia} + (2,0)\, K_{ia} \quad (10)$$

with ε_a the orbital energy of the virtual orbital, ε_i the energy of the orbital from which the electron was taken, J_{ia} the Coulomb integral, given by

$$J_{ia} = \langle ia|ia\rangle = \int dV(1) \int dV(2)\, \phi_i^*(1)\phi_i(1)\, 1/r_{12}\, \phi_a^*(2)\phi_a(2) \quad (11)$$

and K_{ia} the exchange integral, given by

$$K_{ia} = \langle ia|ai\rangle = \int dV(1) \int dV(2)\, \phi_i^*(1)\phi_a(1)\, 1/r_{12}\, \phi_a^*(2)\phi_i(2) \quad (12)$$

The presence of two exchange integrals, always positive, in equation 10, is the frozen orbital approximation to Hund's rule, that suggests that states of highest multiplicity generally lie lower in energy for any given electronic configuration, i.e., for any given electronic assignment to a fixed number of open-shell orbitals.

It is not expected that the energy estimate given by equation 10 is accurate. One of the reasons is that the orbitals used are those appropriate for the ground state, and not the excited state. Another reason is that the ground and excited states might be effected differently by electron correlation, also a criticism of the $\Delta E(SCF)$ model. Nevertheless, it has been found empirically that these estimates are reasonably good for the lowest triplet state of a molecule that is a ground state singlet, and for the lowest $n\text{->}\pi*$ state in planar molecules containing heteroatoms.

The CI procedure then arises from the assumption that all states are of the form (2,7,12)

$$\Psi_I = d_0 \Psi_0(SCF) + \Sigma\, d_{ia}\Psi_i^a + \Sigma\, d_{ijab}\,\Psi_{ij}^{ab} + \cdots\cdots \tag{13}$$

with Ψ_i^a known as a single excitation, Ψ_{ij}^{ab} a double excitation, etc. [2,7,12].

The variation principle then leads to the CI matrix equation

$$(\mathbf{H} - E_I \mathbf{1})\,\mathbf{D}_I = 0 \tag{14}$$

with $\mathbf{H}$ the CI matrix with elements

$$H_{IJ} = \langle \Psi_I | H | \Psi_J \rangle \tag{15}$$

$\mathbf{D}$ are the superposition coefficients of equation 13 as a column matrix, and E_I is the energy of state Ψ_I.

Procedures restricted to only singles (CIS, or mono-excited CI, MECI) are reasonably successful in describing the nature of low lying states (4,16,17) and for this reason are the basis of parametrization of most semi-empirical techniques devoted to electronic spectroscopy. In *ab-initio* calculations, those calculations in which all the integrals of the theory are evaluated *ab-initio*, CIS often overestimates transition energies. When the CI is restricted to both singles and doubles, this is a CISD model. In general CISD is not a successful level of theory for spectroscopy, as the doubles correlate the ground state in a systematic fashion, but not the single excitations that dominate the lowest lying excited states. For this triples are required, CISDT, the important triples being those that correlate the singles that are the principle description of the low lying states of interest. For a good size molecule this is already a difficult calculation.

An example of these points is nicely made in Table I (18,19). Note that the CIS level of theory in this case is remarkably good for the triplets.

Solution of the matrix equation (14) is of major concern in quantum chemistry, for the dimensionality of $\mathbf{H}$ quickly becomes large. Considering n_o occupied mo's and n_v virtual orbitals leads to CIS calculations of dimension $n_v n_o$, CISD procedures approximately

$n_v^2 n_o^2$, CISDT approximately $n_v^3 n_o^3$, an so forth. These equations can be reduced significantly by space and spin symmetries, but the overall scaling remains.

Table I. The calculated spectrum of benzene in eV., using different CI methods, from references 18 and 19.

State	Exp	INDO/S	*Ab-Initio* singles	singles doubles	singles doubles triples
$^3B_{1u}$	3.9	3.90	3.67	5.20	3.83
$^3E_{1u}$	4.7	4.9	5.15	5.78	4.98
$^3B_{2u}$	5.6	5.44	6.01	7.76	7.00
$^3E_{2g}$	6.55	7.06	7.86	8.59	7.28

An example of a CI calculation is given in figure 3 (17). The diagonals of the CI matrix are given for singles by equation 10.

Before proceding, it is interesting to catalog some of the advantages and disadvantages of the ΔE *vs.* CI model.

ΔE

Advantages:

1. Relatively easy
2. Orbital relaxation build in, and because of this
3. the nature of the excited state is easily visualized

Disadvantages:

1. A separate calculation is required for each state of interest
2. It is somewhat difficult to get oscillator strength. This is because the states have different orbitals associated with them, and one needs an intermediate stage of the calculation, for example, a transformation to corresponding orbitals, or the like.
3. Only appropriate for certain excited states, namely those that are of a different spatial or spin symmetry to all states lying below them.
4. The energy of the transition is obtained from the subtraction of two very large

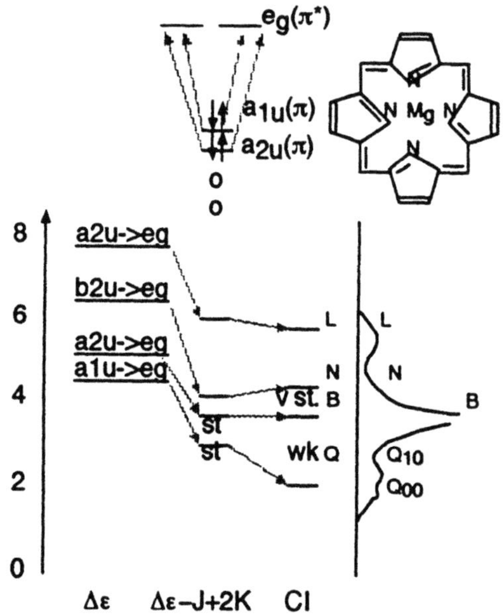

Figure 3. A pictorial representation of the CI procedure applied to the spectroscopy of Mg-porphyrin, from reference 20. Before CI two strong transitions are predicted between the B and Q bands observed. Upon configurational mixing, the lower calculated transition is depressed in energy, and the transition dipole is greatly reduced. In the upper partner the transition dipoles reinforce, giving rise to the very intense B band. Although the Q band is well described by this four-orbital model, the B band is more complex, and interacts with many other configurations, including doubles and some triples. Note that Q_{10} refers to observed vibrational overtones of the Q band.

numbers obtained from separate calculations. For this reason, it is necessary to insure that the calculations are of similar quality. This is not always easy to insure. For example, if the ground state is an RHF calculation, should the excited state be ROHF or UHF. Does it make a difference?

CI:

Advantages:

1. One calculation yields all the states of interest.
2. Relatively easy to calculate transition strengths.
3. Upper bound for all states of interest. Note that this does not imply that there is a boundedness on the energy *differences.*

Disadvantages:

1. The dimensionality of the problem translates into a good deal of computational effort
2. CI expansions are usually very slow to converge
3. Not systematic. The choice of configurations is still an "art"
4. In general, not size extensive. The SCF Hartree-Fock procedure produce energies that grow correctly with the size of the system. For example, if the system is doubled, the energy will roughly double. The same is true for CIS. But it is not true for any other CI procedure other than full CI, CIF, in which all possible excitations for a given model space are included.

An example of a $\Delta E(SCF)$ calculation compared with a CI calculation is given in Table II for the ionization spectrum of Mg Porphin. This system is of importance in photosynthesis.

2.3 DIRECT METHODS

In direct methods the transition energies and the transition moments are obtained "directly". The names usually associated with such techniques are Louiville, Green's functions, and Propagators(1,2,21).

Briefly, we consider the Schrödinger equation

$$H\,|\Psi_\iota> = E_\iota\,|\Psi_\iota> \tag{16}$$

and introduce an excitation operator

$$\Omega_\iota\,|\Psi_O> = |\Psi_\iota> \tag{17}$$

($\iota.\varepsilon.$, $\Omega_\iota = |\Psi_\iota><\Psi_O|$ is an exact excitation operator)

and the Louivillian operator

$$\mathcal{L}_\iota = [\,H, \Omega_\iota\,] = H\Omega_\iota - \Omega_\iota H \tag{18}$$

and note that

$$\mathcal{L}_\iota\ |\Psi_O> = E_\iota\,\Omega_\iota\,|\Psi_O> - E_O\,\Omega_\iota\,|\Psi_O> = \omega_\iota\ |\Psi_\iota> \tag{19}$$

Table II: The calculated ionization spectrum of Mg porphin, in eV., from reference 20.

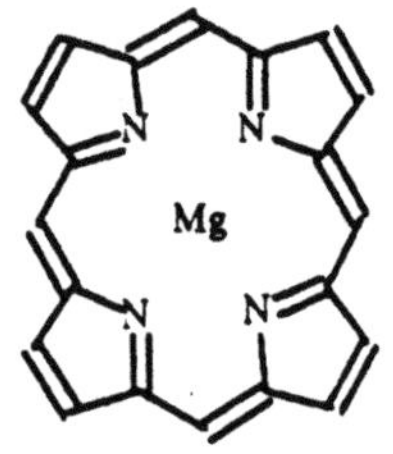

State	ΔE(SCF)	CI	Exp	Comment
$^2A_2$	6.10	5.90(0.96)	6.2	π
$^2A_1$	6.36	6.38(0.95)	6.6	π
^2E	8.76	7.62(0.16)	7.5	mixed
$^2B_1$	8.52	7.76(0.68)		mixed
^2E		7.90(0.38)		π
$^2A_1$		8.09(0.64)	7.9-8.3	π
^2E		8.12(0.47)		mixed
^2E		9.58(0.06)		mixed
$^2B_1$		9.60(0.15)	9.6	mixed
^2E		9.70(0.52)		mostly π
$^2B_1$		9.78(0.82)		mixed
$^2B_1$		10.03(0.07)	10.01	mixed
^2E		10.13(0.05)		lots of N_σ

solves the Schrödinger equation directly for $\omega_i = E_i - E_0$. Note, however, that $\mathcal{L}_\iota$ takes $|\Psi_O>$ directly to $|\Psi_\iota>$, and that in solving equation 19 information is lost on the nature of both states by the excitation process. Defining $|\Psi_O>$ and Ω_ι defines the model. The simplest model is the RPA model in which $|\Psi_O>$ is the SCF wavefunction for the ground state, and

$$\Omega_\iota = \Sigma X_{ia} E_{ia} + \Sigma Y_{ia} E_{ia}^+ \qquad (20a)$$

with **X** and **Y** coefficient matrices to be determined [22], and

$$E_{ia} = (a_\alpha^+ i_\alpha + a_\beta^+ i_\beta) \qquad (20b)$$

Equation 20b is essentially a single excitation only operator, but use of equation 20a also includes de-excitation operators of the same type.

Of considerable interest is that the RPA model is the lowest level of theory that can insure the equivalence of the dipole length and dipole velocity estimates of oscillator strength(1,2). That is, oscillator strength

$$f_{IJ} = 4.32x10^{-9} \int \varepsilon(v)\ dv \qquad (21a)$$

where $\varepsilon(v)$ is the extinction coefficient (mol $^{-1}$ cm. $^{-1}$) and the integral is over the observed band (cm. $^{-1}$) representing the transition, is given theoretically (23) by either

$$f^r_{IJ} = 2/3\ |\langle\Psi_I|\ \Sigma\ r_i\ |\Psi_J\rangle|^2\ (E_I\ - E_J) \qquad (21b)$$

or

$$f^v_{IJ} = 2/3\ |\langle\Psi_I|\ \Sigma\ p_i\ |\Psi_J\rangle|^2 / (E_I\ - E_J) \qquad (21b)$$

Table III: The calculated spectrum of Benzene, in 1000 cm. $^{-1}$. The numbers in parenthesis are oscillator strengths), from reference 19.

State	Observed	Calculated	
		CIS[b] (r , p)	RPA[c] (r)
$^1B_{2u}$	38.0(0.001)	38.0(0, 0)	39.0 (0)
$^1B_{1u}$	49.0(0.1)	48.0(0, 0)	47.0 (0)
^1E_u	55.9(0.690)	55.0(2.16,0.32)[a]	54.6 (0.76, 0.73)

(using atomic units in which the mass of the electron = 1, and h, Planck's constant, $=2\pi$) a consequence of the commutation relationship which relates momentum with position, that is only obeyed in an exact calculation. It is never obeyed, for example, in any truncated CI model, but would be obeyed in the space of a complete basis in an RPA calculation. This might suggest that the RPA model would be particularly good for estimating transition strengths even for poor basis sets, which is the observation. The RPA is a much better model for calculating oscillator strength than is a similar CIS calculation. The CIS level of theory generally exaggerates the transition moments of low lying symmetry allowed excited states. An example of this is presented in Table III for the case of benzene (22).

3. The Intermediate Neglect of Differential Overlap Model Parametrized for Spectroscopy, INDO/s

I have mentioned that the INDO/s model has been used by us, and by very many others, now for thousands of cases. A review of this model Hamiltonian is available (24). This model is based on the original INDO model of Pople, Segal and Santry (3) and is similar to the Complete Neglect of Differential Overlap (CNDO) model adopted for spectroscopy by Jaffé and Delbene (25). The CNDO model does not contain the one-center two-electron terms (Slater Condon Factors[26]) and is therefore not useful for examining the spectroscopy of transition metal complexes, or as accurate in reproducing singlet-triplet splittings in most aromatic systems.

Briefly, this model has been parametrized at the CIS level of theory [4,5,24]. Most parameters of the theory come directly from atomic spectroscopy, and are based on experimental ionization potentials, electron affinities, and the Slater-Condon two-electron integrals. An atom with a valence based s,p basis is given one resonance integral β, those with s,p and d basis, two values $\beta(s) = \beta(p)$ and $\beta(d)$, and those with s,p,d and f basis, three such integrals $\beta(s) = \beta(p)$, $\beta(d)$ and $\beta(f)$. The resonance integrals are determined empirically from a data base of molecules with well established structures and spectroscopies.

Let **F** be the Fock matrix (7,27), **H** the one-electron matrix, and **G** the two-electron matrix, and take the UHF case as an example, then the Hartree-Fock Roothaan method yields (7,10a, 27,28,29)

$$F^{\alpha} = H + G^{\alpha} \tag{22a}$$

and

$$F_{\mu\nu}{}^{\alpha} = h_{\mu\nu} + \Sigma_{\sigma\lambda}\{P_{\sigma\lambda}\,(\mu\nu|\sigma\lambda) - P_{\sigma\lambda}{}^{\alpha}(\mu\sigma|\nu\lambda)\,\} \tag{22b}$$

with $\mathbf{P}^{\alpha}$ the first order Fock-Dirac α spin-density, with elements given by

$$P^{\alpha}{}_{\mu\nu} = \Sigma\, c^{\alpha}{}_{\mu i}\, c^{\alpha}{}_{\nu i}\, n^{\alpha}{}_{\iota} \tag{22c}$$

with similar expressions for β spin. n^{α}_{ι} is the occupation of the mo $\phi^{\alpha}_{\iota} = \Sigma_{\nu} X_{\nu} c_{\nu i}$, either zero or one, and the total density **P** given by

$$\mathbf{P} = \mathbf{P}^{\alpha} + \mathbf{P}^{\beta} \qquad (22d)$$

Then in the INDO model,

diagonal *μ ε atom A*

$$F_{\mu\mu}^{\alpha} = [h_{\mu\mu}] + [G^{\alpha}{}_{\mu\mu}]$$

$$= [U_{\mu\mu} - \Sigma_{B\neq A} Z_B(\mu|R_B{}^{-1}|\mu)] + [\Sigma_{\sigma\lambda}\{P_{\sigma\lambda}(\mu\mu|\sigma\lambda) - P_{\sigma\lambda}{}^{\alpha}(\mu\sigma|\mu\lambda)\}]$$

$$= [U_{\mu\mu} - \Sigma_{B\neq A} Z_B \underline{(\mu\mu|\nu\nu)}] + [\Sigma\{\sigma\lambda \text{ both on A}\}\{P_{\sigma\lambda}(\mu\mu|\sigma\lambda) -$$

$$P_{\sigma\lambda}{}^{\alpha}(\mu\sigma|\mu\lambda)\} + \Sigma\{\sigma \text{ not on A}\}\{P_{\sigma\sigma}\underline{(\mu\mu|\sigma\sigma)}\,] \qquad (23a)$$

where the approximation

$$(\mu|R_B{}^{-1}|\mu) = \underline{(\mu\mu|\nu\nu)} \qquad (23b)$$

is made. Z_A is the atomic number of atom A, and $Z^c{}_A$ is the core charge of atom A, equal to the number of valence electrons in neutral atom A (i.e, 4 for carbon, 8 for iron, etc.) An underline indicates that only the spherically symmetric part of the integral is maintained (3,4).

$$U_{\mu\mu} = U_{\mu\mu}(IP_{\mu}) \qquad (23c)$$

is the core integral taken from ionization potentials. The one-center two-electron integrals are taken from experimental Slater-Condon radial integrals [26]

$$(\mu\sigma|\nu\lambda) = f(F^k, G^k, R^k) \qquad (23e)$$

off-diagonal ν on A, ν on atom B

$$F_{\mu\nu}{}^{\alpha} = [h_{\mu\nu}] + [G^{\alpha}{}_{\mu\nu}]$$

$$= [1/2(\beta_{\mu}{}^{A} + \beta_{\nu}{}^{B})\Delta_{\mu\nu}] - [P_{\mu\nu}{}^{\alpha}(\mu\mu|\nu\nu)] \qquad (23f)$$

$\beta_{\mu}{}^{A}$ = the *resonance integral*, and $\Delta_{\mu\nu}$ = weighted overlap. For further details see references 3, 4 and 24.

Given in this form, the Hamiltonian can be written as

$$H = \Sigma_{\mu\upsilon} \langle\mu|h|\nu\rangle N(\mu^{+}\nu) + 1/4\, \Sigma_{\mu\upsilon}\, \Sigma_{\sigma\lambda} \langle\mu\nu||\sigma\lambda\rangle\, N(\mu^{+}\nu^{+}\lambda\sigma) \qquad (24)$$

and all *ab-initio* theory now carries through for the INDO/s model. $N(\mu^{+}\nu)$ is the normal ordered product, and again μ^{+} and ν are Fermion creation and annihilation operators [2].

If CIS methods are reasonably good for *ab-initio* methods, why perform semi-empirical calculation for spectroscopy? Do not the semi-empirical calculations attempt to mimic the *ab-initio*?

The first comment is that through the use of parameters, the defects of a minimum basis set, MBS, are hopefully improved. *Ab initio* CIS calculations of MBS type are not good. Secondly, the use of parameters obtained through CIS calculations hopeful ly contain some of the dynamic correlation, that part of the correlation that is difficult to obtain without very large CI calculations. All CIS calculations include the *essential* correlation (non-dynamic) providing the excited states of interest are dominated by the single excitations included. Table IV gives an example of *ab initio* calculations on pyridine(16) compared with INDO/s calculations.

Table IV: A Comparison Of *Ab-initio* and INDO/s CIS Calculations for (eV.) Pyridine

State	exp	INDO/s [a]	*ab-initio* [b]	
$^1B_1$	4.32-4.60	4.3	7.24	$n \rightarrow \pi^*$
$^1B_2$	4.76-4.99	4.8	6.9	$\pi \rightarrow \pi^*$
$^1A_2$	5.43	5.45	7.05	$n \rightarrow \pi^*$
$^1A_1$	6.20-6.38	6.16	8.49	$\pi \rightarrow \pi^*$
$^1B_2$	6.82-7.22	7.05	(8.07)[c]	$\pi \rightarrow \pi^*$

a. From Ridley and Zerner, reference 4.
b. From Foresman et al., reference 16. This is using a 6-31G* basis.
c. From the reference 16. This is from a 6-31G+ basis.

The largest calculations to date that have been made using the INDO/s model are on the photosynthetic reaction center involving over 1000 atoms and 3000 electrons, including 4 bacteriochlorophyll molecules, and 2 bacteriopheophytins (Mg free bacteriochlorophyll molecules) (30). The lowest excited state is reported at 10,200 cm^{-1}. The calculations give a transition energy of 11,200 cm^{-1}, and suggest that this state is localized on a pair of chlorophyll molecules known as the "special pair". This finding is totally consistent with experimental findings that require that this state to be the lowest lying absorbing state. Calculated some 1,200 cm^{-1} higher in energy is the first charge transfer state, from the

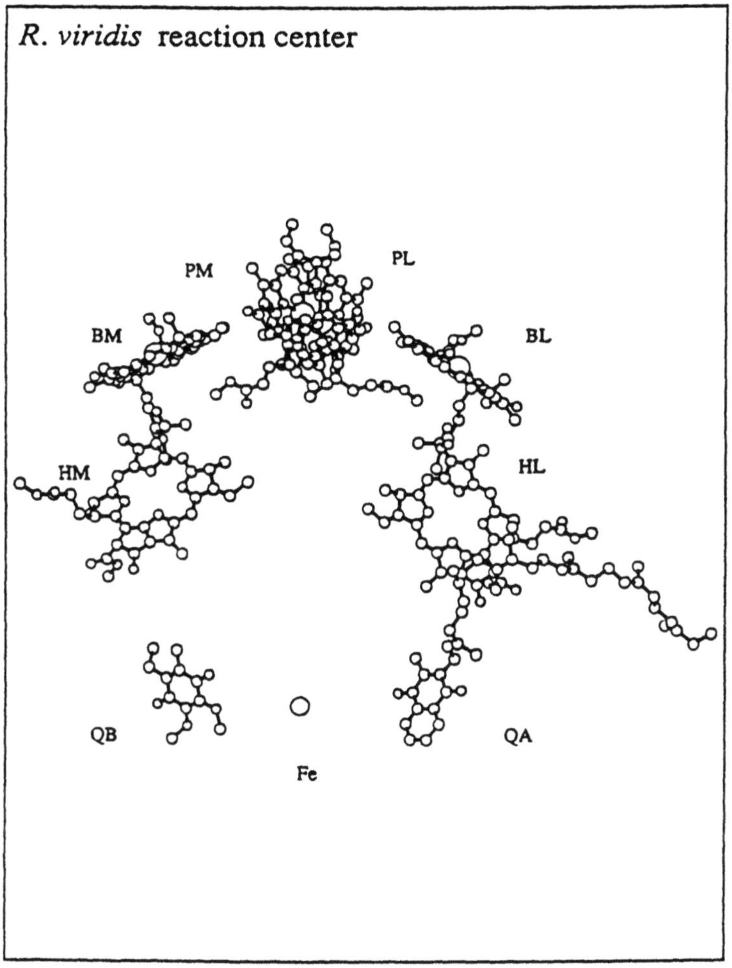

Figure 4. The reaction center of photosynthesis model used to obtain the results in Table V. The hydrogen atoms are not shown to better display the "heavy" atoms.

special pair to the "L-side" pheophytin, a Mg free bacteriochlorophyll molecule. Although the exact location of this state upon absorption is not known, it is required to be the next excited state if the observed electron transfer is to take place. Table V gives a synopsis of these calculations(30,31). The system calculated is given in figure 4. The importance of including an estimate of the effects of relaxation of the protein electronic polarization is important, and is discussed later. If one concentrates only on the special pair, and builds a model of some 600 atoms, the CIS procedure gives a transition energy of 10,080cm^{-1}, in near perfect agreement with the observed value of 10,200 cm-1.

Table V: The calculated Spectroscopy of the Reaction Center from *R. viridis*. Energy in 1000 cm.$^{-1}$ and dipole moment differences $\Delta\mu$ in DeBye.

State	"Gas phase"[a]		Solvent simulated[a]		
	Energy	$\Delta\mu$	Energy	$\Delta\mu$	Ref. **[b] H (I)
Qy1 = P*	11.6	4.6	11.4*	4.6	11.4*
P^+BL^-	19.8	49.9	16.6	46.1	17.0(15.3)
P^+HL^-	19.7	78.9	12.6	75.0	14.2(10.9)
P^+BM^-	20.2	55.5	16.7	48.4	17.1(15.3)
P^+HM^-	20.6	76.9	13.5	74.8	15.0(11.6)

a. From Thompson and Zerner, ref. 30..
b. M.R.Gunner, A.Nicholls, B.Honig, J. Am. Chem. Soc. (1996)100,4277 estimates from Table 4, H and I for $\epsilon = 2$ and $\epsilon = 4$. Their *best* relaxed values are (4:80:4): 14.0, 7.7, 15.4, 8.8 (from Table 5 of that reference)

4. The Calculated Spectra of Transition Metal Complexes

The electronic spectroscopy of transition metal complexes might be cataloged as[32]:

1. The ligand spectroscopy, L->L
2. The Laporte d->d* spectroscopy (ligand field excitations)
3. The metal-to-ligand and ligand-to-metal charge transfer spectroscopy, MLCT, M->L* and LMCT, L->M*
4. Metal ligand exchange coupling, ML->ML*
5. Spin-orbit transitions.

Accurate calculation of the spectroscopy of these systems requires a reasonably democratic treatment of both the ligand orbitals and the d orbitals of the transition metal. This is not always easy to insure, as the former generally participate in delocalized molecular orbitals.

whereas d (and f) orbitals are generally localized on the metal. These transitions are sketched in Figure 5.

The ligand spectroscopy is often much the same as it would be in the absence of the transition metal, with three exceptions. There is often a shift due to inductive effects, that

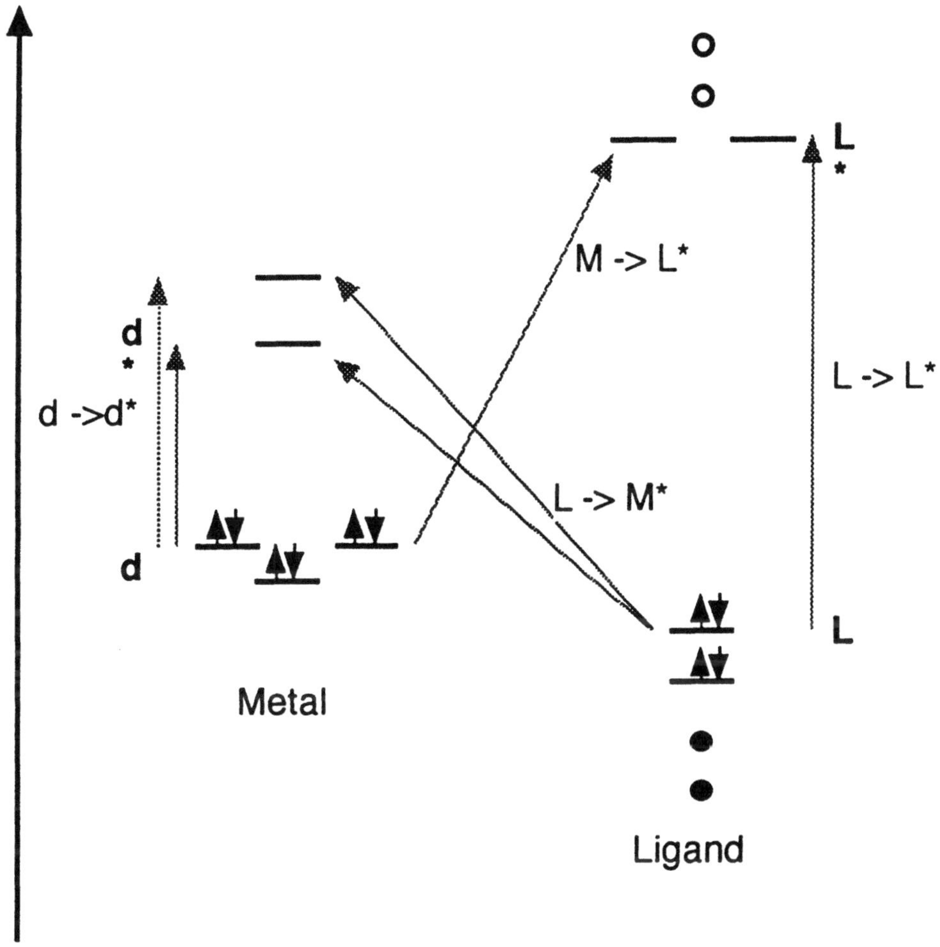

Figure 5. A schematic representation of the transitions found in transition metal compounds.

is, the ligand may lose or gain a fraction of an electron in the complex and this effects the transition energies. The parentage of such transitions in the isolated ligand is often obvious. A second effect involves shifts of transitions that involve the sigma orbitals of the chelating atom. For example, the n->π* excitations in pyridine are shifted greatly to the blue (higher energy), a consequence of the bonding nature of the lone pair of the nitrogen atom. A third effect, often not obvious, is related to metal-ligand exchange coupling, item 4 in the above list, described in more detail below [33].

The calculated spectroscopy of $Mn(II)(H_2O)_6$ is given in table VI [34]. These are results from INDO/CI calculations, in which a high reliance rests on parameters from atomic spectroscopy. Note especially the accuracy obtained compared to experiment for transitions involving different multiplicities. This is not always easy to achieve in methods that have as their starting point Hartree-Fock SCF calculations, as there is a very strong bias in this method for states of highest multiplicity. Unpairing electrons is generally accompanied with a relative gain over the closed-shell situation of *about* 1 eV in correlation energy, a relative error corrected only through consideration of electron correlation. The parametrization of the INDO/CI scheme includes some of this dynamic correlation through the parameters.

Table VI. The Calculated Spectrum of $Mn(II)(H_2O)_6^{2+}$, in 1000 cm.$^{-1}$, from reference 34. Symmetry labels are those of O_h, although actual symmetry is C_{2v}.

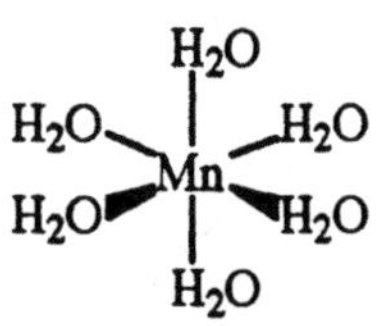

State	Calculated assignment	Observed a.	Observed b.	Configuration
$^6A_1$	0.0	0.0	0.0	$t_1t_2t_3e_1e_2$
$^4T_{1g}$	18.9	18.9	18.9	$t_1t_2t_3^2e^1$
$^4T_{2g}$	21.9	23.1	23.0	$t_1t_2t_3^2e^1$
$^4A_{1g}$	22.4	25.0	25.0	$t_1t_2t_3e_1e_2$
^4E_g	22.6	25.3		$t_1t_2t_3e_1e_2$
$^4T_{2g}$	27.9	28.0	28.0	$t_1^2t_2e_1e_2$
^4E_g	28.5	29.8	29.8	$t_1t_2t_3e_1e_2$
$^2T_{1g}$	29.1			$t_1t_2t_3^2e^1$
$^2T_{1g}$	31.0			$t_1t_2t_3^2e^1$
$^4T_{1g}$	31.3	33.0	32.4	$t_1^2t_2e_1e_2$
$^2A_{1g}$	32.3			$t_1t_2t_3e_1e_2$

As suggested in figure 5, even if the L->L* and d->d* spectroscopy is in good agreement with experiment, that is not a guarantee that charge-transfer transitions will be in good accord with experiment. For this, the ionization potentials, IP, of the fragment loosing the electron, and the electron affinity, EA, of the fragment gaining the electron must also be well represented. The latter is particularly demanding in an *ab-initio* calculation, and requires a reasonably good basis set including diffuse functions. The INDO/CI model can

Table VII: The calculated spectrum of ferrocene, using the INDO/CI model, in 1000 $cm.^{-1}$ The values in parenthesis are oscillator strengths. From reference 4b. Only transitions with calculated intensities are shown after band V

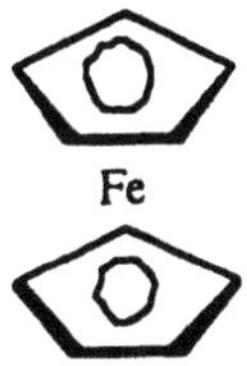

	INDO/S		Obs	Comments
I.	$^3E_1''$	20.5		d->d
	$^3E_1''$	20.6	18.9	d->d
	$^3E_2''$	20.6		d->d
IIa	$^1E_1''$	21.7	21.8	$d_{z}2 \to d_\pi$
b	$^1E_2''$	23.9	24.0	$d_\delta \to d_\pi$
III.	$^1E_1''$	31.9	30.8	$d_\delta \to d_\pi$
IV	$^1E_2''$	36.9	37.7 (0.02)	$d_\delta \to 4s$
	$^1A_1'$	39.7		$d_{z}2 \to 4s$
	$^1A_2'$	39.9		$d_\delta \to L^*$
V	$^1E_1'$	41.2 (0.06)	41.7 (0.01)	$d_\delta \to L^*$
	$^1A_1''$	42.4	42.2 (0.01)	$d_\delta \to L^*$
VI	$^1A_2''$	45.3 (0.29)	46.9 (0.10)	$d_d \to L^*$
	$^1E_1'$	50.2 (0.0)		
VII	$^1A_2''$	50.3 (0.20)	50.9 (0.69)	MLCT
	$^1E_1'$	50.2 (0.30)		

reproduce this, as electron affinities are used in the evaluation of the one-center two-electron integrals, either using the Mataga-Nishimoto formula [24,35]

$$(\mu\mu|\nu\nu) = (a + R)^{-1} \tag{25a}$$

or the Ohno Klopmann formula[36]

$$(\mu\mu|\nu\nu) = (a^2 + R^2)^{-1/2} \tag{25b}$$

with

$$a = (F^0(\mu\mu) + F^0(\nu\nu))/2 \tag{25c}$$

or other similar average and

$$(\mu\mu|\mu\mu) = F^0(\mu\mu) = IP_\mu - EA_\mu \tag{25d}$$

An example of charge transfer excitations calculated using the INDO/CI model is give in Table VII for ferrocene. Most of the difficulty in reproducing the charge transfer spectroscopy of this compound has to do with reproducing the EA of the cyclopentadienyl ring.

Metal-ligand exchange coupling [33,37] produces features in the L->L* spectrum that are not easily observable in the isolated ligand. In a ligand with ground state spin value S, the most obvious L->L* transitions observed are to the spin allowed excited states of the same multiplicity. In coupling with a transition metal of spin state S', the ground state of the complex is one of S+S', S+S'-1,...|S-S'|. Excited states of the ligand with multiplicities other than S, say S*, are usually very weak, but now also couple with the metal spin yielding states of (S*+S'), (S*+S'-1) ..|S*-S'|. One of these states may now be of the same spin-multiplicity as is the ground state, and now be quite observable. The situation is depicted in figure 6 for the case of Fe(II) porphin [38,39].

Metal-ligand exchange coupling is also difficult to reproduce for the same reason as it is difficult to reproduce the energy differences between any states of different multiplicities. Again a reasonably heavy correlation treatment is required to remove the bias that the Hartree-Fock procedure has for states of higher multiplicity. The nice regularity that is seen in Figure 6, easily seen using perturbation theory [33], is lost in the details of the calculation.

Spin-orbit effects can be quite large in all complexes that include heavy elements. In general, the importance of these splittings grow with Z^4, where Z is the atomic number of the element considered. An example of very large spin-orbit effects is given in Table VIII for the complex $Ce(III)(H_2O)_9^{3+}$. This is essentially the metal ion spectroscopy 4f->5d. The ligand field effects can be isolated from the spin orbit effects, and account for only 2000 cm.$^{-1}$ of the observed 10,000 cm.$^{-1}$ total splitting observed in this complex [5].

In addition to being essential in calculating the spectroscopy of complexes containing heavy elements, the inclusion of spin-orbit effects can also be important as a tool in characterizing the excited states even in those cases where the effects are reasonably modest, but are detected in the experiment. Then not only can the location of the bands and the calculated intensity be used for characterizing the excited states, but also the spin-orbit splitting. The calculated transition energies are but one piece of information!

Table VIII: Comparison of the Band Maxima in $Ce(H_2O)_9(EtSO_4)_3$ and Solvent Spectrum of Ce(III) with the Calculated Energies in (cm^{-1})[c]

Ce:crystal[a]	Aqueous[b]	INDO/S-CI
39,060	39,630	40,228
41,950	41,970	42,933
44,740	45,330	45,670
47,400	47,620	48,665
50,130	50,000	50,917

a. S. P. Sinha, reference 40
b. C.K. Jørgensen and J.S. Brinen, reference 41.
c. M. Kotzian, N. Rösch and M.C.Zerner, reference 5.

5. The Effects of Geometry

The spectroscopy of a molecular system *is* sensitive to geometry, as might be anticipated even from the particle in a box model which suggests that the spacing between levels is inversely proportional to the square of the box length. Figure 7 shows the sensitivity of the spectroscopy of pyrimidine to the N-N bond length. Table IX, indicates the sensitivity of spectra calculated using the INDO model when the geometry is obtained from different methods.

The sensitivity of the calculated spectroscopy to geometry could, in fact, be used to help determine the quality of the geometry used. In other words, poor agreement might suggest a poor geometry, and suspect bond lengths of angles that might even be examined using the calculated spectrum as a guide. This sensitivity also suggests that it is not meaningful to ask too great an accuracy for systems in which the geometry is not well determined. In many biological systems, such as the photosynthetic reaction center of Table V, even some of the atoms present are not well established. In this case, the addition of two waters in locations suggested recently leads to a 700 cm.$^{-1}$ red shift in the predicted P^+HL^- transition energy.

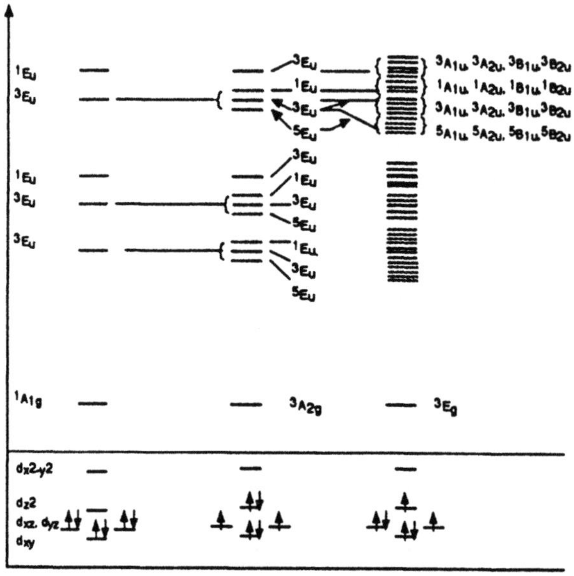

Figure 6: The calculated spectrum of Fe(II) porphin, including the metal-ligand exchange coupling.

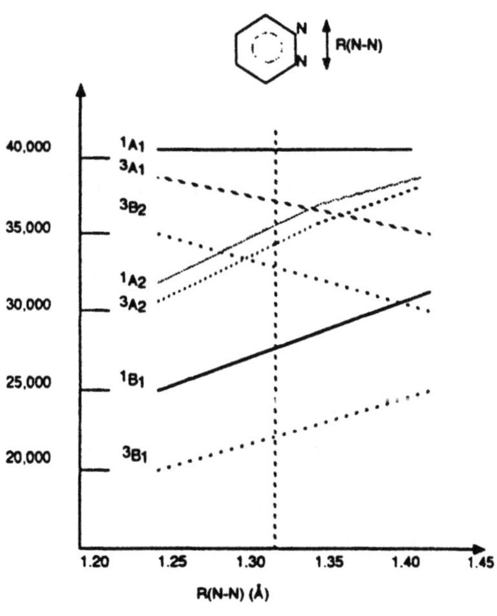

Figure 7. The calculated spectrum of pyridazine as a function of the N-N distance.

Table IX: The sensitivity of Predicted Spectra to the assumed Geometry, from John McKelvey. The data base consists of 20 chromophores each containing 20-30 atoms.

Geometry from:	R^2
//X-Ray	0.94
//B3LYP	0.93
//BLYP	0.95
//6-31G*	0.81
//3-21G*	0.60
//AM1	0.60
//PM3	0.09

6. Solvent Effects

A detailed treatment of solvent effects appears elsewhere in this volume. It is included in this chapter for completeness. Solvation can effect electronic spectroscopy, and there can be major shifts that accompany solvation.

There are several ways to proceed. One way is to perform a classical simulation, and then perform a spectroscopic calculation every fixed number of molecular dynamics or Monte Carlo moves using the structures obtained from the simulation[42-44]. If the interactions are weak, then the solvent molecules might be well represented by point charges in the quantum chemistry used to calculate the spectroscopy, or by the moments of the solvent molecules located at their center of mass. If some of the interactions are strong, or if there is charge transfer [45] then the quantum chemistry must include these tightly bound molecules along with the chromophore. This can be a rather expensive business, but does have the advantage of generating band widths in addition to transition energies.

A simpler procedure employs the self-consistent reaction field approach (SCRF) [46]. In this approach the free energy of solvation is classified according to

$$\begin{aligned} \Delta G_{solution} &= \{\Delta G_{electrostatics} + \Delta G_{polarization}\} \\ &+ \Delta G_{dispersion} + \Delta G_{cavitation} + \Delta G_{specific} \\ &+ \Delta G_s + \mathit{Models\,error} \end{aligned} \quad (26)$$

where $\Delta G_{specific}$ refers to specific solute-solvent interactions that are greater than thermal energy, kT, such as those that involve hydrogen bonding. Such interactions are best treated using a super-molecule approach, although this complicates the simulation by, perhaps, requiring conformational averaging. ΔG_s refers to entropic changes in the solvent caused by the solute.

$\{\Delta G_{electrostatics}+\Delta G_{polarization}\}$ are included in the self-consistent calculation using one of the many self-consistent reaction field models [46,47]. These model generally add to the Fock operator $\mathbf{f^o}$ a term representing interaction with the solvent treated as a dielectric continuum[48]. In its simplest form

$$\mathbf{f} = \mathbf{f^o} - \alpha \mathbf{V} \tag{27a}$$

with

$$\mathbf{V} = \Sigma_l\, g_l(\varepsilon)\Sigma_m\, M_l^m\, \mathbf{M_l^m} \tag{27b}$$

$$g_l(\varepsilon) = a^{-(2l+1)}\, [(l+1)(\varepsilon-1)/(l+(l+1)\varepsilon). \tag{27c}$$

$\mathbf{M_l^m}$ is the m'th component of the l'th moment, as an operator, and $M_l^m(0)$ is the corresponding moment of the reference state, $<\Psi_0|\mathbf{M_l^m}|\Psi_0>$. This latter moment is not known until the SCF is complete, but neither is $\mathbf{f^o}$, which depends on the electron-electron interactions. Both must be updated during the SCF cycling. The parameter α in equation 27 is either 1/2 for the direct reaction filed approach, or it is 1. If $\alpha = 1$, then a correction must be made to obtain the total energy after the SCRF procedure yields the quantum chemical energy which is of the form [47]

$$E_{total}(I) = W_{quantum\text{-}mechanics}(I) + 1/2\ \Sigma_l\, g_l(\varepsilon)\Sigma_m\, M_l^m(I)\, M_l^{-m}(0) \tag{28}$$

for state Ψ_I given that the SCF state is Ψ_0. In equations 27, ε is the bulk macroscopic dielectric constant.

There are several models available for calculating the dispersion contribution to the free energy [48], as there are for estimating the cavitation energy[49,50].

The first term of equation 27b is the famous Born term(51)

$$V^0 = 1/2\,(1-1/\varepsilon)\, a^{-1} \tag{29}$$

and the second term is the Onsager dipole term [51,52]

$$V^1 = 1/2\, \mu\mu/ a^3 \tag{30}$$

In calculating the excited states of molecules, we are interested in the differences in the free energies of two states

$$\Delta(\Delta G_{solution}) = \Delta G_f - \Delta G_i \quad (31)$$
$$= \Delta\{\Delta G_{electrostatics} + \Delta G_{polar}\} + \Delta(\Delta G_{dispersion})$$

I have stressed in equation 31 that model errors, entropic errors and even cavitation errors may well have a minimal effect as they hopefully cancel. (Note however that cavitation contributions might not cancel for excitations into Rydberg orbitals. This cancellation rests on the observation that the ground and excited states of larger molecules are about the same "size".)

There is one further consideration when it comes to estimating spectroscopy, and this has to deal with the time evolution of the process. The dielectric response of a material has a frequency dependence, corresponding to all the different modes that are available for the response to a perturbation. For absorption or emission of light we consider the electronic relaxation quick, comparable to the time required for a chromophore to absorb light, and the

Table X: The calculated spectrum of beta carotene.

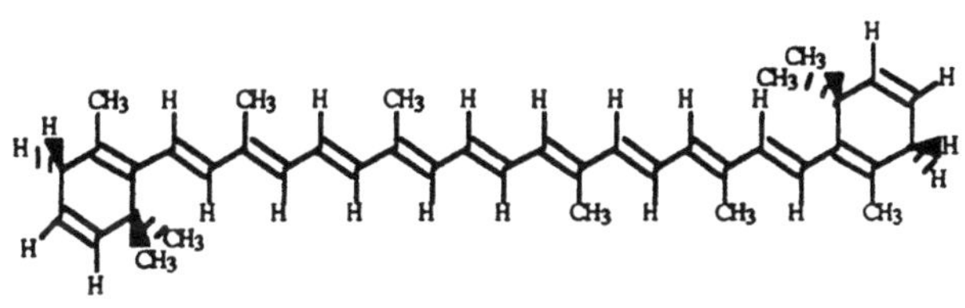

TRANSITION	STATE DIPOLE(D)			TRANSITION MOMENT(D)		
	X	Y	Z	X	Y	Z
In Gas Phase						
(1)-->(2)	0.0000	0.0000	0.0000	1.9856	-3.0928	17.4736
ENGY(CM^{-1})= 24033.9						
W(nm) = 416.1						
OSC = 3.60858						
In Cyclohexane						
(1)-->(2)	0.0000	0.0000	0.0000	-1.9687	2.9058	-17.4065
E(AB,CM^{-1})= 23899.3						
E(DP,CM^{-1})= 22840.7						
W(DP,nm) = 437.8 (exp. is 455nm)						
OSC = 3.39146						

nuclear motion slow (corresponding to a Frank-Condon principle).

The dielectric response is then split into a fast part, $g_1(n^2)$, where n is generally equated to the index of refraction, and a slow (nuclear motion) part, $g_1(D)$, defined as the difference

$$g_1(D) = g_1(\varepsilon) - g_1(n^2) \tag{32}$$

As sketched above, the SCRF procedure assumes that the SCF state is in equilibrium with the solvent, and use of the dialectic constant ε is appropriate. Corrections must then be made for the fact that the excited state upon absorption still realizes the nuclear charge distribution of the ground state, but the electrons of the solvent have followed the electronic excitation of the solute. There are several ways of doing this(47,53].

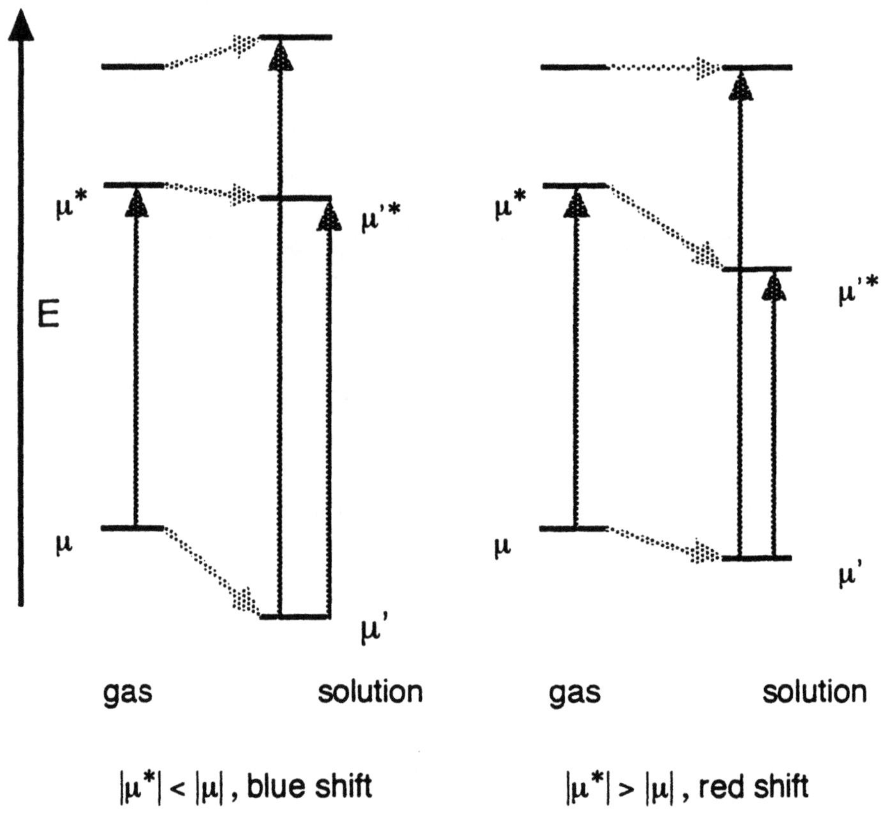

Figure 8. The spectroscopic shifts that might be expected in the case of a molecule that has a dipole moment.

There are some generalities that can be reached from examining these equations. The first is that dispersion always leads to a red shift, and that this red shift is generally greater for higher lying states.

The second is that if the ground state has a larger dipole moment than does the excited state, there will be a greater stabilization of the ground state, and this will lead to a blue shift in the solution spectrum compared to the gas phase one. If the excited state has a greater dipole moment than does the ground state, then the excited state will be more stabilized, and this will lead to a red shift in solution. This is demonstrated in figure 8.

An example of dispersion is given in Table X for beta-carotene. As neither the excited state or the ground state of this system has a dipole moment the shift caused by the electrostatic effects is small, about 135 $cm.^{-1}$ and is mostly due to the quadrapole moments. The dispersion, however, causes a red shift estimated at 1500 $cm.^{-1}$.

An example of the effects of electrostatics is given in table XI for some pi systems. In the

Table XI: Some Calculated Spectroscopic Shifts in $\pi \rightarrow \pi^*$ Transitions, in 1000 $cm.^{-1}$ The quantum chemistry is from ZINDO, reference 46f.

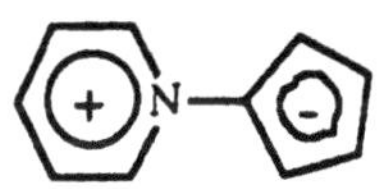

Compound 1

Compound 2

		Calc	Obs
Compound 1			
	1. Gas	20.1	
	2. $CHCl_3$	21.8	19.6[a]
	3. water	24.6	22.1
	$\omega(3)-\omega(2)$	2.8	2.5
Compound 2			
	1. Gas	32.0	
	2. C_6H_{14}	32.8	30.1[b]
	3. water	29.1	26.1
	$\omega(3)-\omega(2)$	-3.7	-4.0

a. Daehne, S .; Schob, F.; Nolte, K.-D. Z. Chem. **1973**, 13,471

b. Kosower, E.M.; Ramsey, B.G J. Am. Chem. Soc. **1959**, 81,856

first of these compounds the dipole moment of the ground state is larger than that of the excited state, and in the second, the first excitation is charge transfer leading to a large excited state dipole.

7. Other Methods

The results reported above are obtained mostly using semi-empirical configuration interaction techniques. Very recently, techniques have been introduce that extend the range of *ab-initio* calculations into the large molecule regime. This is an ongoing development, but many of these models have great promise in estimating molecular electronic spectra. The review presented below is not intended to be comprehensive, but to give a snap-shot of the present activity in this area.

There are several aspects that might be considered in the construction of methods useful for calculating the electronic spectroscopy of very large systems:

1. General, and applicable to all systems, without special considerations to individual systems,
2. Accurate, and systematically improvable,
3. Applicable to the novice as well as the specialist, and
4. Consistent with the tenets of quantum mechanics.

The first of these items is an attempt to remove the need for "art" in the calculation of electronically excited states. The same theory should be useful for all systems. Complete CI's with large enough basis sets meet this criterion, but such calculations are not practical. Multi-reference CI is subject to the choice of the reference functions, and the level of CI performed from each of the reference configurations. MC-SCF depends on the choice of the active space. In all of this, *more* is better, but it is also more expensive.

The second item is generally a criticism of any of the semi-empirical SCF-CI based models. What you get is generally what you get, and one cannot be certain that the very criticism laid above, that more is better, is even true here. Semi-empirical calculations are generally inexpensive, and even "real time" in systems up to hundreds of atoms. They are useful insofar as they make suggestions, and offer explanations where there were none before. They are also excellent to point out the pitfalls that might be met when using more time consuming *ab initio* methodologies. Although controversial, I think they should not be relied on for "kilocalary accuracy" (or in this case wavenumber accuracy).

None of the methods in use today are generally applicable to the true novice. In general, if CIS on a closed-shell molecule, either at the semi-empirical or ab-initio level, does not reproduce the experimental spectrum, invoking experience is required. All of these methods should be used looking at the experimental results and asking the question "is this reasonable?" The use of quantum chemistry in examining electronic spectroscopy is still an iterative process.

There are several consistencies that should be maintained in application of quantum mechanics to problems such as these. We have examine the desirability of insuring, for example, the equivalence between the dipole-length and dipole-velocity formulations for the oscillator strength (equation 24)

$$f^{r}_{IJ} = f^{v}_{IJ}$$

and saw that this generally improved the calculated oscillator strengths using either formulation. The response theory methodologies if carefully formulated ensure this in the case of complete basis sets, but even falling short of this, experience shows that the RPA model yields far better results than either the CIS or CISD methods that are roughly the same amount of work.

Another consistency check is size extensivity, also mentioned above. This might be expected to become more important as the system gets larger. SCF and CIS are size extensive methods, but this, of course, does not always lead to good results. Full CI is also size extensive, but considering the basis sets often needed for good results is not a practical solution for any molecule of size. Rather this methodology today provides benchmarks against which to measure approximate theories of correlation. There does not appear to be any real evidence today that size extensivity is an important constraint in the calculation of spectroscopy, but one *feels* that such evidence will accumulate as larger molecular systems are investigated.

Size consistency generally refers to the inclusion of those configurations that might be required to successfully break bonds, pulling molecules apart into their fragments. Such configurations, when they are needed, are usually detected by contributing to the ground state wave function, and then they are included in a multi-reference treatment. Although important in examining molecules at stretched geometries, they are often important in transition metal complexes, representing other important atomic or ionic configurations.

7.1 THE CASPT2 MODEL

The CASPT2 method has mostly been proposed and examined by Bjorn Roos and his colleagues, and has proven useful in a wide variety of studies [55]. It is a combination of the Complete Active Space SCF (CASSCF) model, with the choice of a Restricted Space (RAS), and then Møller-Plesset second order perturbation theory (PT2).

This model is shown schematically in Figure 9. Some orbitals are chosen frozen, a second set, the RAS1 space, is allowed only a maximum number of holes (excitations). The RAS2 space has arbitrary occupation, set by the conditions on the RAS1 and RAS3 spaces,and the RAS3 space is allowed a maximum occupation. An MC-SCF on this space, which quickly grows quite large, is then used to develop the energy to second order in perturbation theory. The weights of the reference CAS states are checked for balance: they should be about the same for both the ground and excited states of interest. Occasionally there are intruder states, states within the manifold of interesting CAS states, that are poorly treated by perturbation theory. If so, they are removed using level shifting, or, if they are "interesting" they must be included in the active space and the calculation repeated, see figure 10.

In equation 33a, the shaded areas represent matrix elements that need to be calculated. Only the P-P block is diagonalized.

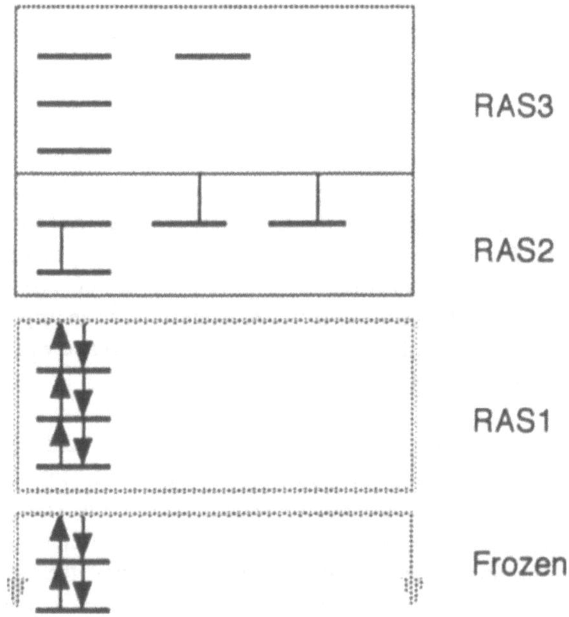

Figure 9. A schematic of the CASPT2 model.

H = 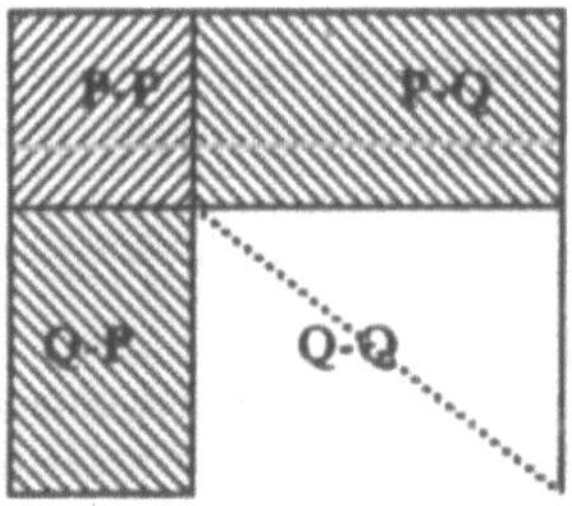

(33a)

PT2 $\Delta E(I) = \Sigma <\Psi_I|H|\Psi_Q> <\Psi_Q|H|\Psi_P> / [E^{CI}(I) - <\Psi_Q|H|\Psi_Q>]$ (33b)

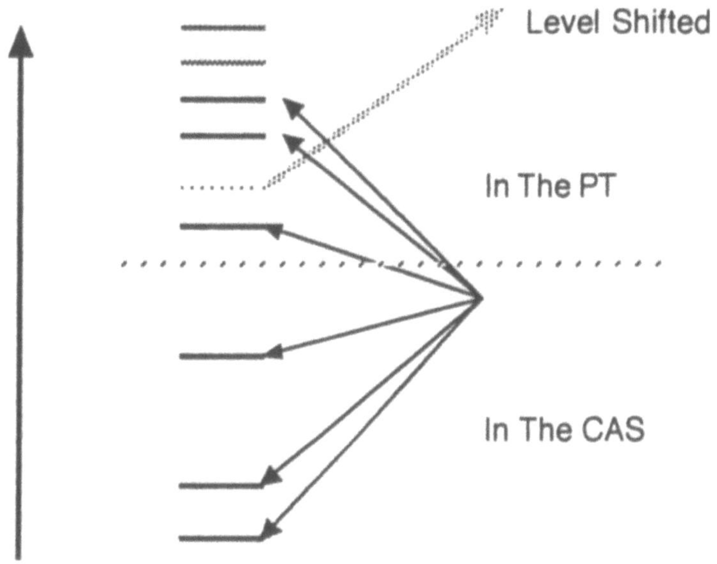

Figure 10. A Schematic of the PT2 refinement to CAS

Table XII: Calculations of the Spectrum of Free Base Porphin

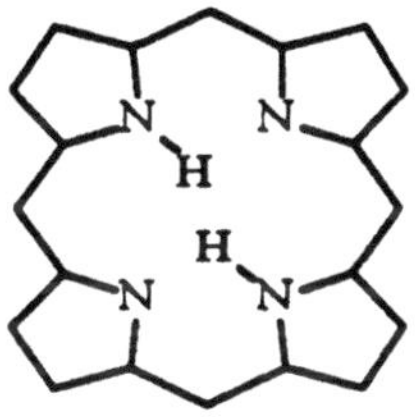

State	CAS[1] (4/4)	CAS[1] (best)	CASPT2[1] (best)	INDO[2] (RPA)	exp[3]
1^1B_{1u}	3.56	2.38[a]	1.70	1.71	1.98-2.02
1^1B_{2u}	3.82	3.20[b]	2.26	1.97	2.33-2.42
2^1B_{1u}	5.14	5.02[c]	2.91	2.99	3.17-3.33
2^1B_{2u}	5.17	5.05[c]	3.04	3.03	ΔE=0.03eV?

a (11/10) b (7/6) c (6/6)

1. Merchán, Orti, Roos,reference 55.
2. Baker and Zerner, Reference 54.

The CASPT2 results for free base porphin are given in Table XII, where they are compared with those obtained from the INDO/s model and experimental results.

The notation [n/m] in the table refers to [orbitals/electrons] in the CAS. Of particular interest in the spectrum of free base porphyrin is the split Soret region at about 3.0 eV. The origins of this split are controversial see, for example, reference [54], sometimes assigned to a vibrational level in the 2^1B_{1u} state, leaving the assignment of the 2^1B_{2u} state to higher energy, but most often the assignment is as in the table above, tracing back to the ^1E_u found in metal D_{4h} porphyrins [55]. This is discussed further below.

7.2 THE SAC-CI MODEL

The symmetry adapted couple-cluster CI model was introduced by Nakatsuji[56,57]. This method has also been applied to several large systems.

The starting point is a ground state wave function of the couple cluster doubles type

SAC $$|\Psi_0\rangle = O_s \exp(S)\,|0\rangle = \Theta_s|\psi_g\rangle \tag{34}$$

$$(H - E)|\Psi_0\rangle = (H - E)\,\Theta_s \exp(S)\,|0\rangle = 0 \tag{35}$$

where the cluster operator

$$S = S_1 + S_2 + \ldots \tag{36a}$$

$$S_1 = \Sigma\, C_{ia}\, S_{ia}^+ \tag{36b}$$

$$S_2 = \Sigma\, C_{ijab}\, S_{ijba}^+ \tag{36c}$$

is spin-symmetry adapted

$$S_{ia}^+ = (a_\alpha{}^+ i_\alpha + a_\beta{}^+ i_\beta)\sqrt{2} \text{ singlet} \tag{37a}$$

$$S_{ia}^+ = (a_\alpha{}^+ i_\beta) \text{ for triplet.} \tag{37b}$$

A CI is now generated using a symmetry adapted excitation operator $\Omega_K{}^+$ on the couple cluster reference wave function $|\psi_g\rangle$.

$$|\Phi_K\rangle = \mathcal{N}_K\, Q\, \Theta_S\, \Omega_K{}^+|\psi_g\rangle \text{ excited state 'k'} \tag{38}$$

where

$$Q = 1 - |\Psi_0\rangle\langle\Psi_0| \quad (39)$$

is a projection operator that keeps the excited state $|\Phi_k\rangle$ orthogonal to the reference state, and $\mathcal{N}_k$ is a normalizing factor. The treatment then proceeds with the assumption of a CI wave function

$$|\psi_k\rangle = |\Phi_k\rangle d_k + |\Psi_0\rangle d_0 \quad (40)$$

$d_0 = 0$ if the symmetry of the excited state and ground state are different, or if the system of equations (the CI matrix) is diagonalized. Most often further approximations are made:

$$|\Psi_0\rangle \sim O_s \exp(S)|0\rangle \sim (1 + S + 1/2!S^2)|0\rangle \quad (41)$$

$$S \sim S_1 + S_2 \quad (42a)$$

$$S_1 = \Sigma\, C_{ia}\ S_{ia}^+ \quad (42b)$$

$$S_2 = \Sigma\, C_{ijab}\, S_{ia}^+ S_{jb}^+ \quad (42c)$$

and

$$\Omega_k^+ = \Omega_{ia}^+ = S_{ia}^+ \quad (43)$$

and then the resulting equations are solved either by projection, in the usual couple-cluster fashion [56,58b],

$$\langle 0|(H-E)|\Psi_0\rangle = 0$$

$$\langle S_{ia}^+ 0|(H-E)|\Psi_0\rangle = 0$$

$$\langle S_{ia}^+ S_{jb}^+ 0|(H-E)|\Psi_0\rangle = 0 \quad (44)$$

or by matrix diagonalization,

$$H_{ij} = \langle \Phi_i | H | \Phi_j \rangle \quad (45a)$$

$$H\ d = d\ E \quad (45b)$$

The latter methodology has the advantage that it creates wavefunctions that are not only orthogonal, but also non-interacting through the Hamiltonian. This procedure, however, reintroduces unlinked clusters into the methodology, and the states are no longer size-extensive, and matrix diagonalization is a more time consuming technology. Nevertheless, the results seem to suggest that diagonalization is the better methodology.

Table XIII presents the results from SAC-CI calculations on free base porphin[58], where they are compared with those obtained from the CASPT2 results from Table XII. Note the reassignment of the second component of the Soret band ($^1B_2$) to the N band. This assignment is based on the fact that the SAC-CI calculation yields no band of calculated intensity in the region of the N band.

7.3 THE EOM-CC MODEL

The Equation of Motion Couple Cluster model (EOMCC) has been proposed by Bartlett and his co-workers[58]. Starting with the generalized Schrödinger equation

$$H\,|\Psi_\iota> = E_\iota\,|\Psi_\iota>$$

Table XIII: The Calculated Spectrum of Free Base Porphin (From Nakatsuji et al., reference 58. (in eV)

	SAC-CI	CASPT2	STEOM[a]	INDO/S CIS	INDO/S RPA	Exp.
Q_x	1.80 (1.1-3)	1.70	1.72	1.70(0.02)	1.71(0.02)	1.98,2.02
Q_y	2.25 (5.7-3)	2.26	2.61(0.02)	2.06 (0.03)	1.97(0.03)	2.33-2.42
B_x	3.59 (1.03)	2.91	3.66(1.03)	3.37(1.62)	2.99 (1.15)	3.10-3.33
B_y		3.04		3.52(2.41)	3.03 (1.23)	
	3.59 (0.0)		3.63 (0.0)	3.59 (0.0)	3.56 (0.0)	
	3.66 (0.0)			3.66 (0.0)	3.59 (0.0)	
N	3.79 (1.73)		3.77 (1.42)	4.09 (1.48)	3.93 (0.42)	3.65
.						
.						
L	4.28 (0.976)		4.24 (0.71)	4.42 (0.37)	4.38 (0.17)	4.25

a. from M. Nuijen and R. J. Bartlett, private communication.

one examines a Louivillian approach much as in the RPA theory described previously,

$$\Omega^R_\iota\,|\Psi_0> = |\Psi_\iota> \text{ (i.e., } \Omega^R_\iota = |\Psi_\iota><\Psi_0|) \tag{46a}$$

$$\mathcal{L}_\iota = [\, H, \Omega^R{}_\iota \,] = H\, \Omega^R{}_\iota - \Omega_\iota{}^R H \qquad (46b)$$

$$\mathcal{L}_\iota \; |\Psi_o\rangle = E_\iota\, \Omega^R{}_\iota\, |\Psi_o\rangle - E_o\, \Omega^R{}_\iota\, |\Psi_o\rangle \qquad (46c)$$
$$= \omega_\iota \; |\Psi_\iota\rangle$$

The starting point is a reference couple cluster wave function (as was the SAC-CI model)

$$|\Psi_o\rangle = e^T\, |\Phi_o\rangle \qquad (47)$$

with $|\Phi_o\rangle$ the Hartree-Fock SCF wavefunction (although it needn't be).

Then

$$T = T_1 + T_2 + T_3 + \ldots \qquad (48a)$$

$$T_1 = \Sigma\, t_{ai}\, a^+ i \qquad (48b)$$

$$T_2 = 1/2!\; \Sigma\, t_{abij}\, a^+ b^+ j i \quad , \text{ etc.} \qquad (48c)$$

As before, in the RPA model,

$$[\, H, \Omega^R{}_\iota \,]\, \{ e^T\, |\Phi_o\rangle \} = \omega_\iota\, \{\, \Omega^R{}_\iota \,\}\{ e^T\, |\Phi_o\rangle \} \qquad (49)$$

Multiplying the above by e^{-T}

$$e^{-T} [\, H, \Omega^R{}_\iota \,]\, e^T\, |\Phi_o\rangle = \omega_\iota\, \Omega^R{}_\iota\, |\Phi_o\rangle \qquad (50)$$

yields

$$[\, \underline{H}, \Omega^R{}_\iota \,]\; |\Phi_o\rangle = \omega_\iota\, \Omega^R{}_\iota\, |\Phi_o\rangle \qquad (51)$$

To obtain the above, $[\Omega^R{}_i \,, T\,] = 0$ constrains $\Omega^R{}_i$ to contain only excitation operators. Although developed as an equation of motion method, this constraint destroys the necessary conditions imposed on a true response model.

Note that

$$\underline{H} = e^{-T} H\, e^T \qquad (52)$$

is not Hermitian.

A big advantage to this method is the generation of a Schrödinger-like equation for the energy differences.

$$H\,\Omega^R = \Omega^R\,\omega \tag{53a}$$

$$\Omega^L\,H = \omega\,\Omega^L \tag{53b}$$

i.e., there is a left and a right sided Schrödinger equation. Only one need to be solved for the excitation energies, but other properties may require the solution of both. The resulting equations do bear a resemblance to those obtained from the SAC-CI model.

The spectrum of acetaldehyde is given in Table XIV, where it is compared with CIS. The CIS model fails, as might be expected, for the higher lying states here. A comparison of the results obtained for free base porphin from an approximate EOMCC model appears in Table XIII, where it is compared with the other models.

Table XIV: The Spectrum of Acetaldehyde[1], in eV. (CH_3CHO)

State	Exp[4]	EOMCC	Osc	$\langle r^2 \rangle$[2]	CIS[3]
1 $^1A'(V)$	-	-	-	57	-
1 $^1A''(V)$	4.28	4.32	6×10^{-5}	57	4.89
2 $^1A'(3s)$	6.82	6.79	0.015	67	*8.51*
3 $^1A'(3p)$	7.46	7.47	0.075	70	*9.22*
2 $^1A''(3p)$	-	7.75	0.011	71	*9.37*
4 $^1A'(3p)$	7.75	7.80	0.020	70	*9.30*

1. S.R. Gwaltney and R.J. Bartlett, Chem. Phys. Let. (1995) 241, 26: [5s3p2d/3s2p]
2. Bohrs[2]
3. K. N. Walzl, C. F. Koerting and A. Kuppermann, JCP (1987) 87, 3796; [6-311(2+)G]
4. M. B. Robin, Higher Excited States of Polyatomic Molecules, Vol 3, Academic Press, N.Y.(1985)

8. Some Further Considerations

Before concluding, it is perhaps desirable to make a few comments about basis sets.

In semi-empirical calculations it is generally assumed that a minimum basis can represent an exact basis. Assuming this, parameters are extracted from atomic and molecular spectroscopy that reproduce experiment. Insofar as the level of CI is sufficient to reproduce the experiment, and the experiment *can* be reproduced by the level of correlation theory used, this has proven to be a successful approach. It is unlikely, however, that models based on the zero differential overlap approximation (ZDO) will ever be able to successfully reproduce Rydberg spectra in a systematic fashion. Although several approaches have been made in this direction (59), the idea that large Rydberg orbitals have zero overlap with other Rydberg orbitals is not tenable. When Rydberg states are important, the present ZDO models will fail. This suggests that whenever one is examining spectra within 1-2 eV. of the ionization limit great care must be exercised [60].

In *ab-initio* calculations larger basis sets are usually required. Since excited states are generally larger than ground states, generally diffuse functions are required. It is rather interesting to note that smaller molecules require the larger basis set per atom. Larger systems have basis sets that are delocalized, and often reasonably modest basis, as DZP or DZP+ are sufficient to obtain reasonable predictions. This is actually suggesting that smaller basis sets will be reasonably successful in examining the spectroscopy of larger systems.

CIS generally is successful in describing the *nature* of the lowest lying excited states, and it is a level of theory that is size-extensive. These observations are some behind the success of the INDO/s semi-empirical model. If one wants greater accuracy, or greater reliability, than that available in the semiempirical models, than one must turn to *ab-initio* models. Since transition energies are generally greatly overestimated at the *ab-initio* CIS level, one needs to go to a rather high level of theory. There does not seem, at present, to be much compromise here. I have reviewed several appropriate techniques in this chapter, based on the configuration interaction method, as well as the RPA, CASPT2, SAC-CI and EOMCC models.

Acknowledgements

It has been interesting to compile all this information in one place. I am grateful for useful information gained in several informal conversations, especially from Mati Karelson, Björn Roos and from Rod Bartlett. Marshall Cory has read this manuscript and made very many useful comments. Our own studies have benefited throughout the years from, especially, the Office of Naval Research and from Eastman Kodak Company.

References

1] a) D.J. Rowe (1968) Rev. Mod. Phys. **40**, 153
b) A.E. Hansen, T.D. Bouman, (1979) Mol. Phys. **37**, 1713
2] P. Jørgensen, J. Simmons (1981) Second Quantization-Based Methods in Quantum Chemistry Academic Press N.Y.
3] a) J. Pople, D. Beveridge, P. Dobosh (1967) J. Chem. Phys., **47**, 2026
b) J. Pople, D. Santry, G. Segal (1965) J. Chem. Phys., **43**, 5129
4] a) J.E. Ridley, M.C. Zerner (1973) Theoret. Chim. Acta , **32**, 111
b) M.C. Zerner, G.H. Loew, R.F. Kirchner, U.T. Mueller-Westerhoff (1980) J. Am. Chem. Soc,**102**, 589 .
5] a) J.C. Culberson, P. Knappe, N. Rösch and M. C. Zerner (1987) Theoret. chim. Acta, **71**,21
b) M. Kotzian, N. Rösch, M.C. Zerner (1991) Intern. J. Quantum Chem., **S25**, 545
b) M. Kotzian, N. Rösch, M.C. Zerner (1992) Theoret. Chim. Acta, **81**, 201
6] J.D. Baker, M.C. Zerner (1991) J. Phys. Chem. **95**, 8614
7] A. Szabo, N.S. Ostlund (1982) Modern Quantum Chemistry, Macmillian Press, N.Y., 1st ed.
8] C.C.J. Roothaan (1960) Rev. Mod. Phys. **32**, 179
9] a) E. Davidson (1973) Chem. Phys. Lett. **21**, 565
b) C. Jackels, E. Davidson (1974) Intern. J. Quantum Chem., **8**, 707
c) R. Caballol, R. Gallifa, J. Ricra, R. Carbo (1974) Intern. J. Quantum Chem., **8**, 373
d) R. Carbo, L. Domingo, J. Gregori (1980) Intern. J. Quantum Chem. , **7**, 725
e) R. Mcweeny, R. Dickerson (1968) J. Chem. Phys. **49**, 4852
f) M. Guest, V. Sanders (1974) Mol. Phys. **28**, 819
g) J. Binkley, J. Pople, P. Dobosh (1974) Mol. Phys., **28**, 1423
h) K. Faegri, R. Manne (1976) Mol. Phys. **31**, 1037
i) H. Hsu, E. Davidson, R. Pitzer (1976) J. Chem. Phys., **65**, 609.
j) W.D. Edwards, M.C. Zerner (1987) Theoret. Chim. Acta **72**, 347
10] a) J.A. Pople, R.K. Nesbet (1954) J. Chem. Phys. **22**, 571
b) P.O. Lowdin (1955) Phys. Rev. **97**, 1509
c) P.O. Lowdin (1956) Advan. Phys. **5**, 161
11] a) A.T. Amos, G.G. Hall (1961) Proc. Roy. Soc. (London) **A263**, 483
b) A. T. Amos and L. C. Snyder (1964) J. Chem. Phys., **41**, 1773:
c) J.G.W. Pratt (1956) Phys. Rev. **102**, 1303
d) B. Sutcliffe (1963) J. Chem. Phys. **39**, 3322
12] R. Pauncz (1979) Spin Eigenfunctions Plenum Press, N.Y.
13]a) P.O. Lowdin (1959) Adv. Chem. Phys. **2**, 207
c)T. Yonezawa, H. Nakatsuji, T. Kawamura, H. Kato (1969) J. Chem. Phys., **51**, 669
d) H. Nakatsuji, H. Kato, T. Yonezawa (1969) J. Chem. Phys. **51**, 3175
e) P.O. Lowdin (1960) Rev. Mod. Phys. **32**, 328
f) A. Hardisson, J.E. Harriman (1967) J. Chem. Phys., **46**, 3639

g) J.E. Harriman (1964) J. Chem Phys. **40**, 2827
h) D. Phillips, J. Schug (1974) J. Chem. Phys., **61**, 1031
i) K.M. Sando, J.E. Harriman (1967) J. Chem. Phys., **47**, 18014)
14] A.D.Bacon and M.C.Zerner (1979) Theoret. chim. Acta, **53**, 21.
15] M.G. Cory, M.C. Zerner, in preperation.
16] J.B. Foresman, M. Head-Gordon, J.A. Pople, M.J. Frish (1992) J. Chem. Phys. **96**, 135
17] R.G.Parr (1963) Quantum Theory of Molecular Electronic Structure, Benjamin, Boston.
18] J.E. Ridley and M.C. Zerner (1976) Theor. Chim. Acta **42**, 223-236.
19]P. J. Hay and I. Shavitt (1974) **60**. 2865.
20] W.D. Edwards and M.C. Zerner (1983) Intern. J. Quantum Chem. **23**,1407.
21] J. Linderberg and N. Y. Öhrn (1973) Propagators in Quantum Chemistry, Academic Press, New York.
22] J. D. Baker and M. C. Zerner (1991) J. Phys. Chem. **95**, 8614--8619.
23] H. Eyring, J. Walter and G.E. Kimball (1944) Quantum Chemistry, John Wiley and Sons
24] M. C. Zerner (1991) "Semi Empirical Molecular Orbital Methods," M.C. Zerner, in Reviews of Computational Chemistry, Vol. 2, Ed. by K.B. Lipkowitz and D.B. Boyd, VCH Publishing, New York, , 313--366.
25] J. DelBene and H. H. Jaffe (1968) J. Chem. Phys., **48**, 1807: *ibid*, 4050.
26] a) J.C. Slater (1960) Quantum Theory of Atomic Structure , vol 1 and 2. McGraw-Hill New York .
b) J. C. Slater, J. B. Mann, T. M. Wilson and J. H. Wood (1969) Phys. Rev., **184**, 672.
27] F. L. Pilar (1968) Elementary Quantum Chemistry, McGraw Hill, New York
28] R. McWeeny, (1989) Methods in Molecular Quantum Mechanics, 2nd edition, Academic Press, London .
29]a) C.C.J.Roothaan (1951) Rev. Mod. Phys., **23**, 69.
b) G.G. Hall (1951) Proc. Roy. Soc. London, **A205**, 541.
c) C.C.J. Roothaan (1960) Rev. Mod Phys., **32**, 179.
30] M.A. Thompson and M.C. Zerner (1991) J. Am. Chem. Soc. **113**, 8210;
31] M.A. Thompson, J. Fajer and M.C. Zerner (1990) J. Phys. Chem.**94**, 3820.
32] M. C. Zerner (1995) "Intermediate Neglect of Differential Overlap Calculations on the Electronic Spectra of Transition Metal Complexes," in Metal Ligand Interactions: Structure and Reactivity, ed. N. Russo and D.R. Salahub, NATO ASI series, Vol. 474 Kluwer Academic Publishers, Dordrecht.
33] M. Cory and M.C. Zerner (1991) Chem. Rev.**91**, 813--822.
34] W.P. Anderson, W. D. Edwards and M. C. Zerner (1986) Inorg. Chem., **25**, 2728.
35] N. Mataga and K. Nishimoto (1957) Z. Physik Chem., **13**, 140. We use a variation of this suggested by K. Weiss, see reference 4a.
36] see, for example, J. A. Pople and D. L. Beveridge (1970) Approximate Molecular Orbital Theory, McGraw-Hill, New York.
37] R. Ake and M. Gouterman (1969) Theoret. Chim. Acta, **15**, 20: *ibid.* (1970) **17**, 408

38] W.D. Edwards , B. Weiner and M. C. Zerner, (1986) J. Am. Chem. Soc. , **108**, 2196.
39] W.D. Edwards, B. Weiner and M. C. Zerner, (1988) J. Phys. Chem., **92**, 6188.
40] S. P. Sinha (1971) Malays. Sci. **6**, 88.
41] C.K. Jørgensen and J.S. Brinen (1963) Mol. Phys. **6**, 629
42] W. L. Jorgensen, "Free energy calculations: A breakthrough for modeling organic chemistry in solution", Acc. Chem. Res. 1989, **22**, 184.
43] a)J. Blair, K. Krough-Jespersen and R. Levy (1989) J. Am. Chem. Soc. **111**, 6948
b) R. Levy, D. B. Kitchen, J. Blair, K. Krough-Jespersen (1990) **94**, 4470.
44] a) J. Gao (1994) J. Am. Chem. Soc. **116**, 9324.
b) G. Pearl, A. Broo and M. C. Zerner, work in progress.
c) M. A. Thompson and G. K. Schenter, J. Phys. Chem., 1995, **99**, 6374.
45] K.K. Stavrev, M.C.Zerner, T.J. Meyer, J. Am. Chem. Soc., (1995) **117**, 8684;
46] a) J. Tomasi and M. Persico,(1994) Chem. Rev., **94**, 2027.
b) A.Warshel and M. Levitt, (1976) J. Mol. Biol. **103**, 227.
c) J. Hylton, R. E. Cristoffersen and G. G. Hall, (1974) Chem. Phys. Lett. **26**, 501
d) J.-L. Rivail and D. Rinaldi, (1976) Chem. Phys., **18**, 233;
e) O Tapia and O. Goscinski, (1975) **29**, 1653;
f) B. T. Thole and P. Th. van Duijnen, (1980) Theoret,. chim. Acta, **55**, 307.
g) S. Miertus, E. Scrocco and J. Tomasi, (1981) Chem. Phys., **55**, 17.
h)M. M. Karelson, A. R. Katritzky, M. Szafran and M. C. Zerner (1989) J. Org. Chem., **61**, 6030.
i) M. M. Karelson, A. R. Katritzky, M. Szafran and M. C. Zerner (1990) J. Chem. Soc. Perkins Trans. **2**, , 161.
47] M.M. Karelson and M.C. Zerner (1992) J. Phys. Chem. **96**, 6949--6957.
47] See M. Karelson and G. Diercksen, this volume
48] C. J. F. Böttcher and P. Bordewijk (1978) Theory of Electron Polarization, 2nd. ed., vol II. Elsevier, Amsterdam, .
49 a)N. Rösch and M. C. Zerner (1994) J. Phys. Chem., **98**, 5817.
b)W. Liptay (1965) Z. Naturforsch., **112**, 9405.
c) G. Rauhut, T. Clark, T. Steinke (1993) J. Am. Chem. Soc., **115**, 9174.
d) F. M. Floris, J. Tomasi, J. L. Pascual-Ahuir (1991) J. Comput. Chem., **12**, 784.
50] a) R. A. Pierrotti (1977) Chem. Revs., **76**, 717.
51) L. Onsager (1936) J. Am. Chem. Soc. **58**, 1486.
52] a) E. G. McRae, (1957) J. Phys. Chem. , **61**, 562.
b) B. Linder, (1967) Adv. Chem. Phys. , **12**, 225
c) H. J. Kim and J. T. Hynes (1992) J. Am. Chem. Soc., **114**, 10508; *ibid*, 10528.
d) H. J. Kim, J. Chem. Phys.,*in press*:
53]a) P. E. Siegbahn, A. Heiberg, B.O. Roos, and B. Levy (1980) Phys. Scr. **21**, 323
b) B. O. Roos, P. R. Taylor and P. E. Siegbahn (1980) Chem. Phys. **48**, 157.
54] Baker and Zerner (1990) Chem. Phys. Let., **175**, 192
55] M. Merchán, E. Orti, B. O. Roos (1994) Chem. Phys. Let. **226**, 27
56] a)H. Nakatsuji , M. Komari, O. Kitao (1987) Chem. Phys. Lett. **142**, 447
b) H. Nakatsuji, O. Kitao and T. Yonezawa (1985) J. Chem. Phys. **83**, 723
57] H. Nakatsuji, J Y Hasegawa and M. Hada (1996) J. Chem. Phys. **104**, 2321

58] a) J. F. Stanton and R. J. Bartlett (1993) J. Chem. Phys. 7029
b) for a review of couple-cluster methods, see R. J. Bartlett and J. F. Stanton (1994) in Reviews of Computational Chem., Vol 5, ed. K. Lipkowitz and D. B. Boyd, VCH, New York
59]a) D. R. Salahub and C. Sandorfy (1971) Theoret. Chim. Acta **20**, 227.
b) W. Haque (1977) J. Chem. Phys. **67**, 3629.
60] M. B. Robin, Higher Excited States of Polyatomic Molecules, Vol 3, Academic Press, N.Y.(1985)

FINE AND HYPERFINE STRUCTURE

Spin properties of molecules

CHRISTEL M. MARIAN
Institute of Physical and Theoretical Chemistry
University of Bonn, Wegelerstraße 12
D-53115 Bonn, Germany

1. Introduction

For chemical purposes the electronic structure of molecules composed of light elements is well described by the Schrödinger equation employing a non-relativistic, spin-independent Hamiltonian. Spin degrees of freedom enter the scene only because electrons — being fermions — obey the Pauli principle, i.e., the total electronic wavefunction has to be antisymmetric with respect to the interchange of two electronic coordinates. At this level of approximation no mixing of spin and spatial symmetries occurs. Eigenfunctions of the electronic Schrödinger equation can simultaneously be made eigenfunctions of the total spin $\vec{S}^2$ with eigenvalues $s(s+1)$.Their spatial parts show characteristic transformation properties under symmetry operations, e.g. rotations and reflections of the nuclear frame. Therefore, we use to classify electronic states according to their spin multiplicity $2s+1$ and according to some irreducible representation of the molecular point group.

A closer look at spectra of molecular radicals reveals, however, that the picture of pure $\vec{S}^2$ eigenfunctions does not hold in detail. High-resolution spectra, even of light molecules, show that electronic spin multiplets are not properly degenerate. Actually, the sublevels are separated energetically by what is called the fine-structure splitting. Furthermore, electronic states like the first excited singlet state of O_2 are long-lived but the probability for a radiative decay to the triplet ground state is observed in night airglows. In addition, non-radiative transitions between states of different spin multiplicities may occur even under collision-free conditions. Predissociations, inter-system crossings, and perturbations in rovibrational ladders belong to the latter type of transitions. Summarizing, there are mechanisms lifting the degenaracy of electronic spin multiplets and coupling states of different spin multiplicities.

S. Wilson and G. H. F. Diercksen (eds.), Problem Solving in Computational Molecular Science, 291–351.

The most important of these, measured by the size of the splittings brought about, is spin-orbit coupling. It is a relativistic effect. Its importance increases with the nuclear charge of the molecular constituents to an extent that molecules containing heavy elements cannot be described even qualitatively correct if spin-orbit coupling is not taken into account. The second important type of interaction contributing to the fine-structure splitting is electronic spin-spin interaction. If there is a first-order spin-orbit splitting, spin-orbit coupling effects generally outweigh spin-spin interaction energies by at least an order-of-magnitude. In spatially non-degenerate electronic states of light molecules, on the other hand, as e.g. the $^3\Sigma^-$ ground state of O_2, first-order electronic spin-spin interaction often is of the same size as second-order spin-orbit interaction.

Classically spin-orbit and spin-spin coupling can be understood as interactions of the magnetic dipole moment associated with the spin of an electron with either the magnetic moment arising from the moving charge of an electron or the spin magnetic dipole moment of another electron. The magnetic dipole moment associated with the electronic spin amounts to

$$\vec{\mu}_e = g_e \mu_\mathrm{B} \vec{S} \tag{1}$$

where g_e is the free-electron spin g factor and $\mu_\mathrm{B} = e\hbar/2mc$ the so-called Bohr magneton. g_e has a value close to 2.0023; the deviation from the value of $g_e = 2$ which is predicted by Dirac theory is mainly caused by radiative corrections. They can be explained by quantum field theory. Similarly, the orbital magnetic dipole moment is given by

$$\vec{\mu}_L = g_L \mu_\mathrm{B} \vec{L} \tag{2}$$

where $g_L = 1$.

As we shall see later on, not all of the degeneracies of the fine-structure levels of an electronic state with a non-vanishing electronic angular momentum can be lifted by spin-orbit or spin-spin coupling. In a field-free surrounding each electronic state of a non rotating diatomic molecule with a projection of the electronic angular momentum on the internuclear axis $\Omega \neq 0$ remains doubly degenerate. Kramers has shown that these degeneracies result from the invariance properties of the Hamiltonian with respect to time inversion. These degenerate pairs of states are therefore often termed Kramers doublets or Kramers pairs.

Kramers pairs remain degenerate for all types of magnetic interaction Hamiltonians involving two angular momenta defined with respect to the molecule fixed coordinate system. Splitting of the parity sublevels requires the interaction of an internal angular momentum like $\vec{L}$ or $\vec{S}$ with an external magnetic field $\vec{H}$ or with the angular momentum $\vec{R}$ brought about by the rotation of the nuclear frame. In the absence of an external field, by far the largest contribution to the splitting of the Kramers doublets is brought about by the simultaneous action (double perturbation) of the electronic spin-orbit Hamiltonian and the rotational Hamiltonian. This allows an off-diagonal coupling of $+\Omega$ and $-\Omega$ sublevels and leads to an energetic splitting of their symmetric and antisymmetric combinations

$$\Psi^{(+)} = (1\sqrt{2})(\Psi(+\Omega) + \Psi(-\Omega)) \tag{3}$$

and

$$\Psi^{(-)} = (1/\sqrt{2})(\Psi(+\Omega) - \Psi(-\Omega)) \tag{4}$$

In diatomic molecules this splitting is denoted by Λ-doubling, Σ-doubling, or Ω-doubling, for Hund's cases a, b, and c respectively. It is especially pronounced in molecules with large rotational parameters, as e.g. hydrides, but it is usually much smaller than fine-structure splitting.

All the interactions of a nucleus with an electron, apart from their strong mutual Coulomb attraction, are classified as hyperfine interactions. One important class among them originates from a coupling of the electric quadrupole moment of a nucleus to the electric field gradient produced by the electronic charge distribution at the nuclear origin. The other important type of interaction arises when the nuclear magnetic dipole moment

$$\vec{\mu}_I = g_I \mu_N \vec{I} \tag{5}$$

associated with a non-zero nuclear spin $\vec{I}$ interacts with electronic magnetic moments. We shall only be concerned here with the nuclear magnetic interaction. The nuclear g factor g_I is of the same order-of-magnitude as g_e and is characteristic of a given nucleus. μ_N, the nuclear magneton, is smaller than the Bohr magneton by a factor of approximately 1836, the ratio of the protonic and electronic masses. Correspondingly, nuclear magnetic interactions are several orders-of-magnitude smaller than pure electronic spin-orbit or spin-spin interactions and often also smaller than Kramers type doubling.

In principle, all the magnetic moments associated with the molecular angular momenta

$\vec{L}$ electronic orbital angular momentum
$\vec{S}$ electronic spin
$\vec{I}$ nuclear spin
$\vec{R}$ rotation of the nuclear frame

can interact with each other and with an external magnetic field $\vec{H}$. We shall focus here on some of the typical internal interactions. All the fine and hyperfine splittings which shall be discussed in the following are so called zero-field splittings occuring in the absence of an external magnetic field.

A final introductory remark: during the course of the summerschool I was repeatedly asked to include more examples where the presented methods have been successfully applied to problems of chemical or spectroscopical interest. I refrain from doing so, because, to my knowledge, there are at least four excellent review articles on this subject which shall merely be referenced here [1, 2, 3, 4].

2. Microscopic coupling operators

2.1. HAMILTONIANS FOR SPIN-ORBIT COUPLING (SOC)

The complete one- and two-electron Breit-Pauli spin-orbit Hamiltonian is given by

$$\hat{\mathcal{H}}_{\mathrm{SO}}^{\mathrm{BP}} = \frac{e^2\hbar}{2m^2c^2}\left\{\sum_i\left(-\vec{\nabla}_i\left(\sum_\alpha \frac{Z_\alpha}{r_{i\alpha}}\right)\times\vec{p}_i\right)\cdot\vec{s}_i \right. \tag{6}$$

$$+ \sum_i\sum_{j\neq i}\left(\vec{\nabla}_i\left(\frac{1}{r_{ij}}\right)\times\vec{p}_i\right)\cdot\vec{s}_i \tag{7}$$

$$+ \sum_i\sum_{j\neq i}\left(\vec{\nabla}_j\left(\frac{1}{r_{ij}}\right)\times\vec{p}_j\right)\cdot\vec{s}_i \tag{8}$$

$$\left. + \sum_j\sum_{i\neq j}\left(\vec{\nabla}_i\left(\frac{1}{r_{ji}}\right)\times\vec{p}_i\right)\cdot\vec{s}_j\right\} \tag{9}$$

where we have chosen a form in which the connection to the Coulombic potential and the symmetry in the particle indices is apparent. Since spin-orbit coupling is a relativistic effect, the operators 6 – 9 are most elegantly derived by reducing the four-component Dirac-Coulomb-Breit operator to two components and separating spin-independent and spin-dependent terms. The original work of Pauli follows a different route though; it starts from the Schrödinger equation of a molecule in an external electric and magnetic field. The Breit-Pauli operator is then obtained by assuming that the scalar potential is purely Coulombic and the magnetic field arises from the electronic spin associated with a magnetic moment $\vec{\mu}_e = 2(e\hbar/2mc)\vec{S}$. From the four-component Dirac-Coulomb-Breit equation the terms 6 – 9 can be deduced without assuming external fields. A Foldy-Wouthuysen transformation of the electron-nuclear Coulomb attraction and collecting terms to order v^2/c^2 yields the one-electron part 6. Similarly the two-electron part 7 of the spin-same-orbit operator stems from the transformation of the two-electron Coulomb interaction. Terms 8 and 9, usually dubbed spin-other-orbit terms, have a different origin. They result, among other terms, from the reduction of the Breit interaction.

The Breit-Pauli spin-orbit Hamiltonian is found in many different forms in the literature; it is mostly written in a short form where spin-same- and spin-other-orbit parts of the two-electron Hamiltonian have been contracted to a single term, either as

$$\begin{aligned}\hat{\mathcal{H}}_{\rm SO}^{\rm BP} &= \frac{e^2\hbar}{2m^2c^2}\left\{\sum_i\sum_\alpha Z_\alpha \vec{s}_i(\frac{\vec{r}_{i\alpha}}{r_{i\alpha}^3}\times\vec{p}_i)\right. \\ &- \left.\sum_{i\neq j}(\frac{\vec{r}_{ij}}{r_{ij}^3}\times\vec{p}_i)\cdot(\vec{s}_i+2\vec{s}_j)\right\} \qquad (10)\end{aligned}$$

or

$$\begin{aligned}\hat{\mathcal{H}}_{\rm SO}^{\rm BP} &= \frac{e^2\hbar}{2m^2c^2}\left\{\sum_i\sum_\alpha Z_\alpha \vec{s}_i(\frac{\vec{r}_{i\alpha}}{r_{i\alpha}^3}\times\vec{p}_i)\right. \\ &- \left.\sum_{i\neq j}\frac{\vec{r}_{ij}}{r_{ij}^3}\times(\vec{p}_i-2\vec{p}_j)\cdot\vec{s}_i\right\} . \qquad (11)\end{aligned}$$

The Breit-Pauli spin-orbit operator has one major drawback. It contains implicitly terms coupling electronic states (with positive energy) and positronic states (in the negative energy continuum) and is thus unbounded from below. Safely it can be employed only in low-order perturbation theory. In molecules containing heavy elements spin-orbit interaction is, however, of the same size as the Coulomb interaction. A perturbation treatment then often is not sufficient and one would instead desire to treat both interactions variationally.

A more appropriate SOC Hamiltonian can be derived if electron-positron pair creation processes are excluded right from the beginning. The four-component no-pair equation can then be reduced to two components by means of a Douglas-Kroll transformation [5]. Without going into details (see e.g. [4]) we just display here the resulting two-component no-pair spin-orbit Hamiltonian

$$\begin{aligned}\hat{\mathcal{H}}_{\rm SO}^{+} &= e^2\hbar\left\{\sum_i\sum_\alpha Z_\alpha \frac{A_i}{E_i+mc^2}\,\vec{s}_i(\frac{\vec{r}_{i\alpha}}{r_{i\alpha}^3}\times\vec{p}_i)\,\frac{A_i}{E_i+mc^2}\right. \\ &- \left.\sum_{i\neq j}\frac{A_iA_j}{E_i+mc^2}\,(\frac{\vec{r}_{ij}}{r_{ij}^3}\times\vec{p}_i)\cdot(\vec{s}_i+2\vec{s}_j)\,\frac{A_iA_j}{E_i+mc^2}\right\} \qquad (12)\end{aligned}$$

with factors E_i and A_i or A_j defined as

$$E_i = \sqrt{p_i^2 c^2 + m^2 c^4} \tag{13}$$

$$A_i = \sqrt{\frac{E_i + mc^2}{2E_i}} \,. \tag{14}$$

As may be seen by comparing Eqs. 10 and 12 the no-pair spin-orbit Hamiltonian has exactly the same structure as the Breit-Pauli spin-orbit Hamiltonian. It differs from the Breit-Pauli operator only by kinematical factors which damp the $1/r_{i\alpha}^3$ and $1/r_{ij}^3$ singularities.

Since spin-orbit coupling is very important in heavy element compounds and the structure of the full microscopic Hamiltonians is rather complicated there have been several attempts to develop approximate Hamiltonians. One of the simplest approaches is to employ an effective one-electron, one-centre spin-orbit operator

$$\mathcal{H}_{\mathrm{SO}}^{\mathrm{eff}} = \frac{1}{2c^2} \sum_i \sum_\alpha \frac{Z_\alpha^{\mathrm{eff}}}{r_{\alpha i_\alpha}^3} \vec{l}_{\alpha i_\alpha} \vec{s}_{i_\alpha} \,. \tag{15}$$

In this operator the effective nuclear charge Z_α^{eff} is parametrized to fit experimental fine-structure splittings of one or several specific states on atom α.

For all-electron calculations, a more sophisticated effective one-electron spin-orbit Hamiltonian can be obtained by starting from the above given microscopic Hamiltonians 10 or 12 and introducing a mean-field approximation. This Fock-type operator can easily be derived from the expressions for matrix elements of singly excited determinants. For convenience it is therefore dealt with in Section 5.1.3.

Finally, several strategies have been pursued for evaluating spin-orbit matrix elements for wavefunctions obtained from pseudopotential or effective core-potential calculations. We shall not go into details here but rather refer to the original literature [6, 7, 8, 9, 10, 11] and a recent review article [12].

2.2. HAMILTONIANS FOR SPIN-SPIN COUPLING (SSC)

In Breit-Pauli theory the magnetic dipole interaction of two electronic spins is described by the following microscopic Hamiltonian:

$$\hat{\mathcal{H}}_{\mathrm{SS}}^{\mathrm{BP}} = \frac{e^2\hbar^2}{2m^2c^2} \left\{ \sum_{i \neq j} -\frac{8\pi}{3}\delta(\vec{r}_{ij})\vec{s}_i\vec{s}_j + \frac{\vec{s}_i\vec{s}_j}{r_{ij}^3} - 3\frac{(\vec{r}_{ij}\vec{s}_i)(\vec{r}_{ij}\vec{s}_j)}{r_{ij}^5} \right\} \tag{16}$$

The term containing the delta function describes the spin-magnetic interaction of two electrons located at the same point in space and does not have a classical analogue. The second and the third terms form the quantum mechanical equivalent to the classical dipolar interaction term. The interaction energy of two classical dipoles $\vec{\mu}_1$ and $\vec{\mu}_2$ with relative location $\vec{r}_{12} = \vec{r}_1 - \vec{r}_2$ is given by

$$E_{\text{int}} = \frac{\vec{\mu}_1 \cdot \vec{\mu}_2}{|\vec{r}_{12}|^3} - \frac{3(\vec{\mu}_1 \cdot \vec{r}_{12})(\vec{r}_{12} \cdot \vec{\mu}_2)}{|\vec{r}_{12}|^5} \tag{17}$$

Like the spin-other-orbit terms 8 and 9, the Breit-Pauli spin-spin Hamiltonian 16 originates from a Foldy-Wouthuysen transformation of the Breit interaction.

2.3. KRAMERS-TYPE DOUBLING: THE NUCLEAR ROTATION HAMILTONIAN

As shown already in the early days of molecular quantum theory a splitting of the Kramers pairs in a rotating diatomic molecule is brought about by the coupling of the rotational angular momentum with the electronic spin and orbital angular momenta. In our formulation of the Hamiltonian we restrict ourselves to basis functions appropriate to Hund's coupling case (a) for which the projection of the electronic angular momenta on the internuclear axis is well defined. This choice of basis is not a severe restriction though. Also molecules which are closer to Hund's coupling cases (b) and (c) can be treated in this way.

The fact that the total angular momentum $\vec{\mathcal{J}}$, apart from the nuclear spin, is equal to $\vec{\mathcal{R}} + \vec{\mathcal{L}} + \vec{\mathcal{S}}$ may be used to express the nuclear rotation Hamiltonian as

$$\begin{aligned} \hat{\mathcal{H}}_{\text{rot}} &= B\{\vec{\mathcal{J}} - \vec{\mathcal{L}} - \vec{\mathcal{S}}\}^2 && (18)\\ &= B\{J(J+1) + S(S+1) - 2\Omega\Sigma - \Lambda^2\} && (19)\\ &- 2B(\mathcal{J}_x\mathcal{S}_x + \mathcal{J}_y\mathcal{S}_y) && (20)\\ &+ 2B(\mathcal{L}_x\mathcal{S}_x + \mathcal{L}_y\mathcal{S}_y) && (21)\\ &- 2B(\mathcal{J}_x\mathcal{L}_x + \mathcal{J}_y\mathcal{L}_y) && (22)\\ &+ B(\mathcal{L}_x^2 + \mathcal{L}_y^2)\ . && (23) \end{aligned}$$

Herein, the first term 19 gives the spacing of the rotational levels. Term 21 represents an additional off-diagonal spin-orbit operator brought about by nuclear rotation. Like the usual spin-orbit Hamiltonian, this term affects the Kramers sublevels equally. The rotational distortion term 20 couples different fine-structure levels of a given electronic state. Its presence leads to a decoupling of the electronic spin from the internuclear axis in high rotational quantum numbers and a transition to Hund's coupling case (b). The electronic angular momentum operators $\mathcal{L}_x$ and $\mathcal{L}_y$ in the Coriolis coupling term 22 mix in electronic states with Λ values differing by ± 1. The last term of the nuclear rotation operator 23 is a two-electron operator which brings about only a shift of the diagonal energies and is usually neglected.

In practical applications the $\vec{\mathcal{L}}$ operator occurring in these formulae is replaced by a sum of one-particle operators $\sum_i \ell_i$. In the computation of the terms containing $\vec{S}$ usually the total spin is employed in this context. This does, of course, not apply to the usual spin-orbit Hamiltonian for which we always choose one of its microscopic forms.

2.4. HAMILTONIANS FOR NUCLEAR MAGNETIC HYPERFINE INTERACTIONS (HFI)

An expression for the Hamiltonian describing the interaction of the nuclear and electronic magnetic momenta can be derived from the Dirac equation in the presence of an external magnetic field [13]. The nucleus is assumed to be a point dipole with magnetic moment μ_I Eq. 5 associated with a non-zero nuclear spin momentum I. Hence, the vector potential at point $\vec{r}_{i\alpha}$ is

$$\vec{A} = \frac{\vec{\mu}_I \times \vec{r}_{i\alpha}}{r_{i\alpha}^3} . \tag{24}$$

When reducing the Dirac equation to two components for this type of external magnetic field, special care has to be taken. At $r_{i\alpha} = 0$, i.e. at the location of the nucleus, the Coulomb potential becomes infinite and the energy of an s electron which has a non-vanishing amplitude at this point might not be small compared to mc^2. Consideration of the latter fact leads to the Fermi contact interaction

$$\hat{\mathcal{H}}_{\text{Fc}} = g_e \mu_B \mu_N \frac{8\pi}{3} \sum_\alpha \sum_i g_{N_\alpha} \vec{I}_\alpha \vec{s}_i \delta(\vec{r}_{i\alpha}) \tag{25}$$

which is non-zero only for s electrons because of the Dirac δ function. The Fermi contact term does not have any classical analogue.

The expression for dipolar interaction of the nuclear and electronic spin magnetic moments is completely equivalent to $\mathcal{H}_{\text{SS}}^{\text{BP}}$ Eq. 16. The microscopic Hamiltonian is given by:

$$\hat{\mathcal{H}}_{\text{IS}} = g_e \mu_B \mu_N \sum_\alpha \sum_i g_{N_\alpha} \left\{ -\frac{\vec{I}_\alpha \vec{s}_i}{r_{i\alpha}^3} + 3 \frac{(\vec{r}_{i\alpha} \vec{I}_\alpha)(\vec{r}_{i\alpha} \vec{s}_i)}{r_{i\alpha}^5} \right\} \tag{26}$$

Finally, a nuclear spin - electronic orbit term is obtained. In most cases only the one-electron term

$$\hat{\mathcal{H}}_{\mathrm{LI}} = g_e\mu_B\mu_N \sum_\alpha \sum_i g_{N_\alpha} \frac{\vec{I}_\alpha(\vec{r}_{i\alpha} \times \vec{p}_i)}{r_{i\alpha}^3} \tag{27}$$

is considered. The corresponding two-electron operator employing $\vec{r}_{ij}$ has a much smaller effect and is usually neglected.

These three terms together form the hyperfine interaction Hamiltonian

$$\begin{aligned} \hat{\mathcal{H}}_{\mathrm{hfs}} = g_e\mu_B\mu_N & \sum_\alpha \sum_i g_{N_\alpha} \left\{ \frac{8\pi}{3} \vec{I}_\alpha \vec{s}_i \delta(\vec{r}_{i\alpha}) \right. \\ + & \left. \frac{\vec{I}_\alpha\left((\vec{r}_{i\alpha} \times \vec{p}_i) - \vec{s}_i\right)}{r_{i\alpha}^3} + 3\frac{(\vec{r}_{i\alpha}\vec{I}_\alpha)(\vec{r}_{i\alpha}\vec{s}_i)}{r_{i\alpha}^5} \right\} \end{aligned} \tag{28}$$

3. Tensor properties of the magnetic interaction terms

An irreducible tensor operator $T^{(k)}$ of rank k has $2k+1$ components $T_q^{(k)}$ with q ranging from $-k$ to k. They form a basis for the irreducible representation $D^{(k)}$ of the full rotation group in three dimensions, $O^+(3)$. This means that the components of $T^{(k)}$ are transformed among themselves upon a coordinate rotation.

Being a part of the Hamiltonian, $\hat{\mathcal{H}}_{\mathrm{SO}}$, $\hat{\mathcal{H}}_{\mathrm{SS}}$, $\hat{\mathcal{H}}_{\mathrm{rot}}$, and $\hat{\mathcal{H}}_{\mathrm{hfs}}$ are, of course, scalar operators, i.e., tensor operators of rank 0 which transform totally symmetric under orthogonal coordinate transformations. On the other hand, they can be thought of as compound tensor operators consisting of vector (first rank tensor) operators and matrix (second rank tensor) operators. As we shall see later, we can make use of this tensorial structure when it comes to the question which matrix elements of the magnetic interaction Hamiltonians are zero by symmetry.

Let us have a look at spin-orbit coupling first. For symmetry considerations it is useful to note that each of the terms in one of the microscopic spin-orbit Hamiltonians can be written in form of a scalar product between an appropriately chosen spatial angular momentum $\vec{\mathcal{L}}$ and a spin angular momentum $\vec{\mathcal{S}}$

$$\hat{\mathcal{H}}_{\mathrm{SO}} = A_{\mathrm{SO}}(r)\vec{\mathcal{L}} \cdot \vec{\mathcal{S}} \tag{29}$$

where $\vec{\mathcal{L}}$ stands for either the usual orbital angular momentum $(\vec{r}_{i\alpha} \times \vec{p}_i)$ or an equivalent momentum $(\vec{r}_{ij} \times \vec{p}_i)$ describing the motion of electron i around electron j, r is the corresponding radial coordinate, and $\vec{\mathcal{S}}$ refers to either particle i or j. Similar considerations apply to the nuclear spin-orbit and the Fermi contact terms of the hyperfine Hamiltonian. With $\mathcal{I}$ representing the nuclear spin magnetic moment they may be written as

$$\hat{\mathcal{H}}_{\mathrm{IL}} = a_{\mathrm{orb}}(r)\vec{\mathcal{L}}\cdot\vec{\mathcal{I}} \tag{30}$$

$$\hat{\mathcal{H}}_{\mathrm{Fc}} = a_{\mathrm{iso}}(r)\vec{\mathcal{I}}\cdot\vec{\mathcal{S}}\ . \tag{31}$$

$\vec{\mathcal{L}}$, $\vec{\mathcal{S}}$, and $\vec{\mathcal{I}}$ are irreducible tensor operators of rank 1. Any first rank tensor operator $\vec{\mathcal{Q}}$ (a vector operator) has three components $\mathcal{Q}_0$, and $\mathcal{Q}_{\pm1}$. Operating with these components on an eigenfunction $|qm\rangle$ of the corresponding operator $\vec{\mathcal{Q}}^2$ yields the following results:

$$\mathcal{Q}_0|qm_q\rangle = m_q|qm_q\rangle \tag{32}$$

$$\mathcal{Q}_{\pm1}|qm_q\rangle = \mp\sqrt{\frac{1}{2}[q(q+1)-(m_q\pm1)m_q]}\ \ |qm_q\pm1\rangle \tag{33}$$

It is seen that $\mathcal{Q}_0$ conserves the projection of q on the z axis, $\mathcal{Q}_{+1}$ steps it up by one unit, and $\mathcal{Q}_{-1}$ steps m_q down by one unit.

In order to form a scalar from two tensor operators $\mathcal{P}$ and $\mathcal{Q}$, a step-up operator component of $\mathcal{P}$ has to be combined with a step-down operator component of $\mathcal{Q}$ and vice versa. Two different notations are found in the literature for these tensor products differing only in normalization constants and phase factors. The properly normalized product of two tensor operators forming a scalar operator reads [14]

$$\{P^{(k)}\otimes Q^{(k)}\}_0^{(0)} = \sum_{q=-1}^{+1}(-1)^{k-q}(2k+1)^{1/2}P^{(1)}_{-q}Q^{(1)}_q\ . \tag{34}$$

Since were are mainly interested in the symmetry properties of the compound tensor operator, we use here the simplified expression

$$(P^{(k)}\cdot Q^{(k)}) = \sum_{q=-1}^{+1}(-1)^{q}P^{(1)}_{-q}Q^{(1)}_q\ . \tag{35}$$

In terms of its tensor components the scalar product of $\vec{\mathcal{L}}$ and $\vec{\mathcal{S}}$ then reads

$$\hat{\mathcal{H}}_{\mathrm{SO}} = f(r)\vec{\mathcal{L}}\cdot\vec{\mathcal{S}} \tag{36}$$

$$= f(r)(\mathcal{L}_0\mathcal{S}_0 - \mathcal{L}_{+1}\mathcal{S}_{-1} - \mathcal{L}_{-1}\mathcal{S}_{+1}) \tag{37}$$

$$= f(r)(\mathcal{L}_x\mathcal{S}_x + \mathcal{L}_y\mathcal{S}_y + \mathcal{L}_z\mathcal{S}_z) \tag{38}$$

where the Cartesian and the spherical tensor components are related by

$$\mathcal{S}_{+1} = \frac{-\mathcal{S}_x - i\mathcal{S}_y}{\sqrt{2}} \qquad \mathcal{S}_0 = \mathcal{S}_z \qquad \mathcal{S}_{-1} = \frac{\mathcal{S}_x - i\mathcal{S}_y}{\sqrt{2}} \tag{39}$$

or reversely

$$\mathcal{S}_x = \frac{\mathcal{S}_{-1} - \mathcal{S}_{+1}}{\sqrt{2}} \qquad \mathcal{S}_z = \mathcal{S}_0 \qquad \mathcal{S}_y = \frac{i(\mathcal{S}_{-1} + \mathcal{S}_{+1})}{\sqrt{2}} \tag{40}$$

and the same holds true for $\vec{\mathcal{L}}$ and $\vec{\mathcal{I}}$. Phase conventions have been chosen to be consistent with those of Condon and Shortley [15].

The dipolar spin-spin coupling operators are scalar operators of the form $P^{(1)} \cdot T^{(2)} \cdot Q^{(1)}$. The tensorial structure of $\hat{\mathcal{H}}_{\mathrm{SS}}$ becomes more apparent if we rewrite Eq. 16

$$\hat{\mathcal{H}}_{\mathrm{SS}}^{\mathrm{BP}} = \frac{e^2\hbar^2}{2m^2c^2} \sum_{i \neq j} \left\{ -\frac{8\pi}{3}\delta(\vec{r}_{ij})\vec{s}_i\vec{s}_j + \frac{\vec{s}_i\vec{s}_j}{r_{ij}^3} - \frac{\vec{s}_i(3\vec{r}_{ij} \otimes \vec{r}_{ij})^{(2)}\vec{s}_j)}{r_{ij}^5} \right\} \tag{41}$$

$$= \frac{e^2\hbar^2}{2m^2c^2} \sum_{i \neq j} \left\{ \frac{8\pi}{3}\delta(\vec{r}_{ij})\vec{s}_i\vec{s}_j + \left(\vec{s}_i \mathcal{D}_{ij}^{(2)} \vec{s}_j\right) \right\} . \tag{42}$$

The first term is a tensor of rank zero involving only spin variables. It does not contribute to the multiplet splitting an electronic state but yields only a (small) overall shift of the energy and is henceforth neglected. $\mathcal{D}_{ij}^{(2)}$ is a traceless (irreducible) second rank tensor operator which reads in form of its Cartesian components

$$\mathcal{D}_{ij}^{(2)} = \frac{1}{r_{ij}^5} \begin{pmatrix} r_{ij}^2 - 3x_{ij}^2 & -3x_{ij}y_{ij} & -3x_{ij}z_{ij} \\ -3x_{ij}y_{ij} & r_{ij}^2 - 3y_{ij}^2 & -3y_{ij}z_{ij} \\ -3x_{ij}z_{ij} & -3y_{ij}z_{ij} & r_{ij}^2 - 3z_{ij}^2 \end{pmatrix} \tag{43}$$

This is exactly the same structure as used for the phenomenological electronic spin-spin operator

$$\hat{\mathcal{H}}_{\mathrm{SS}} = \vec{\mathcal{S}} \mathcal{D}_{\mathrm{SS}}^{(2)} \vec{\mathcal{S}} \tag{44}$$

An overall coupling to a scalar poses again certain conditions on the combination of the tensor components. Up to normalization, all products

$$P_p^{(1)} T_t^{(2)} Q_q^{(1)} \qquad \text{which fulfill} \qquad p + t + q = 0 \tag{45}$$

are obtained with p and q ranging from -1 to 1 and t ranging from -2 to 2. Actual prefactors of the terms depend on the order in which the three tensor operators are coupled. In the electronic spin-spin Hamiltonian one first couples the two spin operators to a second rank tensor operator which is then combined with $\mathcal{D}_{ij}^{(2)}$ to form a scalar, i.e.,

$$\{\{\vec{s}_i \otimes \vec{s}_j\}^{(2)} \otimes \mathcal{D}_{ij}^{(2)}\}^{(0)} \tag{46}$$

The two terms in $\hat{\mathcal{H}}_{\text{hfs}}$ depending on $\vec{I}$ and $\vec{s}$ only, i.e. the dipolar interaction and Fermi contact terms, are often put together. For a given nucleus α a microscopic magnetic hyperfine tensor $\mathcal{A}_{\text{hfs}}^{(2)}(i,\alpha)$ is defined by

$$\hat{\mathcal{H}}_{\text{IS}}(\alpha) = \sum_i \vec{\mathcal{I}}_\alpha \mathcal{A}_{\text{hfs}}^{(2)}(i,\alpha)\vec{s}_i \tag{47}$$

in which

$$\mathcal{A}_{\text{hfs}}^{(2)}(i,\alpha) = g_e\mu_B\mu_N g_{N_\alpha} \left[\begin{pmatrix} \frac{8\pi}{3}\delta(\vec{r}_{i\alpha}) & 0 & 0 \\ 0 & \frac{8\pi}{3}\delta(\vec{r}_{i\alpha}) & 0 \\ 0 & 0 & \frac{8\pi}{3}\delta(\vec{r}_{i\alpha}) \end{pmatrix} + \frac{1}{r_{i\alpha}^5} \begin{pmatrix} 3x_{i\alpha}^2 - r_{i\alpha}^2 & 3x_{i\alpha}y_{i\alpha} & 3x_{i\alpha}z_{i\alpha} \\ 3x_{i\alpha}y_{i\alpha} & 3y_{i\alpha}^2 - r_{i\alpha}^2 & 3y_{i\alpha}z_{i\alpha} \\ 3x_{i\alpha}z_{i\alpha} & 3y_{i\alpha}z_{i\alpha} & 3z_{i\alpha}^2 - r_{i\alpha}^2 \end{pmatrix} \right] \tag{48}$$

Unlike $\mathcal{D}_{ij}^{(2)}$ in Eq. 42, $\mathcal{A}_{\text{hfs}}^{(2)}(i,\alpha)$ is a reducible second rank tensor operator composed of an isotropic part (a diagonal matrix) and an anisotropic (dipolar) part, an irreducible tensor of rank 2.

The related phenomenological operators are given by

$$\hat{\mathcal{H}}_{\text{Fc+IS}}(\alpha) = \vec{\mathcal{I}}_\alpha a_{\text{iso}}(\alpha)\vec{\mathcal{S}} + \vec{\mathcal{I}}_\alpha a_{\text{dip}}^{(2)}(\alpha)\vec{\mathcal{S}} \tag{49}$$

In this case two ways of coupling the operators of the dipolar terms are of common usage.

$$\{\{\vec{\mathcal{I}} \otimes \vec{\mathcal{S}}\}^{(2)} \otimes a_{\text{dip}}^{(2)}\}^{(0)} \tag{50}$$

applies to the case of a weak or zero external magnetic field while

$$\{\{\vec{\mathcal{I}} \otimes \{\vec{\mathcal{S}} \otimes a_{\text{dip}}^{(2)}\}^{(1)}\}^{(0)} \tag{51}$$

is used in the strong-field limit. These two coupling schemes can be converted into one another with the help of 6j symbols. For further details see e.g. [14, 16].

4. How to exploit symmetry?

All the terms in the molecular Hamiltonian are scalar operators. They transform in any molecular point group as the totally symmetric irreducible representation (irrep). Thus, for a matrix element of the spin-orbit operator, say, to be different from zero it is in principle sufficient that the direct products of the irreps of space and spin functions on both sides of the matrix element be equal.

This brings us directly to the question of how spin functions transform under the symmetry operations of a molecular point group. The derivation of their transformation properties is lengthy but interesting. In Appendix A the general connection between the representations of rotations in three-dimensional coordinate space and the special unitary group of order 2 is shown up. For the particular case of C_{2v}, and the symmetry properties of singlet, doublet, and triplet spin functions are derived there explicitly. In this section we just make use of this knowledge without further discussion.

The fact that the magnetic interaction Hamiltoninans are compound tensor operators can be exploited to derive more specific selection rules than the one given above. Furthermore, as we shall see later, the number of matrix elements between multiplet components that actually have to be computed can be considerably reduced by use of the Wigner-Eckart theorem.

A final introductory remark: in the examples presented below I frequently use terms like first-order splittings, second-order couplings etc. These terms are borrowed from the context of perturbation theory. In Section 4 a first-order effect is meant strictly to denote the result of a coupling of the various components of a single degenerate electronic state by some magnetic interaction operator. A second-order effect includes also the direct interaction with other electronic states. In molecular calculations it is not always easy to apply these terms in their strict senses. We shall return to this discussion later on (Section 5).

4.1. A WORD OF CAUTION

In most instances phenomenological operators can be used for symmetry considerations. However, this should be done with great care. Just a word of caution: do not oversimplify!

As an example let us look at the spin-orbit coupling between a $^3\Sigma^-$ and a $^1\Sigma^+$ state of a linear molecule, $\langle ^3\Sigma_0^- | \mathcal{H}_{SOz} | ^1\Sigma_0^+ \rangle$, which is symmetry allowed. To see this, we may employ the phenomenological operator 29. $\mathcal{L}_0$ transforms like $\Sigma^-_{(g)}$ in the symmetry group of a linear molecule and thus allows an interaction of Σ^- with Σ^+ states. The $m_s = 0$ component of a triplet has Σ^- symmetry, $\mathcal{S}_0$ transforms like $\Sigma^-_{(g)}$, and singlets are totally symmetric, resulting again in a symmetry-allowed interaction. Details of the symmetry analysis are dealt with in Section 4.3.

If we, on the other hand, tried to evaluate the spin-orbit coupling matrix element using explicitly the phenomenological operator 29, we would obtain $A_{SO}\langle\Sigma^-|\mathcal{L}_0|\Sigma^+\rangle\langle T^0|\mathcal{S}_0|S^0\rangle = 0$, because $\mathcal{L}_0$ operating on a Σ state gives 0 and the same holds true for the action of $\mathcal{S}_0$ on a state with $m_s = 0$.

The resolution of this discrepancy is closely related to another question. Let us try to understand how an operator like $\vec{\mathcal{S}}$, when combined with $\vec{\mathcal{L}}$, is capable of coupling electronic states of different multiplicities while, according to Eqs. 32 and 33, $\vec{\mathcal{S}}$ as a first rank tensor operator is only able to change the m_S quantum number of a state, not its S value.

For that purpose we will have to leave the global $\vec{\mathcal{L}} \cdot \vec{\mathcal{S}}$ expression and rather employ a sum of two one-particle operators, i.e., $\ell(1)s(1) + \ell(2)s(2)$. Consider the spin-orbit coupling of a $^1\Pi$ and a $^3\Pi$ two-electron state of a linear molecule. As we will see later on, it is the $m_S = 0$ level of the triplet state which couples to the singlet state in this case and the coupling is brought about by the z component of the spin-orbit coupling operator. The wave functions can be written as:

$$\begin{aligned} \langle{}^1\Pi| &= \langle(\sigma(1)\pi(2) + \pi(1)\sigma(2))(\alpha(1)\beta(2) - \beta(1)\alpha(2))| \\ &= \langle\sigma\bar{\pi} - \bar{\sigma}\pi + \pi\bar{\sigma} - \bar{\pi}\sigma| \end{aligned} \tag{52}$$

$$\begin{aligned} |{}^3\Pi\rangle &= |(\sigma(1)\pi(2) - \pi(1)\sigma(2))(\alpha(1)\beta(2) + \beta(1)\alpha(2))\rangle \\ &= |\sigma\bar{\pi} + \bar{\sigma}\pi - \pi\bar{\sigma} - \bar{\pi}\sigma\rangle \end{aligned} \tag{53}$$

The (one-electron part of the) spin-orbit matrix element is given by:

$$\begin{aligned} \langle{}^1\Pi|\hat{\mathcal{H}}_{SO}|{}^3\Pi\rangle &= \langle\sigma\bar{\pi} - \bar{\sigma}\pi + \pi\bar{\sigma} - \bar{\pi}\sigma|\ell_0(1)s_0(1)|\sigma\bar{\pi} + \bar{\sigma}\pi - \pi\bar{\sigma} - \bar{\pi}\sigma\rangle \\ &+ \langle\sigma\bar{\pi} - \bar{\sigma}\pi + \pi\bar{\sigma} - \bar{\pi}\sigma|\ell_0(2)s_0(2)|\sigma\bar{\pi} + \bar{\sigma}\pi - \pi\bar{\sigma} - \bar{\pi}\sigma\rangle \end{aligned} \tag{54}$$

At this stage, all the way down at the integral level, we can finally exploit the fact that ℓ and s act on different coordinates: We can split up the integrals into space- and spin-dependent parts. From the sum 54 then only four terms are different from zero. This is the case because s_0 does not change α into β spin or vice versa and because of the fact that $\langle\sigma|\ell_0|\sigma\rangle = 0$. The remaining terms are

$$\begin{aligned} \langle\sigma\bar{\pi}|\ell_0(2)s_0(2)|\sigma\bar{\pi}\rangle &= -\frac{1}{2}\langle\pi|\ell_0|\pi\rangle \\ -\langle\bar{\sigma}\pi|\ell_0(2)s_0(2)|\bar{\sigma}\pi\rangle &= -\frac{1}{2}\langle\pi|\ell_0|\pi\rangle \\ -\langle\pi\bar{\sigma}|\ell_0(1)s_0(1)|\pi\bar{\sigma}\rangle &= -\frac{1}{2}\langle\pi|\ell_0|\pi\rangle \\ \langle\bar{\pi}\sigma|\ell_0(1)s_0(1)|\bar{\pi}\sigma\rangle &= -\frac{1}{2}\langle\pi|\ell_0|\pi\rangle \end{aligned}$$

which finally yields (at this level of approximation)

$$\langle {}^1\Pi|\hat{\mathcal{H}}_{SO}|{}^3\Pi\rangle \quad = \quad -2\langle\pi|\ell_0|\pi\rangle$$

It is thus the change of sign of individual integrals due to spin integration combined with the effect that the space parts belonging to the reverse spin terms integrate to zero what makes the matrix element survive. In other cases where s_{+1} or s_{-1} are involved spin flips of invidual orbitals occur in addition. It is because of these individual spin flips that within the Russel-Saunders coupling scheme not only the change of m_S is restricted to $\Delta m_S = 0, \pm 1$ but also the spin quantum number S may change only by one unit, i.e., $\Delta S = 0, \pm 1$. Likewise, electronic spin-spin coupling, as an operator composed of second rank tensors can change $\Delta m_S = 0, \pm 1, \pm 2$ and $\Delta S = 0, \pm 1, \pm 2$. There are some restrictions for low multiplicites though. A coupling of states with equal multiplicities, i.e., $\Delta S = 0$, requires $S \geq 1/2$ for spin-orbit interaction and $S \geq 1$ for electronic spin-spin interaction to become effective. The origin of these restrictions will become clear in the context of atomic selection rules.

4.2. LOW SYMMETRY: NON-LINEAR MOLECULES

For non-linear molecules no component of the orbital angular momentum is conserved. In the Russel-Saunders coupling scheme we still assume, however, that s is a (fairly) good quantum number and that we can expand the true electronic wavefunction in terms of $\vec{\mathcal{S}}^2$ eigenfunctions. Given two electronic states we should like to know whether electronic spin-orbit or spin-spin coupling is symmetry-allowed and which of the spin sublevels interact.

Let us work this out for an example: we are interested in the fine-structure splitting of the triplet electronic ground state of methylene (CH_2). In its ground state CH_2 has a bent symmetric equilibrium configuration. The molecule is chosen to lie in the yz plane with z as symmetry axis.

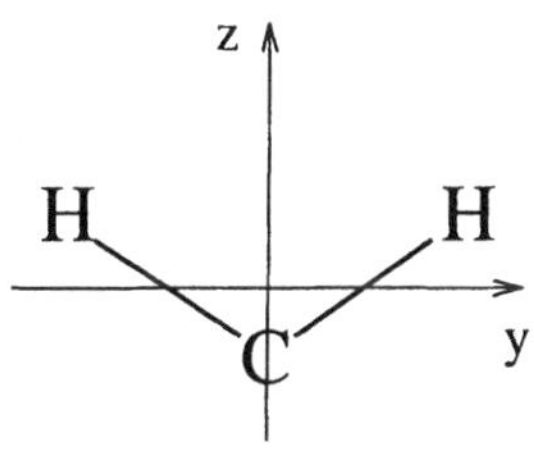

With this convention the triplet ground state is X ${}^3B_1$. Apart from X ${}^3B_1$ three low-lying excited singlet states exist, a ${}^1A_1$, b ${}^1B_1$, and c ${}^1A_1$.

We should like to answer the following questions:

1. Is there any first-order fine-structure splitting in the electronic ground state?
2. Which of the excited states are allowed to contribute to the fine-structure splitting of X $^3B_1$ in second order?
3. Which of the triplet sublevels is lowered in energy due to SOC?

For a matrix element to be non-zero, the direct products of the irreps of space and spin functions on each side have to be equal. Let us check this out first. Spatially A_1 and B_1 states are involved. The irreps according to which singlet and triplet spin functions transform under symmetry operations of the C_{2v} molecular point group can be found in Table 1.

TABLE 1. Transformation properties of the triplet and singlet spin functions, the angular momentum operators, an irreducible second rank tensor operator, and the translational operators in C_{2v} symmetry

function/operator	irreducible representation
T^0	A_2
$T^{\pm1}$	$\{B_1, B_2\}$
S^0	A_1
T^x	B_2
T^y	B_1
T^z	A_2
$\mathcal{L}_x$, $\mathcal{S}_x$	B_2
$\mathcal{L}_y$, $\mathcal{S}_y$	B_1
$\mathcal{L}_z$, $\mathcal{S}_z$	A_2
$\mathcal{D}_{2z^2-x^2-y^2}$	A_1
$\mathcal{D}_{x^2-y^2}$	A_1
$\mathcal{D}_{xy}$	A_2
$\mathcal{D}_{xz}$	B_2
$\mathcal{D}_{yz}$	B_1
$\mathcal{X}$	B_1
$\mathcal{Y}$	B_2
$\mathcal{Z}$	A_1

The direct product representations of the space and spin functions are in this particular case given by

The overall symmetries of the electronic states involved tell us that there is no interaction between b $^1B_1$ and X $^3B_1$ while {a,c} $^1A_1$ and X $^3B_1$ are allowed to couple. However, we can give much more specific answers:

State	spatial symmetry	spin symmetry	$\otimes$
X $^3B_1$	B_1	A_2	B_2
		B_1	A_1
		B_2	A_2
b $^1B_1$	B_1	A_1	B_1
{a,c} $^1A_1$	A_1	A_1	A_1

1. Concerning spin-orbit coupling none of the components of the (spatial) angular momentum operator is found in A_1 symmetry. Therefore, there is no first-order contribution of $\hat{\mathcal{H}}_{SO}$ to the fine-structure splitting of X $^3B_1$. This statement can generally be made for spatially non-degenerate electronic states.
2. On the contrary, both $\mathcal{D}_{2z^2-x^2-y^2}$ and $\mathcal{D}_{x^2-y^2}$ are totally symmetric. Hence, the triplet is split by spin-spin coupling into three distinct multiplet levels.
3. The spatial parts of a,c $^1A_1$ can couple to X $^3B_1$ via the y-component of the spin-orbit operator. S_y couples the singlet spin function S^0 (A_1) to the B_1 triplet function.

Thus, by symmetry, spin-spin interaction is capable of splitting the ground state triplet spin multiplet into three distinct levels. Which of the multiplet levels is lowered in energy and which is raised due to SSC cannot be predicted by means of group theory. This depends on the specific electronic configuration in the considered state. Among the low-lying states only the a $^1A_1$ and c $^1A_1$ states are allowed to perturb the X $^3B_1$ ground state via SOC. One of the triplet components (T^y) is lowered in energy due to this perturbation, the two others remain unaltered.

4.3. HIGH SYMMETRY: LINEAR MOLECULES AND ATOMS

In systems with orbitally degenerate states there is a further advantage of symmetry considerations. In addition to the decision whether two electronic states may couple by some specific operator or not we can exploit the tensorial structure of the magnetic interaction terms by means of the Wigner-Eckart theorem in order to reduce number of matrix elements which have actually to be calculated. Before we turn to the Wigner-Eckart theorem, however, let us have a look at specific selection rules for linear molecules.

It is easily seen from Eqs. 32, 33, and 37 that in linear molecules or atoms where the projection of the angular momenta on the z axis is meaningful, m_L and m_S quantum numbers can be changed by $\hat{\mathcal{H}}_{SO}$ by at most one unit each, while their sum m_J has to remain constant:

$$\Delta\Omega = 0; \qquad \Delta\Lambda = 0, \pm 1; \qquad \Delta\Sigma = 0, \mp 1 \tag{55}$$

TABLE 2. Irreps of singlet and triplet spin functions,the angular momentum operators, an irreducible second rank tensor operator, and the translational operators in $C_{\infty v}$ ($D_{\infty h}$) symmetry

function/operator		irrep.
$\text{Triplet}_{m_S=0}$	$\equiv T^0$	$\Sigma^-_{(g)}$
$\text{Triplet}_{m_S=\pm 1}$	$\equiv T^{\pm 1}$	$\Pi_{(g)}$
$\text{Singlet}_{m_S=0}$	$\equiv S^0$	$\Sigma^+_{(g)}$
$\mathcal{L}_z$, $\mathcal{S}_z$	$\mathcal{L}_0$, $\mathcal{S}_0$	$\Sigma^-_{(g)}$
$\mathcal{L}_{\mp(x\pm iy)/\sqrt{2}}$, $\mathcal{S}_{\mp(x\pm iy)/\sqrt{2}}$	$\mathcal{L}_{\pm 1}$, $\mathcal{S}_{\pm 1}$	$\Pi_{(g)}$
$\mathcal{D}_{(2z^2-x^2-y^2)/\sqrt{6}}$	$\mathcal{D}_0$	$\Sigma^+_{(g)}$
$\mathcal{D}_{\mp(xz+zx\pm i(yz+zy))/2}$	$\mathcal{D}_{\pm 1}$	$\Pi_{(g)}$
$\mathcal{D}_{(x^2-y^2\pm i(xy+yx))/2}$	$\mathcal{D}_{\pm 2}$	$\Delta_{(g)}$
$\mathcal{Z}$	$\mathcal{T}_0$	$\Sigma^+_{(u)}$
$\mp(\mathcal{X} \pm i\mathcal{Y})/\sqrt{2}$	$\mathcal{T}_{\pm 1}$	$\Pi_{(u)}$

Let us look at a common situation: the first-order spin-orbit splitting of an orbitally degenerate electronic state, a $^3\Pi$ state. We shall also make use of the energy expression obtained from the phenomenological operator which in this case reduces to $A_{SO} \cdot m_L \cdot m_S$ since only the z component of the spin-orbit operator is involved.

From the entries in Table 2 it is seen that the $m_s = 0$ component of a triplet state transforms like Σ^- in $C_{\infty v}$ and the $m_s = \pm 1$ components like the Π irrep. The direct product of space and spin functions is irreducible in the case of the $m_s = 0$ component and results in an overall Π symmetry. Reduction of the direct product for $m_s = \pm 1$ components is readily carried out and gives $\Pi \otimes \Pi = \Sigma^+ \oplus \Sigma^- \oplus \Delta$. More specifically one obtains the following splitting pattern:

From the entries in Table 3 we read the following results:

– First-order SOC lifts certain degeneracies but not all of them. In the particular example, the six components of a $^3\Pi$ state are split into

TABLE 3. First-order spin-orbit splitting of a $^3\Pi$ state

	product state	$\|\Omega\| = \|\Lambda + \Sigma\|$	$E[A_{SO}]$
Δ:	$\Pi^{+1}T^{+1};\Pi^{-1}T^{-1}$	2	+1
Π:	$\Pi^{+1}T^{0};\ \Pi^{-1}T^{0}$	1	0
$\Sigma^{\pm}$:	$(1/\sqrt{2})[\Pi^{+1}T^{-1} \mp \Pi^{-1}T^{+1}]$	$0^{\pm}$	−1

three doubly degenerate levels. (Remark: A degeneracy of the $0^{\pm}$ levels is not required by group theory. The $\mathcal{S}_{-1}\mathcal{D}_{+2}\mathcal{S}_{-1}$ and $\mathcal{S}_{+1}\mathcal{D}_{-2}\mathcal{S}_{+1}$ components of the electronic spin-spin coupling operator and higher-order SOC lift their degeneracy. It requires a different type of interaction, i.e., either the coupling to an external magnetic field or the rotation of the nuclear framework to split also the $\Omega = 2$ and $\Omega = 1$ sublevels.)

- There is a first-order splitting pattern common to all states of a specific space+spin symmetry, independent of the physical content. All the molecule dependent physical information is contained in the parameter A_{SO}.

The latter result can be generalized and leads us an important theorem formulated by Wigner and Eckart.

4.4. WIGNER-ECKART-THEOREM

The main reason for working with irreducible tensor operators stems from an important theorem, known as the Wigner-Eckart-Theorem (WET) for matrix elements of tensor operators [17, 18] :

$$\langle \alpha' j' m' | \hat{T}_q^{(k)} | \alpha j m \rangle = \underbrace{(-1)^{j'-m'} \begin{pmatrix} j' & k & j \\ -m' & q & m \end{pmatrix}}_{\text{geometrical part : 3j symbol}} \underbrace{\langle \alpha' j' \| \hat{T}^{(k)} \| \alpha j \rangle}_{\text{physical part : reduced matrix element}} \quad (56)$$

The reduced matrix element (RME) depends on the specific wave functions and the specific operator. It is independent of magnetic quantum numbers.

The 3j symbol depends only on rotational symmetry properties. It is related to the corresponding vector addition coefficient or Clebsch-Gordan

(CG) coefficient by

$$\begin{pmatrix} j' & k & j \\ -m' & q & m \end{pmatrix} = \frac{(-1)^{j'-k-m}}{\sqrt{2j+1}} \langle j'km'q|j'kjm\rangle \tag{57}$$

The general selection rules emerging from the properties of the 3j symbols are

$$m = m' + q \quad ; \quad \triangle(jkj') \tag{58}$$

where the triangle condition $\triangle(jkj')$ is fulfilled if $j + j' \geq k \geq |j - j'|$. Specifically, for tensor operators of rank 0, 1, or 2 one obtains:

$$\langle\alpha' j' m'|\hat{T}_0^{(0)}|\alpha j m\rangle = 0 \quad \text{unless} \quad \begin{cases} j' = j, \\ m' = m. \end{cases} \tag{59}$$

$$\langle\alpha' j' m'|\hat{T}_q^{(1)}|\alpha j m\rangle = 0 \quad \text{unless} \quad \begin{cases} j' = j+1, j, j-1 \\ \text{or}\Delta\text{j} = 0, \pm 1, \\ m' = m - q (q = 1, 0, -1) \\ \text{or}\Delta\text{m} = 0, \pm 1, \\ j + j' \geq 1. \end{cases} \tag{60}$$

$$\langle\alpha' j' m'|\hat{T}_q^{(2)}|\alpha j m\rangle = 0 \quad \text{unless} \quad \begin{cases} j' = j+2, j+1, j, j-1, j-2 \\ \text{or}\Delta\text{j} = 0, \pm 1, \pm 2 \\ m' = m - q (q = 2, 1, 0, -1, -2) \\ \text{or}\Delta\text{m} = 0, \pm 1, \pm 2 \\ j + j' \geq 2. \end{cases} \tag{61}$$

The condition $j+j' \geq 1$ for a matrix element of a first rank tensor operator implies, e.g. that there is no first-order SOC of singlet wavefunctions. Two doublet spin wavefunctions may interact via SOC, but the selection rule $j + j' \geq 2$ tells us that electronic spin-spin interaction does not contribute to their fine-structure splitting in first order.

Furthermore, if a single matrix element $\langle\alpha' j' m'|\hat{T}_q^{(k)}|\alpha j m\rangle$ is known, all possible $(2j'+1)(2k+1)(2j+1)$ matrix elements $\langle\alpha' j' m'''|\hat{T}_q^{(k)}|\alpha j m''\rangle$ can be calculated with the aid of 3j symbols. This is easily achieved by relating

$$\langle\alpha' j' m'''|\hat{T}_{q'}^{(k)}|\alpha j m''\rangle = (-1)^{m'''-m'} \frac{\begin{pmatrix} j' & k & j \\ -m''' & q' & m'' \end{pmatrix}}{\begin{pmatrix} j' & k & j \\ -m' & q & m \end{pmatrix}} \cdot \langle\alpha' j' m'|\hat{T}_q^{(k)}|\alpha j m\rangle \tag{62}$$

Numerical values for the 3j symbols can be looked up in tables [19]. Only the following of their symmetry properties shall be mentioned:

$$\begin{pmatrix} j_1 & j_2 & j_3 \\ m_1 & m_2 & m_3 \end{pmatrix}$$

$$\begin{aligned} &= \begin{pmatrix} j_3 & j_1 & j_2 \\ m_3 & m_1 & m_2 \end{pmatrix} && \text{(even permutations)} \\ &= (-1)^{j_1+j_2+j_3} \begin{pmatrix} j_2 & j_1 & j_3 \\ m_2 & m_1 & m_3 \end{pmatrix} && \text{(odd permutations)} \\ &= \begin{pmatrix} m_1 & m_2 & m_3 \\ j_1 & j_2 & j_3 \end{pmatrix} && \text{(interchange of rows)} \\ &= (-1)^{j_1+j_2+j_3} \begin{pmatrix} j_1 & j_2 & j_3 \\ -m_1 & -m_2 & -m_3 \end{pmatrix} && \text{(m to } -\text{ m)} \end{aligned} \tag{63}$$

In molecular applications the question arises how to use the WET, since L is not a good quantum number! There is a way out: work with Cartesian spatial functions and spherical spin functions and apply relations (39) and (40) for transforming back and forth!

4.5. RECOUPLING OF ANGULAR MOMENTA

Russel-Saunders coupled atomic wavefunctions are eigenfunctions of $\vec{\mathcal{S}}^2$, $\vec{\mathcal{L}}^2$, $\vec{\mathcal{S}}_z$ and $\vec{\mathcal{L}}_z$ and can be represented by the quantum numbers $|LSm_Lm_S\rangle$. If spin-orbit coupling is invoked, it is often more convenient to construct

$$\vec{\mathcal{J}} = \vec{\mathcal{L}} + \vec{\mathcal{S}} \tag{64}$$

and to change the representation from $|LSm_Lm_S\rangle$ to $|LSJm_J\rangle$. We can again use the tensorial properties of the $\hat{\mathcal{H}}_{SO}$ to calculate the spin-orbit matrix elements. They are given by

$$\begin{aligned} \langle LSJm_J|\vec{\mathcal{L}}\vec{\mathcal{S}}|L'S'J'm'_J\rangle &= (-1)^{J+S+L'}\delta_{JJ'}\delta_{m_Jm'_J} \begin{Bmatrix} L & S & J \\ S' & L' & 1 \end{Bmatrix} \\ &\cdot \quad \langle L||\vec{\mathcal{L}}||L'\rangle\langle S||\vec{\mathcal{S}}||S'\rangle \end{aligned} \tag{65}$$

where $\begin{Bmatrix} L & S & J \\ S' & L' & 1 \end{Bmatrix}$ is a 6j symbol and $\langle L||\vec{\mathcal{L}}||L'\rangle$ and $\langle S||\vec{\mathcal{S}}||S'\rangle$ are RMEs. 6j symbols represent recoupling coefficients, up to phase factors and normalization constants. In order to illustrate their meaning, consider three angular momenta j_1, j_2, and j_3. There is no unique way to combine them to a total angular momentum J. 6j symbols allow a conversion between two possible coupling schemes, e.g. $\{\{j_1 \otimes j_2\} \otimes j_3\} = J$ and $\{j_1 \otimes \{j_2 \otimes j_3\}\} = J$ which are related by $\begin{Bmatrix} j_1 & j_2 & j_{12} \\ j_3 & J & j_{23} \end{Bmatrix}$.

This leads us to the following selection rules for $\hat{\mathcal{H}}_{SO}$ in atoms where the bullets (•) symbolize the positions of the angular momenta in the 6j symbol which are related by triangle conditions.

$$\left\{\begin{array}{ccc} L & S & J \\ S' & L' & 1 \end{array}\right\} \quad , \delta_{JJ'}, \delta_{m_J m'_J} \tag{66}$$

$$\begin{array}{ccc} \bullet & \circ & \circ \\ \circ & \bullet & \bullet \end{array} \qquad L + L' \geq 1 \geq |L - L'| \tag{67}$$

$$\begin{array}{ccc} \circ & \bullet & \circ \\ \bullet & \circ & \bullet \end{array} \qquad S' + S \geq 1 \geq |S' - S| \tag{68}$$

$$\begin{array}{ccc} \bullet & \bullet & \bullet \\ \circ & \circ & \circ \end{array} \qquad L + S \geq J \geq |L - S| \tag{69}$$

$$\begin{array}{ccc} \circ & \circ & \bullet \\ \bullet & \bullet & \circ \end{array} \qquad S' + L' \geq J \geq |S' - L'| \tag{70}$$

5. Evaluation of matrix elements

Before turning to the different types of interactions I should like to give a schematic overview over the computational steps which are required to calculate fine- and hyperfine-structure effects in molecules. Various appraoches can be pursued to reach this goal. I will confine the discussion to two of them, a perturbational and a variational procedure, both at the level of electronic states. A treatment at the purely electronic level is applicable as long as no (sharply avoided) intersystem crossings are present or other vibronic (vibrational/electronic) coupling effects are involved.

The various steps of a perturbational treatment of fine- and hyperfine-structure are sketched in Figure 1. The procedure starts with the computation of CI wavefunctions and energies for pure spin states. These are taken as the zeroth-order states $\Psi^{(0)}$ and energies $E^{(0)}$ of the perturbation expansion. Because it is difficult to evaluate transition matrix elements in non-orthogonal orbital bases, in particular for two-electron operators, the determinants in the CI expansions of all interacting electronic states are constructed from a common set of molecular orbitals. Having evaluated the matrix elements for all possible pairs of interacting states the perturbed wavefunctions and energies in a Rayleigh-Schrödinger perturbation expansion can be determined in principle.

A first-order contribution to the energy is obtained only for degenerate states. Let us assume that the unperturbed state $|\Psi_k^{(0)}\rangle$ is n-fold degenerate.

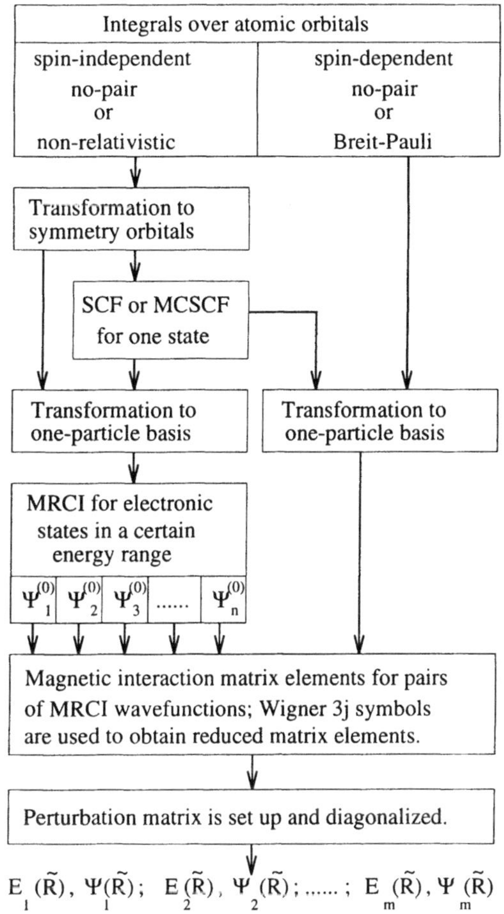

Figure 1. Schematic representation of the computational steps required for the perturbational determination of magnetic interaction effects of electronic states.

According to the rules of degenerate perturbation theory the first-order energies $E_{k,m}^{(1)}$ are obtained by diagonalizing the interaction matrix in a basis spanned by the n components of $|\Psi_k^{(0)}\rangle$. If the degeneracy is lifted completely in first order or if at least second-order effects do not introduce an additional splitting of degenerate levels, the second-order energy $E_{k,l}^{(2)}$ can be expressed as

$$E_{k,l}^{(2)} = \sum_{i\neq k}^{\infty} \sum_{j}^{n'} \sum_{l'}^{n} \frac{\langle c_{k,l,l'} \Psi_{k,l'}^{(0)} | \hat{O}p | \Psi_{i,j}^{(0)} \rangle \langle \Psi_{i,j}^{(0)} | \hat{O}p | c_{k,l,l'} \Psi_{k,l'}^{(0)} \rangle}{E_i^{(0)} - E_k^{(0)}} \tag{71}$$

Up to first order the corresponding wavefunction reads

$$|\Psi_{k,l}^{(0)}\rangle + |\Psi_{k,l}^{(1)}\rangle = \sum_{l'}^{n} c_{k,l,l'}|\Psi_{k,l'}^{(0)}\rangle \left(1 + \sum_{i\neq k}^{\infty}\sum_{j}^{n'} \frac{\langle c_{k,l,l'}\Psi_{k,l'}^{(0)}|\hat{O}p|\Psi_{i,j}^{(0)}\rangle}{E_i^{(0)} - E_k^{(0)}}\right) . \quad (72)$$

If these conditions are not fulfilled, symmetry properties of the involved states and operators may be helpful in predetermining which linear combinations of the $|\Psi_{k,l}^{(0)}\rangle$ are judiciously chosen such that the formulae can be applied nevertheless. In general, this procedure is too complicated, however. Moreover, in a molecule first and second orders of perturbation theory are not easily defined in a stringent way. Consider for example the dissociation of a diatomic molecule. Near the equilibrium geometry the states corresponding to a specific dissociation channel are in many cases energetically well separated. Their mutual interaction is then described properly by second-order perturbation theory. In the separated atom limit first-order degenerate perturbation theory applies. In between, but still close to the dissociation limit, the states are nearly degenerate — or quasi-degenerate as we might say. Neither of the procedures is then strictly applicable and orders of perturbation theory are not well defined. An often chosen way out of this dilemma is to set up a so-called perturbation matrix. It contains unperturbed energies in the diagonal and in addition all possible coupling matrix elements, both diagonal or off-diagonal. Its diagonalization yields directly perturbed energies and wavefunctions. Although this approach has been named quasi-degenerate perturbation theory or also near-degenerate perturbation theory, it is applicable also to states with large energy separations.

In compounds containing heavy atoms spin-orbit and electron correlation energies are approximately of the same size. In this case one cannot expect these effects to be independent of each other. A variational approach is then preferable which treats both interactions at the same level. The first steps in this procedure represented schematically in Figure 2 are completely analogous to the perturbational treatment. Instead of calculating matrix elements for CI wavefunctions, in the so-called spin-orbit CI a Hamiltonian matrix is set up in the basis of configuration state functions or determinants. It contains simultaneously matrix elements of the usual electrostatic Hamiltonian and of the spin-orbit operator. Diagonalization of this matrix (due to its size one usually looks only for a few roots by means of a Davidson procedure) yields directly wavefunctions and energies of spin-orbit mixed states. A schematic representation of computational steps is shown in Figure 2. Special care is advisable in the choice of the spin-orbit operator in this case. The variational determination of spin-orbit

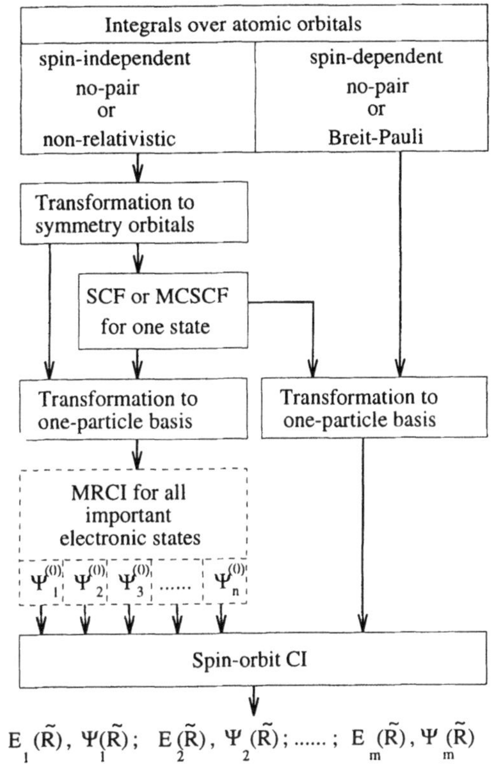

Figure 2. Schematic representation of the computational steps required for the variational determination of spin-orbit effects in electronic states.

coupling requires a spin-orbit operator which is bounded from below like the no-pair Hamiltonian. Otherwise one has to constrain the basis set (e.g. by an energy selection) to prevent a variational collaps.

5.1. SPIN-ORBIT COUPLING

5.1.1. *Integrals: Peculiarities of the Spin-Orbit Operator*

The spatial parts of one- and two-electron spin-orbit operators are purely imaginary operators. If real basis functions are employed, which is usually the case in molecular calculations, the common phase factor i can be left aside in the integral evaluation step; in most applications it is reintroduced then in the very last step when the real spatial integrals are multiplied with spin integration factors. The application of $\hbar\vec{\nabla}_i$ instead of $\vec{p}_i$ leads to an antisymmetric real one-electron integral matrix

$$\langle i|j\rangle = -\langle j|i\rangle \tag{73}$$

where we have introduced the abbreviation

$$\langle i|j\rangle \equiv \langle i(1)|\frac{e^2\hbar^2}{2m^2c^2}\vec{\nabla}_1\left(\sum_\alpha \frac{Z_\alpha}{r_{1\alpha}}\right) \times \vec{\nabla}_1|j(1)\rangle \ . \tag{74}$$

Clearly, all the diagonal elements are zero in this basis.

$$\langle i|i\rangle = 0 \tag{75}$$

The two-electron case has to be examined in more detail. The structure of the integral matrix depends on the way the spin-same- and spin-other-orbit parts are combined. In the programs used in Bonn [4], which we shall refer to, the operators are coded in their forms 10 or 12, i.e., there is only a momentum operator for particle i. If we define

$$\langle ij|kl\rangle \equiv \langle i(1)j(2)| -\frac{e^2\hbar^2}{2m^2c^2}\vec{\nabla}_1\left(\frac{1}{r_{12}}\right) \times \vec{\nabla}_1|k(1)l(2)\rangle \tag{76}$$

one obtains a two-electron integral matrix which is antisymmetric with respect to the interchange of basis functions for particle 1

$$\langle ij|kl\rangle = -\langle kj|il\rangle \tag{77}$$

and in particular

$$\langle ij|il\rangle = 0. \tag{78}$$

By contrast, the matrix is symmetric for an interchange of basis functions related to particle 2

$$\langle ij|kl\rangle = \langle il|kj\rangle \ . \tag{79}$$

Since the two-electron spin-orbit operator is not symmetric in the particle indices, i.e.,

$$\langle ij|kl\rangle \neq \langle ji|lk\rangle \ , \tag{80}$$

twice as many integrals as for the Coulomb operator have to be computed for each of the Cartesian components, i.e., six times as many in total.

Because of these properties special programs have to be used for the transformation of the spatial parts of the spin-orbit integrals to the molecular orbital basis. Furthermore, the usual matrix element formulae derived for the Coulomb interaction (Slater-Condon rules) have to be modified.

5.1.2. *Determinants: Modified Slater-Condon rules*

Symmetrization of the two-electron operator

$$\hat{g}_{12} \rightarrow \frac{1}{2}(\hat{g}_{12} + \hat{g}_{21}) \tag{81}$$

combined with the above mentioned integral matrix properties leads to the following expressions for the matrix element of a pair of determinants:

TABLE 4. Two-electron spin integrals for the $\vec{s}_1 + 2\vec{s}_2$ operator

	$\vert\alpha(1)\alpha(2)\rangle$	$\vert\alpha(1)\beta(2)\rangle$	$\vert\beta(1)\alpha(2)\rangle$	$\vert\beta(1)\beta(2)\rangle$
$\langle\alpha(1)\alpha(2)\vert$	$3/2$	$-2/\sqrt{2}$	$-1/\sqrt{2}$	0
$\langle\alpha(1)\beta(2)\vert$	$2/\sqrt{2}$	$-1/2$	0	$-1/\sqrt{2}$
$\langle\beta(1)\alpha(2)\vert$	$1/\sqrt{2}$	0	$1/2$	$-2/\sqrt{2}$
$\langle\beta(1)\beta(2)\vert$	0	$1/\sqrt{2}$	$2/\sqrt{2}$	$-3/2$

A Diagonal case. D and D' are equal.
Because of the symmetry properties of the spin-orbit Hamiltonian there are no diagonal matrix elements in a basis of real (Cartesian) orbitals.

$$\langle D|\hat{\mathcal{H}}_{\rm SO}|D'\rangle = 0 \tag{82}$$

B Singly excited Slater determinants. D and D' differ in 1 pair of spin orbitals i, j.
Contributions from the one-electron spin-orbit integral and three-index two-electron integrals

$$\langle D|\hat{\mathcal{H}}_{\rm SO}|D'\rangle = \langle i|j\rangle + \frac{1}{2}\sum_k \{\langle ik|jk\rangle + 0 - \langle ki|jk\rangle - \langle ik|kj\rangle\} \tag{83}$$

with $k \in$ core orbitals and common valence orbitals.

C Doubly excited Slater determinants. D and D' differ in 2 pairs of spin orbitals ij, kl.
Contributions only from four-index two-electron integrals

$$\langle D|\hat{\mathcal{H}}_{\rm SO}|D'\rangle = \frac{1}{2}\{\langle ij|kl\rangle + \langle ji|lk\rangle - \langle ji|kl\rangle - \langle ij|lk\rangle\} \tag{84}$$

In these formulae a spin integration has not yet been performed.

It is noteworthy that the spin parts for the interaction of electrons ij, whatever their spins are, with a closed shell orbital do not automatically integrate to zero. Therefore, if a coupling is allowed by symmetry, there is a contribution from the inner shells of an atom (by far the largest two-electron integrals) to a matrix element of two singly excited determinants. The relative importance of the two-electron contributions to a matrix element of singly excited determinants with respect to the one-electron integral varies with the nuclear charge of the constituents. Screening of the nuclear charge by the electrons takes place but to a much less extent than e.g. for a Coulombic interaction. For the first-row elements B to Ne the two-electron part reduces the one-electron matrix element by approximately 50%. For the very light elements it has an even larger compensating effect. The relative size of the two-electron integral contribution decreases with increasing nuclear charge but cannot be neglected completely; for a third-row transition metal like Pt it amounts to approximately 20%. Analyzing the size of the two-electron integrals for heavy elements one makes the interesting observation that those integrals containing only valence orbital indices are small, in the order of 1 cm^{-1}or less.

5.1.3. *Definition of a Spin-Orbit Mean-Field Operator*
If we are only concerned with valence states and are not looking at core hole states, an important simplification of the spin-orbit Hamiltonian can be made without significant loss of accuracy. For heavy elements one can safely neglect the matrix elements coupling doubly excited determinants, Eq. 84. Assuming an average occupation of the orbitals k occuring in the sum 83 and an equal distribution of α and β spins in these k we may define a Fock-like effective one-electron spin-orbit Hamiltonian. A matrix element of this mean-field spin-orbit operator is given by

$$\begin{aligned}\langle i|\hat{\mathcal{H}}^{mf}_{\rm SO}|j\rangle = \langle i|j\rangle + \tfrac{1}{4}\textstyle\sum_k \mathrm{occ}(k) \quad & \{\langle i(k\alpha)|j(k\alpha)\rangle + \langle i(k\beta)|j(k\beta)\rangle - \\ & \langle (k\alpha)i|j(k\alpha)\rangle - \langle (k\beta)i|j(k\beta)\rangle - \\ & \langle i(k\alpha)|(k\alpha)j\rangle - \langle i(k\beta)|(k\beta)j\rangle\} \quad (85)\end{aligned}$$

Since occ(k) reflects a properly chosen average occupation of the orbitals, i.e., 1 for the core orbitals and a value between 0 and 1 for a valence orbital the summation over k in 85 has to be done only once and the result can be added to the usual one-electron integral. The error made by employing this effective one-electron operator was found to be less than 4 cm^{-1} for the case of platinum out of a total of 4000 cm^{-1} [20].

5.1.4. *Matrix Elements for CI Wavefunctions*
So far we know how to calculate spin-orbit matrix elements for pairs of determinants. Because of electron correlation effects one is usually interested in evaluating matrix elements for multi-configurational wavefunctions, i.e., linear combinations of many determinants. Principally, one has to multiply the determinantal matrix elements by the appropriate expansion coefficients of the wavefunction. Technically, however, often a further complication arises: CI wavefunctions are usually determined only for a single, i.e., the $m_S = S$, spin component. This does not always match the selection rules dictated by the spatial part of the matrix element.

This complication does not only arise in spin-orbit coupling calculations but can occur for all types of coupling envolving electronic spin operators. The solution is similar in all instances though and we thus can exemplify it by considering a particular case involving spin-orbit coupling.

Problem: given a molecule which possesses C_{2v} symmetry, calculate $\langle {}^3A_2|\hat{\mathcal{H}}_{SO}|{}^3B_1\rangle$ from CI wavefunctions given with $m_S = 1$. The coupling of a A_2 and a B_1 state requires a spatial angular momentum operator of B_2 symmetry. From Table 1 we read that this is just the x component of $\vec{\mathcal{L}}$. A direct computation of $\langle {}^3A_2, m_S = 1|\mathcal{L}_x\mathcal{S}_x|{}^3B_1 m_S = 1\rangle$ yields zero because the integral of $\mathcal{S}_x$ with two $m_S = 1$ functions is zero.

Answer: calculate instead ME $= \langle {}^3A_2, m_S = 1|\mathcal{L}_x\mathcal{S}_0|{}^3B_1, m_S = 1\rangle$ where $\mathcal{L}_x$ stands symbolically for the spatial part of the microscopic spin-orbit Hamiltonian with x symmetry and $\mathcal{S}_0$ correspondingly for the zero-component of the spin tensor. This is the only non-zero matrix element for the given CI wavefunctions.

The RME (only spin) is then given by

$$\text{RME} = \frac{\langle {}^3A_2, m_S = 1|\mathcal{L}_x\mathcal{S}_0|{}^3B_1, m_S = 1\rangle}{\begin{pmatrix} 1 & 1 & 1 \\ -1 & 0 & 1 \end{pmatrix}} \tag{86}$$

with

$$\begin{pmatrix} 1 & 1 & 1 \\ -1 & 0 & 1 \end{pmatrix} = \frac{1}{\sqrt{6}} .$$

Since the spatial wavefunctions are coupled by $\mathcal{L}_x$, a spin operator of x-symmetry has to used also. $\mathcal{S}_x$ contributes to both $\mathcal{S}_{-1}$ and $\mathcal{S}_{+1}$. Thus, non-zero matrix elements are expected for $\langle m_S = 1|\mathcal{S}_x|m_S = 0\rangle$, $\langle m_S = 0|\mathcal{S}_x|m_S = 1\rangle$, $\langle m_S = 0|\mathcal{S}_x|m_S = -1\rangle$, and $\langle m_S = -1|\mathcal{S}_x|m_S = 0\rangle$. The 3-j symbols for the coupling of these spinfunctions by $\mathcal{S}_{-1}$ and $\mathcal{S}_{+1}$ are easily evaluated as

$$-\begin{pmatrix} 1 & 1 & 1 \\ -1 & 1 & 0 \end{pmatrix} = \begin{pmatrix} 1 & 1 & 1 \\ 0 & 1 & -1 \end{pmatrix}$$
$$= -\begin{pmatrix} 1 & 1 & 1 \\ 0 & -1 & 1 \end{pmatrix} = \begin{pmatrix} 1 & 1 & 1 \\ 1 & -1 & 0 \end{pmatrix} = \frac{1}{\sqrt{6}}$$

Applying relation (40) one obtains the following interaction matrix

5.2. ELECTRONIC SPIN-SPIN COUPLING

5.2.1. *Integrals*

The electronic spin-spin coupling Hamiltonian is a pure two-electron operator, symmetric in the particle indices. The spatial integrals exhibit the same

$<^3A_2\|\mathcal{L}_x S_x\|^3B_1>$	$m_S =$	1	0	-1
	$m_S = 1$	0	$-\mathrm{RME}/\sqrt{2}$	0
	0	$-\mathrm{RME}/\sqrt{2}$	0	$+\mathrm{RME}/\sqrt{2}$
	-1	0	$+\mathrm{RME}/\sqrt{2}$	0

permutation properties as the usual two-electron Coulomb interaction. We therefore refrain from any discussion of this part. In molecules mostly the Cartesian representation of the second rank tensor $\mathcal{D}^{(2)}_{ij}$ (Eq. 43) is employed. Operators of the type xy are combined with yx, etc. Thus, the number of spatial spin-spin interaction integrals to be computed is roughly six times the number of Coulomb integrals.

TABLE 5. Two-electron spin integrals for an $\vec{s}_1 \cdot \vec{s}_2$ operator

	$\|\alpha(1)\alpha(2)\rangle$	$\|\alpha(1)\beta(2)\rangle$	$\|\beta(1)\alpha(2)\rangle$	$\|\beta(1)\beta(2)\rangle$
$\langle\alpha(1)\alpha(2)\|$	$1/2\sqrt{6}$	$-1/4$	$-1/4$	$1/2$
$\langle\alpha(1)\beta(2)\|$	$1/4$	$-1/2\sqrt{6}$	$-1/2\sqrt{6}$	$1/4$
$\langle\beta(1)\alpha(2)\|$	$1/4$	$-1/2\sqrt{6}$	$-1/2\sqrt{6}$	$1/4$
$\langle\beta(1)\beta(2)\|$	$1/2$	$-1/4$	$-1/4$	$1/2\sqrt{6}$

The spin parts of the integrals are shown in Table 5.

5.2.2. *Matrix Elements of Determinants*

In the following, expressions for the spin-spin matrix element of a pair of determinants are given. It is instructive to carry out a spin integration where this leads to a zero contribution of integrals. For equal determinants clearly $\Delta m_S = 0$; the spin projections of singly excited determinants may differ by at most $\Delta m_S = 1$. As seen from the entries in Table 5 in these cases the spin integrals for the interaction of particle 1, say, with an α spin on electron 2 is equal but opposite to the integral for β spin on electron 2. As $\hat{\mathcal{H}}_{SS}$ is a symmetric operator the same is true for an interchange of particles, of course. This has the important consequence that closed shell orbitals do not contribute to spin-spin coupling matrix elements. Otherwise the expressions Eq. 87-89 are identical to the usual Slater-Condon rules.

A Diagonal case. D and D' are equal.
Contributions from two-index two-electron integrals.

$$\langle D|\hat{\mathcal{H}}_{\mathrm{SS}}|D'\rangle = \sum_{i\neq j}^{\text{open shells}} \{\langle ij|ij\rangle - \langle ij|ji\rangle\} \tag{87}$$

with i and $j \in$ common open shell orbitals.

B Singly excited Slater determinants. D and D' differ in 1 pair of spin orbitals i, j.
Contributions from three-index two-electron integrals.

$$\langle D|\hat{\mathcal{H}}_{\mathrm{SS}}|D'\rangle = \sum_{k\neq i,j}^{\text{open shells}} \{\langle ik|jk\rangle - \langle ik|kj\rangle\} \tag{88}$$

with $k \in$ common open shell orbitals.

C Doubly excited Slater determinants. D and D' differ in 2 pairs of spin orbitals ij, kl.
Contributions from four-index two-electron integrals

$$\langle D|\hat{\mathcal{H}}_{\mathrm{SS}}|D'\rangle = \langle ij|kl\rangle - \langle ij|lk\rangle \tag{89}$$

5.3. COUPLING OF ROTATIONAL AND ELECTRONIC ANGULAR MOMENTA

In this section we shall only be concerned with Kramers-type doubling in diatomic molecules and the calculation of Coriolis coupling matrix elements. The latter is a source of interaction between states differing in their Ω values by one unit. In a curve crossing situation this type of interaction may cause inhomogeneous predissociation. The coupling of a vibrational angular momentum and a spin-orbit term occuring in Renner-Teller states of open shell triatomic molecules will not be discussed here. An overview over the latter subject can e.g. be found in [21].

5.3.1. *Integrals and Matrix Elements for CI Wavefunctions*

The integral part for the calculation of rotational coupling is rather simple. The only demanding term, the two-electron operator 23, does not contribute to the splitting of the Kramers doublets and is therefore neglected here. The integrals of $\vec{\mathcal{L}}$ are origin dependent — $\vec{\mathcal{L}}$ is defined with respect to the molecular centre of mass — and have therefore to be computed separately for each of the considered isotopes. Otherwise the evaluation of its matrix elements is very similar to the procedure followed for the spatial part of the one-electron spin-orbit operator.

Molecular orbitals and CI wavefunctions are typically expanded in Cartesian basis functions with real expansion coefficients. Like $\hat{\mathcal{H}}_{\mathrm{SO}}$ the matrix representation of $\vec{\mathcal{L}}$ has no diagonal elements in this basis. For generating a Hund's case (a) basis electronic wavefunctions with definite Λ, S, and Σ quantum numbers are required. This is easily achieved if the perturbational scheme (Figure 1) is employed. The CI wavefunctions, already eigenfunctions of $\vec{\mathcal{S}}^2$ and $\hat{\mathcal{S}}_z$, are transformed to $\vec{\mathcal{L}}_z$ eigenfunctions by forming linear combinations of the Cartesian components. The transformation coefficients can be obtained from the relations between spherical and Cartesion tensor components. A convenient choice is to use the $T^{(|\Lambda|)}_{-\Lambda}$ and $T^{(|\Lambda|)}_{+\Lambda}$ components of a tensor of rank $|\Lambda|$. If, moreover, Σ^- states are multiplied by a phase factor of i, all matrix elements of the perturbation matrix become real.

Matrix elements of $\vec{\mathcal{S}}$ appearing in the rotational Hamiltonian 19–21 are evaluated using the macroscopic (total) spin operator. This does, of course, not apply to the usual spin-orbit matrix elements which are also needed for the evaluation of Kramers splittings. Instead of employing the phenomenological spin-orbit Hamiltonian 29 the matrix elements of $\hat{\mathcal{H}}_{\mathrm{SO}}$ are evaluated using the full microscopic Hamiltonian.

5.3.2. *Interaction Matrix*

Let us consider here only the case where the contributing electronic zeroth order wavefunctions do not undergo curve crossings. The vibrational motion can then to a good approximation be separated off. Perturbation matrices can be built up separately at each bond distance and for each value of the total angular momentum J. In order to avoid confusion the internuclear separation is denoted by Q.

Again, we look at an example situation, a strongly interacting manifold of electronic states consisting of a $^2\Pi$ and a $^2\Sigma^+$ state. The appropriate interaction matrix is shown in Tab. 6.

Herein, E stands for diagonal electronic and rotational excitation energies, $E_i^{(0)}$ and $B[J(J+1) + S_i(S_i+1) - 2\Omega_i\Sigma_i - \Lambda_i^2]$, respectively. SO represents spin-orbit coupling and $\tilde{SO}$ the usual spin-orbit coupling modified by the off-diagonal spin-orbit terms brought about by nuclear rotation term 21. Apart from the diagonal energies $E_i^{(0)}$, also the spin-orbit matrix elements SO_{ij} and $B \cdot L_{ij} \cdot [S(S+1) - \Sigma_i\Sigma_j]^{1/2}$ are J independent. The Coriolis coupling elements designated $\tilde{B}$L and the rotational distortion terms $B \cdot [S(S+1) - \Sigma_i\Sigma_j]^{1/2}$, denoted $\tilde{B}$S, are multiplied by the proper J dependent factors before storing. Further details about the form of the J dependence may be found in textbooks eg. in [22, pages 275–276].

TABLE 6. Structure of the interaction matrix for evaluating the Ω-type doubling within a manifold of a $^2\Pi$ and a $^2\Sigma^+$ state. The following notation has been used: E =: excitation energy (electronic and rotational), SO =: $\langle \hat{\mathcal{H}}_{SO} \rangle$, $\tilde{SO}$=: SO + $\langle \tilde{B}\hat{L}\hat{S} \rangle$, $\tilde{B}\hat{S}$ =: spin-rotation coupling, $\tilde{B}\hat{L}$ =: orbital angular momentum-rotation coupling. All matrix elements refer to a specific bond distance Q and total angular momentum J.

	$^2\Pi_{3/2}$	$^2\Pi_{1/2}$	$^2\Sigma^+_{1/2}$	$^2\Sigma^+_{-1/2}$	$^2\Pi_{-1/2}$	$^2\Pi_{-3/2}$
$^2\Pi_{3/2}$	E+SO	$\tilde{BS}$	$\tilde{BL}$			
$^2\Pi_{1/2}$	$\tilde{BS}$	E+SO	$\tilde{SO}$	$\tilde{BL}$		
$^2\Sigma^+_{1/2}$	$\tilde{BL}$	$\tilde{SO}$	E	$\tilde{BS}$	$\tilde{BL}$	
$^2\Sigma^+_{-1/2}$		$\tilde{BL}$	$\tilde{BS}$	E	$\tilde{SO}$	$\tilde{BL}$
$^2\Pi_{-1/2}$			$\tilde{BL}$	$\tilde{SO}$	E+SO	$\tilde{BS}$
$^2\Pi_{-3/2}$				$\tilde{BL}$	$\tilde{BS}$	E+SO

5.3.3. *Perturbation Expansion of the Interaction Matrix*

If the electronic energy difference is large compared to the rotational energy difference or the spin-orbit splitting, respectively, the characteristic polynomial of the secular determinant corresponding to the interaction matrix (Table 6) can be expanded in a perturbation series. The energy difference between the positive and negative parity sublevels of a $^2\Pi$ state intermediate between Hund's cases (a) and (b) is found to be [23]

$$\begin{aligned} \Delta E(^2\Pi) &= (\frac{1}{2}p + q)\,(1 \pm (2 - Y)\,X^{-1})\,(J + \frac{1}{2}) \\ &\pm\; 2qX^{-1}\,(J - \frac{1}{2})\,(J + \frac{1}{2})\,(J + \frac{3}{2}) \end{aligned} \tag{90}$$

where

$$Y = A(^2\Pi)\,/B(^2\Pi), \tag{91}$$

$$X = \left(Y^2 - 4Y + 4(J + \frac{1}{2})^2\right)^{1/2} . \tag{92}$$

Employing, like Hill and van Vleck [24] and van Vleck [23] in their original publications, the pure precession approximation for evaluating the spin parts of the operators 19–21 but ignoring this type of approximation for the spin-orbit part and the electronic angular momentum, the following formulae for the Λ-doubling parameters p and q of a $^2\Pi$ state are obtained:

$$p(^2\Pi) = \sum_{^2\Sigma} \frac{8\langle ^2\Pi|\hat{\mathcal{H}}_{SO}|^2\Sigma\rangle\langle ^2\Sigma|\hat{B}\cdot\hat{\mathcal{L}}_{-1}|^2\Pi\rangle(-1)^k}{E(^2\Pi) - E(^2\Sigma)} \tag{93}$$

$$q(^2\Pi) = \sum_{^2\Sigma} \frac{8|\langle ^2\Pi|\hat{B}\cdot\hat{\mathcal{L}}_{+1}|^2\Sigma\rangle|^2(-1)^k}{E(^2\Pi)-E(^2\Sigma)} \tag{94}$$

The constants in Eqs. 93 and 94 are chosen to be consistent with the use of spherical tensor conventions for $\vec{\mathcal{L}}$ and $\vec{\mathcal{S}}$. k is even/odd for Σ^+/Σ^- states and the sum runs over all interacting $^2\Sigma$ states. $A(^2\Pi)$ and $B(^2\Pi)$ denote the spin-orbit splitting and the rotational parameters of the $^2\Pi$ state at a given bond distance Q and the $\pm$ sign in front of $(2-Y)X^{-1}$ and $2q$ refers to the upper/lower fine-structure component, respectively.

Even in cases where perturbation theory is not applicable the number of steps required to couple $+\Omega$ and $-\Omega$ sublevels may be used to give a rough estimate of the splitting in a specific state. For an $\Omega = 1/2$ state the coupling of $\Omega = +1/2$ and $\Omega = -1/2$ sublevels is achieved by two steps. For low rotational levels therefore the observed Kramers-type doubling is largest in $^2\Sigma$ and $^2\Pi_{1/2}$ states. The Λ-doubling of a $^2\Pi_{1/2}$ state, for example, can be mediated by either a $^2\Sigma^+$ or a $^2\Sigma^-$ state

$$\langle ^2\Pi_{-1/2}|\hat{\mathcal{H}}_{SO}|^2\Sigma_{-1/2}\rangle\langle ^2\Sigma_{-1/2}|\hat{B}\hat{\mathcal{L}}_{-1}|^2\Pi_{1/2}\rangle \ . \tag{95}$$

Herein, we have labelled the states by $\Psi_{\pm\Omega}$ instead of Ψ_Ω used otherwise. The size of the energy shift depends on the size of the spin-orbit coupling matrix element, the rotational parameter, the matrix element of $\vec{\mathcal{L}}$, and in addition on the energy separation between the states. The direction of the shift depends on two ingredients. The first of them is the relative ordering of the interacting states. A $^2\Sigma$ state lying below the $^2\Pi$ state leads to an upward shift, a $^2\Sigma$ state interacting from above to a downward shift. Which of the two wavefunctions $\Psi^{(+)}$ or $\Psi^{(-)}$ is perturbed depends on the symmetry properties of the electronic and rotational wavefunctions with respect to inversion at the origin. A $^2\Sigma^+$ state has parity $+1$, a $^2\Sigma^-$ state -1. The rotational eigenfunctions are simply spherical harmonics Y_{KM_K} with parity $(-1)^K$. Since only states of equal parity and total angular momentum J can interact, merely one of the sublevels is shifted.

More steps are needed for a $^2\Pi_{3/2}$ state. One of several possibilities in third-order perturbation theory is

$$\langle ^2\Pi_{-3/2}|\hat{B}\hat{\mathcal{L}}_{-1}|^2\Sigma_{-1/2}\rangle\langle ^2\Sigma_{-1/2}|\hat{B}\hat{\mathcal{S}}_{-1}|^2\Sigma_{1/2}\rangle\langle ^2\Sigma_{1/2}|\hat{B}\hat{\mathcal{L}}_{-1}|^2\Pi_{3/2}\rangle \tag{96}$$

The spin-orbit interaction with a Σ state, usually the largest matrix element, contributes only in fourth or higher order

$$\begin{aligned} \langle ^2\Pi_{-3/2}\hat{B}\hat{\mathcal{S}}_{-1}|^2\Pi_{-1/2}\rangle \ &\cdot \ \langle ^2\Pi_{-1/2}|\hat{B}\hat{\mathcal{L}}_{-1}|^2\Sigma_{1/2}\rangle \ \cdot \\ &\cdot \ \langle ^2\Sigma_{1/2}|\hat{\mathcal{H}}_{SO}|^2\Pi_{1/2}\rangle \cdot \langle ^2\Pi_{1/2}|\hat{B}\hat{\mathcal{S}}_{-1}|^2\Pi_{3/2}\rangle \end{aligned} \tag{97}$$

Therefore, the Kramers doubling in $^2\Pi_{3/2}$ state is generally much smaller than in $^2\Pi_{1/2}$ states, at least for low rotational quantum numbers. The Λ-splitting of a Δ state is even smaller since it is a fourth-order effect.

5.3.4. *Diagonalization of the Interaction Matrix*

Since all the coupling matrix elements can be computed it is in general much easier for a quantum chemist to evaluate the Kramers splitting by direct matrix diagonalization. Moreover, the perturbation expansion of the interaction matrix (Table 6) is not justified if the spin-orbit interaction between different states is of the same order of magnitude as their energetic separation.

If no avoided crossings between the interacting states occur one may procede by diagonalizing interaction matrices of the type shown in Table 6 separately for each bond distance Q and total angular momentum J yielding expansion coefficients $a_\kappa(Q, J)$ of the mixed rotronic (rotational/electronic) wavefunctions and the corresponding energy levels $E_\kappa(Q, J)$. The calculated Kramers type splittings will then have to be vibrationally averaged for a comparison with experiment. Note that, although we always start off with Hund's case (a) basis functions, the diagonalization approach can be used whatever Hund's coupling case applies to the considered electronic states because the magnetic interactions between the various angular momenta are all treated at an equal level.

In the presence of sharply avoided crossings between states of equal Ω quantum numbers large non-adiabatic coupling matrix elements tend to occur due to rapid changes of the spin-orbit mixing coefficients with internuclear separation. These difficulties can be avoided to a large extent, if the molecular wavefunction is instead expanded in a basis of uncoupled vibronic functions Ψ_{iv} where each Ψ_{iv} is a product of an eigenfunction $\Phi_i(\vec{r}, \tilde{Q})$ of a spin-free electronic Hamiltonian $\mathcal{H}^{(el)}_{\text{spin-free}}$ at internuclear separation Q and $\chi^{\text{unc.}}_{iv}(Q)$ a corresponding vibrational wavefunction. Like the aforementioned approach this procedure may be applied independent of the actual Hund's coupling cases. For further details see [25].

5.4. NUCLEAR MAGNETIC HYPERFINE COUPLING

5.4.1. *Integrals*

The computation of integrals for the nuclear spin-electronic orbit and the nuclear spin-electronic spin interactions is fairly simple since only one-electron integrals are required.

For $\hat{\mathcal{H}}_{\text{LI}}$ (Eq. 27) only one-electron integrals with the operator $(\vec{r}_{i\kappa} \times \vec{p}_i)/r^3_{i\kappa}$ are needed analogous to the one-particle electronic spin-orbit integrals. We have chosen here the index κ to denote a nucleus in order to avoid confusion with the electronic spin functions α and β. Again, in a basis of real functions there are no diagonal matrix elements with this operator.

No additional integrals have to be computed for the isotropic hyperfine coupling.

One-electron integrals of the type $Y_{2m}(\omega_{i\kappa})/r_{i\kappa}^3$ are required in the dipolar interaction term. Herein the Y_{2m} denote spherical harmonics and $\omega_{i\kappa}$ measures the solid angle between the position of the electron and the nucleus with respect to some centre, usually the nuclear centre of mass. The integrals may also be calculated with respect to a local Cartesian coordinate frame. In polyatomic molecules with symmetry related nuclei one then has to take care that at a later stage the integrals have to be transformed to inertial coordinates, i.e., a coordinate system with axes oriented parallel to the principal axes of the molecular inertia tensor.

5.4.2. *Matrix Elements*

A first-order hyperfine splitting for a nucleus with spin $I \geq 1/2$ is obtained with $\hat{\mathcal{H}}_{\mathrm{LI}}$ only for orbitally degenerate states. It is easily calculated and often yields good results already at the SCF level. Second-order effects occur, because the nuclear spin-electronic orbit operator has the capacity of coupling different electronic states, but their contributions to the hyperfine splitting are generally negligible. Exceptions are e.g. $^1\Sigma$ states where the first-order hyperfine coupling (including the spin-spin terms) vanishes.

$\hat{\mathcal{H}}_{\mathrm{Fc}}$ has diagonal matrix elements only and may be calculated as an expectation value. Despite of this simplicity the computation of the isotropic hyperfine coupling is delicate. The electronic part of the Fermi contact operator (Eq. 25), $\vec{s}_i\delta(\vec{r}_{i\kappa})$, yields the spin density of the electronic wavefunction at nucleus κ, i.e., the difference of the densities for α and β electrons. Only s electrons have a non-vanishing density at a nucleus and may contribute to this entity. This does not necessarily mean that only molecules in electronic states with open shell s configurations experience an isotropic hyperfine interaction. A net spin density at the nucleus is generated in nearly all radicals due to spin polarization of the s electrons by some unpaired spins of the valence electrons. Important mechanisms causing spin polarization are electron correlation and/or spin-orbit coupling. The proper determination of the isotropic hyperfine coupling therefore requires flexible AO basis sets well representing the wavefunctions of inner and valence shell electrons close to the nucleus and an extensive electron correlation treatment including the inner shell electrons. For heavier elements inclusion of relativistic effects, both scalar and spin-dependent, is mandatory.

The dipolar interaction of the nuclear and electronic spin magnetic moments measures the anisotropy of the electronic spin density near a nucleus. Its determination is computationally much less demanding than the isotropic part. For a proper description of the angular charge distribution, polarization functions are needed in the AO basis, but electron correlation effects are less important. A small CI expansion or sometimes even a Hartee-Fock wavefunction may suffice. $\hat{\mathcal{H}}_{\rm IS}$ gives rise to both first- and second-order magnetic hyperfine splittings. Regarding its tensorial structure we find that $\vec{\mathcal{I}}_0\mathcal{T}_0^{(2)}\vec{\mathcal{S}}_0$ may couple different components of an electronic state. In angularly isotropic atomic states, i.e., states of S symmetry, the expectation values of the xx, yy, and zz components of $\hat{\mathcal{H}}_{\rm IS}$ are equal and therefore cancel each other. The same applies to xx and yy components in all states with $\Lambda \neq \pm 1$ of linear molecules. In the particular case of a Π state of a linear molecule also $\vec{\mathcal{I}}_{-1}\mathcal{T}_{+2}^{(2)}\vec{\mathcal{S}}_{-1}$ and $\vec{\mathcal{I}}_{+1}\mathcal{T}_{-2}^{(2)}\vec{\mathcal{S}}_{+1}$ add to the first-order hyperfine interaction. They couple $\langle +\Lambda - \Sigma|$ and $| -\Lambda + \Sigma\rangle$ and vice versa and thus contribute to the Λ doubling. Both $\mathcal{T}_{\pm 2}^{(2)}$ and $\mathcal{T}_{\pm 1}^{(2)}$ can mediate interactions between different electronic states and give rise to second-order contributions.

6. Comparison with experiment

Spectroscopic parameters of a molecule are derived from experimentally determined spectra by fitting term values to a properly chosen model Hamiltonian. In rare cases experimentalists have used a multi-state ansatz like the supermultiplet approach [26] to fit the rovibronic spectra of strongly interacting near-degenerate electronic states. Mostly, however, the model Hamiltonian is an effective one-state Hamiltonian which incorporates the interactions with other electronic states in form of perturbation parameters.

The vibrational dependence of an effective molecular parameter A is usually expressed as

$$A_{v_i} = A_e + \sum_k \alpha_k (v_i + \frac{d_i}{2})^k \tag{98}$$

where A_e is the value of the property at the equilibrium distance and d_i is the degeneracy of the vibrational mode v_i. A general expression for the rotational dependence of a spectroscopic parameter cannot be given. Its functional form varies with the type of basis functions chosen to describe the actual rotation of the nuclear frame which in turn depends on the order in which the angular momenta are coupled (Hund's cases in linear molecules) and the type of rotor (spherical top, symmetric top (prolate, oblate), asymmetric top). Looking at these particular cases in detail goes far beyond the scope of the present article. For more information see e.g. [27, 28, 29].

The safest way of comparing theoretical data to experiment is to compute the spectrum and to fit the calculated term energies to the same model Hamiltonian as has been used by the experimentalists. The "fine-structure" splitting caused by (first-order) spin-orbit coupling is in most instances larger than the energy separation between adjacent rotational levels, at least for low values of J, and rotational excitations can therefore be neglected in a first approximation. Unfortunately, this is not the case for multiplet splittings brought about by electronic spin-spin coupling, rotational coupling, or the nuclear magnetic interactions. This procedure may be cumbersome, therefore, and one often directly compares expectation values computed at the equilibrium geometry of a given electronic state with experimental parameters extrapolated from rovibrational branches. Alternatively, if a property is known for various values of a vibrational coordinate Q, spectroscopic parameters in specific vibrational levels may be calculated. To this end, A is fitted to a function of Q — mostly a polynomial or a cubic spline — and vibrationally averaged, i.e. weighted with the probability density of a particular vibrational state and integrated over the vibrational coordinate Q.

$$A_{v_i} = \langle \chi_{v_i}(Q) | A(Q) | \chi_{v_i}(Q) \rangle \tag{99}$$

A word of caution: in the experimental literature frequently shift operators are used instead of employing the space and spin angular momenta in their tensorial forms as in Eq. 37. The shift operators are defined according to

$$\mathcal{S}_+ = \mathcal{S}_x + i\mathcal{S}_y \tag{100}$$

$$\mathcal{S}_0 = \mathcal{S}_z \tag{101}$$

$$\mathcal{S}_- = \mathcal{S}_x - i\mathcal{S}_y \tag{102}$$

where $\mathcal{S}$ may be replaced with any other angular momentum operator.

6.1. FIRST-ORDER SPIN-ORBIT SPLITTING

Only spatially degenerate states exhibit a first-order zero-field splitting. For a comparison with experiment, the matrix elements of one or the other microscopic spin-orbit Hamiltonian have to be equated with those of a phenomenological operator which reads

$$A_{SO}\vec{\mathcal{L}} \cdot \vec{\mathcal{S}} = a_{\|}\mathcal{L}_0\mathcal{S}_0 + \frac{1}{2}a_{\perp}(\mathcal{L}_+\mathcal{S}_- + \mathcal{L}_-\mathcal{S}_+) \tag{103}$$

if shift operator conventions are employed. One has to be aware of the fact, however, that experimentally determined parameters are effective ones and may contain second order contributions. Second-order SOC may be large, particularly in heavy element compounds, and is not always distinguishable from first-order effects (see also Secs. 6.2).

Using a phenomenological spin-orbit operator, the first-order splitting of two neighbouring atomic fine-structure levels with quantum numbers J and $J-1$ are easily calculated. In a Russel-Saunders (LS) coupling scheme $\vec{\mathcal{L}} \cdot \vec{\mathcal{S}}$ may be expressed as $\frac{1}{2}(\vec{\mathcal{J}}^2 - \vec{\mathcal{L}}^2 - \vec{\mathcal{S}}^2)$ yielding $E(J) - E(J-1) = A_{\mathrm{SO}} J$. This is Landé's interval rule. Unfortunately, for heavy atoms this rule is seldom obeyed well and extraction of an atomic spin-orbit parameter from purely experimental data may be tricky or even impossible.

As an example regard the fine-structure splittings of the $^3D_g(d^9s^1)$ ground state of atomic nickel and the first excited state of atomic copper, $^2D_g(d^9s^2)$.

TABLE 7. Spin-orbit splittings of the ^3D_g (d^9s^1) ground state of atomic Ni and the ^2D_g (d^9s^2) first excited state of the Cu atom.

Ni (^3D_g)	Theory [30] first order	Theory [30] first and second order	Experiment [31]
J	$\Delta E(J, J-1)[cm^{-1}]$		
3	904	672	675
2	602	834	833
Cu (^2D_g)	Theory [32] first order		Experiment [31]
J	$\Delta E(J, J-1)[cm^{-1}]$		
5/2	2006		2042

The ^2D_g state of copper is well separated from other electronic states which by symmetry are allowed to couple in second order. A_{SO} may thus be calculated directly from the splitting by multiplying with 2/5 yielding an expermiental value of 816.8cm^{-1} to be compared with the theoretically determined spin-orbit parameter of 802.4cm^{-1}. Things are not so easy in the case of nickel. The ratio of the measured fine-structure splittings amounts to 1.234 instead of 0.667 as expected from the Landé interval rule. Close lying electronic states are $^3F_g(d^8s^2)$ with components $J = 4, 3, 2$ and $^1D_g(d^9s^1)$ with $J = 2$. Since the coupling of d^8 and d^9 states is negligible in nickel, the $J = 3$ and $J = 1$ levels of ^3D_g are nearly unperturbed and their energetic separation $E(J = 3) - E(J = 1) = 5A_{\mathrm{SO}}$ may therefore be taken to determine A_{SO} =301.6cm^{-1} experimentally in excellent agreement with the thoeretical value of 301cm^{-1}. As the calculations show, the downward shift of the $J = 2$ level with respect to its ideal value is described well by taking into account the second-order interaction with a single perturber, the ^1D_g state.

For a linear molecule only the z-component of $\hat{\mathcal{H}}_{\mathrm{SO}}$ is involved in the evaluation of diagonal matrix elements. In a Hund's case (a) basis the corresponding parameters can therefore be obtained from

$$A_e = \frac{1}{\Lambda\Sigma}\langle \Psi_n^{(0)}(\vec{r}, \tilde{Q}_e)|\hat{\mathcal{H}}_{\mathrm{SO}}|\Psi_n^{(0)}(\vec{r}, \tilde{Q}_e)\rangle_{\vec{r}} \tag{104}$$

where the subscript $\vec{r}$ indicates that the integration is confined to the electronic degrees of freedom. Vibrational averaging is performed by multiplying the Q-dependent electronic integral by the appropriate vibrational wavefunctions and integrating over the vibrational coordinate, i.e.,

$$A_{v_i} = \frac{1}{\Lambda\Sigma}\langle \chi_{v_i}(Q)|\langle \Psi_n^{(0)}(\vec{r}, \tilde{Q})|\hat{\mathcal{H}}_{\mathrm{SO}}|\Psi_n^{(0)}(\vec{r}, \tilde{Q})\rangle_{\vec{r}}|\chi_{v_i}(Q)\rangle_Q \,. \tag{105}$$

Typically, Hund's case (b) applies to molecules with $\Lambda = 0$ for which spin-orbit effects do not contribute to in first order. For states with a non-zero Λ value and a rotational parameter B_v larger than A_v it is most convenient to calculate fine-structure splittings by using the programs described in Sec. 5.3 and to fit them to a function of the form $\gamma \cdot \vec{\mathcal{N}}\vec{\mathcal{S}}$ where $\vec{\mathcal{N}}$ represents the total angular momentum save for from spin. The strong coupling limit, Hund's case (c), cannot be dealt with in first order but requires the inclusion of higher-order (at least second-order) spin-orbit coupling in the calculations.

For the sign of the spin-orbit parameter A the following conventions are used:

$A > 0$ denotes a regular state. In regular states the sublevel with smallest J or Ω quantum number is lowest in energy. Typically, states with less than half filled shells are regular.

$A < 0$ denotes an inverted state. In inverted states the sublevel with largest J or Ω quantum number is lowest in energy. Typically, states with more than half filled shells (hole states) are inverted.

regular state, $^2\Delta_r$

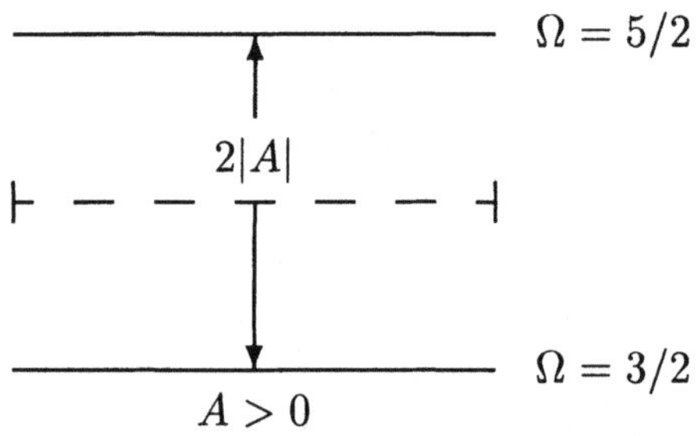

inverted state, $^3\Pi_i$

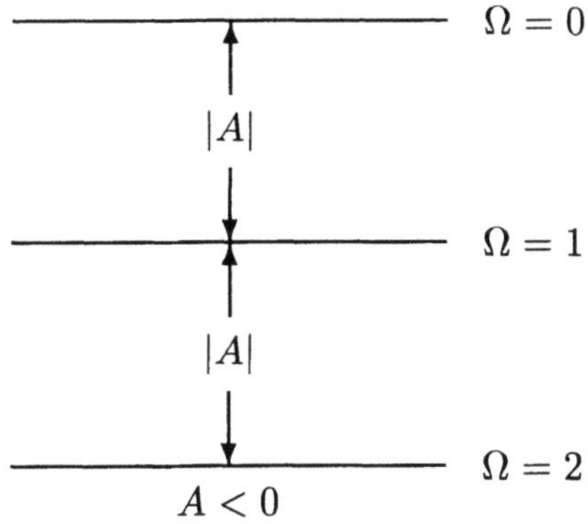

6.2. SECOND-ORDER SPIN-ORBIT SPLITTING

Every electronic state of whatever spin- or spatial symmetry may experience an energy shift because of second-order spin-orbit effects. In doublet states second-order spin-orbit splitting just adds to the first-order effect — in an atom or a non-rotating molecule they can experimentally not be distinguished. In triplets or higher multiplicity states second order SOC causes deviations from the regular first-order splitting patterns addressed in Secs. 4 and 6.1. But also in this case their unambiguous identification is not possible solely on the basis of experimental data because second-order SOC and first-order electronic SSC have identical tensorial structures. In order to see this we rewrite the perturbation sum

$$\sum_{n=1}^{\infty} \frac{\langle\Psi_0|\hat{\mathcal{H}}_{SO}|\Psi_n\rangle\langle\Psi_n|\hat{\mathcal{H}}_{SO}|\Psi_0\rangle}{E_n - E_0} = \langle\Psi_0|\hat{\mathcal{H}}_{SO}\left(\sum_{n=1}^{\infty}\frac{|\Psi_n\rangle\langle\Psi_n|}{E_n - E_0}\right)\hat{\mathcal{H}}_{SO}|\Psi_0\rangle \ . \tag{106}$$

Denoting the spatial part of $\hat{\mathcal{H}}_{SO}$ by the first rank tensor $Q^{(1)}$ one obtains

$$\begin{aligned} &\langle\Psi_0|\sum_i \vec{s}_i Q_i^{(1)}\left(\sum_{n=1}^{\infty}\frac{|\Psi_n\rangle\langle\Psi_n|}{E_n - E_0}\right)\sum_j Q_j^{(1)}\vec{s}_j|\Psi_0\rangle \\ = \ &\langle\Psi_0|\textstyle\sum_i\sum_j \vec{s}_i \underbrace{Q_i^{(1)}\left(\displaystyle\sum_{n=1}^{\infty}\frac{|\Psi_n\rangle\langle\Psi_n|}{E_n - E_0}\right)Q_j^{(1)}}_{\tilde{\mathcal{D}}_{ij}^{(2)}}\vec{s}_j|\Psi_0\rangle \ . \end{aligned} \tag{107}$$

The effective operator in Eq. 107 has exactly the same structure as the last term in Eq. 42. The fine-structure tensor of rank two $\mathcal{D}^{(2)}$ has thus two major contributions

$$\mathcal{D}^{(2)} = \mathcal{D}^{(2)}_{\mathrm{SS}} \oplus \mathcal{D}^{(2)}_{\mathrm{SO}} \,. \tag{108}$$

6.3. FIRST-ORDER ELECTRONIC SPIN-SPIN COUPLING

From the selection rules for tensor operators 61 it follows that only states with $S \geq 1$ are split in first order by SSC. Unlike first-order SOC, first-order SSC is not restricted to spatially degenerate states. For low-symmetry molecules mostly the Cartesian components of the electronic dipolar coupling tensor are specified. However, if symmetry is C_{2v} or higher, only diagonal elements of the tensor operator Eq. 43 contribute to $\langle\Psi_0|\hat{\mathcal{H}}_{\mathrm{SS}}|\Psi_0\rangle$, i.e., the principal axes diagonalizing the SCC tensor coincide with the inertial axes. For triplet and higher multiplicity states one then obtains

$$\langle\Psi_0|\hat{\mathcal{H}}_{\mathrm{SS}}|\Psi_0\rangle = -D_{xx}\mathcal{S}_x^2 - D_{yy}\mathcal{S}_y^2 - D_{zz}\mathcal{S}_z^2 \tag{109}$$

and making use of the traceless property $D_{xx} + D_{yy} + D_{zz} = 0$

$$\langle\Psi_0|\hat{\mathcal{H}}_{\mathrm{SS}}|\Psi_0\rangle = D(\mathcal{S}_z^2 - \frac{1}{3}\mathcal{S}^2) + E(\mathcal{S}_x^2 - \mathcal{S}_y^2) \tag{110}$$

where $D = -\frac{3}{2}D_{zz}$ and $E = \frac{1}{2}(D_{yy} - D_{xx})$. Usually, experimentalists fit spin-spin splittings in terms of these parameters D and E.

For linear molecules $E = 0$ by symmetry, since $D_{yy} = D_{xx}$. In the experimental literature, a different notation is used often in this case, i.e., $\lambda_{\mathrm{SS}} \equiv \frac{1}{2}D$. For similar reasons, atoms in S states do not show a fine-structure splitting caused by electronic SSC at all.

Spin-spin splitting in a triplet state, spatially non-degenerate

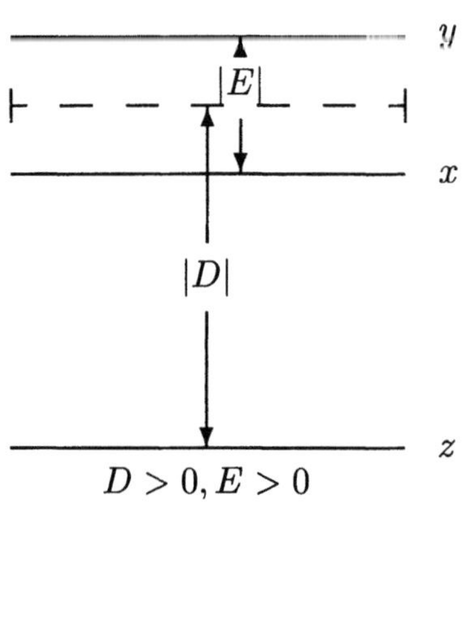

6.4. Λ-DOUBLING

In the experimental literature, labels e and f are frequently used to designate the components of a Kramers doublet rather than the parity eigenvalues $+$ or $-$. Positive and negative parity states and states labelled e and f are related by [33]

$$\left\{ \begin{array}{c} e \\ f \end{array} \right\} \text{designates levels of parity} \left\{ \begin{array}{c} +(-1)^{J-\frac{1}{2}} \\ -(-1)^{J-\frac{1}{2}} \end{array} \right. \tag{111}$$

for half-integral values of the total angular momentum J and by

$$\left\{ \begin{array}{c} e \\ f \end{array} \right\} \text{designates levels of parity} \left\{ \begin{array}{c} +(-1)^{J} \\ -(-1)^{J} \end{array} \right. \tag{112}$$

for integral values of J.

Like the theoretical evaluation of Λ-doubling parameters, the fitting of a model Hamiltonian to reproduce an experimentally recorded spectrum is complicated by the fact that the splittings are determined by off-diagonal couplings. Again, I shall not consider many particular cases in detail but restrict the discussion to $^2\Pi$ states. If high-precision data is experimentally available for many vibrational levels, the Λ-doubling parameters p and q of a $^2\Pi$ state can be fitted to expressions similar to Eqs. 93 and 94. For a particular vibrational level v they read [34]

$$p_v(^2\Pi) = 2\sum_{\Sigma,v'} \frac{\langle ^2\Pi_v | \sum_i a_i \ell_i^+ \mathbf{s}_i^- | ^2\Sigma_{v'} \rangle \langle ^2\Sigma_{v'} | \hat{B} \sum_i \ell_i^- | ^2\Pi_v \rangle (-1)^k}{E(^2\Pi_v) - E(^2\Sigma_{v'})} \tag{113}$$

$$q_v(^2\Pi) = 2\sum_{\Sigma,v'} \frac{|\langle ^2\Pi_v | \hat{B} \sum_i \ell_i^+ | ^2\Sigma_{v'} \rangle|^2 (-1)^k}{E(^2\Pi_v) - E(^2\Sigma_{v'})} . \tag{114}$$

Here, the a_i represent phenomenological orbital spin-orbit parameters and shift operator conventions are used. Parameters p and q computed according to Eqs. 93 and 94 are easily compared to the experimentally determined p_v and q_v by vibrational averaging.

In many cases a much simpler model Hamiltonian is employed by experimentalists. Restricting the perturbation sums to a unique perturber and assuming a pure precession, identical potential approximation, p_v and q_v reduce to [34]

$$p_v(^2\Pi) \approx 2A_v B_v \ell(\ell+1)/(E(^2\Pi) - E(^2\Sigma)) \tag{115}$$

$$q_v(^2\Pi) \approx 2B_v^2 \ell(\ell+1)/(E(^2\Pi) - E(^2\Sigma)) \tag{116}$$

where A_v and B_v are the spin-orbit and rotational parameters in the vth vibrational level of the $^2\Pi$ state. For $A_v \gg B_v$, i.e., $Y \gg 1$ the J dependence of the splitting is simplified [35] to

$$\Delta E(^2\Pi) \approx \begin{cases} p(J+\frac{1}{2}) & \text{for } ^2\Pi_{1/2} \\ (2qY^{-1}+pY^{-2})\,(J-\frac{1}{2})(J+\frac{1}{2})(J+\frac{3}{2}) & \text{for } ^2\Pi_{3/2} \end{cases} \quad (117)$$

which applies in the limit of ideal Hund's case (a) coupling. The Mulliken-Christy formula (117) may serve as a rough estimate of the Λ splitting in $^2\Pi$ states for low J values only and is generally not very well obeyed. Nevertheless, many experimentalists make use of it because of its simplicity.

From Eq. (117) it is seen that for low values of J and large Y (Hund's case (a) coupling) the Λ-splitting is much more pronounced in the $^2\Pi_{1/2}$ fine-structure component than in $^2\Pi_{3/2}$ for which the splitting occurs only in third-order perturbation theory. Due to the approximate J^3 dependence and the increasing coupling between $^2\Pi_{3/2}$ and $^2\Pi_{1/2}$ by rotational distortion terms the splitting increases considerably with J.The Λ-splitting of a Δ state is even smaller since it is a fourth-order effect.

6.5. NUCLEAR MAGNETIC HYPERFINE COUPLING CONSTANTS

The extraction of nuclear magnetic hyperfine coupling constants from electron paramagnetic resonance (EPR) spectra is complicated by the fact that in addition to the internal magnetic moments an external magnetic field is applied. The nuclear spin is then space-quantized with respect to the external field while the electronic matrix elements usually refer to the molecular coordinate system.

Without going into details, let me just mention two examples where measurements of hyperfine coupling constants have to employ external magnetic fields. The first concerns free radicals in solids. Because of the crystalline surrounding their spatial angular momentum is quenched. Neither a rotational movement nor a spatial angular momentum ties the electronic spins then to a molecule-fixed axis. Their electronic spins are therefore also space-quantized. Measuring the angular dependence of the interaction energy allows to determine both the isotropic and anisotropic parts of the magnetic hyperfine tensor. Formulae relating the body-fixed and space-fixed tensor components for many particular cases may be found e.g. in [16].

In the liquid phase molecules are oriented randomly in space. Due to frequent collisions their rotational motion is not quantized and thus all anisotropic components of the magnetic hyperfine tensor average to zero. Only the isotropic hyperfine coupling constants can be measured in liquids.

They can directly be compared to the calculated values of

$$a_{\text{iso}} = g_e \mu_B \mu_N g_{N_\kappa} \frac{8\pi}{3} |\Psi(0)|^2 \tag{118}$$

where $|\Psi(0)|^2 = |\Psi(0)|^2_\alpha - |\Psi(0)|^2_\beta$ is the difference in the spin density at the nucleus κ due to electrons with spin up and spin down.

Things are different for gas phase radicals with a non-zero spatial angular momentum. L and S are then aligned to the internuclear axis. Hyperfine splittings are J dependent and the parameters can experimentally be determined from microwave spectra or by related techniques. In the weak(or even zero)-external-field limit also the nuclear spin can be expressed in terms of molecule-fixed components. Let ϕ measure the azimutal angle of the electron. Applying shift operator conventions the phenomenological magnetic hyperfine coupling Hamiltonian for a diatomic molecule in Hund's (a) coupling case then reads [29]

$$\begin{aligned}
\mathcal{H}_{\text{hfs}} &= a\mathcal{I}_z\mathcal{L}_z && (119)\\
&+ \frac{1}{2}a\left(\mathcal{I}_+\mathcal{L}_- + \mathcal{I}_-\mathcal{L}_+\right) && (120)\\
&+ (b+c)\mathcal{I}_z\mathcal{S}_z && (121)\\
&+ \frac{1}{2}b\left(\mathcal{I}_+\mathcal{S}_- + \mathcal{I}_-\mathcal{S}_+\right) && (122)\\
&+ \frac{1}{2}d\left(\exp(-2i\phi)\mathcal{I}_+\mathcal{S}_+ + \exp(2i\phi)\mathcal{I}_-\mathcal{S}_-\right) && (123)\\
&+ \frac{1}{2}e\left(\exp(-i\phi)(\mathcal{I}_+\mathcal{S}_z + \mathcal{I}_z\mathcal{S}_+) + \exp(i\phi)(\mathcal{I}_-\mathcal{S}_z + \mathcal{I}_z\mathcal{S}_-)\right) && (124)
\end{aligned}$$

Terms 119 and 120 represent the electron orbital - nuclear spin coupling. For states with $\Lambda \neq 0$ the z component makes an important contribution to the diagonal matrix elements of the hyperfine interaction; in singlet electronic states this is even the dominant one. The operator 120 mixes in contributions from other electronic states which are generally too small to be detected, however. Term 121 and 122 contain contributions from both the isotropic and the anisotropic hyperfine coupling tensors. The matrix elements of 121 are non-zero only in the diagonal case, thus bringing about equal shifts of the hyperfine levels in both Kramers components. Term 122 couples different fine-structure levels of an electronic state, e.g. $^2\Pi_{3/2}$ and $^2\Pi_{1/2}$. The operators 123 and 124 originate solely from the dipolar coupling. Like the corresponding electron spin-spin term, the first of them couples the two Kramers sublevels in Π states and leads to an additional Λ-doubling. Finally, 124 may cause interactions with other electronic states which are usually negligible.

If sufficient measurements are available, all these parameters can be determined experimentally. The parameter a can directly be compared with the theoretically determined expectation value

$$a_{\text{orb}}(\Psi) = \frac{g_e \mu_B \mu_N g_{N_\kappa}}{\Lambda} \langle \sum_i \frac{(\vec{r}_{i\kappa} \times \vec{p}_i)}{r^3_{i\kappa}} \rangle_\Psi \, . \tag{125}$$

Equating b with

$$a_{\text{iso}}(\Psi) - T^{zz}_{\text{dip}}(\Psi) = \frac{g_e \mu_B \mu_N g_{N_\kappa}}{\Sigma} \left(\langle \frac{8\pi}{3} \sum_i \vec{s}_i \delta(\vec{r}_{i\kappa}) \rangle_\Psi - \langle \sum_i \frac{z^2_{i\kappa}}{r^5_{i\kappa}} \vec{s}_i \rangle_\Psi \right) \tag{126}$$

and c with

$$3T^{zz}_{\text{dip}} = \frac{g_e \mu_B \mu_N g_{N_\kappa}}{\Sigma} \langle \sum_i \frac{3z^2_{i\kappa}}{r^5_{i\kappa}} \vec{s}_i \rangle_\Psi \tag{127}$$

allows to relate $a_{\text{iso}}(\Psi)$ and the zz component of the dipolar tensor to the parameters b and c. Likewise, its spin-normalized $x^2 - y^2$ and $xz + zx$ components correspond to d and e, respectively.

7. Spin-forbidden transitions

Apart from fine-structure splitting the occurance of spin-forbidden transitions, both radiative and non-radiative, is the most striking feature in which spin-orbit interaction manifests itself. Radiative spin-forbidden transitions in light molecules usually take place at the millisecond time scale, if the transition is dipole-allowed. A dipole- and spin-forbidden transition is even weaker, with lifetimes of the order of seconds. Proceeding down the periodic table spin-forbidden transitions become more and more allowed due to the increase of spin-orbit coupling. For molecules containing elements with principal quantum number 5 or higher (and the late first-row transition metals Ni and Cu) there is hardly any difference between transition probabilities of spin-allowed and spin-forbidden processes.

For (radiationally) long-lived excited states other, non-radiative, depopulation mechanisms are often the lifetime determinating processes. Under collision-free conditions which I shall always assume here a coupling of potential energy hypersurfaces can be brought about, e.g. by nonadiabatic interactions or rotational coupling. If the interacting states differ in spin quantum numbers, electron spin-orbit or spin-spin coupling has to be invoked. A typical example for the latter process is a non-radiative transition between triplet and singlet states in the presence of an intersystem crossing (IC) in a photochemically excited organic molecule.

Let us work out the way to calculate intensities of a spin-forbidden radiative transition by looking at an example: the spin-forbidden radiative transition from the first excited electronic state to the ground state of formaldehyde. We employ perturbation theory here and restrict ourselves to a few terms in the (in principal) infinite perturbation sums Eq. 72 because things are more transparent in this way. In an actual calculation it might be more advantageous to use methods that avoid an explicit summation of the perturbation expressions, as e.g. spin-orbit CI [36, 37], direct perturbation theory [38] or response theory [39].

Formaldehyde is planar in its ground state equilibrium configuration. The molecule is chosen to lie in the yz plane with the z-axis being the symmetry axis.

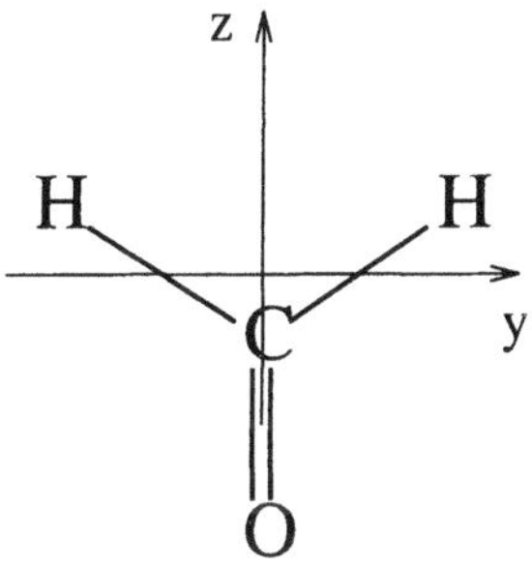

The ground state has $^1A_1$ symmetry, the first excited state, a $^3A_2$, results from an $n \rightarrow \pi^*$ excitation. We are interested in the spin-forbidden radiative transition from X $^1A_1 \leftarrow$ a $^3A_2$. There are various other low-lying valence states of $n \rightarrow \pi^*$ type ($^1A_2$), $\pi \rightarrow \pi^*$ ($^3A_1$,$^1A_1$), $\sigma \rightarrow \pi^*$ ($^3B_1$,$^1B_1$), and $n \rightarrow \pi^*$ type ($^1A_1$) the spin-forbidden radiative transition may borrow intensity from.

1. Consider first the electronic ground state. We shall express its unperturbed wavefunction as a bra-vector.

$$\langle \Psi_0^{(0)} | = \langle \text{X } {}^1A_1^{(0)} | \tag{128}$$

The first-order perturbed wavefunction is analogously written as

$$\langle \Psi_0^{(1)} | = \sum_i^{\text{triplets}} a_i^* \langle {}^3\Phi_i^{(0)} | \tag{129}$$

We now pose the question which of the triplet states are by symmetry allowed to contribute to the spin-orbit perturbed wavefunction of the ground state.

2. The first excited state has three multiplet components. We choose the unperturbed states as

$$\begin{aligned}|\Psi_1^{(0)}, m_{Sx}\rangle &= |\mathrm{a}\ {}^3A_2^{(0)}, m_{Sx}\rangle \\ |\Psi_1^{(0)}, m_{Sy}\rangle &= |\mathrm{a}\ {}^3A_2^{(0)}, m_{Sy}\rangle \\ |\Psi_1^{(0)}, m_{Sz}\rangle &= |\mathrm{a}\ {}^3A_2^{(0)}, m_{Sz}\rangle \end{aligned} \tag{130}$$

and express them as ket-vectors. Accordingly, the first-order spin-orbit perturbed wavefunction may be written as

$$\begin{aligned}|\Psi_1^{(1)}, m_{Sx}\rangle &= \sum_i^{\text{singlets}} b_i|{}^1\Phi_i^{(0)}\rangle \\ |\Psi_1^{(1)}, m_{Sy}\rangle &= \sum_i^{\text{singlets}} c_i|{}^1\Phi_i^{(0)}\rangle \\ |\Psi_1^{(1)}, m_{Sz}\rangle &= \sum_i^{\text{singlets}} d_i|{}^1\Phi_i^{(0)}\rangle \end{aligned} \tag{131}$$

There are also triplet and quintet contributions to $|\Psi_1^{(1)}\rangle$ but they do not have direct dipole transition elements to the singlet ground state and are thus not considered here. Again we ask which of the singlet states may couple to one of the components of the first excited state.

3. In terms of these wavefunctions the electric dipole transition matrix element is

$$\mu_{\text{el}} = \langle\Psi_0^{(0)} + \Psi_0^{(1)}|\sum_j e\vec{r}_j|\Psi_1^{(0)} + \Psi_1^{(1)}\rangle\ . \tag{132}$$

Express the electric dipole transition moment in terms of the contributions from the zeroth order states. Neglect herein the higher-order terms $\langle\Psi_0^{(1)}|\sum_j e\vec{r}_j|\Psi_1^{(1)}\rangle$. Generally, they are only of importance, if the perturbing state is energetically near-degenerate with one of the perturbed states and has a large coefficient in the perturbation expansion. In this case, other methods than a sum-over-states method have to be employed.

(Hint: employ the spin-orbit operator in its Cartesian form 38. Remember that the compound space and spin operator is totally symmetric!)

If you want to practise, skip the following paragraph in the first run. Otherwise, here is the answer.

1. Only states of total (electronic space + spin) symmetry A_1 can interact with the ground state electronic wavefunction. The first-order perturbed ground state wavefunction is apparently given by

$$\langle\Psi_0^{(1)}| = a_1^*\langle\mathrm{a}^3A_2^{(0)}, m_{Sz}| + a_2^*\langle{}^3B_1^{(0)}(\sigma\rightarrow\pi^*), m_{Sy}| + \ldots \tag{133}$$

State	spatial symmetry	spin symmetry	$\otimes$
X $^1A_1$	A_1	A_1	A_1
a $^3A_2$,m_{Sz}	A_2	A_2	A_1
,m_{Sy}		B_1	B_2
,m_{Sx}		B_2	B_1
$^1A_2$	A_2	A_1	A_2
$^3A_1$,m_{Sz}	A_1	A_2	A_2
,m_{Sy}		B_1	B_1
,m_{Sx}		B_2	B_2
$^3B_1$,m_{Sz}	B_1	A_2	B_2
,m_{Sy}		B_1	A_1
,m_{Sx}		B_2	A_2
$^1B_1$	B_1	A_1	B_1

2. The three components of the spin-orbit perturbed excited state wavefunction read:

$$\begin{aligned} |\Psi_1^{(1)}, m_{Sx}\rangle &= b_1|^1B_1^{(0)}(\sigma \to \pi^*)\rangle + \ldots \\ |\Psi_1^{(1)}, m_{Sy}\rangle &= + \ldots \\ |\Psi_1^{(1)}, m_{Sz}\rangle &= d_1|\mathrm{X}\ ^1A_1^{(0)}\rangle + d_2|^1A_1^{(0)}(n \to \pi^*)\rangle + \ldots \end{aligned} \quad (134)$$

Within the set of perturbing states considered here, there is obviously no state contributing to $|\Psi_1^{(1)}, m_{Sy}\rangle$.

3. The evaluation of the transition moment is now straight forward. According to Table 1 components of the dipole operator are found in A_1, B_1, and B_2 symmetries. The direct transition $\langle\Psi_0^{(0)}|\sum_j e\vec{r}_j|\Psi_1^{(0)}\rangle$ is spin-forbidden and the transition element zero. We thus obtain

$$\begin{aligned} \mu_{el} &= \langle\Psi_0^{(0)}|\sum_j e\vec{r}_j|\Psi_1^{(1)}\rangle + \langle\Psi_0^{(1)}|\sum_j e\vec{r}_j|\Psi_1^{(0)}\rangle \\ &= b_1\langle\mathrm{X}\ ^1A_1^{(0)}|\sum_j e\hat{x}_j|^1B_1^{(0)}(\sigma \to \pi^*)\rangle \\ &+ a_2^*\langle^3B_1^{(0)}(\sigma \to \pi^*), m_{Sy}|\sum_j e\hat{y}_j|\mathrm{a}\ ^3A_2^{(0)}, m_{Sy}\rangle \\ &+ d_1\langle\mathrm{X}\ ^1A_1^{(0)}|\sum_j e\hat{z}_j|\mathrm{X}\ ^1A_1^{(0)}\rangle + d_2\langle\mathrm{X}\ ^1A_1^{(0)}|\sum_j e\hat{z}_j|^1A_1^{(0)}(n \to \pi^*)\rangle \end{aligned}$$

$$+ \quad a_1^* \langle \mathrm{a}\ ^3A_2^{(0)}, m_{Sz} | \sum_j e\hat{z}_j | \mathrm{a}\ ^3A_2^{(0)}, m_{Sz} \rangle \ . \tag{135}$$

In the picture of spin-orbit perturbed Russel-Saunders coupled states the dipole transition moment of a spin-forbidden radiative transition is thus a sum of dipole (transition) moments of spin-allowed transitions weighted by spin-orbit coupling coefficients. This is exactly what experimentalists mean when they speak of intensity borrowing.

Two features are remarkable in the sum 135:

- Among the terms not only dipole transition moments of spin-allowed transitions occur but also dipole moments of the ground and excited electronic states involved — their difference to be more precise, because a close look shows that $a_1^* = -d_1$. Employing the usual velocity form of the dipole transition operator one finds that the corresponding diagonal matrix elements of $\vec{p}_i$ are zero. This discrepancy was resolved by Lohr [40] and Goodman and Laurenzi [41]. They argued that in Breit-Pauli theory the appropriate transition dipole operator in the velocity form is given by

$$\vec{\Pi} = \sum_i \left(\vec{p}_i + \frac{e^2\hbar}{2m^2c^2} \vec{s}_i \times \vec{\nabla}_i \left\{ \sum_\alpha \frac{Z_\alpha}{r_{i\alpha}} - \sum_{j \neq i} \frac{1}{r_{ij}} \right\} \right) \tag{136}$$

  They also showed that $\vec{\Pi}$ can be expressed by the commutator $[\hat{\mathcal{H}}_0 + \hat{\mathcal{H}}_{\mathrm{SO}}, \vec{r}] = (i\hbar/m)\vec{\Pi}$ and therefore $e\vec{r}$ is a correct dipole transition operator. Similar considerations apply to the no-pair operator. For the calculation of transition moments of spin-forbidden radiative dipole transitions the length form of the dipole transition operator should thus be employed.

- The contribution of perturbing states can be both positive or negative. In other words, they cannot only borrow intensity to a transition but also take it away.

A further remark concerns the form of the magnetic transition dipole operator. Usually, the contribution of this term in the expansion of the interaction with a radiation field is much smaller than for the electric dipole transition operator. It may play an important role when electric dipole transitions are symmetry-forbidden as e.g. in homonuclear diatomics. Usually, the magnetic dipole transition operator is found in the literature in a form which reads (up to constants) as

$$\vec{\mu}_{\mathrm{L}} = \mu_{\mathrm{B}} \sum_j \vec{\ell}_j \ . \tag{137}$$

In spin-forbidden transitions there is a second, non-negligible term, however, which describes a spin flip. Instead of Eq. 137 one has to employ

$$\vec{\mu}_{\text{mag}} = \mu_{\text{B}} \sum_j (\vec{\ell}_j + g_e \vec{s}_j) \ . \tag{138}$$

The spin-flip term was shown to make a considerable contribution to the transition probability $a^1\Delta_g \rightarrow X^3\Sigma_g^-$ in O_2 [42].

7.2. NON-RADIATIVE TRANSITIONS

Apart from the selection rule for the coupling of states — which should be known by now — the theoretical treatment of spin-forbidden non-radiative transitions does not differ from that of their spin-allowed counterparts. For a non-radiative transition to occur the two coupling states have to be close in energy. We distinguish between two cases:

1. The interaction of two bound electronic states undergoing a potential curve crossing.
2. The coupling of a bound state to the vibrational continuum of another electronic state leading to predissociation.

Let us begin with the first case. The requirement of near-degeneracy brings about that perturbations in the spectrum involving a pair of interacting bound states as initial or final states occur only occasionally, i.e., when their vibrational ladders approximately match in energy. In the time-independent picture which is mostly applied the coupling of two states manifests itself in that the perturbed levels occur as linear combinations of the unperturbed vibrational levels of the pure spin states. The mixing coefficients are easily calculated from the potential energy curves of the involved states and their coupling matrix element at various points of the vibrational coordinate. In a first step the Schrödinger equation for the vibrational motion is solved separately for each of the bound states. Their solutions are taken as the basis for a matrix representation of the interaction Hamiltonian. In a second step the coupling matrix element has to be expressed as a function of the vibrational coordinate, mostly a polynomial or a cubic spline function. The matrix elements of the perturbation matrix are then calculated in a way very similar to the vibrational averaging of spectroscopic parameters described in section 6. Instead of employing the same vibronic level on both sides, now

$$H_{ij} = \langle \chi_{v_i}(Q) | X(Q) | \chi_{v'_j}(Q) \rangle_Q \tag{139}$$

is evaluated where $X(Q)$ is the off-diagonal electronic coupling and v_i and v'_j belong to different electronic states. Diagonalization of this matrix directly yields energies and mixing coefficients of the levels. In practical applications one often combines experimental and theoretical information, if available, such that potential energy curves, deduced from experiment with high accuracy, are employed together with theoretically determined Q dependent coupling matrix elements.

From Eq.139 it is immediately apparent that the size of a particular H_{ij} is not only determined by the magnitude of the electronic coupling matrix element but also by the overlap of the vibrational wavefunctions v_i and v'_j, i.e., their Franck-Condon factors. As a consequence, the low vibrational levels of excited triplet states are often metastable despite a considerable electronic coupling matrix element and a near-degeneracy with a highly excited vibrational level of the electronic ground state.

The mixing of stationary states occuring in a time-independent description is equivalent to a propagation back and forth between the two vibronic states in the language of time-dependent perturbation theory, the relative probabilities of finding the molecule in either state corresponding to its squared expansion coefficient. This picture holds as long as there are no other depletion mechanisms or the radiative lifetimes of the components are long enough for an equilibration to take place. Very often, however, one of the states is short-lived, as e.g. in the case of an IC between two excited states or if one of the states is dissociative. Then the non-radiative transition occurs practically only in one direction.

Concerning bound-continuum interactions we shall restrict ourselves to weak interactions which is mostly fulfilled for spin-forbidden processes. In cases of strong interactions the full apparatus of scattering theory would have to be invoked. In the weak interaction approximation the Wentzel-Rice formula can be applied to calculate the predissociation linewidth Γ and thereby the lifetime for the non-radiative decay $\tau = 1/\Gamma$.

$$\Gamma = \frac{2\pi}{\hbar} |\langle \chi_{v_i}(Q) | X(Q) | \chi_{v'_E}(Q) \rangle_Q|^2 \tag{140}$$

It restricts the coupling of a bound vibrational level to a single continuum state with equal energy

If the potential energy curves of the bound and dissociative states cross approximately at right angles — which is often the case when the dissociative state intersects the bound state at bond distances longer than its equilibrium internuclear separation — then in addition a Franck-Condon-type approximation might be appropriate. Γ may then be estimated by

$$\Gamma \approx |\langle \chi_{v_i}(Q) | \chi_{v'_E}(Q) \rangle_Q|^2 \cdot |\langle X(Q_x) \rangle|^2 \tag{141}$$

where $\langle X(Q_x) \rangle$ is the electronic coupling matrix element at the crossing point.

8. Acknowledgements

I would like to thank Frank Rakowitz and Timo Fleig for helpful discussions and for a careful reading of the manuscript.

A. Symmetry transformation properties of spin functions

A.1. GENERAL CASE

Rotations or reflections of a molecule about a symmetry axis can unambiguously be represented as orthogonal transformations in three-dimensional coordinate space. If we want to study the transformation properties of the spin functions

$$\alpha = \begin{pmatrix} 1 \\ 0 \end{pmatrix} \; ; \; \beta = \begin{pmatrix} 0 \\ 1 \end{pmatrix} \tag{142}$$

under symmetry operations of some molecular point group we will have to find a mapping between these transformation matrices **R** in three-dimensional coordinate space and corresponding matrices **u** in a properly chosen two-dimensional vector space.

In order to find the correspondence let us consider the unitary transformations of a two-dimensional hermitian matrix with trace = 0.

Every two-dimensional traceless hermitian matrix

$$\mathbf{h} = \begin{pmatrix} z & x - iy \\ x + iy & -z \end{pmatrix} \tag{143}$$

can be written as a linear combination of the Pauli spin matrices

$$\sigma_x = \begin{pmatrix} 0 & 1 \\ 1 & 0 \end{pmatrix} \; ; \; \sigma_y = \begin{pmatrix} 0 & -i \\ i & 0 \end{pmatrix} \; ; \; \sigma_z = \begin{pmatrix} 1 & 0 \\ 0 & -1 \end{pmatrix} \tag{144}$$

with real coefficients x, y, and z.

$$\mathbf{h} = x\sigma_x + y\sigma_y + z\sigma_z = (\vec{r}, \vec{\sigma}) \tag{145}$$

At the very right of Eq. 145 **h** is written formally as a scalar product of a (spatial) vector $\vec{r}$ and the three Pauli matrices represented by $\vec{\sigma}$. This is an abbreviation I will commonly use throughout this appendix.

Operating with a unitary transformation on **h** yields a new matrix **h**$'$ that is given by

$$\begin{aligned} \mathbf{h}' &= \mathbf{u}\mathbf{h}\mathbf{u}^\dagger = \mathbf{u}(\vec{r}, \vec{\sigma})\mathbf{u}^\dagger \\ &= x'\sigma_x + y'\sigma_y + z'\sigma_z = (\vec{r}\,', \vec{\sigma}\,') \; . \end{aligned} \tag{146}$$

Expressing the unitary matrix **u** in the form

$$\mathbf{u} = \begin{pmatrix} a & b \\ -b^* & a* \end{pmatrix} \qquad \text{where} \qquad (a \cdot a^* + b \cdot b^*) = 1 \tag{147}$$

Eq. 146 can be rewritten as

$$\begin{pmatrix} z' & x'-iy' \\ x'+iy' & -z' \end{pmatrix} = \begin{pmatrix} a & b \\ -b^* & a* \end{pmatrix} \begin{pmatrix} z & x-iy \\ x+iy & -z \end{pmatrix} \begin{pmatrix} a^* & -b \\ b^* & a \end{pmatrix} . \tag{148}$$

Upon performing the matrix multiplication in Eq. 148 it is found that

$$\begin{aligned} x' &= \frac{1}{2}(a^2+a^{*2}-b^2-b^{*2})x - \frac{i}{2}(a^2-a^{*2}+b^2-b^{*2})y - (ab+a^*b^*)z \\ y' &= \frac{i}{2}(a^2-a^{*2}-b^2+b^{*2})x + \frac{i}{2}(a^2+a^{*2}+b^2+b^{*2})y - i(ab-a^*b^*)z \\ z' &= (ab^*+a^*b)x - i(ab^*-a^*b)y + (aa^*-bb^*)z \end{aligned} \tag{149}$$

which corresponds to a linear transformation of the three-dimensional vector $\vec{r} = (xyz)$ to a new vector $\vec{r}\,' = (x'y'z')$

$$\vec{r}\,' = \mathbf{R}\vec{r} \tag{150}$$

with the transformation matrix $\mathbf{R}$ given by

$$\mathbf{R} = \frac{1}{2}\begin{pmatrix} (a^2+a^{*2}-b^2-b^{*2}) & -i(a^2-a^{*2}+b^2-b^{*2}) & -2(ab+a^*b^*) \\ i(a^2-a^{*2}-b^2+b^{*2}) & i(a^2+a^{*2}+b^2+b^{*2}) & -2i(ab-a^*b^*) \\ 2(ab^*+a^*b) & -2i(ab^*-a^*b) & 2(aa^*-bb^*) \end{pmatrix} \tag{151}$$

By equating $\mathbf{R}$ with the representation matrices of rotations through the three Euler angles, which any rotation in three-dimensional space can be composed of, we can calculate a, a^*, b, b^* and thus the corresponding two-dimensional transformation matrix $\mathbf{u}$. In order to avoid confusion with the spin functions α and β, I shall denominate the Euler angles, usually named α, β, γ, here by τ, υ, ϕ.

The first and the third of the Euler angles, τ and ϕ, refer to rotations about the z axis

$$\mathbf{R}(\tau) = \begin{pmatrix} \cos\tau & -\sin\tau & 0 \\ \sin\tau & \cos\tau & 0 \\ 0 & 0 & 1 \end{pmatrix} . \tag{152}$$

From the matrix elements of $\mathbf{R}(\tau)$ it follows

$$\left.\begin{aligned} (ab \pm a^*b^*) &= 0 \\ (ab^* \pm a^*b) &= 0 \\ \cos\tau &= \tfrac{1}{2}((a^2+a^{*2}) \pm (b^2+b^{*2})) \end{aligned}\right\} \Rightarrow b = b^* = 0 \tag{153}$$

and from

$$\left.\begin{aligned} aa^* &= 1 \\ \cos\tau &= \tfrac{1}{2}(a^2+a^{*2}) \\ \sin\tau &= \tfrac{i}{2}(a^2-a^{*2}) \end{aligned}\right\} \Rightarrow \begin{aligned} a &= +\exp(-\tfrac{i}{2}\tau); a^* = +\exp(+\tfrac{i}{2}\tau) \\ a &= -\exp(-\tfrac{i}{2}\tau); a^* = -\exp(+\tfrac{i}{2}\tau) \end{aligned} . \tag{154}$$

The ambiguity in the signs of a and a^* reflects the fact that no one-to-one correspondence exists between the elements of SU(2), the special unitary group in two dimensions, and $O(3)^+ \equiv SO(3)$, the group of the matrices representing proper rotations in three-dimensional coordinate space. Rather, to each **R** two matrices **u** correspond. In the case of $\mathbf{R}(\tau)$ they are given by

$$\mathbf{u}(\tau) = \pm \begin{pmatrix} \exp(-\frac{i}{2}\tau) & 0 \\ 0 & \exp(+\frac{i}{2}\tau) \end{pmatrix} . \tag{155}$$

A rotation through the Euler angle υ refers to the y axis and is thus represented by

$$\mathbf{R}(\upsilon) = \begin{pmatrix} \cos\upsilon & 0 & -\sin\upsilon \\ 0 & 1 & 0 \\ \sin\upsilon & 0 & \cos\upsilon \end{pmatrix} . \tag{156}$$

This gives the following requirements

$$\left.\begin{array}{r} (ab - a^*b^*) = (ab^* - a^*b) = 0 \\ (ab + a^*b^*) = (ab^* + a^*b) = \sin\upsilon \\ \frac{1}{2}((a^2 + a^{*2}) - (b^2 + b^{*2})) = (aa^* - bb^*) = \cos\upsilon \\ (-(a^2 - a^{*2}) - (b^2 - b^{*2})) = ((a^2 - a^{*2}) - (b^2 - b^{*2})) = 0 \end{array}\right\} \Rightarrow \begin{array}{c} a = a^* \\ b = b^* \end{array} \tag{157}$$

and from

$$\left.\begin{array}{r} (a^2 + b^2) = 1 \\ (a^2 - b^2) = \cos\upsilon \\ 2ab = \sin\upsilon \end{array}\right\} \Rightarrow \begin{array}{c} a = +\cos(\frac{\upsilon}{2}); b = +\sin(\frac{\upsilon}{2}) \\ a = -\cos(\frac{\upsilon}{2}); b = -\sin(\frac{\upsilon}{2}) \end{array} \tag{158}$$

and thus

$$\mathbf{u}(\upsilon) = \pm \begin{pmatrix} \cos(\frac{\upsilon}{2}) & \sin(\frac{\upsilon}{2}) \\ -\sin(\frac{\upsilon}{2}) & \cos(\frac{\upsilon}{2}) \end{pmatrix} . \tag{159}$$

Coupling the three rotations through the Euler angles

$$\pm \begin{pmatrix} \exp(\frac{-i\phi}{2}) & 0 \\ 0 & \exp(\frac{i\phi}{2}) \end{pmatrix} \begin{pmatrix} \cos(\frac{\upsilon}{2}) & \sin(\frac{\upsilon}{2}) \\ -\sin(\frac{\upsilon}{2}) & \cos(\frac{\upsilon}{2}) \end{pmatrix} \begin{pmatrix} \exp(\frac{-i\tau}{2}) & 0 \\ 0 & \exp(\frac{i\tau}{2}) \end{pmatrix} ,$$

the matrix $\mathbf{u}(\tau, \upsilon, \phi)$ corresponding to $\mathbf{R}(\tau, \upsilon, \phi)$ is found to be

$$\mathbf{u}(\tau, \upsilon, \phi) = \pm \begin{pmatrix} \exp(\frac{-i\phi}{2})\cos(\frac{\upsilon}{2})\exp(\frac{-i\tau}{2}) & \exp(\frac{-i\phi}{2})\sin(\frac{\upsilon}{2})\exp(\frac{i\tau}{2}) \\ -\exp(\frac{i\phi}{2})\sin(\frac{\upsilon}{2})\exp(\frac{-i\tau}{2}) & \exp(\frac{i\phi}{2})\cos(\frac{\upsilon}{2})\exp(\frac{i\tau}{2}) \end{pmatrix} . \tag{160}$$

A.2. SYMMETRY OPERATIONS IN C_{2V}

Some of the molecules serving as examples in the text exhibit C_{2v} symmetry. I shall therefore consider the transformation properties of the spin functions under the symmetry operations in this molecular point group.

C_{2v} consists of four symmetry operations:

$\hat{E}$, the identity also corresponding to a rotation through 2π
$\hat{C}_2(z)$, rotation about the z axis through π
$\hat{\sigma}_{xz}$, reflection in the xz plane
$\hat{\sigma}_{yz}$, reflection in the yz plane.

So far, we have been looking only at proper rotations in space. The matrix representations of reflections or improper rotations do not belong to $O(3)^+$ because their determinants have a value of -1. For being able to represent reflections in a plane we observe that these reflections can be thought of as a product of a rotation and the inversion operation, both not necessarily belonging to the considered point group. So, e.g. $\hat{\sigma}_{xz} = \hat{C}_2(y) \cdot \hat{\imath}$. For setting up the two-dimensional unitary matrices for the transformation of the spin functions one then just has to know that spin, like all angular momenta, is invariant with respect to inversion.

Considering the matrix **u** corresponding to the effect of a rotation about the z axis through 2π it is seen that **u** does not equal the identity operation in two dimensions, but rather its negative. For the transformation of the spin functions additional symmetry operations are needed; e.g. a rotation about the z axis through 4π gives the identity.

The so enlarged point group is called the $\bar{C}_{2v}$ double group with eight symmetry operations:

$\hat{E}$, the identity also corresponding to a rotation through 4π
$\hat{\bar{E}}$, a rotation through 2π
$\hat{C}_2(z)$, rotation about the z axis through π
$\hat{\bar{C}}_2(z) = \hat{\bar{E}} \cdot \hat{C}_2(z)$
$\hat{\sigma}_{xz}$, reflection in the xz plane
$\hat{\bar{\sigma}}_{xz} = \hat{\bar{E}} \cdot \hat{\sigma}_{xz}$
$\hat{\sigma}_{yz}$, reflection in the yz plane
$\hat{\bar{\sigma}}_{yz} = \hat{\bar{E}} \cdot \hat{\sigma}_{yz}$.

For the transformation of the spin functions we then get the following two-dimensional representation of $\bar{C}_{2v}$ which is equivalent to the $E_{1/2}$ irrep.

TABLE 8. The $E_{1/2}$ representation of $\bar{C}_{2v}$

Operation	$E_{1/2}$ representation
$\hat{E}$	$\begin{pmatrix} 1 & 0 \\ 0 & 1 \end{pmatrix}$
$\hat{\bar{E}}$	$\begin{pmatrix} -1 & 0 \\ 0 & -1 \end{pmatrix}$
$\hat{C}_2(z)$	$\begin{pmatrix} -i & 0 \\ 0 & i \end{pmatrix}$
$\hat{\bar{C}}_2(z)$	$\begin{pmatrix} i & 0 \\ 0 & -i \end{pmatrix}$
$\hat{\sigma}_{xz}$	$\begin{pmatrix} 0 & 1 \\ -1 & 0 \end{pmatrix}$
$\hat{\bar{\sigma}}_{xz}$	$\begin{pmatrix} 0 & -1 \\ 1 & 0 \end{pmatrix}$
$\hat{\sigma}_{yz}$	$\begin{pmatrix} 0 & -i \\ -i & 0 \end{pmatrix}$
$\hat{\bar{\sigma}}_{yz}$	$\begin{pmatrix} 0 & i \\ i & 0 \end{pmatrix}$

The character table of $\bar{C}_{2v}$ is easily constructed regarding that the barred and unbarred symmetry operations belong to the same class in this case, with the exception of $\hat{E}$ and $\hat{\bar{E}}$.

TABLE 9. Character table of the $\bar{C}_{2v}$ double group

$\bar{C}_{2v}$	$\hat{E}$	$\hat{\bar{E}}$	$(\hat{C}_2(z),\hat{\bar{C}}_2(z))$	$(\hat{\sigma}_{xz},\hat{\bar{\sigma}}_{xz})$	$(\hat{\sigma}_{yz},\hat{\bar{\sigma}}_{yz})$
A_1	1	1	1	1	1
A_2	1	1	1	−1	−1
B_1	1	1	−1	−1	1
B_2	1	1	−1	1	−1
$E_{1/2}$	2	-2	0	0	0

The spin functions $\alpha = \begin{pmatrix} 1 \\ 0 \end{pmatrix}$ and $\beta = \begin{pmatrix} 0 \\ 1 \end{pmatrix}$, i.e. the components of a spin doublet, belong to the $E_{1/2}$ irrep of $\bar{C}_{2v}$ as do all spin functions of an odd number of electrons. But what about singlet and triplet spin functions? For this purpose, let's look at the action of the symmetry operators on typical two-electron spin functions, say $\alpha\alpha$ and $\alpha\beta$. The results, displayed in Table 10, are easily verified.

TABLE 10. Action of the symmetry operators of $\bar{C}_{2v}$ on two-electron spin functions

$\hat{E}$	$\alpha\alpha =$	$\alpha\alpha$	$\hat{E}$	$\alpha\beta =$	$\alpha\beta$
$\hat{\bar{E}}$	$\alpha\alpha =$	$\alpha\alpha$	$\hat{\bar{E}}$	$\alpha\beta =$	$\alpha\beta$
$\hat{C}_2(z)$	$\alpha\alpha =$	$-\alpha\alpha$	$\hat{C}_2(z)$	$\alpha\beta =$	$\alpha\beta$
$\hat{\sigma}_{xz}$	$\alpha\alpha =$	$\beta\beta$	$\hat{\sigma}_{xz}$	$\alpha\beta =$	$-\beta\alpha$
$\hat{\sigma}_{yz}$	$\alpha\alpha =$	$-\beta\beta$	$\hat{\sigma}_{yz}$	$\alpha\beta =$	$-\beta\alpha$

From these results it is immediately clear that the $m_s = 1$ and $m_s = -1$ components of the triplet spin function do not separately transform according to one of the irreps of $\bar{C}_{2v}$. Successively applying the projection operators to $\alpha\alpha$ and $\alpha\beta$ one finds that the singlet function $S^0 = \frac{1}{\sqrt{2}}(\alpha\beta - \beta\alpha)$ transforms totally symmetric, i.e. as A_1, the $m_s = 0$ component of the triplet, $T^0 = T^z = \frac{1}{\sqrt{2}}(\alpha\beta + \beta\alpha)$, as A_2, $T^y = \frac{i}{\sqrt{2}}(\alpha\alpha + \beta\beta)$ as B_1, and finally, $T^x = \frac{1}{\sqrt{2}}(\alpha\alpha - \beta\beta)$ according to B_2.

In general, every spin function corresponding to an even number of electrons transforms as one of the irreps of the usual molecular point group — the boson irreps — while the spin functions with an odd number of electrons belong to the irreps added in the double group — the fermion irreps. Like the spatial angular momentum spin is invariant with respect to inversion. For all molecules with a centre of inversion the spin functions therefore transform according to one of the *gerade* representations. The transformation properties of frequently occurring spin functions are shown in Table 11 for some of the most common molecular double groups. A more complete list is found, e.g. in [43, page 569].

TABLE 11. Transformation properties of spin functions in frequently occurring molecular double groups

$S =$	0	$\frac{1}{2}$	1	$\frac{3}{2}$
$\bar{C}_s$	A'	$E_{\frac{1}{2}}$	$A' \oplus 2A''$	$2E_{\frac{1}{2}}$
$\bar{C}_{2v}$	A_1	$E_{\frac{1}{2}}$	$A_2 \oplus B_1 \oplus B_2$	$2E_{\frac{1}{2}}$
$\bar{D}_2(\bar{D}_{2h})$	$A_{(g)}$	$E_{\frac{1}{2}(g)}$	$B_{1(g)} \oplus B_{2(g)} \oplus B_{3(g)}$	$2E_{\frac{1}{2}(g)}$
$\bar{C}_{3v}/\bar{D}_3(\bar{D}_{3d})$	$A_{(g)}$	$E_{\frac{1}{2}(g)}$	$A_{2(g)} \oplus E_{(g)}$	$E_{\frac{1}{2}(g)} \oplus E_{\frac{3}{2}(g)}$
$\bar{C}_{\infty v}(\bar{D}_{\infty h})$	$\Sigma^+_{(g)}$	$E_{\frac{1}{2}(g)}$	$\Sigma^-_{(g)} \oplus \Pi_{(g)}$	$E_{\frac{1}{2}(g)} \oplus E_{\frac{3}{2}(g)}$

References

1. S R Langhoff and C W Kern. *Modern Theoretical Chemistry*, page 381. H F Schaefer III, Plenum Press, 1977.
2. W G Richards, H P Trivedi, and D L Cooper. *Spin-orbit coupling in molecules.* Clarendon Press, 1981.
3. D R Yarkony. *Int. Rev. Phys. Chem.*, 11:195, 1992.
4. B A Heß, C M Marian, and S D Peyerimhoff. *Advanced Series in Physical Chemistry — Vol.2, Modern Structure Theory Part I*, pages 152 — 278. C-Y Ng and D R Yarkony, World Scientific, 1995.
5. R Samzow. *Die Zweielektronenterme des no-pair-Hamiltonoperators.* PhD thesis, University of Bonn, 1991.
6. C H Teichteil, M Pelissier, and F Spiegelmann. *Chem. Phys.*, 81:273, 1983.
7. W C Ermler, Y S Lee, P A Christiansen, and K S Pitzer. *Chem. Phys. Lett.*, 81:70, 1981.
8. R M Pitzer and N W Winter. *J. Chem. Phys.*, 92:3061, 1988.
9. D Andrae, U Häußermann, M Dolg, H Stoll, and H Preuß. *Theor. Chim. Acta*, 77:123, 1990.
10. M Casarrubios and L Seijo. *Chem. Phys. Lett.*, 236:510, 1995.
11. C M Marian and U Wahlgren. *Chem. Phys. Lett.*, 251:357, 1996.
12. W C Ermler, R B Ross, and P A Christiansen. *Adv. Quantum Chem.*, 19:139, 1988.
13. M Weissbluth. *Atoms and Molecules.* Academic Press, 1978.
14. B L Silver. *Irreducible Tensor Methods.* Academic Press, 1976.
15. E U Condon and G H Shortley. *The Theory of Atomic Spectra.* Cambridge University Press, 1967.
16. W Weltner Jr. *Magnetic Atoms and Molecules.* Van Nostrand Reinhold, 1983.
17. E P Wigner. *Gruppentheorie.* Vieweg, 1930.
18. C Eckart. *Rev. Mod. Phys.*, 2:305, 1930.
19. M Rotenberg, R Bivins, N Metropolis, and Jr. K Wooten. *The 3j and 6j Symbols.* MIT Press, 1959.
20. B A Heß, C M Marian, U Wahlgren, and O Gropen. *Chem. Phys. Lett.*, 251:365, 1996.
21. M Perić, B Engels, and S D Peyerimhoff. *Quantum Mechanical Electronic Structure Calculations with Chemical Accuracy*, pages 261 — 356. S R Langhoff, Kluwer, 1995.
22. M Mizushima. *The Theory of Rotating Diatomic Molecules.* John Wiley & Sons, 1975.
23. J H van Vleck. *Phys. Rev.*, 33:467, 1929.
24. E L Hill and J H van Vleck. *Phys. Rev.*, 32:250, 1928.
25. C M Marian. *Ber. Bunsenges. Phys. Chem.*, 99:254, 1995.
26. R W Field. *Ber. Bunsenges. Phys. Chem.*, 86:771, 1982.
27. L Engelbrecht and J Hinze. *Adv. Chem. Phys.*, 44:1, 1980.
28. A Carrington, D H Levy, and T A Miller. *Adv. Chem. Phys.*, 13:149, 1970.
29. A Carrington. *Microwave Spectroscopy of Free Radicals.* Academic Press, 1974.
30. C M Marian. *J. Chem. Phys.*, 93:1176, 1990.
31. C E Moore. *Atomic Energy Levels, NBS Circular 467.* Natl. Bur. Std., 1949.
32. C M Marian. *J. Chem. Phys.*, 94:5574, 1991.
33. J M Brown, J T Hougen, K P Huber, J W C Johns, I Kopp, H Lefebvre-Brion, A J Merer, D A Ramsay, J Rostas, and R N Zare. *J. Mol. Spectrosc.*, 55:500, 1975.
34. H Lefebvre-Brion and R W Field. *Perturbations in the spectra of diatomic molecules.* Academic Press, 1986.
35. R S Mulliken and A Christy. *Phys. Rev.*, 38:87, 1931
36. B A Heß, R J Buenker, C M Marian, and S D Peyerimhoff. *Chem. Phys.*, 71:79, 1982.
37. S J Hutter. *Methodologische Aspekte der Behandlung der Spin-Bahn-Kopplung*

zweiatomiger Moleküle. PhD thesis, University of Bonn, 1994.
38. D R Yarkony. *J. Chem. Phys.*, 84:2075, 1986.
39. O Vahtras, H Ågren, P Jørgensen, H J A Jensen, T Helgaker, and J Olsen. *J. Chem. Phys.*, 96:2118, 1992.
40. L L Lohr, Jr. *J. Chem. Phys.*, 45:1362, 1966.
41. L Goodman and B J Laurenzi. *Adv. Quantum Chem.*, 4:153, 1968.
42. R Klotz, C M Marian, S D Peyerimhoff, B A Heß, and R J Buenker. *Chem. Phys.*, 89:223, 1984.
43. G Herzberg. *Molecular Spectra and Molecular Structure III. Electronic Spectra and Electronic Structure of Polyatomic Molecules.* Van Nostrand Reinhold, 1966.

MOLECULAR PROPERTIES AND SPECTRA IN SOLUTION

MATI KARELSON
Department of Chemistry, University of Tartu
Jakobi St. 2, Tartu EE 2400, Estonia

1. Introduction. The solvation free energy.

In the previous chapter [1], the quantum chemical approach to the theoretical calculation of molecular properties in solution was reviewed. In this lecture series, we consider the recipes for the calculation of individual physical and chemical properties of the molecules in the disordered condensed media, i.e. in the pure liquids and solutions. Both the recipes for the calculation of the time-independent properties (such as the solute charge distribution, equilibrium constants of chemical reactions, and molecular free energy of solvation) and the time-dependent properties (such as the spectroscopic transitions of molecules and rates of chemical reactions) will be discussed.

For the study of many equilibrium properties of molecules in a condensed medium, it is important to calculate the solvation energy or the solvation free energy. The *ad hoc* calculation of these quantities from the quantum and statistical mechanical first principles is, except of the most simple model systems, still not achievable. Therefore, a variety of different physical models have been developed to describe various aspects of the molecular solvation in condensed media. In the following we consider some recipes to calculate the free energy of solvation of a molecule, $\triangle G_S$ in a disordered condensed medium (liquid or solution). We proceed from the following partitioning of this free energy [1,2]:

$$\triangle G_S = \triangle G_{el} + \triangle G_{disp} + \triangle G_{cav} + \triangle G_{HB} \tag{1}$$

where $\triangle G_{el} = E_{el}$ is the electrostatic interaction energy of the solute charge distribution with the reaction field it creates in the surrounding polarizable dielectric medium, $\triangle G_{disp} = E_{disp}$ is the dispersion interaction energy between the solute and adjacent solvent molecules, $\triangle G_{cav}$ is the free energy of

S. Wilson and G. H. F. Diercksen (eds.), Problem Solving in Computational Molecular Science, 353–387.

the cavity formation in the solution and ΔG_{HB} is the solute-solvent hydrogen bonding energy. The different calculation schemes for individual terms in Equation (1) have been discussed in [1]. Notably, the first three terms are, in most cases, directly related to the size and shape of the molecular cavity.

The self-consistent reaction field (*SCRF*) method [3-5] can be used to calculate the electrostatic interaction energy, E_{el}, between the solute charge distribution and the respective reaction field in a dielectric medium

$$E_{el} = E_{rf} - E_0 \tag{2}$$

where E_0 and E_{rf} are the quantum chemically calculated total energies for the isolated solute molecule and for the solute molecule in the condensed dielectric medium, respectively. If to assume the full relaxation of the polarizable medium in the electrostatic field of the solute dipole (cf. *Model B* in [1,6]), the E_{rf} is obtained directly as the solution of the respective *SCRF* Schrödinger equation

$$\left[\hat{H}^0 + \Gamma\langle\phi|\hat{\mu}|\phi\rangle\hat{\mu}\right]\Psi = E_{rf}\Psi \tag{3}$$

where $\hat{H}^0$ is the Hamiltonian for the isolated solute molecule, $\hat{\mu}$ is the total dipole moment operator of the solute molecule, $\langle\phi|\hat{\mu}|\phi\rangle$ is the *SCRF* expectation value of the total dipole moment, and Ψ is the total wavefunction of the solute in the condensed medium. In the last equation, the factor Γ describes the reaction field response of the solvent. As shown before [1], this factor depends both on the dielectric properties of the solvent and the size and shape of the solute cavity. In the case of the simplest, spherical model for the solute cavity (*SCa SCRF*)

$$\Gamma = \frac{1-\epsilon}{(2\epsilon+1)a_0^3} \tag{4}$$

where a_0 is the cavity radius and ϵ is the dielectric constant of medium. In the case of the ellipsoidal cavity with principal axes **a**, **b** and **c** (*ECa SCRF*), Γ becomes a tensor [1,7,8] but the general form of Eq. (3) remains the same.

Different approaches are available for the calculation of the dispersion interaction energy between the solute and solvent molecules in solutions [1,9]. According to the quantum-mechanical method developed by Rösch and Zerner [10] the dispersion energy, E_{disp}, for a solute molecule in a spherical cavity is given by the following equation

$$E_{disp} = -\frac{2}{3}\frac{1}{a_S^3 a_M^3}\sum_{J\neq I}\sum_{K\neq O}\frac{\left(\mu_{IJ}^M\right)^2\left(\mu_{KO}^S\right)^2}{E_K^S - E_O^S + E_J^M - E_I^M} \tag{5}$$

where the superscript S refers to the solvent molecule and the superscript M to the solute molecule. The μ_{IJ}^M and μ_{KO}^S are the transition dipoles between the respective states of the solute (I and J) and the solvent (K and O) molecules. In the last equation, E_K^S, E_O^S and E_J^M, E_I^M, denote the energies of the K-th and O-th state of the solvent and of the J-th and I-th state of the solute molecule, respectively. The a_S and a_M are the cavity radii for the solvent and solute molecules, respectively. In the case of the ellipsoidal cavity for the solute molecule, the following equation was implemented for the calculation of the dispersion energy [2]

$$E_{disp} = -\frac{2}{9}\left(\frac{1}{\mathbf{a}^3 a_S^3} + \frac{1}{\mathbf{b}^3 a_S^3} + \frac{1}{\mathbf{c}^3 a_S^3}\right)\sum_{J\neq I}\sum_{K\neq O}\frac{\left(\mu_{IJ}^M\right)^2\left(\mu_{KO}^S\right)^2}{E_K^S - E_O^S + E_J^M - E_I^M} \tag{6}$$

where **a**, **b** and **c** are the principal semiaxes of the ellipsoid.

The free energy of the cavity formation, ΔG_{cav}, can be calculated either proceeding from the scaled particle theory (*SPT*) [11,12] or from the microscopic surface tension on the surface of the solute cavity [13,14]. According to the *SPT* theory, the free energy of formation of the solute cavity is

$$\Delta G_{cav} = RT\Big(1 - \ln(1-y) + \left(\frac{3y}{1-y}\right)\frac{a_M}{a_S} + \\ + \left[\frac{3y}{1-y} + \frac{9}{2}\left(\frac{y}{1-y}\right)^2\right]\left(\frac{a_M}{a_S}\right)^2\Big) \tag{7}$$

for the spherical cavity [12] and

$$\Delta G_{cav} = RT\Big(1 - \ln(1-y) + \left(\frac{\alpha y}{1-y}\right)\frac{a_M}{a_S} + \\ + \left[\frac{\beta y}{1-y} + \gamma\left(\frac{y}{1-y}\right)^2\right]\left(\frac{a_M}{a_S}\right)^2\Big) \tag{8}$$

for the ellipsoidal cavity [2]. In the last formulae, y denotes the reduced number density of the solvent [12], a_M and a_S are the intrinsic radii of the solute and solvent molecules, respectively, α, β, and γ denote the geometrical coefficients and a_M is the characteristic length of the ellipsoid (the main semiaxis). According to the surface tension model, the free energy of the cavity formation is given as follows [9]

$$\triangle G_{cav} = \sigma S_M \tag{9}$$

where σ is the surface tension of the solvent and S_M is the solvent-accessible surface area of the solute. For the *AM1-SM1* and *AM1-SM1a* solvational models by Cramer and Truhlar [13], the use of different surface tension values for different atomic types in the solute molecule has been suggested

$$\triangle G_{cav} = \sum_{k=1}^{N} \sigma_k A_k \tag{10}$$

where A_k are the solvent-accessible surface areas of different atoms in the solute molecule and σ_i denote the microscopic surface tension for the cavity type. In the case of *AM1-SM1* method, each atom has been assumed to have an individual σ value whereas in the case of the *AM1-SM1a* method it depends also on the functionality of a given atom (e.g., the alcoholic and carbonyl oxygen atoms have different σ values).

Significantly, all "cavity" methods are parametric as they depend on the solute cavity shape and size in solution. Those have been calculated proceeding from the experimental mass densities of compounds, from fitting the experimental data [14] or from the quantum-chemical electron isodensity surfaces of molecules [15]. The components of the solvation free energy for the acetonitrile in aqueous solutions as calculated for different (spherical or ellipsoidal) cavities of different sizes (cf. Eqs (2)-(10)) are given in Table 1. The electrostatic (*SCRF*) contributions were calculated using semiempirical *AM1* (*Austin Model 1* [16,17]) method for the quantum-chemical motif. The dispersion energies were calculated using the spectroscopic *INDO/1* parameterization [18] and *AM1* optimized molecular geometries in solution.

The general appearance of the dependence of different terms of the solvation free energy on the cavity size is rather similar for different solutes and thus the data for acetonitrile are representative to discuss the general trends. First of all, the free energy of the cavity formation has always the positive sign (destabilization) whereas both the electrostatic and dispersion interactions have always the negative sign (stabilization of the system). Also, it is interesting to notice that at the cavity sizes corresponding to the realistic estimates from the mass density of the compound or from the electronic isodensity surface, the dispersion energy and the cavity formation energy are nearly cancelling, i.e.

$$\triangle G_{cav} \approx -E_{disp} \tag{11}$$

The hydrogen bonding effects in aqueous solution may be considered small if to assume that the formation of an hydrogen bond between the solute and the water molecule causes the cleavage of the hydrogen bond between two water molecules of nearly the same energy. Therefore, if to neglect the hydrogen bonding effects, the free energy of solvation of an electroneutral solute should basically be determined by the electrostatic term, E_{el}.

TABLE 1. AM1 SCRF calculated electrostatic solvation energies, E_{el}, $INDO/1$ calculated dispersion energies, E_{disp}, SPT and the surface tension cavity formation free energies, ΔG_{cav}, for acetonitrile in aqueous solution for different cavity sizes and shapes. All energies are given in $kcal/mol$.

$V(\mathring{A}^3)$	$E_{el}^{SPH(a}$	$E_{el}^{ELL(b}$	$E_{disp}^{SPH(c}$	$E_{disp}^{ELL(d}$	$\Delta G_{cav}^{SPH(e}$	$\Delta G_{cav}^{ELL(f}$	$\Delta G_{cav}^{ST\,(e}$
65.4	4.92	3.99	5.87	5.42	3.87	3.52	3.12
73.6	4.24	3.38	5.40	5.10	4.11	3.70	3.38
82.4	3.69	2.74	4.97	4.77	4.37	3.86	3.65
92.0	3.23	2.29	4.59	4.48	4.63	4.04	3.92
102.1	2.85	1.88	4.24	4.18	4.90	4.22	4.20
113.1	2.54	1.56	3.02	3.89	5.18	4.40	4.50
124.8	2.27	1.27	3.62	3.62	5.46	4.62	4.81
137.3	2.04	1.03	3.35	3.38	5.75	4.82	5.12

$a)$$AM1SCRF$ electrostatic solvation energy for a spherical cavity
$b)$$AM1SCRF$ electrostatic solvation energy for an ellipsoidal cavity
$c)$$INDO/1SCRF$ dispersion energy for a spherical cavity (Eq. (5))
$d)$$INDO/1SCRF$ dispersion energy for a an ellipsoidal cavity (Eq. (6))
$e)$$SPT$ cavity formation free energy fora spherical cavity (Eq. (7))
$f)$$SPT$ cavity formation free energy for an ellipsoidal cavity (Eq. (8))
$f)$cavity formation free energy from the surface tension model (Eq. (9))

The PCM and $SCRF$ calculated electrostatic solvation energies for a number of small electroneutral solutes have been compared with the experimental free energies of solvation, $\Delta G(g \to w)$ [19, 20]. These results, presented in Table 2, demonstrate that the SCa $SCRF$ electrostatic solvation energies are insufficient for the description of experimental data. Somewhat better correlation can be observed in the case of PCM calculated electrostatic solvation energies. Interestingly, the best correlation between the calculated and experimental free energies of solvation was obtained in the case of the multi-cavity self-consistent reaction field (MCa $SCRF$) method [20]. The basic idea behind this method is the coupling of the rotational and vibrational motions in the solute molecule with the solvent orientation and electronic relaxation. Therefore, these groups in the molecule which are rotationally free or have small rotation or inversion barriers are considered to interact independently with the dielectric medium. Each of these fragments can be characterized by its partial charge q_A, partial dipole moment μ_A, and higher electrical moments, and is embedded into a spherical or ellipsoidal cavity surrounded by a dielectric continuum characterized by its macroscopic dielectric permittivity (ϵ). The latter is a function of the electric field frequency ω and, strictly speaking, this dependence should also be accounted for in the coupling of the intramolecular nuclear motions in the solute with the rotational and librational movement of the solvent molecule(s) [21-24]. For processes that have time scales sufficiently different from the time scale of the solvent dielectric (orientational) relaxation, two values of ϵ, corresponding to the full solvent relaxation ($\epsilon_0 = \epsilon(\omega = 0)$) and only to the solvent electronic relaxation ($\epsilon_\infty = \epsilon(\omega \rightarrow 0)$) can be used.

The solute molecules studied by the $MCa\ SCRF$ method have been divided into different number of rotationally flexible groups depending on the presence of these groups in the molecule. Thus, one group of the molecules that were studied (CH_3Cl, CH_3Br, CH_3I, CH_3OH, CH_3NH_2, CH_3CN, CH_3CHO, CH_3SH), were divided into two fragments, one of which is the methyl group and the other involves the substituent attached to it. Another group of molecules (CH_3COOH, CH_3CONH_2, CH_3COCH_3, CH_3SCH_3, CH_3CH_2OH) had three fragments, with two terminal substituents ($-CH_3$,$-OH$,$-NH_2$) connected with a bridge group ($> C = O$, $-CH_2-$,$-S-$). Ethylene glycol and ethylene diamine were divided into four fragments (two methylene groups and two terminal substituents, either $-OH$ or $-NH_2$). However, the results of the calculation indicated that methyl fluoride was still better represented by a single cavity and the respective value was used in the correlation. Also, for acetonitrile, which has a rigid $CN-$ group connected to the methyl group, the single cavity representation gives closer value to the experimentally observed solvation energy. For the rest of the molecules presented in Table 2, a good agreement between experimental and $AM1\ MCa\ SCRF$ energies can be observed. The linear regression between the calculated energies of solvation and the experimental values of $\Delta G(g \rightarrow w)$ [25] gives the following relationship [20]:

$$\Delta E_{el}(MCa\ SCRF) = (0.33 \pm 0.30) + (0.953 \pm 0.058)\,\Delta G(g \rightarrow w) \quad (12)$$

$$\text{Correlation coefficient } R = 0.975$$
$$\text{Standard deviation } \sigma = 0.71 (kcal/mol)$$

The slope of this relationship is close to 1 and the intercept is a small number which is in accordance with the assumption about the mutual cancellation of the dispersion and cavity-formation effects for this set of compounds.

However, such one-component calculation scheme of the solvation energies is applicable only for some small solutes. One must certainly not forget the other terms required in the total solvation energy; namely the cavity-formation, solvent restructurization and dispersion effects. An effective method for a semiempirical treatment of these effects in combination with the general Born model for electrostatic interactions has been proposed by Cramer and Truhlar, and the results obtained by them for a wide range of compounds look very promising [26-29]. Also, by allowing for the contribution, corresponding to the hydrogen bonding, the following equations for the solvation free energies of a set of electroneutral solutes in aqueous solution, obtained for the spherical (SCa) and ellipsoidal cavity (ECa) models were reported [2]

$$\delta\triangle G_s = \triangle G(g \rightarrow w) - E - el - E_{disp} - \triangle G_{cav} \tag{13}$$

$$\delta\triangle G_s(SCa) = (-0.27 \pm 0.25) - (1.94 \pm 0.33)N_{HBA} - (1.37 \pm 0.25)N_{HBD}$$

$$R = 0.94 \quad \sigma = 0.98$$

$$\delta\triangle G_s(ECa) = (-0.40 \pm 0.29) - (2.20 \pm 0.38)N_{HBA} - (1.33 \pm 0.28)N_{HBD}$$

$$R = 0.93 \quad \sigma = 1.28 \tag{14}$$

where N_{HBA} and N_{HBD} are the numbers of the hydrogen bonding acceptor and hydrogen bonding donor centres in the solute molecule, respectively. The coefficients in front of the hydrogen bonding descriptors correspond to the average difference between the energies of the solute-water and water-water hydrogen bonds. In Eq. (13), the cavity formation free energy, $\triangle G_{cav}$, was calculated proceeding from the SPT theory (Eqs. (7) and (8), respectively), and the dispersion energy, E_{disp}, from the Rösch-Zerner theory (Eqs. (7) and (8), respectively).

In the case of ionic solutes, the ion-molecule electrostatic interactions should be accounted for in the calculation of free energies of solvation. Numerous theoretical schemes have been developed for the calculation of ionic solvation energies [30-35]. Here we notice that the application of the MCa $SCRF$ calculation scheme together with the Born equation leads to accurate solvation energies of simple inorganic cations and anions in aqueous solvation provided that the first solvation sphere is treated explicitly quantum mechanically in the framework of supermolecule approach [36].

A critical parameter in the calculation of ionic and molecular solvation energies is the solute cavity size. It can be found by minimizing the difference between the calculated and experimental solvation free energies at different cavity sizes. However, a common scheme for the calculation of the cavity size for different compounds is necessary in order to minimize the

TABLE 2. *AM1 PCM* [19] and *AM1* single cavity and multi-cavity *SCRF* (*Model B*) [20] calculated electrostatic solvation energies E_{el} and experimental free energies of transfer of compounds from the gas hase to the aqueous solution $\Delta G(g \rightarrow w)$ $(kcal/mol)$

Compound	E_{el} $(SCa\ SCRF)$	E_{el} $(MCa\ SCRF)$	ΔE_{el} (PCM)	$\Delta G(g \rightarrow w)$
CH_3CH_3	0.00	-0.10	—	1.81
CH_3F	-0.25	-5.04	—	-0.22
CH_3Cl	-0.84	-2.07	-1.8	-0.56
CH_3Br	-0.79	-0.70	—	-0.80
CH_3OH	-0.62	-1.24	-2.4	-0.90
CH_3NH_2	-2.20	-4.66	-1.2	-4.56
CH_3COOH	-1.27	-5.79	-7.1	-6.70
CH_3CN	-3.32	-5.30	-3.8	-3.89
CH_3CHO	-2.68	-3.66	-5.3	-3.55
CH_3CONH_2	-0.21	-10.8	-9.5	-9.70
CH_3COCH_3	-2.17	-4.38	-5.6	-3.81
CH_3SH	-1.87	-1.91	—	-1.26
CH_3SCH_3	-0.58	-2.11	—	-1.56
CH_3CH_2OH	-0.84	-4.00	—	-4.96
$(CH_2OH)_2$	-1.56	-7.87	—	-7.75
$(CH_2NH_2)_2$	-0.98	-9.28	—	-9.88

number of empirical parameters involved in the theory. For convex cavity shapes (sphere and ellipsoid), it is feasible to construct the solute cavity by adding the contributions from both the solute and solvent molecules. Such scheme allows to account for the dielectric saturation at the vicinity of the solute molecule and for the change of the cavity size in the solvents with molecules of different size and is thus advised to be used in the calculation of solvation effects.

On the basis of the above given discussion on the theoretical calculation of solvation free energies for electroneutral and ionic solutes, it is recommended, in general, to develop the experiment in a manner or restrict the set of compounds to be studied in a way that only a single solvation interaction is predominant. In such case, the results of theoretical treatment can be more confidently interpreted and understood in terms of basic intermolecular interactions (reduction of intermolecular interactions).

2. Theoretical modelling of solvent effects on the physical prop-

erties of molecules in condensed media.

The traditional chemical thinking proceeds from the axiom of the constancy of the chemical structure, independently of the medium into which it is submerged. This axiom has been extremely useful and consists one of the cornerstones of chemistry. However, the distances between the neighbouring molecules in condensed (liquid) media are so small that significant alterations in their electron distribution and even geometric structure can be expected. Therefore, it is of principal interest to study theoretically the influence of the solvent effects on the geometric and electronic structure of chemical compounds. Such analysis is advisable in all studies of phenomena in condensed media since it allows to distinguish between the mechanical mixing effects and the mutual electronic and nuclear polarization of the solute and solvent molecules, leading thus to a deeper understanding of solvation.

First, it is instructive to investigate the reaction field effects on the charge distribution of the solute molecule. In Table 3, the *AM1 SCF* and *SCRF* calculated dipole moments are presented for various substituted benzoic acids, phenols and anilines [37] in two media, corresponding to the gas phase ($\epsilon = 1$) and to the polarizable medium with the dielectric permittivity of water ($\epsilon = 78.5$). First, it is obvious that the relative increase in the dipole moment due to the surrounding polarizable medium is highly individual for a given combination of the substituents at the benzene ring. However, this increase is not directly correlated with the absolute value of the dipole moment but rather depends on the relative position of the substituents. The largest increase is observed in the case of molecules with the presence of direct polar resonance between the substituents (p-nitrophenol and p-nitroaniline). A similar effect, the solvent assisted polar resonance, has also been observed in push-pull ethylenes [38-41] and other conjugated systems [42-44].

The changes in the electron distribution due to the electrostatic solvation are also reflected in the respective calculated Mulliken atomic partial charges. For instance, the polarity of the $C = O$ double bond in the 2-H-oxo form of the 3-hydroxypyrazole, as estimated from the Mulliken charges, is substantially increased by the dielectric medium

TABLE 3. *AM1 SCF* and *SCRF* calculated total dipole moments μ (*Debye*) for some substituted benzoic acids, phenols and anilines in the gas phase and in a polarizable dielectric medium

Compound	Substituent	$\mu(\epsilon = 1)$	$\mu(\epsilon = 80)$	$\Delta\%^{(a}$
Benzoic acid	$-H$	2.040	2.951	45
	$m - CN$	2.267	2.880	27
	$p - CN$	1.753	2.342	34
	$m - NO_2$	4.113	5.399	31
	$p - NO_2$	3.543	5.201	47
Phenol	$-H$	1.236	1.535	24
	$m - CN$	2.096	2.827	35
	$p - CN$	3.286	4.744	44
	$m - NO_2$	3.963	5.605	41
	$p - NO_2$	5.234	8.051	54
Aniline	$-H$	1.537	1.697	10
	$m - CN$	4.053	5.486	35
	$p - CN$	4.916	7.519	53
	$m - NO_2$	5.839	7.839	34
	$p - NO_2$	7.244	12.590	74

(*a* Relative increase in the dipole moment
$\Delta\% = 100[\mu(\epsilon = 80) - \mu(\epsilon = 1)]/\mu(\epsilon = 1)$

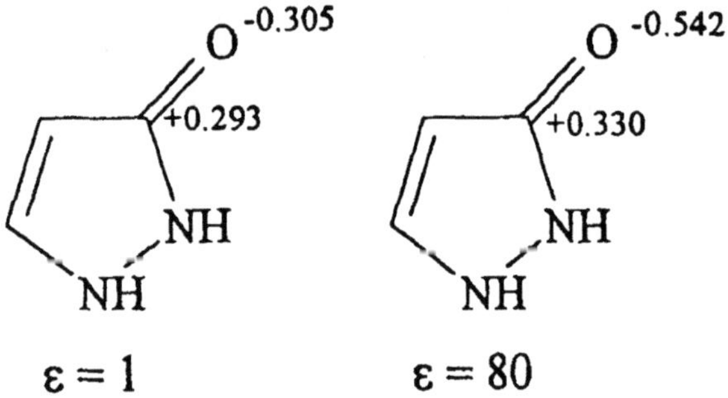

(II.1)

In fact, two different resonant forms of this compound may be written out as being predominant in each of these media

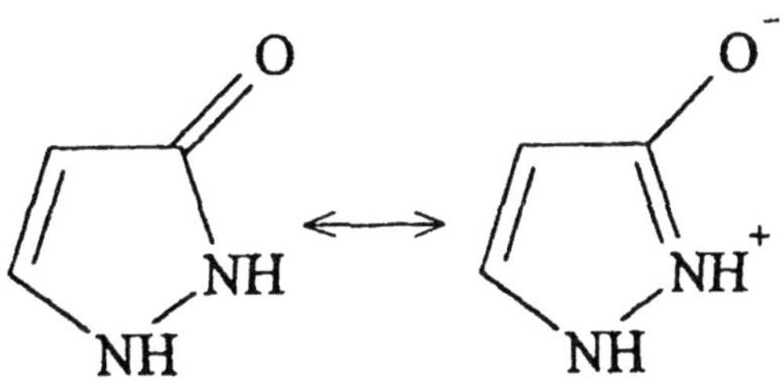

(II.2)

Notably, the left-hand structure has a dienic electron system whereas the right-hand structure, corresponding to the high dielectric constant medium, has a 6 π-electron aromatic system. Therefore, the electrostatic solvation can interfere with the aromaticity of a cyclic compound. This question will be analyzed more closely in the next Chapter.

Another solvent-assisted intramolecular electronic effect is the merostabilization of radicals [45-50]. This effect, also called the "captodative" effect [51,52] is defined as an additional stabilization of the carbon radical with an electron-donor and an electron-acceptor group simultaneously connected to the radical centre. The respective merostabilization energy is defined as follows [48,55]

$$\triangle E_M = E_{XY} - \frac{1}{2}(E_{XX} + E_{YY}) \tag{15}$$

where E_{XY} is the energy of the electron-donor and electron-acceptor substituted radical $^{\cdot}CHXY$, E_{XX} is the energy of the doubly electron-donor substituted radical $^{\cdot}CHX_2$ and E_{YY} is the energy of the doubly electron-acceptor substituted radical $^{\cdot}CHY_2$. Earlier calculations by P. von R. Schleyer [53] supported the presence of the extra stabilization of $^{\cdot}CH(CN)NH_2$ and $^{\cdot}CH(CN)BH_2$ radicals in their isolated form. However, this conclusion was later disputed on the basis of more elaborate quantum-chemical calculations [54]. Our studies have shown that the merostabilization is largely supported by the presence of the polarizable dielectric continuum. In Table 3, the results of the *AM1 UHF SCRF* calculations on the $^{\cdot}CH(ON)_2$, $^{\cdot}CH(CN)_2$ and $^{\cdot}CH(CN)OH$ radicals, corresponding to three dielectric media are presented [55]. In the gas phase ($\epsilon = 1$), no additional stabilization was obtained for the $^{\cdot}CH(CN)OH$ radical. However, even in the low dielectric constant medium, a substantial stabilization energy $\triangle E_M$ was predicted (3.53 $kcal/mol$) for this radical which was further enhanced by the high dielectric constant medium (to 12.57 $kcal/mol$). Also, a dramatic change in the value of the dipole moment was obtained for the $^{\cdot}CH(CN)OH$ radical while transferred from the gas phase to the high dielectric constant medium (cf. Table 3). Such strong dependence of the dipole moment on the dielectric constant of the medium causes a principally nonlinear relationship between the electrostatic solvation energy E_{el} and the Kirkwood-Onsager factor Γ (Eq. 4). Correspondingly, the calculated merostabilization energy ΔE_M was predicted to have a nonlinear dependence from this factor (cf. Fig. 1). Notably, the classical Kirkwood-Onsager formula using a constant value of the dipole moment of the compound in different media leads to a qualitatively wrong prediction of the slope of this relationship (Fig. 3).

The solvent reaction field may also have significant effects on the solute molecular geometry, conformational transitions and equilibria in solutions. In most cases, the bond lengths and valence bond angles of the solute are little affected by the surrounding dielectric medium. However, larger effects may be observed in systems with large solvent assistance to the polar resonance.

TABLE 4. *AM1 UHF SCRF* calculated enthalpies of formation ΔH_f^0 *(kcal/mol)*, dipole moments μ *(Debye)* and merostabilization energies ΔE_M *(kcal/mol)* in the media of different dielectric constant, ϵ.

Radical	ϵ	ΔH_f^0	μ	ΔE_M
$^{\cdot}CH(ON)_2$	1	-80.03	1.289	—
	2	-84.38	1.411	—
	80	-86.56	1.514	—
$^{\cdot}CH(CN)_2$	1	82.14	2.247	—
	2	79.34	2.583	—
	80	76.51	2.015	
$^{\cdot}CH(CN)OH$	1	0.65	1.419	1.09
	2	-6.05	4.875	-3.53
	80	-17.59	6.193	-12.57

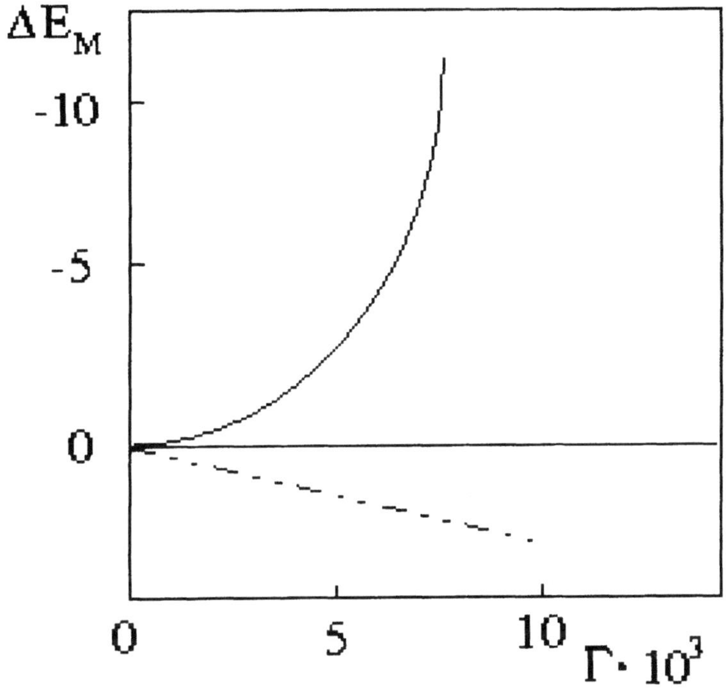

Figure 1. The dependence of the *AM1 UHF SCRF* calculated merostabilization energy $\triangle E_M$ $(kcal/mol)$ of the $^{\cdot}CH(CN)OH$ radical on the Kirkwood-Onsager factor Γ. The solid line corresponds to the *SCRF* theory, the dashed line to the classical theory.

For instance, in the above discussed 2-H-oxo form of the 3-hydroxypyrazole,

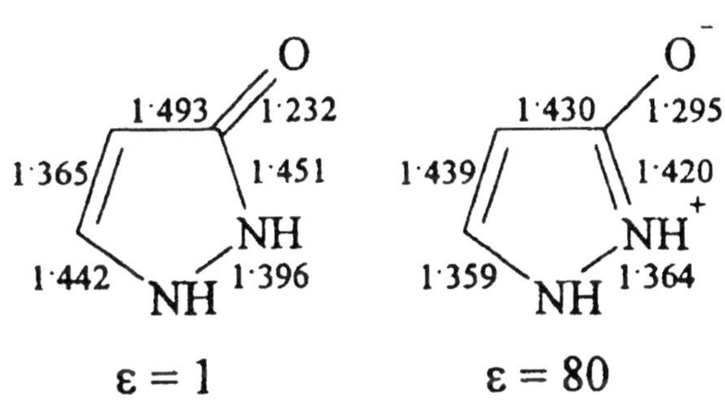

(II.3)

the *AM1 SCRF* calculated bond lengths are significantly changed by the high dielectric constant medium as compared to those in the gas phase. First of all, the $C = O$ bond length is substantially increased, from the length typical for a double bond (1.232 $\mathring{A}$) to a closer value for the single bond (1.295 $\mathring{A}$). This is also in accordance with the formal representation of predominant resonance structures for this compound in the corresponding media (cf. Scheme II.2). In the heterocyclic ring system, the bond lengths are significantly more equalized in the high dielectric constant medium as expected for an aromatic cyclic structure.

The solvent effects on the conformational dynamics and equilibria of the molecules in liquids and solutions have been rather thoroughly studied, particularly in connection with the conformational structure of biomolecules [56-65]. For instance, even such simple system as the $^{\cdot}CH(CN)OH$ may have its conformational structure affected by the dielectric medium. The conformational angle of interest in this radical is defined between the two planes given by the sp^2 radical C center and the OH bond (α) (cf. Scheme II. 4)

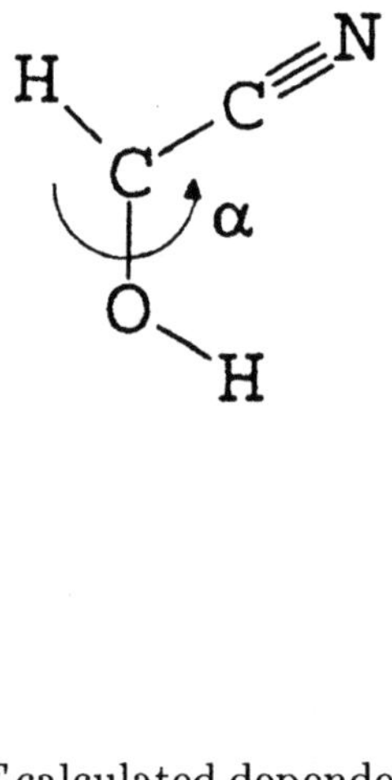

(II.4)

The *AM1 SCF* and *SCRF* calculated dependence of the relative energy (zero corresponds to the energy at $\alpha = 0°$) on the conformational angle in the $CH(CN)OH$ radical is given on Fig. 2. It is clear that the conformational minimum is shifted from $\alpha = 0°$ in the gas phase to the $\alpha = 180°$ in the high dielectric constant medium. This is primarily due to the large increase of the dipole moment of this radical in the latter conformation as compared to the most stable conformation in the gas phase (the *AM1 SCRF* calculated dipole moment in the medium with the dielectric constant $\epsilon = 80$ is $\mu = 1.569\,D$ at $\alpha = 0°$ and $\mu = 4.875\,D$ at $\alpha = 180°$ [55]) and the corresponding large increase in the respective reaction field.

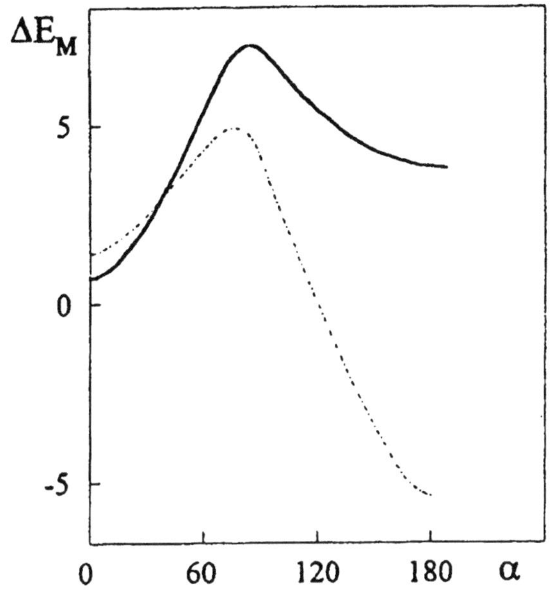

Figure 2. Dependence of the merostabilization energy $\triangle E_M$ $(kcal/mol)$ of $^{\cdot}CH(CN)OH$ radical on the conformational angle α (deg) defined on Scheme (II. 4). Solid line corresponds to the gas phase and the dashed line to the dielectric medium with $\epsilon = 80$.

Analogously, proceeding from the *SCRF* model calculations, the electrostatic solvent effect of approximately 1 $kcal/mol$ has been predicted for the relative energies of different conformational minima of some alkanes [56], and 1,2-dichloroethane and furfural [57] in aqueous solutions.

The solvent effect on the rotational barrier and relative energy of different conformational minima of formamide [64] and other small amides [62,63,65] has been studied using the combination of the *ab initio* molecular orbital method with the *SCRF* electrostatic solvation model. The rotation about the $C - N$ bond in amides is of fundamental importance for understanding the essence of conformational preferences in proteins and the development of the respective accurate potentials is prerequisite for the molecular dynamics simulation of the structure of these macromolecules.

It has been established [64] that the conformational barriers at the $C - N$ bond of formamide are substantially influenced by the presence of the high dielectric constant minimum. Two transition states of the C_s symmetry were found for the conformational change of this molecule, one of which corresponds to the rotation about the $C - N$ bond (*TS1*) and another to the inversion at the nitrogen atom ($TS2$). The 6-31G**//6-31G** calculated rotational barrier relative to the planar structure (*TS1*) was increased (by 4.8 $kcal/mol$ according to the *SCRF* in dipole approximation, by 1.4 *kcal/mol* according to the *SCRF* in 32-pole approximation and by 2.1 *kcal/mol* according to the *PCM* model) and the inversion barrier reduced (by 1.2 *kcal/mol* according to *SCRF* in dipole approximation, by 1.2 *kcal/mol* according to *SCRF* in 32-pole approximation and by 0.9 *kcal/mol* according to the *PCM* model) when the molecule was transferred from the gas phase to a dielectric medium with $\epsilon = 78.54$ [64]. These trends can be rationalized in terms of the increase of the weight of the more polar resonance form of formamide in the high dielectric constant medium

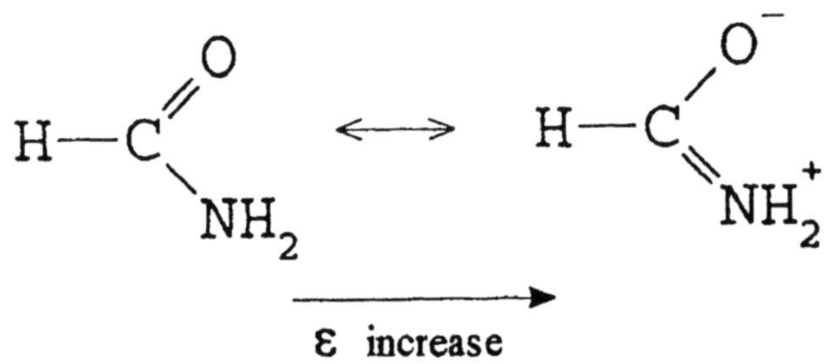

(II.5)

Obviously, the double bond character of the $C = N$ bond is increased by the higher dielectric constant of the medium and the respective rotation barrier about this bond should be also increased. At the same time, the nitrogen atom is shifted from the sp^3 valence hybridization given by the left-hand side resonance form to a more sp^2-like valence state presented by the right-hand side resonance form. Accordingly, the inversion barrier at this atom should be reduced by the high dielectric constant medium. As a result of these trends, the two conformational transition states of formamide, separated by about 2.7 *kcal/mol* in the gas phase in favour of the rotational barrier, *TS1*, are predicted to be practically the same energy in the high dielectric constant medium. This conclusion is of large importance for the accurate prediction of conformational transformations and equilibria in polypeptides and proteins at intracellular conditions.

Important properties of molecules are connected with their response to external perturbations. One of such properties is the hyperpolarizability of the molecules for which the theoretical methods of study in solution have been developed recently [66-69]. The frequency dependent first hyperpolarizabilities $\beta_{vec}(-2\omega, \omega, \omega)$ (second harmonic generation) of push-pull substituted conjugated organic molecules were studied by Yu and Zerner [66] using the time-dependent Hartree-Fock theory with the inclusion of the self-consistent reaction field theory. The importance of the inclusion of solvent effect was confirmed by the comparison of experimental values of β_{vec} with those calculated for the isolated molecules and for the molecules submerged into a dielectric continuum. The fitted slope between the *SCRF* calculated and the experimental β_{vec} values was 1.07 whereas the same slope for the calculated data without inclusion of the solvent effects was only 0.49. Thus, it is mandatory to include these effects in the theoretical treatment of hyperpolarizabilities of molecules in condensed media.

K.V. Mikkelsen et al. [67] have developed a multiconfiguration self-consistent reaction field response method for calculating frequency dependent molecular properties as well as electronic excitation energies and transition moments of solvated molecules. The sample calculations of a water molecule surrounded by different dielectric media using the complete active space (*CAS*) wavefunctions indicate the increase of polarizabilities when increasing the dielectric constant. At the same time, for a given component of the frequency dependent polarizability, the relative difference between the four different frequency dependent polarizabilities increases also with increasing the dielectric constant of the surrounding medium.

J. Tomasi et al. [68] have presented a theory of the analytical calculation of the *SCF* dipole polarizability and hyperpolarizabilities for molecular solutes described with the polarizable continuum model (*PCM*). The first numerical results, obtained using the respectively modified version of the *HONDO-8* program package [70], suggest a significant dependence upon the basis set used of the calculated solvent effects on the different components of the hyperpolarizability. Therefore, likewise to the isolated molecules, a special attention has to be paid to the use of proper basis sets when calculating the molecular response properties in condensed media.

3. Theoretical modelling of solvent effects on the chemical reactivity of molecules in condensed media

One of the most important applications of the quantum theory of molecules in condensed media is the prediction of their chemical reactivity. It is often necessary to predict the rate or equilibrium constant of a chemical reaction in different organic and inorganic media. Some of the simplest chemical reactions are the Brønsted acid-base equilibria consisting the proton transfer between two reagents, either electroneutral or charged

$$\begin{aligned} AH &\rightleftharpoons A^- + H^+ \\ B + H^+ &\rightleftharpoons BH^+ \end{aligned} \tag{16}$$

Many of these reactions have been experimentally thoroughly studied both in the gas phase and in solution (cf., e.g. [71,72]). Due to the large solvation energies of ions involved, the solvent effects on equilibria (16) are dramatic. Also, the proton is always specifically solvated by available lone electron pairs of the solvent and thus the complete theoretical treatment of the solvent effects on the acid-base equilibria in solutions should include both the supermolecule study of the specific hydrogen-bonding effects in the first coordination sphere of different species involved in these equilibria and the dielectric polarization of the bulk solvent.

One of the first attempts of the quantum-chemical description of the solvent effect on the acid-base equilibria was made at the *ab initio* level of quantum theory using the combination of the supermolecule approach with the self-consistent reaction field theory [33]. The *SCF* and *SCRF* calculations corresponding to four media $\epsilon = 1$ (gas phase), $\epsilon = 2.209$ (1,4-dioxane), $\epsilon = 20.74$ (ethanol) and $\epsilon = 80.1$ (water) were performed for small OH-acids (H_2O, FOH and CH_3OH) and ammonium ions (NH_4^+, NFH_3^+, $NF_2H_2^+$, NF_3H^+). The results demonstrated the importance of the electrostatic solvation in higher dielectric constant media (cf. Fig. 3 for OH-acids). However, the electrostatic solvation accounts only for about $50-70\%$ of the total experimental solvent effect on the free energy of acid-base equilibria. Therefore, the specific solvation effects have to be added in the framework, for example, of the supermolecule approach.

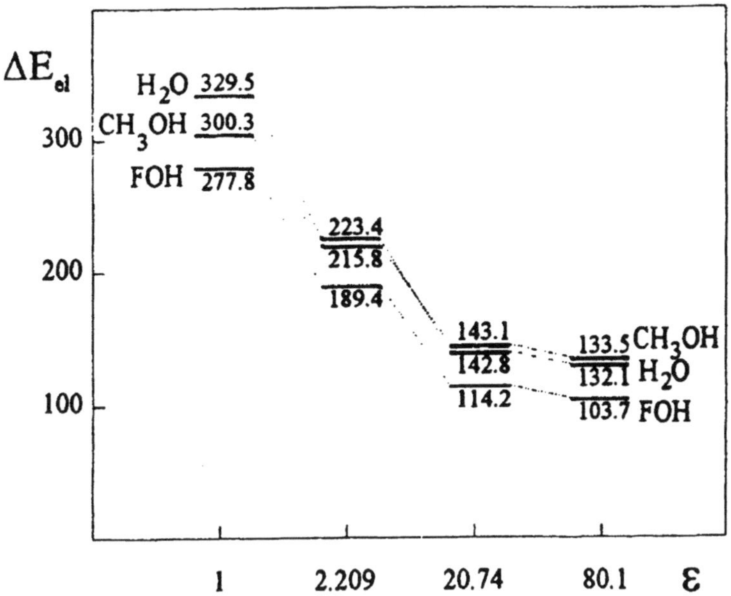

Figure 3. The *ab initio* calculated electrostatic components ΔE_{el} $(kcal/mol)$ of acid-base equilibrium free energies for H_2O, FOH and CH_3OH in different dielectric media [33].

For instance, the account for the specific solvation in the first coordination sphere of the species participating in the acid-base dissociation of water

$$H_2O \rightleftharpoons H^+ + OH^-$$

gives the difference with the energy of this equilibrium in gas phase ΔE_{SM} = 105.6 $kcal/mol$. The calculated difference arising from the $SCRF$ electrostatic polarization is somewhat larger, ΔE_{SCRF} = 197.4 $kcal/mol$. The calculated free energy of the dissociation of water in water using both the supermolecule and the $SCRF$ corrections, is ΔG_{calc} = -26.5 $kcal/mol$ which is in a remarkably good agreement with the experimental free energy of this equilibrium at the room temperature (ΔG_{exp} = -24.1 $kcal/mol$).

T. Furuki et al. [73] have calculated the hydration enthalpies of carboxylic acids and the respective carboxylate anions in water using the polarizable continuum model (*PCM*) method to account for the electrostatic solvation of these compounds. The results are in satisfactory agreement with the experimental hydration energies, however, no direct comparison was made with the respective acid-base dissociation free energies.

J.J. Urban et al. have successfully used the *AM1-SM2*, *PM3-SM3* [54,55] and *GB/SA* [74] methods for the theoretical calculation of the aqueous acidity of $o-$, $m-$, and $p-$fluorophenols relative to the unsubstituted phenol [75]:

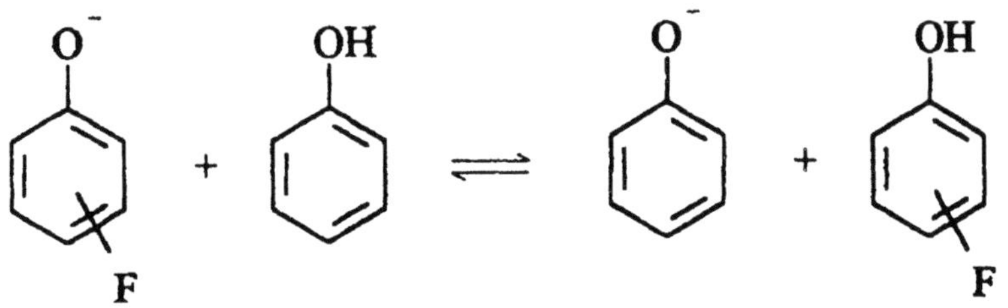

(III.1)

These solvation models produce calculated free energies of hydration that are in good agreement with the available experimental data and give confident predictions of the relative acidities of different fluorophenols with respect to phenol (cf. Table 4.). The solvent effects were rationalized in terms of the charge distribution differences of fluorophenols in the gas phase and in the aqueous solution. It was found that while the fluorine substitution increases the hydrophobicity of the phenols and phenoxides, the perturbation of ring charge density imparted by fluorine substitution is quite similar in the gas and aqueous phases.

TABLE 5. Calculated and experimental acidity (kcal/mol) of substituted fluorophenols relative to phenol in the gas phase and in the aqueous solution [75]

Medium	Method	ortho-F	meta-F	para-F
Gas phase	AM1	4.8	6.4	5.9
	PM3	5.8	7.6	5.9
	HF/6-31+G(d)	2.7	7.0	2.5
	MP2 6-311+G(2d,p)	3.1	5.9	2.8
	experiment	3.2	5.3	2.3
Water	AM1-SM2	4.5	2.8	1.9
	PM3-SM3	5.2	2.8	1.1
	experiment	1.6	1.1	0.1

From the point of view of the complexity of the solvation phenomena, the prototropic tautomeric equilibria are simpler to model theoretically than the acid-base equilibria. In a prototropic tautomeric equilibrium, the isomers of the same molecule varying only by the position of a proton in their molecular structure, are participating. It is feasible to assume that the solute-solvent dispersion energy, the cavity formation free energy and the hydrogen bonding energy should be very similar for different tautomeric forms participating in these equilibria.

Therefore, the theoretical tautomeric equilibrium constant can be calculated from the following equation:

$$\log K_T = \frac{\triangle G_T}{2.3RT} \approx \frac{E_{el}}{2.3RT} = \frac{\delta \triangle H}{2.3RT} \tag{17}$$

where $\triangle G_T$ is the free energy of the tautomeric equilibrium and $\delta\triangle H$ is the difference in the calculated heats of formation of the tautomeric species participating in this equilibrium.

The tautomeric equilibria of heterocyclic compounds have been extensively studied using both the *SCRF* and *PCM* solvation models at the semiempirical and *ab initio* level of quantum theory [76-89]. Already the simple *SCa SCRF* method, implemented in combination with the semiempirical *AM1* parameterization [16] has given a quantitative description of the tautomeric equilibrium of different substituted pyridines in aqueous solutions [76]. The same method has also given the correct prediction of the predominant tautomers of various five-membered rings with two heteroatoms in solutions [78]. The tautomeric equilibria of nucleic acid bases may have a large importance in predicting the probability of mutagenic transformations in living cell [90]. The *AM1 SCRF* method has given good quantitative predictions of the relative stability of uracil, cytosine, guanine and adenine tautomers in aqueous solutions [79].

Most of the *SCa SCRF* calculations of equilibrium constants have been performed for cyclic molecules characterized by the intramolecular rigidity of the structure. Thus, not unexpectedly, this method has failed in the case of internally flexible molecular structures. However, recently it has been demonstrated [89] that the use of the multi-cavity *SCRF* (*MCa SCRF*) method [20] may significantly improve the prediction of the tautomeric equilibrium constants for aliphatic systems. For 14 aliphatic ketones and aldehydes, the *AM1 SCa SCRF* and *MCa SCRF* calculated $\log K_T$ values were correlated with the experimental data by the following regression equations:

$$\begin{aligned} \log K_T(SCa) &= (0.03 \pm 0.61) + (0.95 \pm 0.09) \log K_T(exp) \\ &\qquad R^2 = 0.896; \quad s = 1.06 \\ \log K_T(MCa) &= (-0.03 \pm 0.22) + (0.99 \pm 0.03) \log K_T(exp) \\ &\qquad R^2 = 0.986; \quad s = 0.39 \end{aligned}$$

The *MCa SCRF* results are significantly superior to those obtained using the *SCa SCRF* method (cf. also Fig. 4) and provide a satisfactory explanation for the variation found.

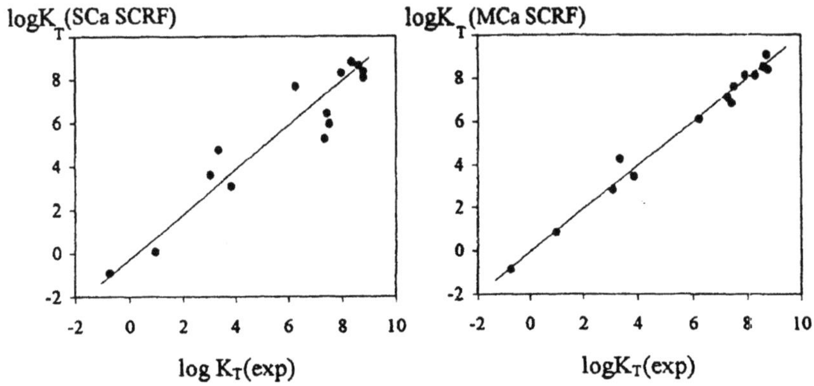

Figure 4. The correlation between the *AM1 SCa SCRF* and *MCa SCRF* calculated, and experimental log K_T values for a series of aliphatic ketones and aldehydes in aqueous solution [89].

In most cases, the theoretical calculations have been limited to the gas phase and to the aqueous solutions. Therefore, it would be important to study the applicability of the solvation theories in different dielectric media. An appropriate reaction for such study is the tautomeric equilibrium between acetylacetone (**1**) and its enol form (**2**)

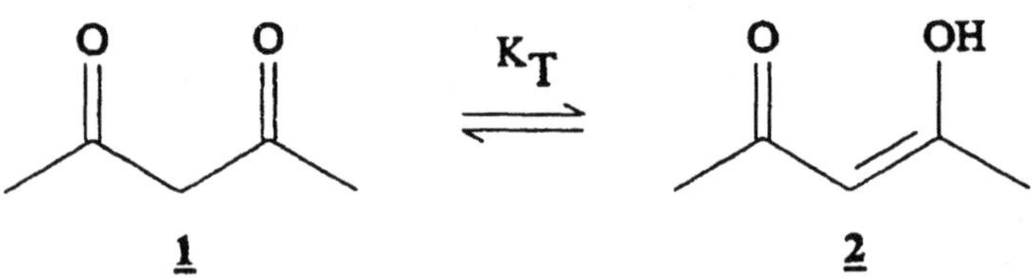

(III.2)

The experimental equilibrium constant of this reaction has been precisely measured using bromatometry and NMR spectroscopy in the dioxane-water mixtures with a large variation of the dielectric constant of the medium, from pure dioxane ($\epsilon = 2.209$) to water ($\epsilon = 78.4$) [91]. An excellent agreement between the calculated and experimental log K_T values in different media was obtained using self-consistent reaction field methods (cf. Figure 5.) [2]. Notably, the *AM1 SCa SCRF* and *ECa SCRF* methods lead to practically identical results.

C.J. Cramer and D.G. Truhlar have performed extended basis set electronic structure calculations with high levels of electron correlation for heterocyclic tautomerizations, augmented by a detailed analysis of the aqueous solvation free energy differences including both the electronic and geometric relaxation in aqueous solution [87], using the *AM1-SM2* solvation model. They have also shown that in flexible systems, it is important to account for the local (bond) dipole moments rather than the total dipole moment of the molecule.

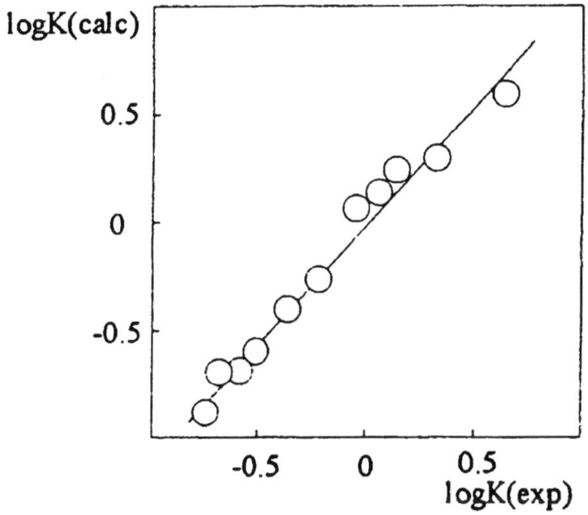

Figure 5. *AM1 SCa SCRF* Calculated vs. Experimental Tautomeric Equilibrium Constants log K_T for Acetylacetone in Different Dioxane/Water Mixtures.

The calculation of absolute rate constants of chemical reactions in solution from the first principles is a tremendously difficult task. In the present lecture series, we do not attempt to give an overview on the extensive literature available on the respective methodology but will only give some examples of the quantum-chemical study of the solvent effects on the reaction profiles in solution.

One of the most well-known nucleophilic substitution reactions with a large solvent effect on the reaction rates and yields is the Menshutkin reaction. In variance from the ordinary S_N2 reactions, this reaction proceeds between the electroneutral reagents (alkyl halide and amine, e.g. ammonia) and results in ionic products (the respective alkylammonium cation and halide anion)

$$NH_3 + RX \rightarrow RNH_3^+ + X^- \tag{18}$$

It had been established already by Menshutkin over 100 years ago that this reaction strongly depends on the polarity of the solvent employed. The reaction is fast in the solvents of high polarity but it is dramatically slowed down in lower dielectric constant media. This is opposite to the behaviour of ordinary S_N2 reactions involving ionic reagents which are accelerated by the decrease of the solvent polarity. Therefore, the Menshutkin reaction provides an interesting example of the solvent influence on the chemical reactivity. Recently, several reports on the quantum-chemical study of the Menshutkin reaction at the *ab initio* level have been published [92-94]. These involve the reactions $NH_3 + CH_3X$ $(X = F, Cl, Br, I)$. The solvent effect on the reaction was usually studied by using appropriate model Hamiltonians for the calculation of reaction potential energy surfaces in different media. A substantial lowering of the potential energy barrier was predicted in the high polarity solvents as compared to the gas phase, in accordance with the experimental evidence. In all these studies, a linear arrangement of the nitrogen, carbon and halogen atoms was assumed along the reaction path (cf. complex **1** in Scheme (III. 3)).

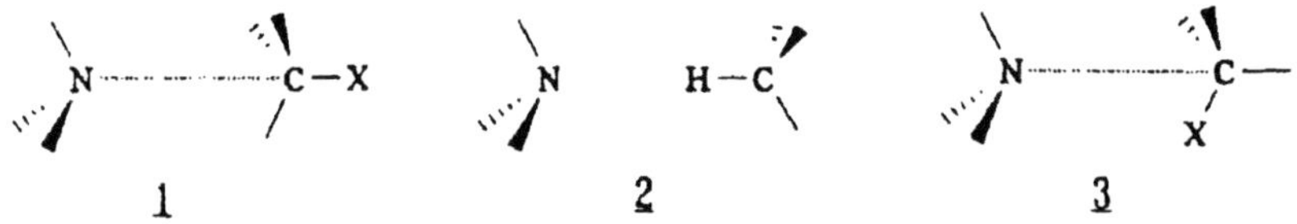

(III.3)

Such arrangement of reacting atoms is characteristic for almost all S_N2 reactions. However, our *ab initio* and semiempirical calculations [95,96] have indicated that this is true for the Menshutkin reaction only in highly polar media. In the gas phase, the relative position of the above-mentioned three atoms at the reaction centre has been predicted to be forcibly stretched along most of the reaction path (cf. complexes **2** and **3** on Scheme (III. 3)). These results indicate that the shape of the reaction potential surface may strongly depend on the reaction medium and therefore the study of the full surface is requested in each individual medium.

An important characteristics of the chemical reactivity is the aromaticity of carbo- and heterocyclic compounds. Numerous aromaticity scales have been invented either on the basis of experimental data or proceeding from some theoretical model calculations. Three different types of aromaticity scales have been distinguished on the basis of their origin, i.e. the energetic, geometric and magnetic aromaticity criteria [97]. A suggestion that aromaticity is at least a two dimensional phenomenon has been widely accepted [98-101]. Recently, the attention has been turned to the possible variation of the degree of aromaticity of a compound with the environment of the molecule [102].

The aromaticities can be expected to, and do, vary in magnitude relatively, and in many cases absolutely, with molecular environment. This can be illustrated with reference to benzene and azulene molecules.

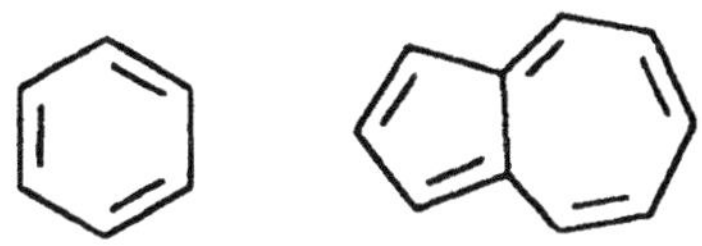

(III.4)

Benzene, a centrosymmetric molecule, with consequently no dipole moment, is expected to possess a geometry and therefore an aromaticity which is largely invariant with the medium: indeed the $C-C$ bond length as determined by electron diffraction in the gas phase of 1.393 [103] is well within the experimental error of that found for the crystal by neutron diffraction of 1.392 [104]. Azulene, by contrast, has a dipole moment with the five membered ring as the negative and the seven membered ring as the positive pole. Microwave measurements have given 0.796 D as the gas phase dipole moment [105]. The dipole moment has also been measured in dioxane solution as 1.00 [106]. Calculations suggest that this difference is significant: the dipole moment is predicted by *AM1* to be 1.496 in the gas phase and 1.757 for solution in dioxane. Unfortunately, the microwave spectrum and X-ray crystal determinations of azulene do not give accurate bond lengths due to low ground state population (approximately 3% at 25° C) in the gas phase and a disordered crystal, respectively. The bond lengths calculated by *AM1* for azulene in the gas phase, in dioxane and water solution (at 20° C) are shown in Table 5 [102]. It is immediately obvious that formally single bonds tend to become shorter as the dielectric constant is increased and that formally double bonds to tend to become longer. The Pozharskii and Bird aromaticity criteria can be used to express these variations quantitatively: $\Delta N\%$ decreases from 53.4 to 49.8% from water to the gas phase and the Bird index decreases from 69.5 to 67.4%. It can be therefore concluded that azulene is significantly less aromatic in the gas phase than in the media of higher dielectric constant.

Aromaticity variation with molecular environment is of particular significance in heterocyclic chemistry.

The measured and calculated bond lengths for imidazole, pyrrole, pyrazole, 1*H*-1,2,4-triazole and benzimidazole have an analogous tendency of change to that of the azulene when transferred from the gas phase to a high dielectric constant medium. Also, the specific hydrogen bonding of the solvent with the lone pairs of cyclic heteroatoms has a further influence on the aromaticity of the heterocyclic compounds. The Pozharskii and Bird aromaticity indices derived from *AM1 SCF* and *SCRF* bond lengths increase successively on progressing from the gas phase to dioxane, to dioxane with specific solvation, to water, and to water with a specific solvation by a single solvent molecule, and finally to water with maximum possible

TABLE 6. *AM*1 *SCF* and *SCRF* calculated bond lengths and aromaticity indices for azulene in different media [102]

Bond	Bond type	gas phase	dioxane	water
1-2	Formally double bond	1.381	1.382	1.384
3-10	"	1.390	1.396	1.397
4-5	"	1.358	1.363	1.364
6-7	"	1.361	1.363	1.363
8-9	"	1.355	1.360	1.362
2-3	Formally single bond	1.445	1.444	1.444
10-4	"	1.412	1.407	1.402
5-6	"	1.416	1.417	1.414
7-8	"	1.416	1.417	1.413
9-10	"	1.484	1.483	1.482
9-1	"	1.459	1.458	1.456
$\Delta N^{(a}$	—	0.246	0.234	0.228
$I^{(b}$	—	67.4	68.9	69.5

(*a* Pozharskii's aromaticity index [98]
(*b* Bird's aromaticity index [107]

number of specifically solvating water molecules. There are especially large increases in the aromaticity indices derived from experimental bond lengths in going from the gas phase to the crystal, presumably because of the very strong association in the crystal where the hydrogen bonds are evidently nearly symmetrical. Thus it had been concluded that the aromaticity of various heterocycles increases significantly with the polarity of the medium [102].

4. Calculation of molecular spectra in solutions

Much attention has been paid to the development of theoretical models for the interpretation of the solvent effects on the electronic states of molecules in solution and on the frequency shifts in absorption and emission spectra. For early theoretical elaborations, the reviews [108,109] can be suggested.

A complete treatment of the solvent effects on the electronic absorption spectra in solution had been proposed by us in the framework of the electrostatic *SCRF* model and quantum chemical configuration interaction (*CI*) method [6]. The absorption of the light by matter was considered as an instantaneous process. In other words, during the photon absorption no solvent orientational polarization is expected to happen, but rather only the electronic polarization of solvent will respond to the changed electron density of the solute molecule in its excited state. That is, the solvent orientation for the excited state remains the same as it was for the ground state, the solvent electronic polarization, however, must reflect the excited state dipole and other electric moments of the molecule. Considering the *SCRF* Hamiltonian

$$\hat{H} = \hat{H}^0 + \Gamma\langle\phi|\hat{\mu}_{tot}|\phi\rangle\hat{\mu} \tag{19}$$

we have for state $|\Psi_I\rangle$

$$\begin{aligned} H_{II} = \langle\Psi_I|\hat{H}^0 + \Gamma\langle\phi|\hat{\mu}|\phi\rangle\hat{\mu}|\Psi_I\rangle = \\ \langle\Psi_I|\hat{H}^0|\Psi_I\rangle - \Gamma\langle\Psi_0|\hat{\mu}|\Psi_0\rangle\langle\Psi_I|\hat{\mu}|\Psi_I\rangle \end{aligned} \tag{20}$$

which is the zeroth order estimate of the energy of the state $|\Psi_I\rangle$. Then, for a single excitation, $I \rightarrow A$, the excitation energy is given as follows

$$\begin{aligned} \triangle E_{ia} = \epsilon_a - \epsilon_i - J_{ia} + \left\{ \begin{matrix} 2 \\ 0 \end{matrix} \right\} K_{ia} = \epsilon_a^0 - \epsilon_i^0 - -J_{ia} + \\ + \left\{ \begin{matrix} 2 \\ 0 \end{matrix} \right\} K_{ia} - \Gamma\langle\Psi_0|\hat{\mu}|\Psi_0\rangle \left[\langle\Psi_i^a|\hat{\mu}|\Psi_i^a\rangle - \langle\Psi_0|\hat{\mu}|\Psi_0\rangle\right] \end{aligned} \tag{21}$$

where J_{ia} and K_{ia} are the respective Coulomb'; and exchange matrix elements, Γ is the reaction field tensor at the dipole level, and

$$\hat{f}_0(k)|\phi_i(k)\rangle = \epsilon_i^0|\phi_i(k)\rangle \tag{22}$$

with $\hat{f}_0(k)$ being the Fock operator for the k-th electron in the isolated solute molecule. The off-diagonal *CI* matrix elements are given by

$$\begin{aligned} H_{IJ} = \langle\Psi_I|\hat{H}^0 + \Gamma\langle\phi_0|\hat{\mu}|\phi_0\rangle\hat{\mu}|\Psi_J\rangle = \\ \langle\Psi_I|\hat{H}^0|\Psi_J\rangle - \Gamma\langle\Psi_0|\hat{\mu}|\Psi_0\rangle\langle\Psi_I|\hat{\mu}|\Psi_J\rangle \end{aligned} \tag{23}$$

Equations (20) - (23) demonstrate that some part of the solvent effect is already included in the ordinary *CI* treatment when proceeding from the *SCRF* Fock matrix. However, several concerns should be addressed to obtain the proper treatment of the excitation energies in solution.

First, $\langle\phi_0|\hat{\mu}|\phi_0\rangle$ should be the ground-state dipole moment after *CI*, and this implies an iterative procedure. At the *CIS* level, commonly used for the spectroscopic calculations, this is no concern because of Brillouin's theorem, which implies that the *CI* does not change the dipole moment of the molecule. It has been discussed [6] that even at higher levels of excitation in *CI*, this effect should not be large and might be estimated from the respective perturbation operator.

If the *Model A* is implemented for the *SCRF* theory (cf. the previous lecture notes [1]), the solvent cost has to be added to the energy obtained after diagonalization for the excited state

$$W_I^A = \langle\Psi_I|\hat{H}|\Psi_I\rangle + \frac{1}{2}\Gamma\langle\Psi_0|\hat{\mu}|\Psi_0\rangle\langle\Psi_I|\hat{\mu}|\Psi_I\rangle \tag{24}$$

No such correction is needed if theory *B* is applied for the calculation of the *SCRF* wave function and energy.

There are two approaches to address the instantaneous electronic polarization of the solvent during the excitation of the solute molecule. In the first case, the following correction term has to be added to the *CI* excitation energy

$$\triangle E_I = \frac{1}{2}\Gamma\left(n_D^2\right)\left[\langle\phi_0|\hat{\mu}|\phi_0\rangle\langle\phi_I|\hat{\mu}|\phi_I\rangle - |\langle\phi_I|\hat{\mu}|\phi_I\rangle|^2\right] \tag{25}$$

where $\Gamma\left(n_D^2\right)$ is the reaction field tensor for the optical relative dielectric permittivity of the solvent, n_D^2. In the last equation, the first term removes the incorrect term arising from the *SCRF* orbitals and energies in forming the *CI* matrix, and the second adds the response of the electronic polarization of the solvent to the dipole of the excited state. Equation (25) is first order in electron relaxation. Higher orders can be examined by defining the perturbation

$$X^{(I)} = \frac{1}{2}\lambda\Gamma\left(n_D^2\right)\left[\langle\phi_0|\hat{\mu}|\phi_0\rangle - \langle\phi_I|\hat{\mu}|\phi_I\rangle\right] \tag{26}$$

which is clearly different for each excited state and would, if pursued, lead to a set of excited states that were nonorthogonal. In principle, these corrections need not to be small.

Summarizing, the excitation energy from the ground state $|\Psi_0\rangle$ to excited state $|\Psi_I\rangle$ of a solute molecule in a dielectric medium is given as follows

$$\begin{aligned} W_I^A - W_0^A &= \langle\Psi_I|\hat{H}|\Psi_I\rangle - \langle\Psi_0|\hat{H}|\Psi_0\rangle + \\ &+\frac{1}{2}\Gamma\langle\Psi_0|\hat{\mu}|\Psi_0\rangle\left[\langle\Psi_I|\hat{\mu}|\Psi_I\rangle - \langle\Psi_0|\hat{\mu}|\Psi_0\rangle\right] + \\ &+\frac{1}{2}\Gamma\left(n_D^2\right)\langle\Psi_I|\hat{\mu}|\Psi_I\rangle\left[\langle\Psi_0|\hat{\mu}|\Psi_0\rangle - \langle\Psi_I|\hat{\mu}|\Psi_I\rangle\right] \end{aligned} \tag{27}$$

for *Model A*, and

$$W_I^B - W_0^B = \langle\Psi_I|\hat{H}|\Psi_I\rangle - \langle\Psi_0|\hat{H}|\Psi_0\rangle +$$

$$\frac{1}{2}\Gamma\left(n_D^2\right)\langle\Psi_I|\hat{\mu}|\Psi_I\rangle\left[\langle\Psi_0|\hat{\mu}|\Psi_0\rangle - \langle\Psi_I|\hat{\mu}|\Psi_I\rangle\right] \tag{28}$$

for *Model B*. The last two equations are first order in electron polarization of the solvent.

A second approach to the calculation of spectra in solutions is based on the assumption that the ground and excited states are intimately coupled in an instantaneous absorption process [110,111]. In this model, the solute ground state electron distribution responds to the electron distribution in the excited state through the instantaneous polarization of the solvent. In such a case, the energy of the absorbing (ground) state is shifted by

$$\frac{1}{2}\Gamma\left(n_D^2\right)\langle\Psi_0|\hat{\mu}|\Psi_0\rangle\langle\Psi_0|\hat{\mu}|\Psi_0\rangle -$$
$$-\frac{1}{2}\Gamma\left(n_D^2\right)\langle\Psi_0|\hat{\mu}|\Psi_0\rangle\left[\frac{1}{2}\langle\Psi_I|\hat{\mu}|\Psi_I\rangle - \langle\Psi_0|\hat{\mu}|\Psi_0\rangle\right] \tag{29}$$

In the last equation, the first term removes the first order in the electron polarization part of the dielectric relaxation included in the *SCRF* of the ground state, and the second term adds back the appropriate interaction of the ground state with the "mean" reaction field, created by the excited state, $|\Psi\rangle$.

This leads to a *Model A1* (notations from Ref. [6]), according to which the excitation energy is given as follows

$$\begin{aligned} W_I^{A1} - W_0^{A1} &= \langle\Psi_I|\hat{H}|\Psi_I\rangle - \langle\Psi_0|\hat{H}|\Psi_0\rangle + \\ &+\frac{1}{2}\Gamma\langle\Psi_0|\hat{\mu}|\Psi_0\rangle\left[\langle\Psi_I|\hat{\mu}|\Psi_I\rangle - \langle\Psi_0|\hat{\mu}|\Psi_0\rangle\right] + \\ &+\frac{1}{4}\Gamma\left(n_D^2\right)\left[\langle\Psi_0|\hat{\mu}|\Psi_0\rangle - \langle\Psi_I|\hat{\mu}|\Psi_I\rangle\right]^2 \end{aligned} \tag{30}$$

and to a *Model B1*, according to which

$$\begin{aligned} W_I^{B1} - W_0^{B1} &= \langle\Psi_I|\hat{H}|\Psi_I\rangle - \langle\Psi_0|\hat{H}|\Psi_0\rangle + \\ &\frac{1}{4}\Gamma\left(n_D^2\right)\left[\langle\Psi_0|\hat{\mu}|\Psi_0\rangle - \langle\Psi_I|\hat{\mu}|\Psi_I\rangle\right]^2 \end{aligned} \tag{31}$$

In the cases, examined in [6], all four models (*A, B, A1*, and *B1*) yield very similar results when $\pi \rightarrow \pi^*$ and $n \rightarrow \pi^*$ spectroscopic transition solvatochromic shifts are compared between the nonpolar and polar solvents. The use of models *A1* and *B1*, however, can occasionally lead to red shifts in $n \rightarrow \pi^*$ transitions when calculations in nonpolar solvents with reference to gas phase calculations are compared. In Table 6., the results of the *INDO/1* [18,112,113] *SCRF CIS* calculated spectroscopic transition energies are given for some solvatochromic dyes

1a 2a 3a (IV.1)

The relative shifts due to the solvent are reproduced theoretically in most cases. Even the absolute values of the spectroscopic transition energies are in satisfactory agreement with the respective experimental values which demonstrates the applicability of the spectroscopic *INDO/1* parameterization for the spectra in solutions. In these cases, the specific solvation by the hydrogen-bonding solvents was neglected.

TABLE 7. The *INDO/1 SCRF CIS* calculated spectral transitions for some dyes in different media

Molecule	Solvent	$\nu_{calc}(cm^{-1})$	$\nu_{exp}(cm^{-1})$
1a	gas phase	29,700	—
	cyclohexane	26,300	27,400
	water	22,500	23,300
2a	gas phase	36,900	—
	n-hexane	34,800	30,200
	water	31,100	26,100
3a	gas phase	20,200	—
	chloroform	21,800	19,600
	water	24,600	22,100

Our earlier calculations of the solvatochromic shifts in the spectra of azoles have, however, indicated the importance of such solvation [114] (cf. Table 7). The *SCRF CIS* theory alone explains all the shifts, except that observed in water. Water and acetonitrile are predicted to have practically the same shift which is not the case in experiment. The explicit bonding of two water molecules to the nitrogen lone pairs leads in the case of pyrimidine and pyridazine to the calculated large red shift instead of the experimentally observed solvatochromic blue shift for these compounds. However, by treating the whole complex of the azole and two water molecules in the surrounding reaction field leads to quantitatively correct blue shifts.

The same theoretical approach has been successfully used for the prediction of the emission spectra of p-N,N-dimethylaminobenzonitrile

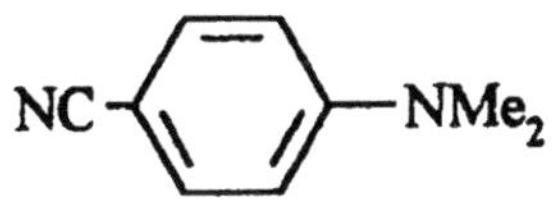

(IV.2)

and its analogues in solution [115]. The dual fluorescence of this molecule was discussed in terms of internal twisting of the excited state as well as in terms of N-inversion of the dimethylamino group. The results gave a definite indication that the N-inversion mechanism alone cannot explain the dual fluorescence without twisting the dimethylamino group.

TABLE 8. The *INDO/1 SCRF CIS* calculated and experimental spectral transition energies of azoles in different media

Molecule	Solvent	$\nu_{calc}(cm^{-1})$	$\nu_{exp}(cm^{-1})$
Pyrimidine	gas phase	32966	—
	isooctane	33559	34200
	diethyl ether	34127	34400
	acetonitrile	34697	34800
	water	34743	36900
	2 H_2O	30982	36900
	water+2 H_2O	36572	36900
Pyridazine	gas phase	28329	—
	isooctane	29460	29740
	diethyl ether	30382	30150
	acetonitrile	31296	31080
	water	31368	33570
	2 H_2O	26490	33570
	water+2 H_2O	33927	33570
Pyrazine	gas phase	30387	—
	isooctane	30387	31610
	diethyl ether	30387	31610
	acetonitrile	30387	31740
	water	30387	33160
	2 H_2O	32900	33160
	water+2 H_2O	33301	33160

The multiconfigurational self-consistent reaction field (*MC SCRF*) and the *SCRF* methods have been applied for the calculation of the solvent shifts of molecular photoelectron spectra [116]. The solvation shift was defined as the difference between two vertical ionization energies for the vacuum and solvated systems

$$\delta E = \triangle E_{BE} = \left(E_T^f - E_T^i\right) - \left(E_C^f - E_C^i\right) \tag{32}$$

where T denotes the total solvated system, which can be a single solute molecule or a cluster embedded into dielectric continuum, and C denotes the isolated molecule, i the initial state and f the final ionic state. The calculations were performed for water, benzene, methanol and formamide surrounded by dielectric media corresponding to various solvents. The wave functions for single- or multiconfigurational closed- and open-shell states were optimized with their reaction fields, using either a continuum approach with one solute molecule embedded into dielectric medium or a semicontinuum approach with one solute molecule and a solvation shell of solvent molecules surrounded by the dielectric medium. It was established, that comparatively large multipolar expansions of the solute charge distribution are needed in order to adequately treat the solvent effects on the photoionization spectra [116].

Finally, we mention the combined approaches to molecular absorption spectra in solutions. N. Rösch and M.C. Zerner have implemented the Abe's semiclassical solute-solvent dispersion model into *INDO/1 CI* theory of molecular electronic spectra (cf. also Chapter 1) [10]. This theory, when combined with the *SCRF CI* theory for electrostatic solvent effects, has enabled to predict correctly the $n \rightarrow \pi^*$ transition energies of acetone in cyclohexane and the solvent shifts of different spectral transitions of benzene, naphthalene, chrysene and azulene in cyclohexane. The calculation of molecular absorption spectra in condensed media has been performed also by modelling the medium using the molecular dynamics (*MD*) method and the spectroscopically active solute quantum-mechanically [117,118]. Such approach involves, of course, very long computational time as the quantum calculation has to be performed for a large number of *MD* configurations.

5. Conclusions

As a summary, we may conclude that a large variety of methods are available for the theoretical study of different chemical and physical processes and phenomena in disordered condensed media (liquids and solutions). However, one has to be careful in the application of a given method for the description of an experimental property or observable of interest. Due to the enormous complexity of solvation, specific methods are often needed for specific purposes. However, this encourages further investigations in two directions - first, the development of new theoretical methods and approaches to solvent effects, and secondly, the validation of the limits of applicability of available methods.

References

1. Karelson, M. , Diercksen, G.H.F in: *Problem Solving in Computational Molecular Science, NATO ASI Series*, Wilson, S. and Diercksen, G.H.F. (Eds.) Kluwer, Dordrecht. 1997.
2. Karelson, M. *Adv. Quant. Chem.* **1997**, in press.
3. Karelson, M.M. *Org. React.* **1980**, 17, 357.
4. Tapia, O. Goscinski, O. *Mol. Phys.* **1975**, 29, 1683.
5. Rivail, J-L; Rinaldi, D. *Chem. Phys.* **1976**, 18, 233.
6. Karelson, M.; Zerner, M.C. *J. Phys. Chem.* **1992**, 92, 6949.
7. Böttcher, C.J.F.; Bordewijk, P. *Theory of Electric Polarization*, 2nd ed., vol. II, Elsevier Co, Amsterdam, 1978.
8. Ågren, H.; Knuts, S.; Mikkelsen, K.V.; Jensen, H.J. Aa *Chem. Phys.* **1992**, 159, 211.
9. Tomasi, J.; Persico, M. *Chem. Rev.* **1994**, 94, 2027.
10. Rösch, N.; Zerner, M.C. *J. Phys. Chem.* **1994**, 98, 5817.
11. Reiss, H.; Frisch, H.L.; Lebowitz, J.L. *J. Chem. Phys.* **1959**, 31, 369.
12. Reiss, H. *Adv. Chem. Phys.* **1966**, 9, 1.
13. Cramer, C.J.; Truhlar, D.G. *J. Am. Chem. Soc.* **1991**, 113, 8305.
14. Bachs, K.M.; Luque, F.J.; Orozco, M. J. *Comput. Chem.* **1994**, 15, 446.
15. *Gaussian 94, Revision B.1*, M.J. Frisch, G.W. Trucks, H.B. Schlegel, P.M. W. Gill, B.G. Johnson, M.A. Robb, J.R. Cheeseman, T. Keith, G.A. Petersson, J.A. Montgomery, K. Raghavachari, M.A. Al-Laham, V.G. Zakrzewski, J.V. Ortiz, J.B. Foresman, J. Cioslowski, B.B. Stefanov, A. Nanayakkara, M. Challacombe, C.Y. Peng, P.Y. Ayala, W. Chen, M.W. Wong, J.L. Andres, E.S. Replogle, R. Gomperts, R.L. Martin, D.J. Fox, J.S. Binkley, D.J. Defrees, J.Baker, J.P. Stewart, M. Head-Gordon, C. Gonzalez, and J.A. Pople, Gaussian, Inc., Pittsburgh PA, (1995).
16. Dewar, M.J.S.; Zoebisch, E.G;. Healy, E.F.; Stewart, J.J.P. *J. Am. Chem. Soc.* **1985**, 107, 3902.
17. Stewart, J.J.P. *MOPAC Program Package*, QCPE, No 455 (1989).
18. Ridley, J.; Zerner, M.C. *Theor. Chim. Acta* **1973**, 32, 111.
19. Ford, G.P.; Wang, B. *J. Am. Chem. Soc.* **1992**, 114, 10563.
20. Karelson, M.; Tamm, T.; Zerner, M.C. *J. Phys. Chem.* **1993**, 97, 11901.
21. Su, S.-G.; Simon, J.D. *J. Phys. Chem.* **1987**, 91, 2693.
22. Castner, E.W., Jr.; Maroncelli, M.; Fleming, G.R. *J. Chem. Phys.* **1987**, 86, 1090.
23. Castner, E.W., Jr.; Fleming, G.R.; Bagchi, B. *Chem. Phys. Lett.* **1988**, 143, 270.
24. Bagchi, B.; Chandra, A. *J. Chem. Phys.* **1990**, 93, 1956.
25. Hine, J.; Mokerjee, P.K. *J. Org. Chem.* **1975**, 40, 292.
26. Cramer, C.J.; Truhlar, D.G. *J. Am. Chem. Soc.* **1991**, 113, 8305.
27. Cramer, C.J.; Truhlar, D.G. *J. Am. Chem. Soc.* **1991**, 113, 8552.
28. Cramer, C.J.; Truhlar, D.G. *J. Comput. Chem.* **1992**, 13, 1089.
29. Cramer, C.J.; Truhlar, D.G. *Science* **1992**, 256, 213.
30. Gurney, R.W. *Ions in Solution*, Dover Press, New York, 1962.
31. Marcus, Y. *Ion Solvation*, Wiley Interscience, Chichester, 1986.
32. Krestov, G.A. *Thermodynamics of Ion Solvation*, Ellis Horwood, New York, 1991.
33. Karelson, M.M. *Org. React.* **1983**, 20, 127.
34. Honig, B.; Sharp, K.; Yang, A.-S. *J. Phys. Chem.* **1993**, 97, 1101.
35. Sitkoff, D.; Sharp, K.; Honig, B. *J. Phys. Chem.* **1994**, 98, 1978.
36. Diercksen, G.H.F.; Karelson, M.; Tamm, T.; Zerner, M.C. *Int. J. Quant. Chem.* **1994**, S28, 339.
37. Tamm, T.; Karelson, M. *Proc. Estonian Acad. Sci. Chem.* **1991**, 40, 122.
38. Tamm, T.; Karelson, M. *Org. React.* **1989**, 26, 211.
39. Pappalardo, P.R.; Sanchez Marcos, E. *J. Chem. Res.* **1989**, 290.
40. Pappalardo, P.R.; Sanchez Marcos, E. *J. Chem. Soc. Faraday Trans.* **1991**, 87, 1719.

41. Pappalardo, P.R.; Sanchez Marcos, E.; Ruiz-Lopez, M.F.; Rinaldi, D.; Rivail, J.L. *J. Phys. Org. Chem.* **1991**, 4, 141.
42. Wong, M.W.; Frisch, M.J.; Wiberg, K.B. *J. Am. Chem. Soc.* **1991**, 113, 4776.
43. Pappalardo, P.R.; Sanchez Marcos, E.; Ruiz-Lopez, M.F.; Rinaldi, D.; Rivail, J.L. *J. Am. Chem. Soc.* **1993**, 115, 3722.
44. Parrondo, R.M.; Karafiloglou, P.; Pappalardo, P.R.; Sanchez Marcos, E. *J. Phys. Chem.* **1995**, 99, 6461.
45. Balaban, A.T. *Rev. Roum. Chim.* **1971**, 16, 725.
46. Balaban, A.T.; Istratoiu, R. *Tetrahedron Lett.* **1973**, 1879.
47. Baldock, R.W.; Hudson, P.; Katritzky, A.R.; Soti, F. *J. Chem. Soc. Perkin Trans. 2* **1974**, 1422.
48. Katritzky, A.R.; Zerner, M.C.; Karelson, M.M. *J. Am. Chem. Soc.* **1986**, 108, 7213.
49. Bordwell, F.G.; Lynch, T.Y. *J. Am. Chem. Soc.* **1989**, 111, 7558.
50. Bordwell, F.G.; Zhang, X-M.; Alnajjar, M.S. *J. Am. Chem. Soc.* **1992**, 114, 7623.
51. Viehe, H.G.; Merenyi, R.; Stella, L.; Janousek, Z. *Angew. Chem. Int. Ed. Engl.* **1979**, 18, 917.
52. Viehe, H.G.; Janousek, Z.; Merenyi, R.; Stella, L. *Acc. Chem. Res.* **1985**, 18, 148.
53. Crans, D.; Clark, T.; Schleyer, P.v.R. *Tetrahedron Lett.* **1980**, 21, 3681.
54. Pasto, D.J. *J. Am. Chem. Soc.* **1988**, 110, 8164.
55. Karelson, M.M.; Tamm, T.; Katritzky, A.R.; Szafran, M.; Zerner, M.C. *Int. J. Quant. Chem.* **1990**, 37, 1.
56. Hylton McCreery, J.; Christoffersen, R.E.; Hall, G.G. *J. Am. Chem. Soc.* **1976**, 98, 7191.
57. Urban, J.J.; Cramer, C.J.; Famini, G.R. *J. Am. Chem. Soc.* **1992**, 114, 8226.
58. Vásquez, M.; Nemethy, G.; Scheraga, H.A. *Chem. Rev.* **1994**, 94, 2183.
59. Pincus, M.R. *Biopolymers* **1992**, 32, 347.
60. Kostrowicki, J.; Scheraga, H.A. *J. Phys. Chem.* **1992**, 96, 7442.
61. Cramer, C.J.; Truhlar, D.G. *Theoret. Comput. Chem.* **1994**, 1, 9.
62. Wiberg, K.B.; Wong. M.W. *J. Phys. Chem.* **1992**, 96, 668.
63. Luque, F.J.; Orozco, M. *J. Org. Chem.* **1993**, 58, 6397.
64. Burton, N.A.; Chiu, S.C.-L.; Davidson, M.M.; Green, D.V.S.; Hillier, I.H.; McDouall, J.J.W.; Vincent, M.A. *J. Chem. Soc. Faraday Trans.* **1993**, 89, 2631.
65. Wiberg, K.B.; Rablen, P.R.; Rush, D.J.; Keith, T.A. *J. Am. Chem. Soc.* **1995**, 117, 4261.
66. Yu, J.; Zerner, M.C. *J. Chem. Phys.* **1994**, 100, 7487.
67. Mikkelsen, K.V.; Jørgensen, P.; Jensen, H.J.Aa. *J. Chem. Phys.* **1994**, 100, 6597.
68. Cammi, R.; Cossi, M.; Tomasi, J. *J. Chem. Phys.* **1996**, 104, 4611.
69. Sun, H.; Mumby, S.J. *J. Chem. Phys.* **1996**, 104, 1018.
70. Dupuis, M.; Farazdel, A.; Karna, S.P.; Maluendes, S.A. *HONDO-8*, IBM Corporation, Kingston, New York, 1990.
71. Löffler, J.E.; Grunwald, E. *Rates and Equilibria of Organic Reactions*, J. Wiley & Sons, New York, 1965.
72. Taft, R.W.; Topsom, R.D. *Progr. Phys. Org. Chem.* **1987**, 16, 1.
73. Furuki, T.; Sakurai, M.; Inoue, Y.; Chûjô, R.; Harata, K. *Chem. Phys. Lett.* **1991**, 188, 84.
74. Still, W.C; Tempczyk, A.; Hawley, R.C.; Hendrickson, T. *J. Am. Chem. Soc.* **1990**, 112, 6127.
75. Urban, J.J.; von Tersch, R.L.; Famini, G.R. *J. Org. Chem.* **1994**, 59, 5239.
76. Karelson, M.M.; Katritzky, A.R.; Szafran, M.; Zerner, M.C. *J. Org. Chem.* **1989**, 61, 6030.
77. Karelson, M.M.; Tamm, T.; Katritzky, A.R.; Cato, S.J.; Zerner, M.C. *Tetrahedron Comput. Methodol.* **1989**, 2, 295.
78. Karelson, M.M.; Katritzky, A.R.; Szafran, M.; Zerner, M.C. *J. Chem. Soc. Perkin Trans 2* **1990**, 161.

79. Katritzky A.R.; Karelson, M. *J. Am. Chem. Soc.* **1991**, 113, 1561.
80. Strandberg, M.; Schiemenz, G.P.; Karelson, M. *Proc. Estonian Acad. Sci. Chem.* **1994**, 43, 41.
81. Parchment, O.G.; Hillier, I.H.; Green, D.V.S.; Burton, N.A.; Morley, J.O.; Schaefer, H.F., III *J. Chem. Soc. Perkin Trans. 2* **1992**, 1681.
82. Gould, I.R.; Green, D.V.S.;Young, P.; Hillier, I.H. *J. Org. Chem.* **1992**, 57, 4434.
83. Wong, M.W.; Wiberg, K.B.; Frisch, M.J. *J. Am. Chem. Soc.* **1992**, 114, 1645.
84. Freitas, L.C.G.; Longo, R.L.; Simas, A.M. *J. Chem. Soc. Faraday Trans.* **1992**, 88, 189.
85. Woodcock, S.; Green, D.V.S.; Vincent, M.A.; Hillier, I.H.; Guest, M.F.; Sherwood, P. *J. Chem. Soc. Perkin Trans. 2* **1992**, 2151.
86. Gould, I.R.; Hillier, I.H. *J. Chem. Soc. Perkin Trans. 2* **1993**, 1771.
87. Cramer, C.J; Truhlar, D.G. *J. Am. Chem. Soc.* **1993**, 115, 8810.
88. Contreras, J.G.; Alderete, J.B. *J. Phys. Org. Chem.* **1995**, 8, 395.
89. Karelson, M.; Maran, U.; Katritzky, A.R. *Tetrahedron* **1996** (in press).
90. Katritzky, A.R.; Lagowski, J.M. *Adv. Hetrocycl. Chem.* **1963**, 1, 339.
91. Lochmüller, C.H.; Maldacker, T.; Cefola, M., *Anal. Chim. Acta* **1969**, 48, 139.
92. Solá, M.; Lledos, A.; Duran, M.; Bertrán, J.; Abboud, J-L.M. *J. Am. Chem. Soc.* **1991**, 113, 2873.
93. Gao, J. *J. Am. Chem. Soc.* **1991**, 113, 7796.
94. Shaik, S.; Ioffe, A.; Reddy. A.C.; Pross, A. *J. Am. Chem. Soc.* **1994**, 116, 262.
95. Maran, U.; Pakkanen, T.A.; Karelson, M. *J. Chem. Soc. Perkin Trans. 2* **1994**, 2445.
96. Maran, U.; Karelson, M.; Pakkanen, T. *J. Mol. Struct. (Theochem)* **1996**, (submitted).
97. Katritzky, A.R.; Barczynski, P.; Musumarra, G.; Pisano, D.; Szafran, M. *J. Am. Chem. Soc.* **1989**, 111, 7.
98. Katritzky, A.R.; Feygelman, V.; Musumarra, F.; Barczynski, P.; Szafran, M. *J. Prakt. Chem.* **1990**, 332, 853.
99. Katritzky, A.R.; Feygelman, V.; Musumarra, F.; Barczynski, P.; Szafran, M. *J. Prakt. Chem.* **1990**, 332, 870.
100. Katritzky, A.R.; Barczynski, P. *J. Prakt. Chem.* **1990**, 332, 853.
101. Katritzky, A.R.; Karelson, M.; Malhotra, N. *Heterocycles* **1991**, 32, 127.
102. Krygowski, T.M. *J. Chem. Inf. Comput. Sci.* **1993**, 33, 70.
103. Jug, K.; Koster, A.M. *J. Phys. Org. Chem.* **1991**, 4, 163.
104. Bastiansen , O. *Acta Crystallogr.* **1957**, 861.
105. Bacon, G.E.; Curry, N.A.; Wilson, S.A. *Proc. R. Soc. London A* **1964**, 279, 98.
106. Tobler, H.J.; Bauder, A.; Gunthard, Hs.H. *J. Mol. Spectrosc.* **1965**, 18, 239.
107. Wheland, G.W.; Mann, D.E. *J. Chem. Phys.* **1949**, 17, 264.
108. Bird, C.W. *Tetrahedron* **1987**, 43, 4725.
109. Basu, S. *Adv. Quant. Chem.* **1964**, 1, 145.
110. Amos, A.T.; Burrows, B.L. *Adv. Quant. Chem.* **1973**, 7, 289.
111. Marcus, R. *J. Chem. Phys.* **1956**, 24, 979.
112. Jortner, J. *Mol. Phys.* **1962**, 5, 257.
113. Ridley, J.; Zerner, M.C. *Theor. Chim. Acta* **1976**, 42, 223.
114. Bacon, A.D..; Zerner, M.C. *Theor. Chim. Acta* **1979**, 53, 21.
115. Karelson, M.; Zerner, M.C. *J. Am. Chem. Soc.* **1990**, 112, 9405.
116. Broo, A.; Zerner, M.C. *Chem. Phys. Lett.* **1994**, 227, 551.
117. Medina-Llanos, C.; Ågren, H.; Mikkelsen, K.V.; Jensen, H.J. Aa. *J. Chem. Phys.* **1989**, 90, 6422.
118. Blair, J.T.; Krogh-Jespersen, K.; Levy, R.M. *J. Am. Chem. Soc.* **1989**, 111, 6948.
119. Luzhkov, V.; Warshel, A. *J. Am. Chem. Soc.* **1991**, 113, 4491.

Molecular Simulation – A Primer

Florian Müller-Plathe, Max-Planck-Institut für Polymerforschung,
D-55128 Mainz, Germany

1. Classical Molecular Simulation – Why Bother?

Quantum chemistry allows us to calculate molecular properties with arbitrary accuracy: potential surface, geometry, moments, spectra, etc. However, there are properties which do not depend on a single molecule alone but can only be defined for an ensemble of many molecules: melting point, diffusion coefficient, density, for example. "Many molecules" can refer to a few dozen molecules in a molecular cluster but can also mean Avogadro's number.

There are properties which are thermodynamical. That implies that they are temperature dependent, take e.g. the thermal expansion coefficient. Quantum chemistry is mostly performed at 0 Kelvin, and expansions around 0 K (for example, the "thermochemistry" from harmonic vibrational analysis) are frequently poor approximations.

In addition, there are properties which are dynamical, in other words time-dependent. Most transport coefficients belong to this class: diffusion coefficient, thermal and electrical conductivity, viscosity. Quantum chemistry does not include time in its formulation.

To make things worse, there are properties which involve the free energy, i.e. entropic as well as enthalpic terms (e.g. solubility, partition coefficient, binding constant, pK_a). Entropy, on the other hand, is strongly related to the available conformation space. This means that a representative sample of relevant conformations has to be taken into account. A large enough sample in practice often consists of thousands or millions of conformations.

In summary, there is a large number of interesting properties whose calculation needs a system of many molecules in many different conformations. This leads to the sad fact that these properties cannot be dealt with by traditional quantum chemistry.

The various methods of molecular simulation establish the link between a microscopic model of the system and its macroscopic properties. They all contain two important ingredients: A simplification of the molecular interaction, called an empirical, phenomenological or effective force field, and a solid body of statistical mechanics to connect microscopic atomic motions to macroscopic observables.

In this short overview, we will start with the potential energy function, then present the molecular dynamics method as one example of molecular simulation, then look at the statistical mechanics for calculating a few bulk properties, and, finally, (re)introduce electrons and quantum mechanics, in order to address chemical reactions. Since all chapters but the last contain mainly textbook knowledge, few references are given. Instead, the appendix provides an annotated list of references which may be useful for further study.

S. Wilson and G. H. F. Diercksen (eds.), Problem Solving in Computational Molecular Science, 389–412.

2. Force Fields

2.1 Forms

The basis of all molecular simulation methods is a potential energy function $U(\mathbf{R})$. It gives the potential energy of the entire system of N atoms as a function of the positions of all atoms. The vector formed by all atomic co-ordinates, is denoted by $\mathbf{R}$ without subscript, whereas a subscripted vector $\mathbf{R}_i$ will refer to the co-ordinates of atom i. $U(\mathbf{R})$ could be the result of a quantum-chemical calculation after invoking what is (historically incorrect, see Brian Sutcliffe's contribution) called the Born-Oppenheimer approximation. However, in molecular simulation, $U(\mathbf{R})$ is usually described by simple analytical terms.

The choice of the form is mostly guided by chemical common sense. The form for the "bonded" interactions often looks something like

$$U_{\text{bonded}} = \sum_{\text{bonds}} \frac{K_b}{2}(b-b_0)^2 + \sum_{\text{bond angles}} \frac{K_\phi}{2}(\phi-\phi_0)^2 + \sum_{\text{dihedral angles}} \frac{K_\tau}{2}(1-\cos(3\tau-\tau_0)). \quad \text{Eq. 1}$$

Here, b, ϕ and τ are the actual value of a bond distance, a bond angle and a dihedral angle, respectively; b_0, ϕ_0 and τ_0 are the corresponding "equilibrium values" (where this particular term has a minimum); K_b, K_ϕ, and K_τ are the corresponding force constants. For clarity, I have omitted additional subscripts denoting the different terms in each sum. It goes without saying that chemically different bonds will have different parameters. The form above is the one we are mostly using [1], but it is in no way special. In order to illustrate this, Eq. 2 gives alternatives for the three terms which are also frequently used.

$$\begin{aligned} U_{\text{bonded}} &= \sum_{\text{bonds}} D\left[1-\exp\left\{-a(b-b_0)^2\right\}\right]^2 \\ &+ \sum_{\text{bond angles}} \frac{K_{\cos\phi}}{2}(\cos\phi-\cos\phi_0)^2 + \sum_{\text{dihedral angles}} \left[\sum_{n=0}^{5} C_n \cos^n\tau\right] \end{aligned} \quad \text{Eq. 2}$$

The alternative forms obviously contain different parameters.

As an aside: In many applications, one replaces some of the bonded interactions by constraints, i.e. one keeps parts of molecules or whole molecules rigid during the simulation. In particular, bond distances are often constrained, rather than, described by a harmonic or Morse term as in Eq. 1 and 2, respectively. Although in practice constraints are being used in place of potential energy functions they are by no means equivalent. Using constraints removes degrees of freedom from the system and is equivalent to solving the system's equations of motion in a lower dimensional space. This difference in dimensionality leads, for example, to correction terms for calculated properties like the free energy, the pressure and – in certain cases – conformational distributions.

The terms of Eqs. 1 and 2 include interactions involving 2 atoms (bonds), 3 atoms (bond angles) and 4 atoms (dihedral angles) which are connected to each other. Atoms of a molecule which are further apart in terms of bond topology and atoms belonging to different molecules are normally treated by "nonbonded" interactions, usually taken to be pairwise additive. The choice of the terms is again motivated by physical and chemical considerations.

$$U_{\text{nonbonded}} = \sum_{\text{nonbonded pairs } ij} 4\varepsilon_{ij}\left[\left(\frac{\sigma_{ij}}{r_{ij}}\right)^{12} - \left(\frac{\sigma_{ij}}{r_{ij}}\right)^{6}\right] + \frac{1}{4\pi\varepsilon_0}\frac{q_i q_j}{r_{ij}} \quad \text{Eq. 3}$$

The r^{-12} term in Eq. 3 approximates a short-range repulsion between filled electron shells (Pauli principle), the r^{-6} term is the leading term in an expansion of the dispersive attraction between atoms (London forces), the r^{-1} term is caused by a Coulombic attraction or repulsion of the two atomic partial charges q_i and q_j. The r^{-12} and r^{-6} terms together are referred to as a Lennard-Jones potential, and ε_{ij} and σ_{ij} are the well depth and the zero point, respectively. Again, variations of Eq. 3 are common: The repulsion is often taken to be r^{-9} or exponential (Buckingham potential). The dispersive attraction may contain other powers in addition to r^{-6}, like r^{-8}. There may be special forms for hydrogen bonds. The Coulombic terms can be augmented by atomic polarisabilities. Some force fields, for example those for silicon melts and crystals, include three-atom terms in addition to pair terms.

2.2 Parameters – Caveats

From the discussion so far it should have become clear that there is nothing unique about a force field. In most cases, it matters little which functional form is chosen as long as it is physically reasonable and satisfies the applicable boundary conditions. Choosing an angle-bending term harmonic in $\cos\phi$ (Eq.2), rather than, harmonic in ϕ (Eq. 1) may have certain technical advantages, it will, however, not change the essentials of the description. Consequently, there is no form which is a priori "better" than another. In stark contrast, the quality of a force field depends crucially on its parametrisation. That is, once an analytical form has been chosen, one must find appropriate values for its parameters. This step is of utmost importance and it is here, where mistakes are easily made. In the following, I give a few caveats. You should be aware that these are not necessarily sound scientific results but my personal opinion. Many people would strongly debate some of them.

- *A parameter is a parameter is a parameter.* A force field parameter such as in Eqs. 1-3 should be regarded as a number which is chosen so as to make the simulated system reproduce certain properties of the real system. A parameter does not have to equal a microscopic descriptor of the molecule. As an example, the force constant of the bond term in Eq. 1 *can* be chosen so that the experimental vibrational frequency of this bond is matched exactly, but it does not have to be. In fact, by allowing a different value for the force constant one might improve the calculated value for some other property. Not even the molecular geometry is sacred: b_0 and ϕ_0 need not equal the experimentally determined equilibrium bond length and bond angle. This does not mean that one is allowed to use a completely unphysical C-C bond length of 10 Å or –1.3 Å, but it can be very justified to use a value of 1.6 Å instead of measured 1.5134 Å.
- *There is no such thing as a universal force field.* The most complicated force field is still a function with comparatively few adjustable parameters; 10 parameters per atom type is a good average which includes bonded as well as non-bonded terms. One cannot expect this simple function to correctly describe all properties of all systems under all conditions. This means that an OH group will not only have to be parametrised differently when it is part of a water molecule, an aliphatic alcohol, a phenol or a

carboxylic acid. One may need different parameters for the same alcohol depending on whether it is in a liquid, in a crystal, in aqueous solution, in a mixture with a hydrocarbon and so on. In addition, one may need different parametrisations for different properties of the bulk alcohol that one is trying to calculate.

- *Assembling force fields is dangerous.* Taking the bonded interactions from one force field, the Lennard-Jones parameters from a second one and adding the partial charges from a semiempirical quantum chemical calculation is common practice. Using such a hotchpotch of terms without serious testing often leads to disastrous results. The reason for this is that the parameters cannot be selected independently; they have to harmonise with each other to give a reliable description of the system. One wrong parameter can mess it all up.

- *Ab initio force fields are no better than any other force fields.* There are relatively few molecular properties which quantum chemistry can provide in a such way that they can be used directly and profitably in the construction of a force field. Quantum chemistry does very well for molecular geometries. However, in many cases these can be guessed with sufficient accuracy: a C-C single bond is 1.53 ± 0.02 Å except in exotic circumstances. Similarly, vibrational force constants can often be transferred from similar molecules and need not be calculated. Quantum chemistry is impractical for Lennard-Jones parameters of most molecules of interest to simulators, since the description of dispersive interactions needs huge basis sets and a high-order treatment of electron correlation. Atomic partial charges can be derived from wave functions. However, the fundamental problem (the atomic partial charge is not an observable and, hence, there is no well-defined quantum-mechanical operator for it) leads to technical problems (there are many recipes for extracting partial charges from a wave function, all of which give different answers) as well as language problems (the quantum chemists uses partial charges as a data reduction concept to approximately characterise the distribution of electrons in a molecule; the molecular simulator views them as effective force field parameters, i.e. mere numbers). One additional important reason why nonbonded parameters from quantum chemistry cannot be used directly, even if they could be calculated accurately, is that they have to implicitly account for everything that has been neglected: three-body terms, polarisation, etc. (One should add that this applies to experimental parameters as well: a set of parameters describing a water dimer in vacuum will, in general, not give correct properties of bulk liquid water.)
  On the bright side, there is an application area where quantum chemistry is very successful, namely in the parametrisation of dihedral angle (torsional) interactions. Except for small molecules there is no way to determine torsional barriers as well as the energetic differences between conformations (say, *trans* and *gauche*) which is as quick and reliable as ab initio quantum chemistry. Often Hartree-Fock or density functional calculations in a polarised split-valence basis do a very nice job.

- *Off-the-shelf force fields must be carefully tested before being used in anger.* There is a number of ready-made force fields. At the push of a mouse button, all commercial simulation packages will provide a force field for the molecule drawn on the screen. However, many of these force fields combine several of the above vices. Most importantly, they have been assembled in order to cover all possible applications of the software, from drug molecules to zeolites. They can be used, but only after thorough testing.

- As a consequence of all this: *A "well-tested" force field is one whose application range and limitations are known and understood.* In other words, one should look at the performance of a force field for a variety of properties under a variety of conditions, in order to determine whether or not it suits one's own application. Serious force field developers provide such data with their force fields.

- And finally: *The parameters used must be listed where they are used.* Statements like "We used version X of force field Y from company Z." mean that the paper's results are not reproducible, so it cannot be published. Even if the parameters have been published, it is often in the appendix of some manual, without any reference of how they were obtained and how they were checked. In ten years, not even the force field's own mother will be able to tell what the parameters were at the time of the calculation.

2.3 Parametrisation – An example

Let us look at a specific example of how a force field can be derived. A while ago, we needed a model of liquid dimethyl sulfoxide (DMSO) which is often used as a solvent [2]. We began by defining the properties which were most important to us. These were: the density at room temperature (the molecular "size" of the model), the heat of vaporisation (the "cohesion" of the liquid) and the diffusion coefficient as a representative of the dynamical properties. Next, we made a fundamental choice by selecting a united-atom model (methyl groups are represented by a single interaction site, not 4 atoms) rather than, an all-atom model. This is a simplification which causes certain properties to be described less accurately. However, the computer time for the nonbonded interactions increases with the atom number density squared. This means, using 4 instead of 10 atoms in a molecule makes the calculation faster by a factor of more than 6 (100/16).

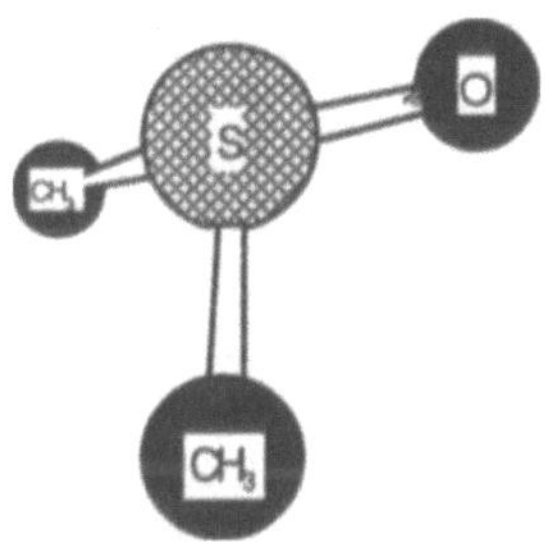

Figure 1. The structure of dimethyl sulfoxide DMSO.

The structure of the DMSO molecule is shown in Figure 1. Note that, due to the sulphur lone pair, it is not planar. We opted for a totally rigid molecule. Rigidity is enforced by applying distance constraints not only for the 3 chemical bonds but also for the two O–CH_3 distances and the CH_3–CH_3 distance. The molecular geometry (Table 1) is the experimental one, except for the S–CH_3 bonds which are found to be 0.18 nm in X-ray crystallography. This bond length was chosen to improve the short-range behaviour of the atomic radial distribution function.

Table 1. The rigid united-atom force field for DMSO, from [2].

atom type	S	O	CH_3
ε (kJ/mol)	1.2970	1.7154	0.9414
σ (nm)	0.356	0.263	0.366
q (e)	0.139	-0.459	0.160
m (amu)	32.0600	15.9994	15.0350
r(S–O) = 0.153 nm		∠(O–S–C) = 106.75°	
r(S–CH_3) = 0.195 nm		∠(C–S–C) = 97.4°	

With the geometry fixed, we adapted the nonbonded force constants ε, σ and q, so the model gave the desired density, heat of vaporisation and diffusion coefficient (Table 2). This was essentially done by trial and error. It is interesting to note that the dipole moment of our final DMSO model (calculated from the atomic partial charges and the molecular geometry) is about 4.5 Debye (in contrast to atomic partial charges, the molecular dipole moment is an observable). The dipole moment of a single DMSO molecule calculated by ab initio quantum chemistry is about 4.0 Debye. However, immersing a DMSO molecule into a liquid of other DMSO molecules causes an increase of the dipole moment to 4.4 Debye, as has been verified in ab initio self-consistent reaction field calculations. This example nicely illustrates that parameters of an effective force field for a bulk liquid have to incorporate implicitly environmental effects. Using partial charges appropriate for a single molecule in vacuum would have lead to an incorrect description of the liquid.

Table 2. Bulk properties of DMSO calculated with the model of Table 1 (at room temperature and ambient pressure). The boxed quantities were used in the parametrisation of the model.

property		MD	exper.
density (kg/m^3)		1.099	1.095
heat of vaporisation (kJ/mol)		52.87	52.88
diffusion coefficient (10^{-5} cm^2/s)		11	8
rotational correlation time (ps)		3.9	5.2
thermal expansion coefficient (10^{-3} K^{-1})		0.91±0.11	0.928
isothermal compressibility (10^{-7} $Torr^{-1}$)		0.901±0.099	0.70
specific heat (J mol^{-1} K^{-1})		108.1±7.7	118.28
excess Helmholtz free energy (kJ mol^{-1})		-29.7±0.8	-29.7
static dielectric constant		30	46
shear viscosity (cP)	Einstein	1.26±0.24	1.991
	Green-Kubo	1.29±0.27	

Table 2 shows how well the model reproduces bulk properties of DMSO (see Section 4 for how to calculate these properties in a simulation). Both the density and the heat of vaporisation leave little to be desired. The diffusion coefficient is off by 30% which is rather typical for liquid force fields.

Certain connections among different liquid properties are evident in Table 2. For example, the excess free energy (the free energy for turning 1 mol of liquid DMSO into an ideal gas) is reproduced rather well. This is not too surprising, since the model was parametrised to give the correct heat of vaporisation which is the enthalpy for turning liquid DMSO into DMSO vapour of the same temperature and pressure. Similar connections exist, e.g. between the density and the thermal expansion coefficient (its derivative with respect to temperature) and between the diffusion coefficient D and the viscosity η (cf. the Stokes-Einstein relation $\eta = k_B T / 6\pi a D$). It is worth keeping in mind such relations when parametrising a force field. They make it possible to parametrise against the most convenient property among a set of connected properties whilst being reasonably confident that certain other properties will be reproduced as well. This is exactly what we did in the present example: the heat of vaporisation converges much more rapidly than the excess free enthalpy, and the same is true for the diffusion coefficient and the viscosity.

3. Molecular Simulation Methods

Suppose we have the potential energy function $U(\mathbf{R})$. Its simple analytical form allows us to easily evaluate its gradients with respect to the position of every atom i.

$$\nabla_i U(\mathbf{R}) = \frac{\partial U(\mathbf{R})}{\partial \mathbf{R}_i} \qquad \text{Eq. 4}$$

Different molecular simulation methods use in different ways the potential energy and/or its gradient, whose negative is the force. Figure 2, which is similar to a ternary phase diagram tries to illustrate this point.

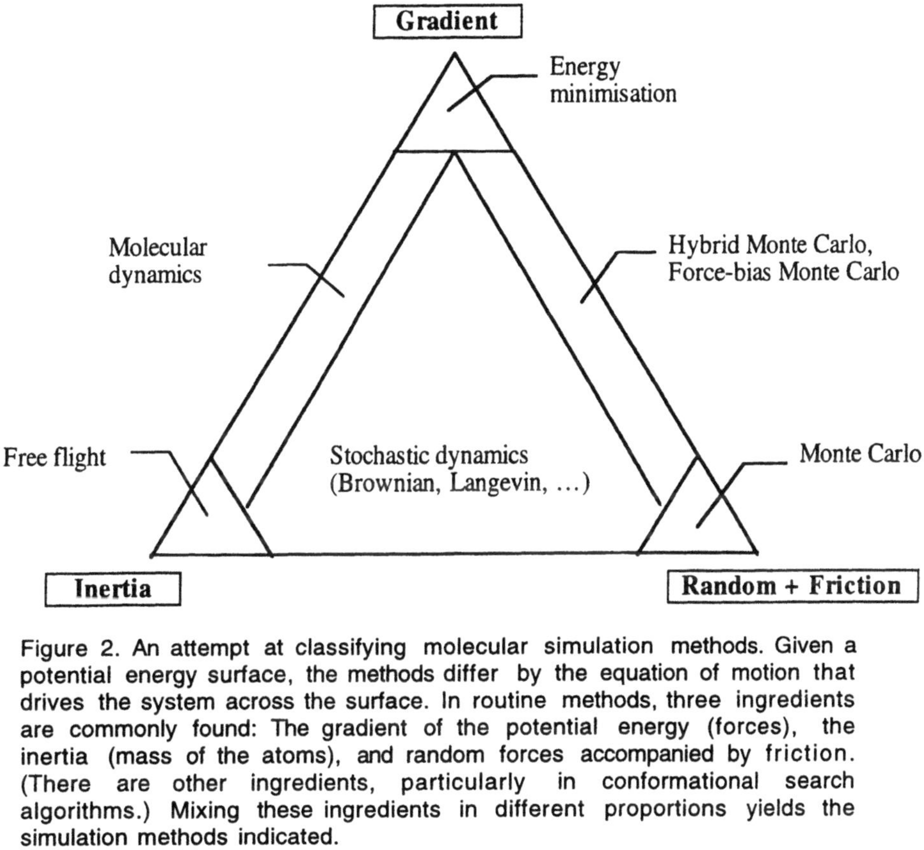

Figure 2. An attempt at classifying molecular simulation methods. Given a potential energy surface, the methods differ by the equation of motion that drives the system across the surface. In routine methods, three ingredients are commonly found: The gradient of the potential energy (forces), the inertia (mass of the atoms), and random forces accompanied by friction. (There are other ingredients, particularly in conformational search algorithms.) Mixing these ingredients in different proportions yields the simulation methods indicated.

If only the gradient information is used the system is driven into the nearest local minimum. This is called energy minimisation, geometry optimisation or *molecular mechanics* (NB: some optimisation methods use higher derivatives, such as the Hessian, as well).

If the atoms feel their mass as well as the potential energy gradient, their motion will be partly determined by inertia. In the simplest case, the atoms will follow Newton's second law (in Cartesian co-ordinates)

$$\mathbf{f}_i = m_i \mathbf{a}_i \quad \text{or, more formally,} \quad \frac{\partial^2 \mathbf{R}_i}{\partial t^2} = -\frac{1}{m_i} \nabla_i U(\mathbf{R}). \qquad \text{Eq.5}$$

This equation of motion leads to the method of *molecular dynamics*. (NB: If the atoms follow Newton's equation of motion without feeling any forces, they perform a free flight. For an ensemble of atoms such a system is known as the ideal gas. However, it is irrelevant in the context of molecular simulation, since everything can be calculated analytically.)

A third component is that of random motions. It is possible to move atoms over a potential energy surface in a sequence of random steps and still obtain distributions which correspond to well-defined statistical-mechanical ensembles. This procedure which does not need the gradient but only the potential energy itself, is known as Metropolis *Monte Carlo*. However, Monte Carlo can be run more efficiently if some gradient information is used in guiding the random moves. Such methods are known as force-biased Monte Carlo.

It is possible to devise equations of motion which contain all three ingredients: gradient, inertia and random forces. These are collectively referred to as *stochastic dynamics*. These are best thought of as atoms moving through a viscous liquid: The atoms are accelerated by systematic forces (gradient), they try to keep going in one direction (inertia), but they undergo collisions with the (presumed) atoms of the viscous medium. The medium produces both random forces which accelerate the atoms and a friction which decelerates them.

3.1 Molecular dynamics

Conceptually the most straightforward molecular simulation method is molecular dynamics. Its fundamental equation is Newton's second law (Eq. 5). It is a differential equation and to solve it, we need (i) an integration algorithm and suitable (ii) initial and (iii) boundary conditions.

In principle, most standard numerical algorithms for differential equations can be used. In practice, people tend to use higher-order Gear predictor-corrector algorithms for smooth potential energy functions (say the harmonic oscillator or the Lennard-Jones fluid) and Verlet integrators for more irregular potential functions (like chemically realistic models of molecular systems). The latter (which outside molecular simulation are known as Stürmer's method and have been know for a hundred years) have the additional advantage of blending relatively well with bond constraints and other changes to pure Newtonian dynamics.

To derive the simplest Verlet scheme, we Taylor-expand the atom's position at the next time step $t+\Delta t$ around its current position (time t).

$$\mathbf{R}_i(t+\Delta t) = \mathbf{R}_i(t) + \Delta t\left(\frac{\partial \mathbf{R}_i}{\partial t}\right)_t + \frac{\Delta t}{2}\left(\frac{\partial^2 \mathbf{R}_i}{\partial t^2}\right)_t + \ldots$$

We can identify the first derivative with the velocity and the second derivative with the acceleration which is related to the force (Eq. 5).

$$\mathbf{R}_i(t+\Delta t) = \mathbf{R}_i(t) + \Delta t\mathbf{v}(t) + \frac{\Delta t}{2m_i}\mathbf{f}_i(t) + O(\Delta t^3) \qquad \text{Eq. 6}$$

Similarly, we can perform a Taylor expansion backward in time to the position at the previous time step.

$$\mathbf{R}_i(t-\Delta t) = \mathbf{R}_i(t) - \Delta t\mathbf{v}(t) + \frac{\Delta t}{2m_i}\mathbf{f}_i(t) - O(\Delta t^3) \qquad \text{Eq. 7}$$

Adding Eqs. 6 and 7 and rearranging leads to cancellation of the first and third-order terms and gives the Verlet scheme for the positions

$$\mathbf{R}_i(t+\Delta t) = 2\mathbf{R}_i(t) - \mathbf{R}_i(t-\Delta t) + \frac{\Delta t}{m_i}\mathbf{f}_i(t) + O(\Delta t^4). \qquad \text{Eq. 8}$$

Thus, we have an explicit scheme accurate through third order in the time step which propagates the atoms from their positions at time t to their new positions at $t+\Delta t$. What is needed in this formulation of the Verlet algorithm (there are others) to perform a time step are the forces acting on the atoms at time t, as well as the positions at the current *and the previous* time step.

3.2 Initial conditions

The initial conditions in a molecular simulation are the initial positions and velocities of all atoms. The initial velocities are typically picked at random from a Maxwell-Boltzmann velocity distribution at the desired temperature. Finding starting positions for all atoms can be considerably trickier. The easiest case are liquids where the positions can be taken, for example, as lattice positions and the system can be left to equilibrate itself to a disordered structure typical for a liquid. This is possible because liquids have short relaxation times so that equilibration will take place in a few dozen picoseconds. However, when the intrinsic time scales of the system become longer (larger flexible molecules, proteins, polymers) additional information is necessary. For proteins, one typically starts from their experimental X-ray or NMR structures. For amorphous polymers, special methods are needed.

3.3 Boundary conditions

There are several kinds of boundary conditions. Very important are the spatial boundary conditions. The system studied is typically small (10^2-10^5 molecules) compared to an experimental system (10^{23} molecules). Therefore, *vacuum boundary conditions* (used in most standard quantum chemistry) should only be chosen when the physics studied involves a vacuum surrounding the system. This can be the case, e.g., for a small molecular cluster or a liquid droplet. For a bulk system, vacuum boundaries are inappropriate because a small system consists of all surface and no bulk volume, so that surface artefacts completely overshadow any bulk behaviour. The most common boundary conditions for mimicking bulk behaviour are *periodic boundary conditions*: The parent system is surrounded ad infinitum with copies of itself in all dimensions, a trick borrowed from solid state theory. Figure 3 illustrates the idea. Periodicity applies not only to the interactions between atoms but also to the motion of atoms: if an atom leaves the parent cell its periodic image appears at the opposite side. In the case of molecules, periodicity can also apply to the chemical connectivity. An interesting consequence of periodic boundary conditions (which need not be cubic as in the example but may be based on any space-filling polyhedron) is the artificial periodicity (superlattice) imposed on the system, even if the system is inherently non-periodic like a liquid. It is clear that in order to rule out artefacts, the system has to be "large enough" for effects of periodicity to vanish in the noise of other influences. What "large enough" means often has to be determined empirically by running simulations of increasing sizes. For some properties (phase transitions, anomalous diffusion, etc.) large enough can mean prohibitively large. It is maybe worth mentioning that there are other spatial boundary conditions for eliminating surface artefacts such as *stochastic boundary conditions*. These are, however, beyond the scope of this text.

In addition to boundary conditions regarding the spatial layout of the system, there are other boundary conditions. We have already mentioned constraints as an example of reducing the number of degrees of freedom. Other boundary conditions concern the equations of motion themselves. In practice, one hardly ever solves Newton's equations of motion (Eq. 4). Newton's equations conserve the total energy of the system, leading to the microcanonical or constant-*NVE* (number of atoms, volume or density, total energy) statistical-mechanical ensemble. Since experiments are much more often performed under constant temperature *T* (rather than, total energy), constant pressure *P* (rather than, volume) and/or constant chemical potential μ (rather than, number of atoms), one often needs to simulate *NVT* (canonical), *NPT* (isothermal-isobaric), μVT (grand canonical) or other ensembles. For this, Eq. 4 needs to be modified accordingly.

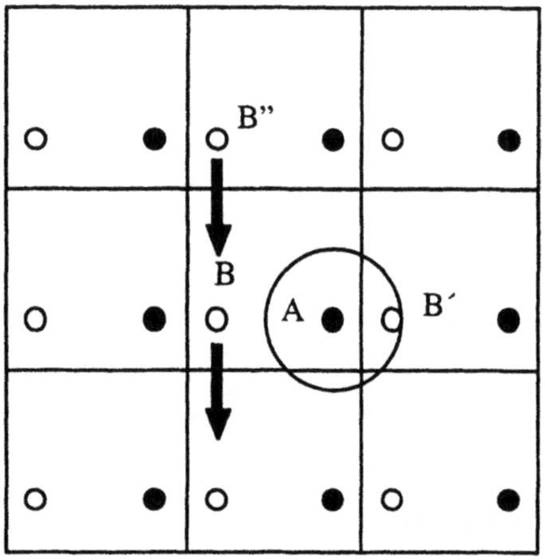

Figure 3. Illustration of periodic boundary conditions (for clarity, in 2 dimensions): (i) Atom *A* interacts with the nearest periodic image of atom *B*, *B´* ("minimum image convention") but not with atom *B* itself which is outside the interaction cut-off distance. (ii) When atom *B* moves out of the parent simulation cell (shaded), its periodic image *B´´* enters through the opposite face.

4. Calculation of Properties

4.1 Ergodicity

Ultimately, one does not solve Newton's equations of motion for the fun of it but in order to calculate properties of the system. These properties are usually calculated as averages of some kind. In a statistical mechanical sense, they are ensemble averages. In order for a simulation to be able to yield correct averages, it has to be *ergodic*. This simply means that the simulated system has to sample all of its available phase space. Ideally, during the simulation the system has to assume every atomic conformation permitted by the potential energy function at the respective temperature with all combinations of atomic momenta.

This is, of course, impractical and, fortunately, for many properties it is sufficient to sample only a small relevant part of phase space. However, care must be taken that all phase space is in principle accessible. This is less of a problem in molecular dynamics where one can rely on Newton's equations of motion to sample all relevant phase space: They are the same equations Mother Nature uses, and Nature does sample phase space. In Monte Carlo simulations, on the other hand, the moves are designed by the simulator, and it is possible to inadvertently come up with moves which do not allow all conformation space to be reached. An example is given in Figure 4.

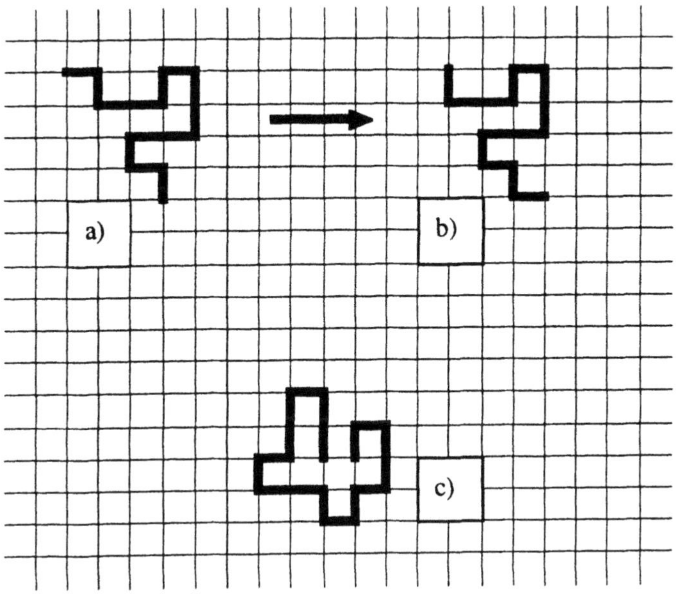

Figure 4. Monte Carlo of a polymer on a 2-dimensional lattice. The moves are reptation or "slithering snake" moves (*a* and *b*): The "head" of the polymer move to a neighbouring free lattice position and the "tail" follows. A configuration is shown in *c* that can never be formed or destroyed by a reptation move. Hence, if only reptation moves are allowed this simulation is not ergodic.

4.2 Thermodynamic averages

To an observable A corresponds an ensemble average $\langle A \rangle$. Nature obtains the ensemble average by averaging both over time (the characteristic duration of the measurement) and over many particles. In a simulation, both time and the number of particles are usually (much) smaller than in Nature. In molecular dynamics, the ensemble average (averaging over many systems) becomes a time average

$$\langle A \rangle = \lim_{t \to \infty} \frac{1}{t} \int A(t)dt = \lim_{m \to \infty} \frac{1}{m} \sum_{i=1}^{m} A_i$$

(averaging over conformations of the same system at different points in time). The second form is already suitable for simulations. One calculates A_i at every time step i (or every so many time steps) and forms the average. How many times A needs to be sampled depends on its features. There are good-natured quantities which converge rapidly (like the radial distribution function of a liquid, see below), others have natural fluctuations much larger than their average (like the pressure), so many samples are needed to obtain a meaningful arithmetic mean. In addition, the inherent relaxation time of the quantity is important. Potential energies calculated at two subsequent time steps will be correlated, i.e. they will not be all that different. Hence, in order to obtain the average potential energy, one has to allow sufficient time to elapse between samples for them to be independent in a statistical sense.

There are many properties which can be obtained as thermodynamic averages. Take the temperature as an example. The equipartition theorem tells us that each degree of freedom in a system stores, on average, $k_BT/2$ kinetic energy and $k_BT/2$ potential energy. The total kinetic energy is therefore

$$E_{\text{kin}} = \frac{N_{\text{DF}} k_B T}{2}, \qquad \text{Eq. 9}$$

where N_{DF} is the number of the degrees of freedom. In the simplest case, there are 3 Cartesian co-ordinates per atom, and $N_{\text{DF}} = 3N$. (If constraints are used there are fewer degrees of freedom!)

On the other hand, classical mechanics tells us that the kinetic energy of a moving atom (no constraints) is $mv^2/2$. Therefore, the total kinetic energy is

$$E_{\text{kin}} = \frac{1}{2}\sum_{i=1}^{N} m_i v_i^2. \qquad \text{Eq. 10}$$

Eliminating E_{kin} from Eqs. 9 and 10 and rearranging gives

$$T = \frac{2}{3Nk_B}\sum_{i=1}^{N} m_i v_i^2, \qquad \text{Eq. 11}$$

the temperature of a system without constraints. Well, this is not completely correct. I have been deliberate sloppy here: the symbol E_{kin} really has two different meanings in Eqs. 9 and 10. In Eq. 10, it denotes the *instantaneous* kinetic energy, whereas in Eq. 9 it denotes the *average* kinetic energy (the equipartition theorem does not say how much kinetic energy is contained in a degree of freedom at any one moment). Therefore, the two cannot really be equated and, consequently, T in Eq. 11 is some kind of instantaneous temperature. In order to calculate the real temperature, we have to correct Eq. 11 by taking the ensemble average.

$$T = \left\langle \frac{2}{3Nk_B}\sum_{i=1}^{N} m_i v_i^2 \right\rangle \qquad \text{Eq. 12}$$

In a very similar way, an expression for the pressure can be derived.

$$P = \frac{1}{3V}\left[\left\langle \sum_{i=1}^{N} m_i v_i^2 \right\rangle + \left\langle \sum_{i=1}^{N} \mathbf{f}_i \cdot \mathbf{R}_i \right\rangle\right], \qquad \text{Eq. 13}$$

where the first term is related to the kinetic energy (pressure of the ideal gas) and the second term is the so-called virial, which involves the atoms' positions and forces and which has the unit of an energy.

In addition to temperature and pressure, one often calculates the average potential energy U and its components. These components are a consequence of the model we use for the potential energy function (Eqs. 1-3), Nature does not know about bond bending terms or hydrogen bond terms but only about the total potential energy. Nonetheless, it can be very useful for interpretation to separate out the components. Moreover, the intermolecular part of U can be related to the enthalpy of vaporisation of molecular liquids after accounting for some corrections (the PV term, different molecular conformations in the gas and the liquid, changes in vibrational frequencies and, consequently, zero-point energy, etc.).

4.3 Fluctuations

I have already indicated that most observables fluctuate with time. Fluctuations are not only a nuisance when one is trying to converge an ensemble average (see last section). There are many macroscopic properties which can be calculated from microscopic fluctuations. An example is the isothermal compressibility κ which can be calculated from the fluctuations of the volume V of the system.

$$\kappa = \frac{\left\langle (V - \langle V \rangle)^2 \right\rangle}{k_B T \langle V \rangle}$$

Other properties involving derivatives, like the specific heat and the thermal expansion coefficient, can be calculated in a similar manner. Incidentally, these can often be estimated more reliably by performing several simulations at different state points and taking the derivative numerically. However, fluctuations are the method of choice for the dielectric constant (relative permittivity) ε_r of fluids. Fluctuations of the total dipole moment **M** of the system are used.

$$(\varepsilon_r - 1)\left(\frac{2\varepsilon_{RF} + 1}{2\varepsilon_{RF} + \varepsilon_r}\right) = \frac{\langle M^2 \rangle - \langle M \rangle^2}{3\varepsilon_0 V k_B T}$$

Here, ε_{RF} is the effective dielectric constant used in the reaction field treatment of Coulombic interactions (not discussed in this article [1]), and ε_0 is the vacuum permittivity.

4.4 Correlation coefficients, time correlation functions and transport coefficients

In statistics, the measure of the correlation of two quantities A and B is often the correlation coefficient C_{AB}.

$$C_{AB} = \frac{\langle (A - \langle A \rangle)(B - \langle B \rangle) \rangle}{\left[\left\langle (A - \langle A \rangle)^2 \right\rangle \left\langle (B - \langle B \rangle)^2 \right\rangle \right]^{1/2}} \qquad \text{Eq. 14}$$

If A and B have a tendency to be simultaneously above their respective means C_{AB} will be close to 1; if A is always above its mean when B is below its mean and vice versa (anti-correlation) C_{AB} is close to -1; if A and B assume their values completely independently (no correlation) C_{AB} is around 0.

In simulations, all sorts of quantities can be correlated, in order to find out whether they are independent or not. A common example are atomic positions (A=$\mathbf{R}_i$ and B=$\mathbf{R}_j$): If their correlation coefficient is 1, the atoms move synchronously together; if it is -1 they may vibrate against each other; if it is 0 their motions are independent. Positional cross correlations have been used extensively to analyse protein dynamics and a lot of nonsense has been published there because frequently elementary requirements were not satisfied. Note that Eq. 14 is full of angle brackets. The definition of C_{AB} assumes that these averages and fluctuations are converged which implies that A and B are in equilibrium and are fluctuating around their equilibrium values. If this is not the case, for example because they are still relaxing towards equilibrium, C_{AB} is meaningless.

A time delay can be incorporated into C_{AB} by extending it to a time correlation function $C_{AB}(t)$.

$$C_{AB}(t) = \frac{\langle (A(t) - \langle A \rangle)(B(0) - \langle B \rangle) \rangle}{\left[\left\langle (A - \langle A \rangle)^2 \right\rangle \left\langle (B - \langle B \rangle)^2 \right\rangle \right]^{1/2}} \qquad \text{Eq.. 15}$$

Now, the correlation is not measured instantaneously, but between B at some (arbitrary) time zero and A at a time t later. (The example mentioned to me on one occasion in Brazil was that of Carnival and the birth rate where $C_{AB}(t)$ shows a positive peak at t= 9 months.)

Time correlation functions can, again, be calculated for many different properties and the same information can be obtained as with the correlation coefficients augmented by the time dependence. (Note that Eq. 14 is just a special case of Eq. 15 with t=0.) However, modern statistical mechanics allows us also to obtain macroscopic properties from microscopic time correlation functions. Many transport coefficients can be calculated as time integrals of correlation functions (so-called Green-Kubo relations). As an example, take the diffusion coefficient D which can be defined in terms of the unnormalised velocity auto-correlation function (in an *auto*-correlation function, A=B).

$$D=\frac{1}{3}\int_0^\infty \langle \mathbf{v}(t)\,\mathbf{v}(0)\rangle dt \qquad \text{Eq. 16}$$

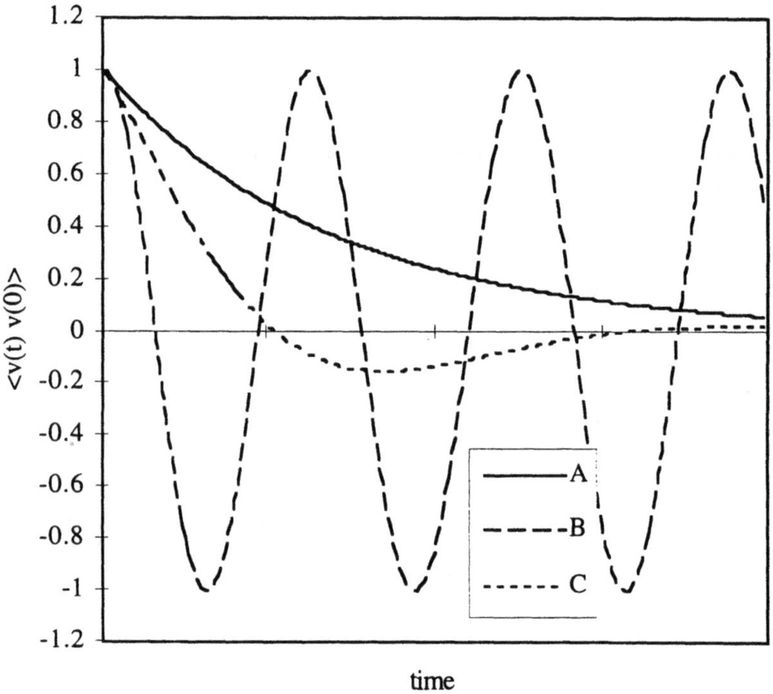

Figure 5. The shape of the (normalised) velocity auto-correlation function gives some qualitative description of the atomic or molecular motion. Curve A is that of a dilute gas where decorrelation is effected by infrequent random collisions with other gas molecules. Curve B is the periodic vibration of an atom in an ideal harmonic solid (no damping). Curve C is typical for a molecular liquid: The negative lobe occurs because – on average – the direction of motion of a molecule is reversed (it hits the wall of the solvent cage) before it is randomised by further collisions.

The velocity auto-correlation function can be regarded as a measure of how long an atom will continue to travel in a given direction before this direction is randomised or reversed. (The shape of the velocity auto correlation function also tells us something about the qualitative characteristics of the molecular motion, see Figure 5.) It is clear that the longer an atom travels in one direction, the larger is the displacement it achieves, and the larger becomes the diffusion coefficient.

Statistical mechanics also provides an alternative route to transport coefficients. To every Green-Kubo relation corresponds an Einstein relation. In the case of the diffusion coefficient it involves the atomic or molecular mean-square displacement.

$$6Dt = \left\langle \left|\mathbf{R}(t) - \mathbf{R}(0)\right|^2 \right\rangle \qquad \text{Eq. 17}$$

The mean-square displacement is the squared distance the particle has travelled between times 0 and t. For a truly diffusing particles (there exist other types of motion!) the mean-square displacement is linear in time and Eqs. 16 and 17 can be shown to be completely equivalent. The pattern found here for the diffusion coefficient is general in pairs of Green-Kubo and Einstein relations: In the Green-Kubo relation appears the correlation function of a quantity (here: the velocity) whose time integral (here: the position) squared appears in the corresponding Einstein relation. Other examples of transport coefficients which can be calculated alternatively via the Einstein or the Green-Kubo route are: ionic conductivity, thermal conductivity, viscosity.

Before leaving the time correlation functions let us quickly mention that correlation functions can also be analysed for their frequency spectra. This is done by fast Fourier transforming them. The Fourier transform of Curve B in Figure 4, for example would reveal one frequency typical for this harmonic solid, whereas the transform of Curve C (liquid) would give a whole spectrum of frequencies found in a liquid, the so-called phonon density of states.

4.5 Structural properties

In principle, all geometric quantities can be monitored in a simulation: distances, bond angles, dihedral angles and so on. This is a major advantage of simulation compared to experiment. There are few experiments that actually measure molecular geometry directly. Mostly, the geometry has to be inferred from the measurement using some physical model (an example is the Karplus relation which connects NMR coupling constants and dihedral angles), and in many cases, experiment is restricted to certain systems (crystallography needs crystals) or ranges (NMR-NOE experiments cannot measure distances beyond 0.6 nm). Since in simulations the positions of all atoms are known, all geometric parameters can be calculated directly. By suitable post-processing one can obtain their averages, fluctuations, time evolution, time correlation functions, characteristic frequencies and so forth.

In addition to local geometric descriptors, also more global variables can be calculated like the end-to-end distance (important for polymer chains) or the mean-square radius of gyration R_G^2 (important for macromolecules, can be measured by e.g. light scattering).

$$R_G^2 = \left\langle \frac{1}{M} \sum_i m_i \left|\mathbf{R}_i - \mathbf{R}_{\mathrm{CM}}\right|^2 \right\rangle$$

Here, M is the total mass of the molecule and $\mathbf{R}_{\mathrm{CM}}$ the position of its centre of mass; the unweighted form is also sometimes used.

A structural descriptor often used for liquids and for amorphous solids is the radial distribution function $g(R)$ (also called the pair correlation function; do not confuse it with a time correlation function). By a Fourier transform it is related to the static structure factor which may be obtained from X-ray or neutron scattering. It is essentially a histogram of interparticle distances found in the simulation normalised by the distribution of interparticle distances expected for a system of uniform density. In a liquid, it has some structure at short distances corresponding to first and second solvation shells and maybe other effects; at long distances it goes to one (Figure 6).

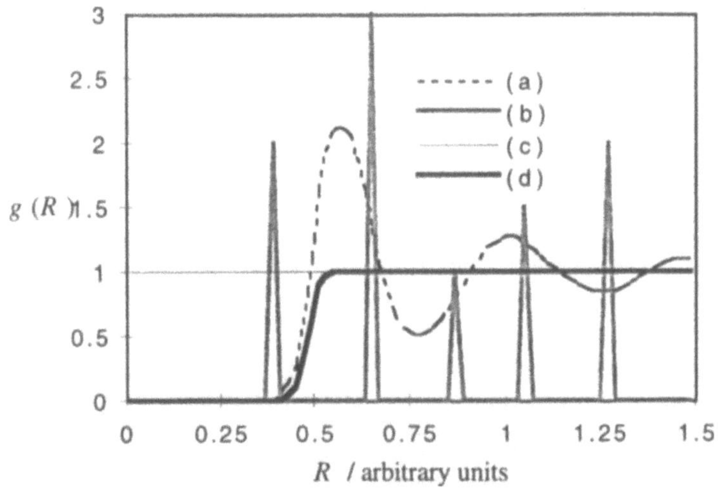

Figure 6. Typical radial distribution functions for (a) a liquid, (b) an ideal solid, (c) an ideal gas, and (d) a real gas.

Another intermolecular feature are hydrogen bonds. They are typically defined by some geometric criterion, like the two oxygens have to be within 0.3 nm and the O-H-O angle has to be greater than 130°. Their existence can be monitored, as can be their time development. Residence times of, say, water molecules near the surface of a protein can be determined and compared to results from NMR exchange experiments.

In addition to atomic and molecular positions, one can also calculate molecular orientation, both with respect to a laboratory frame (e.g. director of a liquid-crystalline phase, orientation of molecules with respect to a surface or interface) and towards each other (e.g. the packing of benzene molecules in the liquid is T-shaped and not coplanar). To this end, one has to define unit vectors **u** within the molecule and calculate them throughout the simulation. Experiments often measure the projection of these vectors onto other directions (i.e. scalar product with another unit vector) or higher Legendre polynomials of the projections. The time behaviour of orientation can also be conveniently studied by correlating an orientation vector at time *t* with the same vector at time *t*=0.

$$\langle \cos\phi(t) \rangle = \langle \mathbf{u}(0) \cdot \mathbf{u}(t) \rangle$$

In analogy to Eq. 16, the time integral defines a transport coefficient for the molecular reorientation, the rotational correlation time τ.

$$\tau = \int_0^\infty \langle \mathbf{u}(0) \cdot \mathbf{u}(t) \rangle \, dt$$

This correlation time can be related to infrared linewidths. Similar correlation times involving the second, rather than, the first Legendre polynomial can be obtained by NMR, Raman, and a variety of fluorescence methods.

5. Chemistry in Condensed Media – Hybrids of Quantum Chemistry and Molecular Simulation

5.1 Hybrid (QM/MM) schemes

From what was said in the preceding sections one could assume that molecular simulation is the universal tool for inferring macroscopic behaviour from a microscopic model. Well, it is not. Apart from technical difficulties like the inaccuracy of force fields and insufficient simulation times there are major systematic limitations. One of them we will briefly address in this section.

Force fields have serious deficiencies when it comes to describing properties and processes that explicitly involve the electronic degrees of freedom. As a consequence, the following systems and processes cannot be studied by force-field-based molecular simulation:

- *Chemical reactions.* Force fields can, in fact, be parametrised to describe the making and breaking of chemical bonds. However, one first has to know the potential energy curve along a breakable bond before it can be modelled by a suitable function (no predictive power). Secondly, one has to specify which bonds can be formed or destroyed, whereas ideally one would like the system to decide which chemistry takes place in a simulation, since it might be different from the simulator's preconceived ideas.
- *Electronically excited states.* Again, it would be possible to derive parameters for the interaction of a molecule in its first excited state with a surrounding solvent, if there only were data to parametrise against.
- *Transition metals.* With their numerous oxidation states and co-ordination numbers they are a pain for force fields and most applications have used ad-hoc patches rather than systematically derived parameters.

Hence, there have been many attempts to combine the best of quantum chemistry and molecular simulation. Continuum solvent models (reaction field and variations thereof, see Mati Karelson's contribution) are very popular now in quantum chemistry but they do not solve all problems, since the environment is treated in a static mean-field approximation. The Car-Parrinello method has found its way into chemistry and it is probably the most rigorous of the methods presently feasible. However, its computational cost allows at present only the study of systems of a few dozen atoms for periods of a few dozen picoseconds. In this section, our own studies with so-called hybrid or quantum-mechanical/molecular-mechanical (QM/MM) methods are described. These concentrate on the part of the system which is of primary interest (the reactants or the electronically excited solute, say) and treat it by semiempirical quantum chemistry. The rest of the system (solvent, surface, outer part of enzyme, etc.) is described by a classical force field. With this, we hope to incorporate the essential influence of the in itself uninteresting environment on the structure and dynamics of the primary system. The approach lacks the rigour of the Car-Parrinello scheme but it allows us to surround a primary system of up to a few dozen atoms by an environment of several ten thousand atoms and run the whole system for several hundred thousand time steps which is equivalent to hundreds of picoseconds.

The hybrid scheme used by us has been described in detail [3]. Therefore, only a short summary is given here. The system is divided into a primary (QM) part treated quantum-chemically and the environment treated by a force field (MM part). For the QM part, a semi-empirical calculation is carried out. We have used the MOPAC program [4] for this task and have, consequently, employed several NDDO methods (MNDO, AM1, PM3) in the course of our work. The interactions among atoms of the MM part are described entirely by a classical force field. They were evaluated using the GROMOS87 program [5], and the GROMOS force field or close variations of it were used in our work.

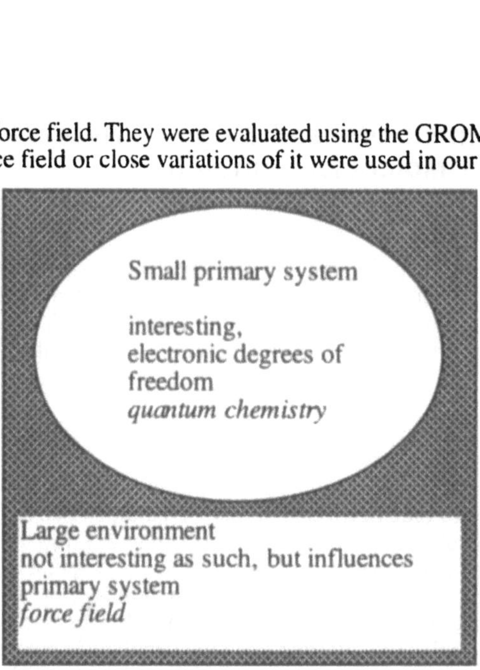

Figure 7. The hybrid (QM/MM) method: the system is broken up into a primary system and an environment. These are treated differently, but interact with each other.

The interactions between the QM part and the MM part are treated by a combination of quantum-chemical and force-field contributions. If the system can be separated into a QM solute and an MM solvent, the partitioning of interactions arises naturally: The electrons of the QM part feel the electric field due to the partial charges of the solvent molecules which enter the one-electron Hamiltonian as additional point charges. (In this way, also the polarisation of the solute by the solvent is accounted for.) The dispersion attraction and the short-range repulsion are modelled by a Lennard-Jones 6-12 potential. If, on the other hand, a molecule extends over the QM/MM boundary, parts of it are described quantum-chemically and parts by the force field, and there are chemical bonds crossing the boundary. On the QM side, the dangling bonds are saturated by hydrogen atoms. The interaction across the boundary is then described by force field terms (bond stretching, bond angle bending and dihedral angle deformation). The partial charges of the first three atoms on the MM side (fewer than 4 chemical bonds away from the last QM atom) do not enter the QM Hamiltonian.

In this way, forces on all atoms are calculated which may come from quantum chemistry, the force field or both. These forces are used in a conventional MD algorithm to propagate the atoms to the configuration at the next time step. Additional techniques of molecular simulation such as umbrella sampling (not explained here) for the calculation of potentials of mean force are used as needed.

5.2 The conformations of dimethoxy ethane in aqueous solution

Conformational equilibria of the dimethoxy ethane molecule (ethylene glycol dimethyl ether, glyme) have puzzled researchers for a while. This molecule has three rotatable dihedral angles involving heavy atoms only, which can assume *trans* (180°), *gauche*$^+$ or *gauche*$^-$ (±60°) conformations, giving rise to ten different molecular conformers. Of these, the all-*trans* (*ttt*), *trans-gauche-trans* (*tgt*) and the *trans-gauche*$^+$*-gauche*$^-$ (*tgg*$^-$) have the lowest energies (Figure 8). While all quantum chemical calculations in vacuum unanimously have found *ttt* to be the minimum, all experimental measurements in condensed phases (crystal, melt, solution) found mainly *tgt*. It could be shown that this effect has nothing to do with the level of the quantum-chemical treatment, but is due to the different environment (vacuum vs. condensed phase). It could be reproduced by a simple self-consistent reaction-field treatment [6] which just assumed a polarisable dielectric continuum around the glyme molecule (cf. the two MBPT2 curves in Figure 9). The shift in relative stability is caused by the large difference of the dipole moments of *ttt* (zero by symmetry) and *tgt* (1.4 D). The dipole moment causes polarisation of the continuum and, consequently, stabilisation of the polar conformation.

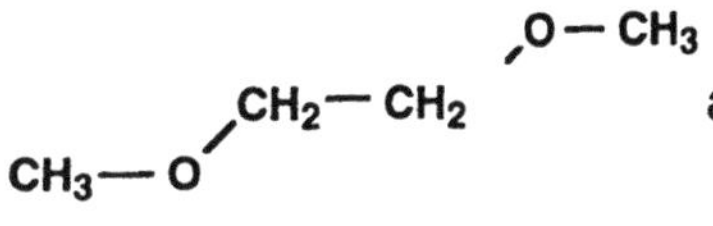

all-trans (*ttt*)

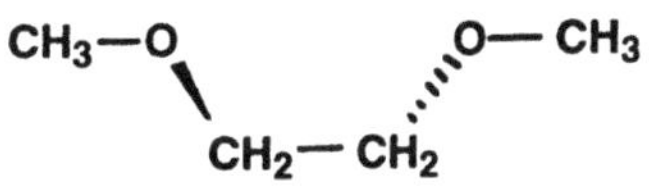

trans-gauche-trans (*tgt*)

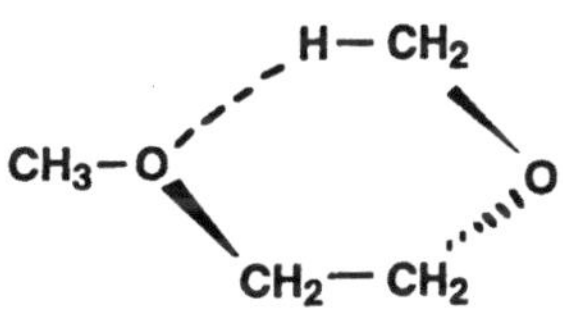

trans-gauche$^+$-gauche$^-$ (tg^+g^-)

Figure 8. Low-energy conformations of dimethoxy ethane.

The stabilisation of polar conformations apparently is a static effect. The environment only has to respond to the solute's dipole in a mean-field way. The hybrid MD calculation (the two MNDO curves in Fig. 9) shows the same effect (vacuum vs. condensed-phase environment), but is an overkill. Describing the environment by explicit water molecules does not improve the picture at a qualitative level. (The magnitude of the stabilisation of ≈5 kJ/mol, rather than, the 2 kJ/mol found for the reaction field is more in line with experimental observations. However, it is difficult to determine unambiguously how much of this is due to deficiencies of the MNDO method; cf. the dismal performance of MNDO for the rotational barriers in Figure 9.)

However, the reaction field treatment cannot account for the third low-energy conformer (tg^+g^-). This is within 0.1 kJ/mol of *ttt* in the gas phase as found by ab initio calculations, the stabilisation being ascribed to a weak intramolecular hydrogen bond between a methyl hydrogen and an ether oxygen at the other end of the molecule (cf. Fig. 8). Because of the large dipole moment (1.7 D), reaction-field treatments predict this conformer to be by far the most stable in solution [7], completely at odds with experiment which never found evidence for this conformer in the condensed phase. On the other hand, the presence or absence of this conformer has serious implications in the physics of poly(ethylene oxide) and polymer electrolytes derived from it [8]. We have addressed this problem with a hybrid calculation which described the glyme solute by MNDO and the aqueous environment by the SPC water model [3]. The MD calculation confirmed not only the experimental finding (nearly complete absence of the tg^+g^- conformer) but also provided the explanation: Water molecules compete very effectively with the intermolecular methyl group for positions near the ether oxygens, since they are much better hydrogen donors.

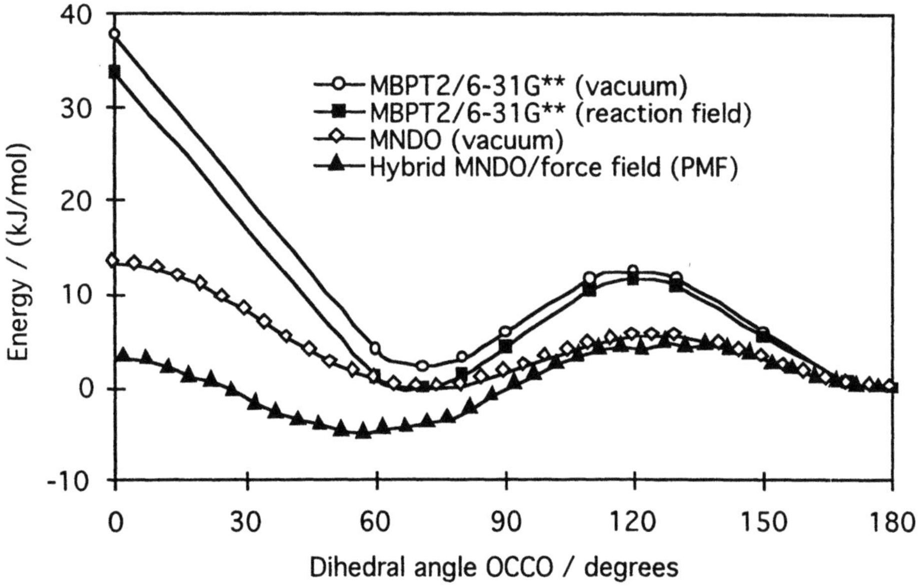

Figure 9. The potential curve for the inner dihedral angle (OCCO) of dimethoxy ethane. The two outer dihedrals (COCC) are in *trans* position (≈180°).

5.3 An S_N2 reaction in solution

We have also studied the symmetric reaction

$$Cl^- + NH_2Cl \rightarrow NH_2Cl + Cl^-$$

in dimethyl ether (DME) solution at 248 K [9]. First, the above reaction was studied in vacuum using both semiempirical and ab initio methods, after which it was decided that the PM3 method reproduced best the transition state energy found by MBPT2/6-311+G**. It was subsequently used in a hybrid scheme to treat the reactants, whereas for DME a rigid model was developed within the framework of the GROMOS force field. Umbrella sampling was used to calculate the free energy along the reaction path. The calculation yielded a number of important results on the modification of the reaction energetics as one goes from the gas phase into solution. Firstly, there is a considerable screening of the ion-dipole interaction between the reactants at large distances. Secondly, the energy of the encounter complex (Cl^- hydrogen-bonded to one of the hydrogens of chloro amine) with respect to infinite separation changes from -18.9 kcal/mol to -12.5 kcal/mol due to the better solvation of the separated reactants. Thirdly, the transition barrier is raised from 6.4 kcal/mol to 15.0 kcal/mol, for the same reason. Fourth, comparison of Mulliken populations qualitatively shows an increased charge separation in the encounter complex as well as in the transition state.

5.4 The mechanism of HIV protease

This enzyme hydrolyses proteins and plays an important role in the replication of the AIDS virus. It belongs to the family of aspartic proteases. It catalyses the reaction

$$R_1\text{-NH-CO-}R_2 + H_2O \rightarrow R_1\text{-NH}_2 + \text{HOOC-}R_2\ ,$$

where R_1 and R_2 denote the rests of the peptide chain. HIV protease achieves catalysis by a combined acid-base mechanism: There are two aspartic acid residues in the active site, only one of which is protonated. At the time of the reaction, the substrate protein is firmly bound in the active site as is the water molecule which is to be split ("lytic water"). The role of the unprotonated aspartate is clear from its position (X-ray structure): It removes a proton from the lytic water so the latter can attack the carbonyl carbon in an activated form, often regarded as a quasi-free OH^-. The protonated aspartic acid has to transfer its proton either to the amide nitrogen (to form a better leaving group) or to the carboxyl oxygen (to activate the C=O bond and facilitate the water attack). From the available experimental information alone, it cannot be decided which of the two mechanisms prevails, basically because X-ray crystallography cannot determine which of the two aspartic acids initially carries the proton. Moreover, it is not clear in how far the mechanism of the nucleophilic water attack and the two proton transfers is step-wise or concerted.

We have used the hybrid method to study both possible pathways of the enzyme mechanism [10]. A part of the substrate peptide, the lytic water molecule and relevant parts of the two aspartic acids are described by the PM3 method, after we had determined that this method reproduces best ab initio (MBPT2/6-31G**) energetics of the uncatalysed, acid-catalysed and base-catalysed hydrolysis of formamide in vacuum. The calculations clearly favoured one mechanism over the other (carboxyl oxygen is protonated, not nitrogen). Moreover, the proton transfers and the nucleophilic attack appear to be concerted: We used umbrella sampling to drive the oxygen of the lytic water towards the carbonyl carbon, and, within a few picoseconds of the oxygen going over the transition state, the proton transfers took place spontaneously.

5.5 Hybrid methods – A critique

The combination of a semiempirical kernel and a force-field environment has been very successful in the above applications and has yielded information that could – at present – be obtained by other means (experimental or computational) only with considerable difficulty or not at all. One has to realise, though, that in practice there is a window of usefulness of the method: If the energy differences are large enough the energetics will be mainly determined by vacuum values, i.e. environment effects will be negligible. In other words, if a reaction barrier is 200 kJ/mol, this is mainly the intrinsic potential energy barrier of the system, and the environmental effects (solvation, polarisation, entropy terms) of maybe 10 kJ/mol will not alter the physics of the system significantly. If, on the other hand, the energy differences are small compared to the intrinsic inaccuracies of semiempirical method and force field the results cannot be trusted.

However, the method also has more serious problems. They mainly arise from the often limited accuracy of the semiempirical schemes. For the time being, one is forced to compromise between accuracy of the quantum-chemistry part and the convergence of statistical-mechanical quantities which is determined mainly by the number of time steps sampled. Our calculations are biased toward longer simulation times and better statistical sampling. As a minimum safety measure, we have always calculated the system or a truncated version of it in vacuum both by ab initio methods (usually MBPT2 and polarised bases) and the semiempirical schemes at our disposal. This calibration helps to answer three vital questions:

- Does a semiempirical scheme make any sense for the system under study? Often, none of the semiempirical methods is able to describe, say, a reaction profile correctly. For example, originally we had planned to study the reaction of fluoride with fluoro amine (for which there were ab initio data in the literature). This had to be given up when the ab initio transition state structures were not reproduced in semiempirical calculations. Similarly, our first choice of an enzyme was a thioredoxin. Here, the semiempirical estimates of the transition state energy were too far off to be useful.
- Which of the semiempirical schemes works best (or least badly) for the system under study? For example, only MNDO did distinguish between the *ttt* and *tgt* conformers of glyme (Figure 9). Neither AM1 nor PM3 gave a barrier between these states.
- What are the residual errors? For example, the *ttt-tgt* barrier in glyme is 5 kJ/mol by MNDO and 12 kJ/mol by ab initio (Figure 9). This means that calculated transition rates would be too high, whereas the equilibrium concentrations (relative energies of the minima) are still correct.

As long as it is not feasible to use more accurate quantum-chemical schemes (probably ab initio or DFT) inside the hybrid method, we believe it is mandatory to carry out test calculations before embarking on hybrid MD calculations. Otherwise, one's results can easily be wrong quantitatively as well as qualitatively.

Acknowledgements

The work of our own presented in Sections 2.3 and 5 was carried out in collaboration with Haiyan Liu (now Hefei, PR China) and Wilfred van Gunsteren (ETH Zürich). Two students of the Bad Windsheim NATO ASI, Christian Berweger and Uwe Micka, are gratefully acknowledged for their critical reading of the manuscript.

References

1 F. Müller-Plathe, Comput. Phys. Commun. **78**, 77 (1993).
2 H. Liu, F. Müller-Plathe and W.F. van Gunsteren, J. Am. Chem. Soc. **117**, 4363 (1995).
3 H. Liu, F. Müller-Plathe and W.F. van Gunsteren, J. Chem. Phys. **102**, 1722 (1995).
4 J.J.P. Stewart, Comput.-Aided Mol. Design **4**, 1 (1990).
5 W. Scott and W.F. van Gunsteren, in: E. Clementi and G. Corongiu (Eds.) *Methods and Techniques in Computational Chemistry METECC-95* (STEF, Cagliari, 1995).
6 F. Müller-Plathe and W.F. van Gunsteren, Macromolecules **27**, 6040 (1994).
7 F. Müller-Plathe, Braz. J. Phys. **24**, 965 (1994).
8 F. Müller-Plathe, H. Liu and W.F. van Gunsteren, Comput. Polym. Sci. **5**, 89 (1995).
9 H. Liu, F. Müller-Plathe and W.F. van Gunsteren, Chem. Eur. J. **2**, 191 (1996).
10 H. Liu, F. Müller-Plathe and W.F. van Gunsteren, J. Mol. Biol. **261**, 454 (1996).

Appendix: Further Reading

Disclaimer

The aim is to provide useful references for beginners in the field of molecular simulation. The list is probably incomplete and certainly biased according to my personal preferences and ignorances. I assume no responsibility whatsoever for anything. If you need to do molecular simulation of some kind you are still responsible for bringing yourself up to date with the literature. Prices are approximate and as I happen to remember them.

Molecular simulation – general techniques

M.P. Allen and D.J. Tildesley, *Computer Simulation of Liquids* (Oxford Science Publishers, Oxford, 1987). 0-19-855645-4, ca. DEM 60.- (GBP 26.-).

This is the Bible of molecular simulation and the paperback is affordable. Do not start without it.

D. Frenkel and B. Smit, *Understanding Molecular Simulation* (Academic, San Diego, 1996), 0-12-267370-0, ca. DEM 140.-

This textbook covers basics and also some of the more recent algorithms in some detail. In some sense, it is complementary to Allen&Tildesley. Emphasis is on ensembles, phase equilibria, free energy and a few advanced topics. Chapters on techniques are illustrated with applications. Highly recommended.

W.F. van Gunsteren, P.K. Weiner and A.J. Wilkinson (Eds.), *Computer Simulation of Biomolecular Systems* (ESCOM, Leiden, 1993) 90-72199-15-4, ca. DEM 350.-.

This is a very good collection of review articles and it is much more general than the title suggests. If it were not for the price I would recommend it unconditionally.

K. Binder and G. Ciccotti, *Monte Carlo and Molecular Dynamics of Condensed Matter Systems* (Società Italiana de Fisica, Bologna, 1996) 88-7794-078-6, ca. DEM 240.- (LIT 240 000.-).

Contains the proceedings of a four-week-long summer school in Como in 1995. There are many excellent articles on most aspects of simulation. The volume has 1000 pages. Highly recommended.

M.P. Allen and D.J. Tildesley (Eds.), *Computer Simulation in Chemical Physics* (Kluwer, Dordrecht, 1993) 0-7923-2283-5, ca. DEM 300.- (NLG 357.-)

Proceedings of a NATO summer school. The selection of articles encompasses a comprehensive list of application areas of computer simulation.

D.C. Rapaport, *The Art of Molecular Dynamics Simulation* (Cambridge University Press, Cambridge, 1995) 0-521-44561-2, DEM 110.- (GBP 45.-)

Teaches you how to run MD calculations (including writing programs). It contains much material on ensembles, thermostats, non-equilibrium etc. It mostly deals with atomic fluids, though.

W.F. van Gunsteren and H.J.C. Berendsen, Angew. Chem. Int. Ed. Engl. **29**, 992 (1990).

A review article covering much of the MD methodology. For those who cannot afford books.

Computer simulation – more specialised

K. Binder and D.W. Heermann, *Monte Carlo Simulation in Statistical Physics – An Introduction* (Springer, Heidelberg, 1988) 3-540-19107-0, 0-387-19107-0, ca. DEM 65.-

An introduction to the more statistical-mechanical side of simulation: lattice models, random walks, percolation. It contains thorough treatments of error analysis and finite-size effects.

K. Binder (Ed.), *Monte Carlo and Molecular Dynamics Simulations in Polymer Science* (Oxford Science Publications, Oxford, 1995) 0-19-509438-7, ca. DEM 180.- (GBP 72.50).

Some very good introductions to aspects of polymer simulations and a good account of current research.

W.F. van Gunsteren, S.R. Billeter, A.A. Eising, P.H. Hünenberger, P. Krüger, A.E. Mark, W.R.P. Scott and I.G. Tironi, *Biomolecular Simulation: The GROMOS96 Manual and User Guide* (vdf, Zürich, 1996) 3-72812-422-2, ca. DEM 120.- (CHF 100.-)

This 1000-page manual contains not only the instructions for the use of van Gunsteren's simulation program GROMOS96 (including the GROMOS force field) but also a lot of physical, mathematical and simulational background material. It is available independently of the program.

L. Monnerie and U.W. Suter (Eds.), *Atomistic Simulations of Polymers*, Advances in Polymer Science Vol. **116** (Springer, Berlin, 1994). DEM 230.-

Comprehensive overview of current simulation activities in the polymer area.

C.R.A. Catlow, S.C. Parker and M.P. Allen, *Computer Modelling of Fluids Polymers and Solids* (Kluwer, Dordrecht, 1990) 0-7923-0549-3, DEM 250.- (NLG 332.-).

The proceedings of the 1988 NATO summer school at Bath. Some articles are very good. Expensive.

G. Ciccotti, D. Frenkel and I.R. MacDonald (Eds.), *Simulation of Liquids and Solids* (North-Holland, Amsterdam, 1987) 0-444-87061 X, ca. 100.- DEM (NLG 119.-).

This is an annotated collection of reprints of classical papers in the field, from Metropolis' 1953 to Car &Parrinello 1986. It is not only of historical interest but contains various seminal papers which are still useful to read, especially on free energies, non-equilibrium methods and ensembles. Saves many a trip to the library.

Analytical theory

J.-P. Hansen and I.R. MacDonald, *Theory of Simple Liquids 2nd Ed.* (Academic Press, London, 1986) 0-12-323852-8, ca. DEM 80.-

The ultimate book on liquid theory, difficult.

M. Doi and S.F. Edwards, *The Theory of Polymer Dynamics* (Clarendon, Oxford, 1988) 0-19-852033-6, DEM 60.- (GBP 25.-)

Polymer theory at its best.

Statistical mechanics

D.A. McQuarrie, *Statistical Mechanics* (Harper Collins, New York, 1976) 06-044366-9, ca. DEM 200.-.

All you ever want to know about statistical mechanics with a particular emphasis on correlation functions. Very well written and expensive.

R. Kubo, M. Toda and N. Hashitsume, *Statistical Physics II* (Springer, Berlin, 1985) 3-540-53833-X, 0-387-53833-X, ca. DEM 90.-.

From the inventors of the Kubo relation. A well-written account of what can be done with correlation functions.

D. Chandler, *Introduction to Modern Statistical Mechanics* (Oxford University Press, New York, 1987) 0-19-504277-8, ca. DEM 55.- (GBP 25.-).

A very compact book containing many important things, for example how to treat chemical reactions. It is well written but sometimes so terse that you need someone to explain it to you. Best used in a course.

CHEMICAL REACTIONS IN BULK AND ON SURFACES

RISTO NIEMINEN
Laboratory of Physics,
Helsinki University of Technology,
02150 Espoo, Finland

Introduction

These lectures deal with computational modelling of chemical bonding, electronic and atomic structures and reaction kinetics in solid-state and surface physics problems. A brief review of the relevant computational methods is given in the first lecture, including density-functional theory molecular-dynamics techniques and Monte Carlo simulations. The second lecture deals with applications to metallic surfaces, and focuses on the basic concepts and some recent work on their electronic and atomic structure and related properties. The third lecture surveys current work on semiconductor surfaces, with particular emphasis on elemental defects in semiconductors as well as atomic structures and band offsets at interfaces. The topic of the fourth lecture is the Monte Carlo approach to modelling of non equilibrium kinetics of heterogeneous catalysis, based on discrete lattice methods as well as rules and rate constants incorporating the energetics old individual reaction steps.

Below, a short synopsis of each of the four lectures is given togther with some references, mainly to recent relevant work of the Helsinki University of Technology group which was discussed in the lectures.

1. Methodology

A brief review of the relevant computational methods is given in the first lecture

evolution of computational techniques [1]
density-functional theory: local-density approximation and beyond [2]
quantum Monte Carlo techniques [3]

S. Wilson and G. H. F. Diercksen (eds.), Problem Solving in Computational Molecular Science, 413–416.

first-principles molecular dynamics: Car-Parrinello methods and all that [4]

quantum dynamics [5]

classical molecular dynamics and Monte Carlo simulation [6]

2. Metallic Surfaces

The second lecture deals with applications to metallic surfaces, and focuses on the basic concepts and some recent work on their electronic and atomic structure and related properties

atomic and electronic structure of metal surfaces [7]

adatom diffusion and growth morphology: islanding [7] , [8]

adsorption of molecules at surfaces: dissociation of diatomic molecules [9]

reactivity of metals: role of hybridisation [10]

3. Semiconductor Interfaces and Surfaces

The third lecture surveys current work on semiconductor surfaces, with particular emphasis on elemental defects in semiconductors as well as atomic structures and band offsets at interfaces.

atomic and electronic structure of semiconductor surfaces [11]

the quest for new semiconductor materials defects, doping and metastability in compound semiconductors [12] - [18]

thin film growth and STM imaging semiconductor heterostructures and band offsets [19] , [20]

4. Monte Carlo Studies of Nonequilibrium Kinetics

The topic of the fourth lecture is the Monte Carlo approach to modelling of non equilibrium kinetics of heterogeneous catalysis, based on discrete lattice methods as well as rules and rate constants incorporating the energetics old individual reaction steps.

kinetic phase transitions at surfaces: discrete lattice models [21] - [23]

oscillations in surface reactions [24] - [26]

Monte Carlo simulation of Fischer-Tropsch synthesis: model-buildin5 and rate-limiting steps [27] - [29]

References

[1] A.P. Seitsonen, M.J. Puska and R.M. Nieminen: Real space electronic structure calculations: combination of the finite difference and conjugate gradient methods, Phys. Rev. B 51, 14057 (1995).
[2] J. P. Perdew, K. Burke and M. Enzerhof: Generalized gradient approximation made simple, Phys. Rev. Lett.. 77, 3865 (1996).
[3] L. Mitas and R.M. Martin: Quantum Monte Carlo of nitrogen: atom, timer, atomic, and molecular solids, Phys.Rev.Lett. 72, 2438 (1994).
[4] M.J. Puska, A.P. Seitsonen and R.M. Nieminen: Electron-positron Car-Parrinello methods: self-consistent treatment of charge densities and ionic relaxations, Phys. Rev. B 52, 10947 (1995).
[5] P.L. Silvestrelli, A. Alavi, M. Parrinello and D. Frenkel: Ab initio molecular dynamics of laser melting of silicon, Phys. Rev. Lett. 77, 3149 (1996).
[6] L. Lewis and R.M. Nieminen: Defect-induced nucleation and growth of amorphous silicon, Phys.Rev. B 54, 1459 (1996).
[7] R. Stumpf and M. Scheffler: Theory of self-diffusion and growth of A1(111), Phys.Rev. Lett. 7 2, 254 (1994).
[8] G. Boisvert, L.J. Lewis, M.D. Puska and R.M. Nieminen: Energtics of diffusion on the (100) and (111) surfaces of Ag, Au and Ir from first pnnciples, Phys.Rev. B 52, 9078 (1995).
[9] B. Hammer, K.W. Jacobsen and J.K. Nørskov: Dissociation path for H_2 on Al(110), Phys.Rev.Lett. 69, 1971 (1992).
[10] B. Hammer and J.K. Nørskov: Electronic factors determining the reactivity of metals, Surf.Sci. 343, 211 (1995).
[11] D.J. Chadi: Energy-minimisation approach to the atomic geometry of semiconductor surfaces, Surf. Sci. 299/300, 311 (1994).
[12] S. Pöykkö, M.J. Puska and R.M. Nieminbn: Ab initio study of fully relaxed divacancies in GaAs, Phys. Rev B 53, 3813 (1996).
[13] T. Mattila and R.M. Nieminen: Direct antisite formation in electron irradiation ol GaAs, Phys.Rev.Lett. 74, 2721 (1995).
[14] T. Mattila, A.P. Seitsonen and R.M. Nieminen: Large atomic displacements associated to the nitrogen anti site in GaN, Phys.Rev. B 54, 7909 (1996).
[15] S. Poykko, M.J. Puska, M. Alatalo and R.M. Nieminen: Metastable defcct complexes in GaAs, Phys.Rev. B 54, 7909 (1996).
[16] S. Poykko, M.J. Puska and RM. Nieminen: Metastability of the anti-strllcture pair in GaAs, Phys. Rev. B (in press, 19973.
[17] T. Mattila and R.M. Nieminen: Ab initio study of oxygen point deiects in GaAs, GaN and AlN, Phys. Rev. B (in press, 1996).
[18] T. Manila and R.M. Nieminen: Point-defect complexes and broad-band luminesence in GaN and AlN, Phys. Rev. B (in press, 1997).
[19] F. Bernardini and R.N. Nieminen: Valence-band offset at lattice-matched interface between ZnSe and III-V wide ternary alloys, Phys. Rev. B (in press, 1997).
[20] A. Linden, M. Pessa, A. Salokatve, F. Bernardini, R.M. Nieminen and M.. Paalanen: Band onsets at the GaInP/GaAs heterojunction, J. Appl. Phys. (in press, 1997).
[21] H.-P. Kaukonen and R.M. Nieminen: Simulation studies of the dissociation of H_2 on stepped surfaces, Surf. Sci. 247, 43 (l991) .
[22] H.-P. Kaukonen and R.M. Nieminen: Computer simulation of the catalytic oxidation of carbon monoxide on platinum, J.Chem.Phys. 91, 4380 (1989).
[23] J.-P. Hovi, J. Vaari, H.-P. Kaukonen and R.M. Nieminen: Simulations of the

effect of preadsorbed poisons and promoters on irreversible surface reactions, Comp. Mat.Sci. 1, 35 (1992).
[24] A.P. J. Jansen and R.M. Nieminen: Oscillations induced by an inert adsorbate in CO oxidation, J. Chem.Phys. (in press, 1997).
[25] J.-P. Hoyt, A.P.J. Jensen and R.M. Nieminen: Oscillatory temporal behavior in an auto catalytic surface reaction model, Phys.Rev. E (in press, 1997).
[26] A.P. J. Jansen and R.M. Nieminen: Oscillations induced by an inert adsorbate in CO oxidation, J. Chem.Phys. (in press, 1996).
[27] J.P. Hovi, J. Lahtinen, Z.S. Liu and R.M. Nieminen: Monte Carlo study of CO hydrogenation on cobalt model catalysts, J. Chem.Phys. 102, 7674 (1995).
[28] J. Lahtinen, J.-P. Hoyt and R.M. Nieminen: Monte Carlo simulations combined with UHV-atmospheric pressure reaction studies of CO hydrogenation on cobalt, Topics in Catalysis 2, 173 (1995).
[29] J.-P. Hoyt, J. Lahtinen, J. Vaari and R.M. Nieminen: Monte Carlo simulations of CO hydrogenation on a (0001) surface, Surf.Sci. 311, 331 (1994).

MIX
Papier aus verantwortungsvollen Quellen
Paper from responsible sources
FSC® C105338

If you have any concerns about our products,
you can contact us on
ProductSafety@springernature.com

In case Publisher is established outside the EU,
the EU authorized representative is:
Springer Nature Customer Service Center GmbH
Europaplatz 3, 69115 Heidelberg, Germany

Printed by Libri Plureos GmbH
in Hamburg, Germany